# MARS COLONIES

## Plans for Settling the Red Planet

Edited by
Dr. Frank Crossman
The Mars Society

Published by Polaris Books
Lakewood, Colorado

ISBN - 978-0-9741443-8-2

# TABLE OF CONTENTS

# INTRODUCTION _WHY WE MUST COLONIZE MARS

**Robert Zubrin**

## COLONIZING MARS

Mars is the decisive step in humanity's outward migration into space. The Red Planet is hundreds of times further away than the Moon, but it offers a much greater prize. Indeed, uniquely among the extraterrestrial bodies of our solar system, Mars is endowed with all the resources needed to support not only life but the development of a technological civilization. In contrast to the comparative desert of Earth's moon, Mars possesses oceans of water frozen into its soil as permafrost, as well as vast quantities of carbon, nitrogen, hydrogen, and oxygen, all in forms readily accessible to those clever enough to use them. Additionally, Mars has experienced the same sorts of volcanic and hydrologic processes that produced a multitude of mineral ores on Earth. Virtually every element of significant interest to industry is known to exist on the Red Planet. With its 24-hour day-night cycle and an atmosphere thick enough to shield its surface against solar flares, Mars is the only extraterrestrial planet that will readily allow large scale greenhouses lit by natural sunlight.

Mars can be settled. For our generation and many that will follow, Mars is the New World.

With this in mind, the fall of 2018, a donor to the Mars Society offered $20,000 to sponsor a international contest asking people all over the world to lay out their plans for a 1000 person Mars colony. The contest required participants to explain how to make such a settlement feasible and attractive, with attention not only to its technical design, but its economic system, social and political organization, and aesthetics as well. Exactly 100 teams from around the globe submitted plans, and a team of ten expert judges sorted them out to down-select to 25 semi-finalists, and then ten finalists, who had it out in person before the Mars Society convention in Los Angeles in October 2019. But winner, finalist, or not, all the semi-finalists were of such merit as to deserve publication, collectively offering a wealth of ideas – not necessarily in agreement – that will serve as a treasure trove for those who will endeavor to take this next giant leap.

We present them to you in this book.

## HOW WE CAN

*"We hold it in our power to begin the world anew ..."*
*-Tom Paine, 1776*

The question of colonizing Mars is not fundamentally one of transportation. If we were to use a vehicle comparable to the SpaceX Starship now under development to launch habitats carrying settlers to Mars on one-way trips, firing them off at the same rate we launched the Space Shuttle when it was in its prime, we could populate Mars at a rate comparable to that which the British colonized North America in the 1600's—and at much lower expense relative to our resources. No, the problem of colonizing Mars is not that of moving large numbers to the Red Planet, but of the ability to use Martian resources to support an expanding population once they are there. The technologies required to do this will be developed at the first Mars base, which will thus act as the beachhead for the wave of immigrants to follow. It will develop techniques for extracting water out of the soil, for conducting increasingly large scale greenhouse agriculture, for making ceramics, metals, glasses and plastics out of local materials, and constructing large pressurized structures for human habitation and industrial and agricultural activity.

Over time, the base could transform itself into a small town. The high cost of transportation between Earth and Mars will put a strong financial incentive to find astronauts willing to extend their surface stay beyond the basic one-and-a-half-year tour of duty, to four years, six years, and more. Experiments have already been done showing that plants can be grown in greenhouses filled with CO2 at Martian pressures—the Martian settlers will thus be able to set up large inflatable greenhouses to provide the food required to feed an expanding resident population. Mobile microwave units will be used to extract water from Mars' abundant permafrost, supporting such agriculture and making possible the manufacture of large amounts of brick and concrete, the key materials required for building large, pressurized structures. While the base could start as an interconnected network of "tuna can" habitats, by its second decade, the settlers might live in underground vault pressurized domains the size of shopping malls. Not too long afterwards, the expanding local industrial activity will make possible a vast expansion in living space by manufacturing large supplies of high strength plastics like Kevlar and spectra that could allow the creation of inflatable domes encompassing Sun-lit pressurized areas up to hundreds of meters in diameter. Each new reactor landed will add to the power supply, as will locally produced photovoltaic panels and solar thermal power systems. However, because Mars has been volcanically active in the geologically recent past, it is also highly probable that underground hydrothermal reservoirs exist on the Red Planet. Once such reservoirs are found, they can be used to supply the settlers with abundant supplies of both water and geothermal power. As more people steadily arrive and stay longer before they leave, the population of the town will increase. In the course of things, children will be born, and families raised on Mars, the first true colonists of a new branch of human civilization.

We don't need any fundamentally new or even cheaper forms of interplanetary transportation to send the first teams of human explorers to Mars. However, meeting the logistical demands of a Mars base will create a market that will bring into being low-cost, commercially developed systems for interplanetary transport. Combined with the base's own activities in developing the means to use Martian resources to allow humans to be self-sufficient on the Red Planet, such transportation systems will make it possible for the actual colonization and economic development of Mars to begin.

***Fig 1. In the course of things, children will be born, and families raised on Mars, the first true colonists of a new branch of human civilization. (Credit: Robert Murray, Mars Society)***

While the initial exploration and base building activities on Mars can be supported by government or corporate largess, a true colony must eventually become economically self-supporting. Mars has a tremendous advantage compared to the Moon and asteroids in this respect, because unlike these other destinations, the Red Planet contains all the necessary elements to support both life and technological civilization, making self-sufficiency possible in food and all basic, bulk, and simple manufactured goods.

That said, Mars is unlikely to become autarchic for a very long time, and even if it could, it would not be advantageous for it to do so. Just as nations on Earth need to trade with each other to prosper, so the planetary civilizations of the future will also need to engage in trade. In short, regardless of how self-reliant they may become, the Martians will always need, and certainly always want, cash. Where will they get it?

While it may be possible to export plentiful Martian deuterium (fusion power fuel) to Earth, or food to and other necessaries to miners in the asteroid belt, in my view the most likely export that Mars will be able to send to Earth will be patents. The Mars colonists will be a group of technologically adept people in a frontier environment where they will be free to innovate — indeed, *forced* to innovate — to meet their needs, making the Mars colony a pressure cooker for invention. For example, the Martians will need to grow all their food in greenhouses, strongly accentuating the need to maximize the output of every square meter of crop-growing area. They thus will have a powerful incentive to engage in genetic engineering to produce ultra-productive crops, and will have little patience for those who would restrict such inventive activity with fear-mongering or red tape.

Similarly, there will be nothing in shorter supply in a Mars colony than human labor time, and so just as the labor shortage in nineteenth-century America led

Yankee ingenuity to a series of labor-saving inventions, the labor shortage on Mars will serve as an imperative driving Martian ingenuity in such areas as robotics and artificial intelligence. Recycling technology to recapture valuable materials otherwise lost as wastes will also be strongly advanced. Such inventions, created to meet the needs of the Martians, will prove invaluable on Earth, and the relevant patents, licensed on Earth, could produce an unending stream of income for the Red Planet. Indeed, if the settlement of Mars is to be contemplated as a private venture, the creation of such an inventor's colony — a Martian Menlo Park — could conceivably provide the basis for a fundable business plan.

Martian civilization will become rich because its people will be smart. It will benefit the Earth not only as a fountain of invention, but as an example of what human beings can do when they rise above their animal instincts and invoke their creative powers. It will show all that infinite possibilities exist, not to be taken from others, but to be made.

Such ideas, and many others, are discussed in this book. They also discuss many other aspects of creation of what must serve as the seeds of new branches of human civilization. In this, they are not in agreement. The economic and social systems advocated, for example, range from socialist to libertarian. I see such diversity of possibilities as a great strength. Mars is big enough for many, many colonies, founded by many different kinds of people who will have their own ideas as to what forms of social organization offer the greatest scope for people to realize their hopes and full potentials. Indeed, the chance to be a maker of one's own world, rather than just an inhabitant of one already made, is a fundamental form of freedom, whose attraction may well prove to be the primary driver for many to accept the risks and hardships that settling another planet must necessarily involve. It is hardly to be imagined that human social thought has reached the final and best answers possible in the early 21st century. There will always be people with new ideas, who will need a place they can go where the rules haven't been written yet to give them a fair try. That said, not all of their novel ideas will prove workable. Some may prove impractical, and cause the colonies that adopt them to stagnate or fail altogether. But, as was the case with the "noble experiment" of testing the ideas of enlightenment liberalism embraced by America's founders, sometimes the new path can lead to better ways. Those colonies that demonstrate them will succeed, draw immigrants, grow, and be an example of all others, on Earth, Mars, and the thousands of new worlds that await us beyond.

So shall we continue to progress.

**WHY WE MUST**

Ideas have consequences.

Science, reason, morality based on individual conscience, human rights; these are the central ideas of the humanist heritage. Whether expressed in Hellenistic,

Christian, Deist, or purely Naturalistic forms, it all drives toward the assertion of the fundamental dignity of man. As such, it rejects human sacrifice and is ultimately incompatible with slavery, tyranny, ignorance, superstition, perpetual misery, and all other forms of oppression and degradation. It asserts that humanity is capable and worthy of progress.

This last idea – progress – is the youngest and proudest child of Western humanism. Born in the Renaissance, it has been the central motivating idea of our society for the past four centuries. As a civilizational project to better the world for posterity, its results have been spectacular, advancing the human condition in the material, medical, legal, social, moral, and intellectual realms to an extent that has exceeded the wildest dreams of its early utopian champions.

Yet now it is under attack. It is being said that the whole episode has been nothing but an enormous mistake, that in liberating ourselves we have destroyed the Earth. As influential Malthusians Paul Ehrlich and John Holdren put it in their 1971 book *Global Ecology*: "When a population of organisms grows in a finite environment, sooner or later it will encounter a resource limit. This phenomenon, described by ecologists as reaching the 'carrying capacity' of the environment, applies to bacteria on a culture dish, to fruit flies in a jar of agar, and to buffalo on a prairie. It must also apply to man on this finite planet."

Note the last sentence: *It must also apply to man on this finite planet.* Case closed. The only thing left to decide is who gets death and who gets jail.

We need to refute this. The issue before the court is the fundamental nature of humankind. Are we destroyers or creators? Are we the enemies of life or the vanguard of life? Do we deserve to be free?

Again, ideas have consequences. Humanity today faces a choice between two very different sets of ideas, based on two very different visions of the future. On the one side stands the antihuman view, which, with complete disregard for its repeated prior refutations, continues to postulate a world of limited supplies, whose fixed constraints demand ever-tighter controls upon human aspirations. On the other side stand those who believe in the power of unfettered creativity to invent unbounded resources and so, rather than regret human freedom, demand it as our birthright. The contest between these two outlooks will determine our fate.

If the idea is accepted that the world's resources are fixed with only so much to go around, then each new life is unwelcome, each unregulated act or thought is a menace, every person is fundamentally the enemy of every other person, and each race or nation is the enemy of every other race or nation. The ultimate outcome of such a worldview can only be enforced stagnation, tyranny, war, and genocide. Only in a world of unlimited resources can all men be brothers.

On the other hand, if it is understood that unfettered creativity can open unbounded resources, then each new life is a gift, every race or nation is fundamentally the friend of every other race or nation, and the central purpose of government must not be to restrict human freedom, but to defend and enhance it at all costs.

And that is why we must take on the challenge of colonizing Mars. For in doing so, we make the most forceful statement possible that we are living not at the end of history, but at the beginning of history; that we believe in freedom and not regimentation, in progress and not stasis, in love rather than hate, in peace rather than war, in life rather than death, and in hope rather than despair.

# 1: THE TEAM "LET IT BE" MARS SETTLEMENT HOW THE UTOPIA COULD MATERIALIZE

**P. Brisson**
Mars Society Switzerland
*pierre_brisson@yahoo.com*
**R. Heidmann**
Association Planète Mars
**T. Volkova**
Swiss Space Center, EPFL

A 200 residents district (© R. Heidmann ) :
*__1)__ Line of 13 ice- sheltered apartments; __2)__ Greenhouses for their 26 residents; __3)__ Access airlock; __4__/20m diam. Dome; __5)__ 30m diam. Dome; __6)__ Corridor; 7) Solar panels (if basic energy source)*

**INTRODUCTION**

The disinterest of political deciders in a manned Mars exploration program is challenged by the on-going revolution of low-cost space transportation, which opens a wide spectrum of new opportunities. The Moon and Mars both present scientific, touristic and economic attractiveness, as well as the prospect of in situ resources exploitation. But, concerning this last point, the Red planet is much more promising for a vision of space settlement. In fact, the sole meaningful disadvantage of Mars, its remote location, is made much less significant by the wide spectrum of resources available there, which can soon lead to a large self-sufficiency.

The Mars Society Contest deals with a first settlement of 1000 residents, self-supporting to the maximum possible, and economically viable. We totally concur with this basis as it framed our past studies. But besides these overall specifications, the conditions under which the project could be realistically funded and lead to success, impose constraints of their own. That said, having chosen to **rely mainly on proven technologies**, the challenge that appears most compelling is **the mere scale of the settlement**.

Three phases should be considered in this endeavor. After a **preparatory phase** (3 to 5 years, several billions $), dedicated to qualification and transfer of the required building equipment (500 mT), the **building phase** will start, spanning 20 years, deemed the maximum that private investors could accept. This growth rate implies 6 Starship-like flights at each launch window, hence a minimum of 12 spaceships to be built, since a spaceship cannot return to Earth in less than a synodic period. In a stabilized cruise regime, this fleet size allows to ferry around 300 mT and 370 passengers per synodic period, which is the transportation capacity needed for the settlement. The **operational phase** will begin when planned facilities are totally set-up and functioning, even if their progressive availability will authorize a partial use and qualification before. In parallel, the Colony will have to prepare the development of its activities, through a Martian **research & technology effort** which will need dedication of personnel and further financial support.

Our survey of the different facets of the project, including a check of its economic viability through our economic model, allows to consider the project as conceivable, should nevertheless two decisive preliminary assumptions be fulfilled:

-Success in **gathering a pool of investors** capable of bringing some $ 30 to 40 billions;
-**Availability of a Super Heavy** transport system, with a transfer price of no more than 2 million$/mT and 1 million$/passenger.

To finish with, let us underscore some general points which framed our work:
-**most attention should be given to the well-being of residents** (safety first, but not only);

- in order to be properly funded and conceived as leading to a perennial undertaking, the first settlement setting-up should be **structured and managed as an enterprise**;
- thus, though **not motivated by the will to establish a new branch of humanity**, this valuable goal being premature, the settlement should nevertheless be considered as its proof of concept;
- contrary to a widespread view, during this starting phase of human presence, **the vast majority of people traveling to Mars will return to Earth**, having fulfilled the dream of their life or accumulated a big enough amount of money; thus, we'll speak of **residents rather than colonists**;
- for transportation, we selected the **SpaceX SuperHeavy / Starship** scheme, even though we acknowledge some of its shortfalls, as it has the merit of putting forward performance and attractive cost figures and, furthermore, to be effectively pushed forward by its proponent.

## 1 – SAFETY AND HEALTH

Humankind has always been attracted to climb out of the cradle, as Konstantin Tsiolkovsky (1857-1935) once said but surviving in Space remains a challenge. Mars settlement creation related human health threats include hostile environment (reduced gravity, near-vacuum and radiation), habitats inside environment (noise, closed life support, confinement, isolation), and human body necessary adaptation (neurovestibular, musculoskeletal, and cardiovascular changes) to this new environment. Residents' survival must be considered a mission primary success criterion and residents' physical and mental well-being is crucial for the creation of a successful settlement sustainability; it will be the primary responsibility of the medical team.

To date, all accidents that occurred during space missions have occurred because of design and manufacturing decisions that were made and were within the technological know-how and limitations that currently existed. To reduce the risk of accidents, worst case scenarios should therefore be considered when setting resident survival design requirements. This in mind, we underline that making a preliminary design for residents' survival is more effective than postproduction modification of a settlement design.

Utmost attention should be paid to radiations (solar particle events as well as a continuous high level of GCR ) protection design development, by implementing several options:

- For buildings, use of high-performance structural materials, such as carbon composite with a high hydrogen content having effective radiation shielding properties;
- Adequate planning for best location of the base (low altitude) and timing of EVAs;
- Seeking shelter and using localized shielding with personal warning systems (Cucinotta, Kim, and Ren 2006a).

The usually high specific ionization of high-energy nuclei included in the galactic cosmic radiation is the ultimate limiting factor for long-term space operations, because although their relative dose contributions are comparable to those of light particles, their biological effects which as of yet are poorly understood, are far more serious (Cucinotta and Durante 2006). Residents of a Mars settlement will face other challenges such as a low-fat diet that can conflict with the need for a high calorie diet which can provoke the possibility of cardiovascular disease over the years. Diminished immune function because of confined environment and the use of recycled air, which increases the risk of reactivation of latent viruses, could decrease performances and increase the risk of accidents. In addition to the risks imposed by a generally reduced immune response and anti-viral defense shedding, it also appears that spaceflight induces some species of bacteria to proliferate (Manco et al. 1986; Takahashi et al. 2001).

As a solution, different design strategies can be implemented to solve such problems and take them correctly into consideration. One of them is “design for Minimum Risk”. Such design takes into account all requirements and relies on safety factors and safety margins established by analysis and tests, past experiences, and international standards to ensure an acceptable level of risk. Also fault-tolerant design criteria should be implemented. Safety factors should be carefully adapted to the Mars settlement design to ensure that the risk of failure is reduced to an acceptable level (Safety factor = Actual capability/Required capability). All settlement failures should be reviewed to gain appreciation of their consequences, to understand the operating environment and the performance of the structure under these conditions. And finally, hazard control should be included in the process design with a strategy of taking into account the maximum failure tolerance, that can provide more capability to continue operations in a safe manner after experiencing failures.

Containers must be designed to be leak-tight for all environments expected during the lifetime of hardware. In particular this concerns pressure loads, structural loads, temperature extremes, exposed materials, and life cycle. Other environments can be relevant for particular hardware. Then microbiological risks can be mitigated by early development and implementation of effective countermeasures starting with the settlement design phase. Specifically, selection of antimicrobial materials, use of air filtration, control of humidity and condensates, setting up of acceptability limits, verification monitoring, and remediation technologies, will all play important roles in a successful human settlement of Mars.

Other specific design requirements, such as contaminants consideration, should be studied carefully. For human health, sizes $< 10$ μm in aerodynamic diameter, which comprise respirable particles, is of utmost interest (Hines et al. 1993). The Mars dust load, which overlays the human-generated load, presents a big challenge for air quality control system designers.
Further design considerations should be devoted to noise control strategy and acoustic analysis. Designers should develop reasonably quiet noise sources. But a specific acoustic design development strategy should be thought out in order to keep some places noisy so as to provide some minimum noise so that people can feel safe by understanding that systems keep working.

Also, we have to work to reduce the high risk of human factor on Mars. Computer control can help solve this problem. For example in Stanley Kubrick's film "2001: A Space Odyssey" we see that the "Artificial Intelligence" controlled the whole space mission and never made mistakes...until a certain point. It's definitely a good design strategy for supporting Mars residents during their daily life. Computer control systems can help controlling hazardous functions because tasks are either very complex or require a very fast response to control parameters. But to avoid problems, we see in this film that astronauts are assisted by AI, not replaced by "him". It means that "cognification" of AI will lead to alteration of job roles rather than their elimination. (Volkova T., IAC 2018).

Fire is a very serious threat in any confined space under pressure because in such environments chosen partial pressure of oxygen is usually higher than in the air on Earth. As a result, many materials that are non-combustible or self-extinguishing in the "normal" air are easily ignitable in the conditions of an habitat. In general, smoke detector designs can help prolong the escape time if detectors are installed at each level and in each bedroom. In addition, success in preventing fires in a Mars settlement can be achieved by eliminating the use of flammable materials as much as possible. Low and reduced gravity also plays a role in the overall effectiveness of fire extinguishing agents. Although much remains to be learned about the extinction of fire within reduced gravity, today we can consider using key fire suppression agents which were tested under low gravity conditions: Foam (AFFF and FFFP), Halon 1211, water mist, inert gas agents.

## 2 - TECHNICAL DESIGN

Our survey of the settlement functional needs and of the corresponding infrastructure shows that we master most of the required technologies. In fact, from a technical point of view, the actual challenges are the following ones: **rising the RAMS figures** (*Reliability, Availability, Maintainability, Safety*) to an acceptable level, despite the specific difficulty due to the settlement remoteness and being able to **sustain the high rate of facilities building** resulting of the setting-up duration constraint (20 years), mainly in three production domains: energy, food, habitats.

### 2.1. General layout of the colony

Urban planning considerations will be of great importance given the living conditions of the inhabitants: EVA only in pressurized spacesuit or vehicle; living inside only within limited "shirtsleeves" areas. Despite those conditions, residents will need to get the availability of a complete range of services: they should "lack nothing" and should find in public areas the ambience and diversity allowing them to get rid of any enclosure feeling. Urban planners and architects will have a beautiful ground to exercise their talents (cf. chapter 6). They will however be subject to limitations: **maximum use of local resources; drastic saving of a limited workforce; specific workers safety constraints**. We can imagine two main types of general layout: linear, where the various infrastructures are located along a few major communication axis, by zones conforming to specific uses; or radiating, with the different zones distributed along "spokes" centered on an interconnection "hub", where the social infrastructures can be located. This second solution is more rational for communication. In any case, since we should expect a gradual development of the colony, with construction resources (material and human) under strict limiting constraints, we are condemned to follow a network pattern of a large number of modules of relatively small size, and consequently to an arrangement preferably based on "linear". The building of larger modules, whether domes or cylinders, presents serious difficulties. Also, such structures present low efficiency in several respects: volume occupancy, quantity of atmosphere to provide, envelope thickness... Nevertheless, it will be absolutely necessary to provide a few of them, in order to avoid giving the colony the monotonous aspect, asocial and depressing, of a series of identical modules.

### 2.2. Main factors for site selection

A series of varied criteria will have to be applied for the site selection of the settlement. The most important is probably the abundance of and easy access to water (ideally water ice), which is required for the return transfers propellant production (and daily life!). Then come the parameters influencing the easiness of spaceships access to the planet: mainly latitude, but also landing terrain altitude (the lower is the better for EDL). For a better installation, consideration should also be given to the following factors: proximity of other resources (silica, carbonates, hematite...) mining sites, terrain smoothness (for an easy driving around), presence of slopes procuring natural partial shielding against radiations. Finally, a very important factor is the attractiveness of the landscape and the accessibility to various types of scenic geological sites; tourists, who should be an important part of the customers population, deserve to really feel being on the magnificent planet of their dreams and to enjoy the sheer beauty of its wilderness. As for planetary science activities, the settlement will be rather regarded as a support base, with terrain work ongoing at secondary outposts cleverly located (mostly through robots under direct guidance).

**2.3. Synthesis of the main needs governing the settlement design and size**

Different studies provide a rather homogeneous set of data related to the volume of infrastructures and consumables needed to sustain the operations and maintain the living environment of such a settlement. Of course, it is desirable that the needs of both setting-up and full-level operation phases, be coherent: what has been deployed for building should not be left unused afterwards. We tried our best to conform with this constraint in elaborating our scenario and found a good balance with the following main characteristics:

| | |
|---|---|
| -Duration of the construction period: | 20 years |
| -Lodging surface/resident - Total pressurized surface/resident | 30 m² - 50 m² |
| -Greenhouses surface/resident | 80-100 m² |
| -Mass to land beforehand, then at each window (6 spaceships) | 500 mT - 310 mT |
| -Electric power | |
| -for propellant synthesis | 250 kW/spaceship |
| -for materials processing | 500 kW |
| -city needs (lighting, heating, air & water processing...) | 20 kW/resident |
| -greenhouses, complementary artificial lighting | 15 kW/resident |

Then we tried to evaluate needed material productions for the largest infrastructures, i.e. those linked to the great number of residents; namely, the **habs, domes and greenhouses**.

**2.4. Pressurized structures**

In order to get an idea of the quantity of materials needed, we considered the following designs, elaborated by R. Heidmann in his study "Martian Habitats: Molehills or Glass Houses", published in 2016 on the Web site of the Mars Society French chapter, planete-mars.com (English language pages). Note that ***they should be considered only as evaluation tools, not as optimized concepts***, and that pertinent aspects discussed in chapter 6 were not yet integrated at this stage of the study.

-For the **habs**: 6 m diameter hexagonal modules offering 60m² on 2 stories, made of glass plates (1.5+1.5 cm thickness, 3x1.5 m²) covered by plastic foam or glass fiber isolation plates and mounted on a steel frame. Those "apartments" are either semi-buried (for 18 months "short" stays), with a large window that can be occulted by a protective water ice shutter, or totally covered with piled up regolith (for long stays, offices, workshops). This technology looks well fit for mass production, emplacement on site and it also minimizes soil moving. This is not to say that this is the best design; many other valuable solutions have been proposed, even if a lot should be discarded as not fit to a mass production scheme. **The main problem is the accessibility of carbonates**, required to synthetize NaOH and CaO, significant ingredients of glass manufacturing…

*Scheme of a pack of 20 apartments, for 40 persons (semi-buried) © R. Heidmann*

As for the frames, strong longitudinal beams are needed to withstand the force applied to the ends of each row of modules (200 kN on each of the 6 beams). And the hexagonal frame itself needs to be duplicated to accommodate the limited 3x1.5 m² size of the glass plates. Choosing steel raises the question of corrosion, with the risk of loss of tightness on the long run. Using stainless steel would induce the need of nickel for 10% of the frame mass (from collected meteorites, or imported?).

-For **greenhouses**, the same basic modules should be used, with isolation plates affixed on the sides of the modules set. Thermal contact with the ground is restrained to bottom fins. Glass plates between successive rows of cells are replaced by simple steel beams linking floor and ceiling (to withstand the 50 kN/m² stress).

*One of the apartment's hexagonal faces is a window.*
*A water ice curtain provides isotropic shielding when flipped against the window.*

-**Socialization hubs** (restaurants, sports halls, goods shops, libraries, meeting halls, performance halls, city services offices...) should be added. For that we relied on geodesic domes made of triangular glass plates, using the same materials as for the habs, and anchored on foundations made of Martian regolith cement (1m thickness for a 20m diameter dome, 2m for a 30m one), rigidified by steel beams. The basements of the domes are topped by circular bleachers, thus providing 0 to 100% radiation shielding to the floor and, by the same token, different degrees of visibility of the sky and landscape; it's up to the visitor to choose his exposure at any time. We have supposed the assembly of **ten 20m diameter and two 30m diameter** units, requiring a total of 10x1600 mT + 2*3500 = **23000 mT of duricrete** (hence a lot of water!). The required glass and steel quantities are small compared to those dedicated to the numerous habs and greenhouses cells, but the assembly, by people in spacesuits, will be a daunting task, for which assistance of robotic acrobats will be unavoidable. Particularly, the enormous number of plates to frame beams bondings, presents a severe quality challenge. That's why we consider that a 30m diameter is a maximum, not to speak of the amount of linear stress between dome and foundation, which grows with diameter... And, always, a pressure of 50 kN/m²... Forget huge plastic domes!
-To that should be added the materials needed for the implementation of **scientific and industrial facilities**, with a different set of specifications, as they will not be replicated in such a large number and probably set up in bored tunnels. From a production point of view, they look not so significant, except for the **astroport**, which should need to get a dozen of hardened landing platforms, built of a concrete layer (for a total of about 2000 mT). What is significant is the quantities of machines and equipment that will be installed within them.
-To finish with, we should not forget the **circulation corridors**. It is highly desirable that most of the facilities be linked together, so that residents can access almost anywhere without donning a spacesuit. Ideally, those links should permit the crossing of two vehicles. And, since this is a place where nobody stays for long, it will not a big problem to have one of the lateral walls largely made of glass, thus providing the residents when they move, an opportunity to admire the landscape and reinforce the feeling that they really are on Mars! Examining the overall layout of the settlement, we concluded that we need, as a mean value, a total of 1.2 km of corridors. This looks a lot, but represents only 6.5% of what is needed for habs plus greenhouses...

On this basis we found the following production needs:

| | Total (mT) in 20 years | Rate (mT /sol) |
|---|---|---|
| Glass | 27500 | 3.8 (~11 plates) |
| Steel | 6000 | 0.8 |
| Duricrete | 25000 | 3.5 |

The means required to reach these production rates have been evaluated on the basis of a survey of machines available on the market that are close to match these figures, and compared with the in-depth data base produced by the Mars

Homestead project. For steel, for instance, we found an oven (Nabertherm 700/12) weighting 2.6 mT and with a capacity of 6 mT of steel/sol. And for glass, an oven (N2214) with the following characteristics: 1400°C, 3.9 mT, 140 kW, interior 1x1.4 $m^2$, adapted to a discontinuous process; the dimensions of our panes (3x1.5 $m^2$) should be attainable with an extrapolation of this kind of machine. It thus appears that the ovens (and ancillary equipment) could all be landed by only one spaceship. In conclusion, despite its vast extent,

***the first Martian colony setup does not require enormous industrial means.***

**2.5. Habs and corridors equipment**

A question related to both categories of buildings is that of hatches. More precisely, three main points should be considered: safety in case of a major leak, easy circulation, dimensions of opening (related to mass). Should every apartment be tight? How many cumbersome door frames is it acceptable to cross along daily walks? Which format to choose, knowing that the hatch mass increases rapidly with size, due to the required rigidity of the frame and door wings?

Another less critical but mass-intensive item is the plumbing equipment of habitats, for water, air, heating and waste. We can assume that tubes can be manufactured in situ, from plastic mainly, but, at the beginning, more elaborate parts will not be manufactured on Mars. The same for electrical equipment and electronics, except for cables, if aluminum is accepted. We should also mention lighting equipment, especially if artificial lighting is installed in the greenhouses. The total mass has been estimated at 400 kg/hab module, meaning a load of **20 mT** /synodic period. Concerning furniture, we propose that pieces be manufactured, mainly by 3D printing, on demand from individual visitors, and that pieces can be bought or exchanged in a specialized shop.

**2.6. Greenhouses design**

Food production has to take place in greenhouses, water tanks or small specialized habitats (for animals). The same basic modules as for habitats are used (cf.2.4). Food producers will have three objectives: quantity, variety, organoleptic quality, and three special considerations: maintaining a healthy environment for their products (all are living beings), a maximum energetic and dietetic quality for the consumer, a minimum no-reusable waste amount.

Sun irradiance at Mars distance of the Sun is such that it will be worth **taking advantage of natural light**; its sols, with nights of an acceptable length, will facilitate growing superior plants and raising small animals. However since natural light might be insufficient during the long months of the austral Martian winter (irradiance stays around 400 W/m2 during several months) or during dust storms, it will have to be **complemented by artificial light**. Greenhouses will also have to be **thermally isolated** as much as possible, **and** heated. Based on past studies, we estimate the mean power required for the auxiliary lighting for 1000 residents at 15 MW (at 30% of the required power for full artificial lighting, several thousands of light tubes!). Heating power could be extracted from the "cold" source of nuclear generators, the main power source we selected.

In order to ease circulation to and from the other parts of the settlement, greenhouses will get the same atmosphere pressure, albeit this should translate in a heavier structure than strictly necessary (plants could grow under a lower pressure) and the extraction of much more nitrogen (the buffer gas) from the Martian atmosphere.

The greenhouses volume will be limited and we'll need to grow small plants or animals with the highest nutritive content. The first organisms we will have to grow will be spirulina, green algae. They breath $CO_2$ and rejects $O_2$. They otherwise provide rich organic molecules, which will be used indirectly to feed new living beings which in turn will be harvested (or killed!). The ultimate goal is to recreate a closed environment with minimal outside replenishment for maintaining it. Besides oxygen release, the second interesting property of spirulina is that it is very rich in useful proteins and it can be eaten. Other staple food will be vegetables and small fruits (berries), fish (tilapia) and, along the years, we might bring in poultry (eggs!) and small animals like rabbits. The main concern with terrestrial animals will be the control of their microbial environment.
The minimum vegetable growth surface to feed one man is estimated to be 60 to 100 m², depending on the cultivation mode. We'll have to use the greenhouses volume as much as possible, that is **several levels within the same volume**. A good example of what we could do is what is being experimented in some city farms like "Sky greens vertical farming system", in Singapore. On Mars, the nutrients will be fed directly to the plants through **hydroponic** in the trough (minimum waste and maximum control!), the support being neutral mineral beads from Martian regolith (cleaned of its perchlorates) and easily controllable for its microbial content. Microbial control will be essential, and contacts between living-beings (microbiomes) limited to the maximum possible.

*Socialization will occur inside domes, where residents can balance the delight of a scenic outside view with their radiation exposure budget*

## 2.7. Energy production

### 2.7.1. The needs

The large array of heavy equipment for the settlement set-up will require, from the beginning, availability of electrical power at a significant level:
-Propellant production for 6 Spaceships 1500 kWe (primarily electrolysis)
-Materials extraction and processing 500 kWe (see table below)
-Greenhouses lighting (30% of time, @500W/m²) +1500 kWe per synodic period

***Total for the building phase: from 3.5 MWe to 17 MWe***
*(heating (habs and greenhouses) is supposed to rely on wasted heat from nuclear generators).*

| PROCESS | mT/day | We |
|---|---|---|
| Hematite reduction & Steel production | 0.8 | 900°C heating 160 kW |
| Glass production | 3.8 | 1550 °C 210 kW (if elec.heating) |
| Duricrete preparation | 3.5 | |
| $CO_2$ decomposition | CO: 0.4 $O_2$: 0.23 | 1100°C: 100 kW breaking down: 40kW |
| $N_2$ separation (fertilizer not incl.) | $N_2$:0.013 | 5 kW |
| | | |
| TOTAL | | **0.5 MWe** |

*Materials processing power needs (see below)*

For the steady, “cruise”, state operation of the settlement, we suppose that the building effort will be kept at the same level, to cope both with facilities maintenance and continuous growth of capacities. But to that should be added the day-to-day energy needs, which will grow with the population size. Evaluations from existing literature span a rather quite large spectrum (1 to 3), loosely related with population size. For instance, the Mars Homestead project gives 90 kWe/resident (but at the scale of 12 persons), while Martin Fogg gives **20 kWe/resident** for bigger settlements (like ours). We took this value, for a population growing from 100 to 1000 people, which leads from 2 MWe to 20 MWe along the period. Hence:
***needs span from 5.5 to 37 MWe, with an increment of 3.5 MWe / synodic period.***

2.7.2. The case for photovoltaics
Taking into account all limiting factors (distance from the sun, seasons, height of the sun above the horizon, day-night cycle, dust storms) the mean power from fixed panels should reach no more than 50W/m². Therefore, generating 3.5 MW implies deploying a panel surface area of 70,000 m². Even with predictable improvements, the mass will still be around 100 to 50 mT. But the main problems will arise on the planet: robots should be imported for deployment, and dust cleaning; there is a problem of life duration and, last but not least, long global dust storms will shut down the power supply: the settlement should then enter a “dormant” mode and rely on a potent emergency system such as fuel cells (taping the propellant stocks). On the other hand, this solution is simpler to produce, to transport and even to deploy than nuclear fission. It is also politically “clean”.

Would it be reasonable to consider later in situ mass production of solar panels? This point of view is widespread, because silicon abounds in the Martian ground.

But the industrial process is very complex (as an illustration, « solar » quality silicon requires a purity of 99.9999 % !).

2.7.3. The nuclear fission generator, best fit

A 3 MWe fission-based generator, with two independent Brayton loops (for redundancy), can be designed with a mass of less than 10 mT, and core dimensions and fuel load similar to those of the 40 kWe low power projects contemplated for exploration missions. It is in fact the "cold source" which needs to be scaled with power, remembering that on Mars, no cold sinks such as rivers or seaside are available: the atmosphere is much too thin for this purpose; radiative heat removal would demand a cumbersome assembly of very high temperature radiator panels; using the ground as a cold source implies a large amount of soil moving…That said, some essential facilities functions (e.g. habs heating) could be fulfilled through direct use of generators wasted heat. But **transportation of high temperature calories on distances presents limits**:
-**temperature** would not exceed about 600 K; this is largely convenient for heating up living areas and greenhouses, as well as for some parts of materials processing, but for higher temperatures and most energy-demanding processes, electricity or gas combustion will be necessary;
-**distances** between generators and heat-consuming facilities should be minimized, because of the mass of the transfer ducts and of the unavoidable resulting thermal losses; nevertheless, proximity raises concerns for inhabited areas.

Each nuclear generator core will be put in a pit topped with a metallic shield, and the whole system should be safe even in case of a failure (in the core or in the fluid circuits), leading ultimately to a definitive burying of the nuclear part of the device. **Maintenance** operations will be restricted to the emerging parts: turboalternators and fluid circuits connections, with a surface radiological environment conforming to nuclear industry rules. At the end of life of the core (20 years?), the best practical option is to bury the reactor in its pit and mark its presence for the far future.

This technology presents several other drawbacks: political acceptability, continuous replacement of fuel-depleted generators, long-term fate of the used generators… But despite these drawbacks, given the technology mastered so far, if solar power may be considered in the long run (and from the beginning as an emergency resource), the only reasonable option to power up a first settlement of the size discussed here (1000 residents) is definitely nuclear fission (the NASA/DOE Megapower project leads to interesting openings).

## 3 - THE SETTLEMENT AS AN ENTERPRISE

To discuss how to make a Martian economy successful, we first have to consider the costs of building and servicing an inhabited base on Mars. We then have to see how the financing of the construction and the period before profitability can be organized, what the Colony can sell and to whom; how long we will need, on the one hand, to cover variable costs and, on the other hand, to begin remunerating investors who funded the not subsidized starting costs (some will be) and the fixed assets. Within the fixed costs we should consider the last stage of the development of the interplanetary transport system, the first elements of a fleet and the equipment allowing an active and comfortable life on Mars. The variable costs will include all expenses which have to be renewed and depend of usage and time.

**3.1. Elements of cost**

The amount representing the development cost of the vessel will depend on the stage at which the decision is made to create a Colony on Mars. For the time being, expenses already made are not specifically assigned to the project. Assuming a global amount of some 15 billion dollars for the overall development of the MCT (Super-Heavy + Starship), we may eventually need only some 10 billion more when the decision is made. Then we will need a minimal fleet servicing the interplanetary transport system. This should be 11 BFR and 14 Starship (6 x 2 Starships * for the trip, 6 Super-Heavy for the launching, 3 tankers to fill a Starship before its interplanetary injection; 2 Starships + Super-Heavy for back-up). This could be secured for about 15 billion. We have to add up the cost of building the first "City", including the cost of material and equipment transferred (about 500 Tons; 3 million $ per metric T, including 2 million for the transport), the maintenance of machines and people needed for this initial work. This could be made for some 20 billion. At the end we see that we'll need at most 50 billion dollars investments to validly commercially run a Mars base.

**Unfortunately, the Starships sent during one launch window cannot come back on Earth in time to be re-used during the next launch window (but the booster can!).*

**3.2. Need for a strong financial structure**

Assuming conservatively that we will need to do most of these 50 billion expenses in the first 8 years, we can start receiving paying guests on the third synodical voyage. The sooner and the faster we succeed in getting income, the better. In order to start a long-lasting Colony, we should not rely, on the long run, on governmental subsidies or generous philanthropic donations. We should consider the Colony as a corporation which, on the long run, should live upon its own resources. Feasibility will first of all depend on the structuring of the financing of the development, construction and launching periods. We propose the commitment of the following partners:

-**Space Agencies**. NASA could possibly get the support of others agencies, as they did for the ISS. They could together be expected to gather some 5 to 10 billion per year at the beginning of the project. To get interested in the economic success of the Colony, they could be one of the main shareholders of the Operating Company (see below).

-a **Mars Foundation**. Such Foundation could be created with donations made by rich entrepreneurs interested in the realization of a Mars Colony (Elon Musk of course, but also people like Jeff Bezos, Larry Page, Robert Bigelow…). They could also provide a large amount (in the range of 3 to 5 billion per year during 8 years) and become other main shareholders of the Operating Company. The Foundation could also use its funds as guarantees to help raise money elsewhere.

-a **Mars Colony Operating Company** (the "Company", possibly dubbed "The New India Company"). Created to operate, manage and develop the Colony, its main shareholders (the Agencies and the Foundation) will issue stocks on the world markets in order to associate the Public to the Venture. This could be done after some construction progress has been made and a large part of the initial net worth of the Company spent (creating confidence in the project). We could expect a few billion dollars from this equity issues. In case of needs and as a function of its projected returns, the Company could also issue debt with different seniority levels, function of the length of the repayment periods, of the grace periods, the interest rates and the guarantees received. Such guarantees could be issued by the Agencies or the Foundation.
-the **Public**. Depending on the success of the first realizations on Mars and on the confidence that can thus be built, the Public could provide an important amount of funding, as said above. Once on the market the stock of the Company will be traded and people will speculate on its future success (on the basis of prospects projected by financial analysts). This could improve the value to the stock and facilitate new issues of equity after the IPO. It is also possible that, in order to foster the collection of equity, the Foundation commits to match the money raised on the market.
**-industrial suppliers of the Company**. They could accept a deferred payment of some of their supplies as retention money (10%?), to be released upon evidence that their equipment is functioning on Mars. They could also be motivated to accept to be paid such retention money with options to purchase stocks of the Company (thus saving liquidities). It might be also that the suppliers, attracted by the prospect of higher gains, prefer to lease or rent some equipment.
-a **Space Bank**. Such Bank could help structure and organize the financing, open the world stock markets to the Company and act as the market-maker of their stocks. Its shareholders would be the same as the Company's but the Bank could also call for public money for itself. Besides raising equity, the Bank could also organize loans to relay equity issues or pre-finance the voyage and sojourn of individuals or corporations wanting to participate in the Venture.
-an **insurance company**, the "Space Insurance Company" would also be valuable for the whole structure. This company would have expertise in the Mars settlement process as well as in the insurance business and could insure people going to Mars, equipment shipped to Mars and get access to the world capacities of reinsurance. This would allow financing and credits otherwise impossible. Its main shareholders would be the Foundation and the Space Agencies.

### 3.3. Making money by building a margin

Now, raising money is one thing, making the project floating and keeping it going is another one. In order to last, the Colony will need resources. The first obvious service (but not the only one as we will see) which can be sold is residency on Mars.

**-Residency services**

This leads to get a look at the making up of the population. Considering all the professional qualifications resulting of the special conditions of life on Mars, we estimate the needs to amount to 550 staff for a whole population of 1000 people.

Among the 550 staff ("paid residents") we will have all kind of technicians, a few administrators, either direct employees of the Company or contractors. They are likely to stay several synodical cycles on Mars. On the side of the 450 customers or "paying guests", we will have three categories of people: 1) researchers, studying the planet and its environment; 2) people commissioned by their corporations to take advantage of the Martian environment to study or develop specific technologies; 3) tourists, any people wanting to experience life on Mars (retirees or not) provided they can afford it. Retirees might stay longer than one synodical cycle on Mars, most others won't.

**-Freelance-entrepreneurs, key to make the Colony successful**

Beyond this basic population distribution and in another dimension (but within the paying guests category, including the "tourists"), a fourth category of people will come and might stay on Mars more than one synodical cycle, we call them "freelance-entrepreneurs". They will be people who candidate to travel to Mars, to try and test and maybe succeed in creating "something new" which could be a process, a technology, a new way to do things within a very stimulating environment. They will submit their project with a business plan to the Company. The latter could subsidize part of their trip or their stay, asking for some return in case of success; the rationale being that one out of one hundred such freelance-entrepreneurs might create something on Mars which could be sold with an important benefit for the Company. This could also be a way to bring competition and improve the functioning or the management of the Company. The Mars Colony will be a proactive start-up incubator.

### 3.4. The problem of price

Elon Musk tells us that he could ship people to Mars and offer them a free return to Earth for only 500.000 dollars. Conservatively, at this stage, we prefer to rather take 1.000.000 dollars for a return ticket. Besides, the paying-guests will have to pay for the maintenance of their living quarters on Mars, all their consumables and those of the staff servicing the Colony, twice as numerous as they are, and well paid (most likely in the order of three times what they would be paid on Earth). Therefore we consider that the cost of transport both way and of one synodical stay could reach up to 5 million dollars (let's hope less!). But "cost" is not "price" and price is to be set by supply and demand. Everything above 5 million would be for the Company to start amortizing its fixed assets, supporting its financial charges and develop further the Colony.

Is there a market for these amounts? This is indeed a crucial question because a supply must meet a demand; if not, no income is possible. The Company may decide to subsidize stays or trips and could do so to initiate the fluxes but it could do it for the first or second trip, no longer. It will have to make money as soon as possible.

### 3.5. Reasonable hopes

Right from the beginning, residents will have to try to "do better", to invent and innovate. The staff will work on it, freelance-entrepreneurs will put pressure on them and sometimes find, by themselves, new ways to do things, and do it cheaper. We may hope that this process widens the difference between cost and price paid, so as to generate a higher exploitation margin for the Company or allow it to lower prices so as to get more candidates to come and hopefully stay on Mars. As the margin widens, it will induce an expectation of profit and that will be enough to generate a rise in the value of stocks. People would start to make money on trading them and invest more. In this context, the Company will offer residents only the basic services for the lowest ticket price and will let the residents free to consume what they want atop of that minimum. They could anticipate by paying for a package of facilities before leaving Earth or just wait until they arrive on Mars and decide what to consume but whatever their decision, they will have to decide on most (or a large part) of what they consume once on Mars, with their own money. This is the best way to insure the best adaptation of supply to demand, avoid wastes and guide investments. We expect the beginning of rentability to occur after 20 years of commercial exploitation (such exploitation starting, let us say, on the second or third synodical cycle). If we add an 8 years investment period before starting exploitation, we get close to 30 years. Therefore there should neither be payment of financial costs (dividends or interests) nor repayment of loans before close to 30 years. We will need a "grace period" during which these payments be waived. But, as said before this will not make impossible the expectation of profits and increase in value of the stocks before the Company turns profitable, indeed right from the first years, on the basis of the technological success of the Colony.

We are now waiting for the demonstration of the operating capacity of the Super-Heavy+Starship. This will allow to seriously consider starting the project i.e. structuring the various entities necessary for the financing and allowing the first call for money.

**3.6. Economic model**

Building a scenario seemingly reasonable, technically feasible and financially realistic, is not sufficient to give life to the idea of a Martian settlement endeavor. As already stated above (§3.4), once the salable activities have been imagined and defined, it is necessary to check if the services offered could find a market. More precisely, is there a sufficient number of potential customers, from the beginning and on the long-term, to match the capacities of the company, with pricing conditions leading to a sufficient operational margin. There exists much less literature about this aspect of the project than on technical or human ones. We therefore built a spreadsheet model aimed at establishing a desirable price list, which we then confronted with the histogram of household incomes (in the USA, then extrapolated to the world), in order to get a documented judgement about this market. To run the spreadsheet we had to enter a great number of data, most of them being, true to say, quite putative. Among those data, the most influential are the following ones:

-Transfer costs: we considered 2 M$/mT for cargo and 1 M$/passenger (totaling a mass of 0.5 mT with spacesuit, food, water, personal equipment and belongings); that is twice the values assumed by SpaceX, which we consider too optimistic. This is of course, together with the long travelers stay duration, the dominant cost factor. Besides, this high transport costs make the cargo items prices much less determining (about a third of the total cost); it also allows to allocate interesting remuneration to the service staff (we need it in order to get volunteers and compensate for risk).

-Remuneration costs: the first aim of the model was to determine what actually should be the percentage of paying-residents (customers); by considering the list of functions to be fulfilled and how many people each specialized staff can attend, we found, for a population of 1000, a total paying-residents number of 460 people; for the 540 remaining paid-residents, we picked three salary categories of 35000, 15000 and 9000 $/month, leading to 29% of the staff total cost.

| | **FEUILLE Budget** : Fares data and margin computation (over a synodnic period). | | | | | | | | | |
|---|---|---|---|---|---|---|---|---|---|---|
| COSTS | | | | INCOMES | | | | | | MARGIN |
| PERSONNEL | CARGO | PAYING RESIDENTS FLIGHTS | TOTAL | TOURISTS | LONG-TERM IMMIGRANTS | MISSIONNED PERSONNEL | MEDIA, PATENTS, MISC. | LOCAL VAT | TOTAL | |
| 619 | 912 | 737 | 2268 | 1751 | 89 | 555 | 100 | 32 | 2527 | 259 |
| | | | **FARES** (total for the whole stay) | | | | | **VAT Rate :** | | |
| | | | 1 syn.per. | 6,0 | 5,5 | 8,0 | | 0,05 | | |
| | | | 2 syn.per. | 6,8 | 6,5 | 9,0 | | | | |
| | | | 3 syn.per. | 7,3 | 7,0 | 9,5 | | | | |
| | | | 4 syn.per. | 7,8 | 7,5 | 10,0 | | | | |
| | | | permanent | 17,0 | 12,0 | 12,0 | | | | |

The final results (with our set of data!) appear in the above last sheet of the model, showing:

-that the three operational costs categories are of comparable magnitude;

-that a minimal margin requires fares between 6 to 9 M$ (depending on customer category). Such prices level, when put in relation with the wealthy population volume (annual income above $ 250,000)*, weighted by a ratio of would-be Martian travelers of 1/10000 (deduced from the high ratio of interest expressed for the Mars One project), leads to:

***a potential touristic customers base of 6000 people worldwide***

** In 2011, 2% of US tax households declared an annual income above $ 250,000, suggesting a capacity of funding such an extraordinary expense.*

If this number cannot be taken as such, in absolute terms, we anyway don't get 100. The consideration of this niche of wealthy individual clients in quest of exceptions is therefore legitimate in this exercise. This customer base could make a sizable part of the backbone of a Martian economic model, at least initially.

## 4 - SOCIAL AND CULTURAL ASPECTS

After an incident in 1970s, when a pair of cosmonauts had to be brought home from the space station earlier than planned because "They fought like cat and dog" and finally one of them said, "If you don't bring us down to Earth now, I am not going to work with this corpse any more" (quoted in Ignatius A., 1992) a greater attention to the social and psychological issues has been paid and must be especially paid during the Mars settlement setting up.

To reduce the risk that such a problem happens, proper residents selection, training, inflight-monitoring, aptitude to support, ability to work under reduced gravity conditions, are recommended. Also, all residents should be professional, sociable, responsive, have self-control and a sense of humor, in order to be able to bear with others. Leadership characteristics studies made by scientists in four space-analog environments (aviation, polar bases, submarines, and expeditions) concluded that an effective leader profile included a focus on mission objectives and the ability to take charge during critical situations and sensitivity to other people expertise, optimism, hard work as well as attention to group harmony and cohesion.

With the help of Earth-based habitats in extreme environments simulators, a number of psychological issues that are related to long-duration space missions have been studied and definitely should be taken into account. However, living on Mars can seriously raise the risks associated with psychological problems and interpersonal conflicts. All previous experiences show that a group of people living and working in isolated and confined environments go through stages that are time-dependent. Some people believe that psychological changes occur after the halfway point of a mission, especially in the third quarter (Sandal, G.M. et al., 1995; Stuster et al., 2000). Others define it in terms of three successive time phases: initial anxiety, mid-mission boredom, and terminal euphoria. However, not all space-analog studies have noted such stages (Kanas et al., 1996). It means that special measures, e.g. independent help decision-making systems must be integrated into the current life residents program to prevent such effects.

As a key solution to this problem, our best practices of integration of education, cultural, medical, sports programs should be used. When people are busy with these activities, they are not focused on their mood and apply their energy to their self-development and regular work. From this point of view a several hundred people Mars Colony, in comparison with a limited astronauts number, will be easier to sustain and make autonomous. Leaders selection could be done as it is for the Swiss governing bodies, one of the most stable government systems in the world. The Swiss Federal Council is a seven-member executive council that heads the federal administration, operating as a combination cabinet and collective presidency. In the Mars Colony, a college of several leaders can likewise be selected to head the Colony administration. Any Colony resident should be eligible to participate. It is also quite obvious that the Colony will be multinational and therefore that the local government should reflects that plurality. For others features of the governance see (**5 – How should the Colony be governed**?).

Since, as said above, the Colony should be multinational, it is important to pay attention to developing training programs to increase the awareness of resident candidates to cultural differences and other issues that may lead to conflicts. But of course, conflicts are unavoidable. In this case, continual speech analysis techniques for detecting stress, workload mishandling and cognitive impairment, regular communication with Earth and other architectural solutions (see "**6- Aesthetics and Architectural aspects**") can also help control the situation.

## 5 - HOW SHOULD THE COLONY BE GOVERNED?

### 5.1. Principles

Civilization (living together) implies that people behave according to rules so as not to encroach on the rights of others or on the interests of the community to which they belong. Residents of the Martian Colony must therefore respect the rules ("laws") established by the Operating Company ("the Company", legal structure created by its shareholders to manage and develop the Colony) and that they must have accepted before leaving Earth. This is essential for an efficient management and for the coherence of activities within the Colony.

### 5.2. As much freedom as possible

Most likely, these rules, to be applied to the relations between residents and between the residents and the Company, will be incorporated into the articles of association of the Company and will result from a compromise between the different countries participating in the Company as shareholders, either directly as States (through their Space Agencies), or through some of their citizens (investors)*. Of course, these rules will be drafted in the framework of generally acknowledged human rights declarations but, on account of the risks inherent to the extreme and dangerous Martian environment, of the scientific interest of the planetary specificities and of the ecological vulnerability of Mars, some of the fundamental individual rights may be curtailed if their exercise could jeopardize either the Colony safety or scientific research (for example, health-inspectors should have the right to conduct urgently, without prior authorization, a thorough cleaning of the private premises of anyone residing in the Colony). Indeed, the very particular Martian environment implies that strict safety rules be respected without discussion in order to allow the population to survive from one synodic window of departure to the next one. This also implies that vital and scarce resources, such as energy, oxygen or water, be allocated on the basis of demand and supply only to the extent that the community's vital survival needs can be safeguarded. Finally, this means that people cannot be allowed to dispose freely of their waste and that they have to limit to the strict minimum their ecological footprint on the planet. This means that, for the sake of common safety, the interplay of supply and demand cannot go uncontrolled, that expertise must always be respected, that important decisions must always be carefully taken, but at the same time that emergencies must always be dealt with efficiently.

** It is assumed that there will be private interests among the shareholders. Their participation should be welcome because the development will require important financial resources, and the participants should request and obtain the right to take part in the decision process according to the weight of their investment.*

### 5.3. Different people, different rights

In the population of a nascent society of 1,000 people on Mars, we will have to differentiate between different categories of residents because they will have different interests and responsibilities. On the one hand, we will have (1) individual residents ("paying guests"), either tourists, researchers or private long-term residents; (2) freelance-enterprises, whether companies or individuals ("freelance-enterprises") pursuing an independent economic objective; and on the other hand, (3) the staff of the Company ("staff", "paid-residents") responsible for the administration or the satisfaction of the needs deemed necessary for the proper functioning of the Colony. The staff category will include companies operating at the request (of) and under contract with the Company ("contractors"). It will be organized into various "operational departments", each one responsible for a range of specific services necessary for the proper functioning of the Colony. It will exercise statutory management rights. On the other hand, paying residents will have the right to get in return of their money, a counterpart that they value as much as what they paid for, and the freelance-enterprises that have invested capital to get a profit, should have the right to maximize this profit. In most businesses carried out on Mars, the shareholders of the Company will be the main actors because they will, very likely, have provided some part of the financing and consequently, they will expect, rightly, a return on their commitment. As owners of the Company they will collectively be the ultimate decision makers for the use of the assets of the Company and the evolution of the Colony. They will be represented on Mars by a "Colony Directory" (three people, in any case an odd number).

**5.4. Organization, looking for a balance**

The staff will be placed under the authority of a governing body which could be called the "Colony Executive Council", responsible for coordinating and controlling the various activities developed in the Colony. Around the Colony Directory (in charge of the day to day management), this Council will include the heads of the operating departments concerned with the decisions to be made and five representatives of all residents, the "Council of Martian Resident Representatives" ("CMRR"), elected every six months by the paying guests (including two by those who have been on Mars for more than one synodic cycle).

Decisions concerning a specific activity will be made by the Executive Council only after consulting the people in charge of the operational service concerned. The leaders of the "vital departments" (control of energy, data processing, water, atmosphere, air conditioning, food, safety, health) and the head of global scientific research, shall have the right to participate in all meetings of the Council. Any bearer of a dissenting minority opinion in the Executive Council shall have the right to submit a referendum to the entire population of Martian residents, except in the event of opposition from the head of one of the vital departments. A reasonable percentage (10%?) of the resident population should be allowed to make proposals to their fellow citizens provided that it does not interfere with the safety of the population of the Colony, that its material resources permit it, and that they are accepted by the Company (the Directory will have a right of veto). Disputes will be arbitrated by three arbitrators, two appointed by the Company and one elected by the CMRR. They will be independent from the Executive Council, except for safety matters. Law enforcement and arbitration decisions will be controlled / executed by a police force of five people (who will also be in charge of health inspections) under the authority of the Directory.

### 5.5. Adaptability to change, more than ever a must

A permanent adaptation to a changing situation will be necessary to allow the correct development of the Colony, but an authority to arbitrate the needs of residents will also be necessary, given the constraints related to the scarcity of resources, while preserving the interests of the Company's shareholders and respecting the planet. Difficult piloting... but we still have time to think about it!

## 6 – AESTHETICS AND ARCHITECTURAL ASPECTS

*Figure 1:Photo of a Moscow's interior coach showing bodies of water*

*Figure 2: From the Stanley Kubrick's film: "2001, a Space Odyssey»*

To identify and protect environmentally sensitive areas on Mars there should be a prime consideration for settlement planning. The unique Mars landscape will nurture the architectural ideas for the first Martian settlement.

On account of the length of their stay, residents individual adaptability and appropriate performance capabilities are necessary. At the same time, the architecture of the whole structure or facility has to provide systems and inhabitants with security, sustainability and good living standards. An optimized, compact, modular and flexible design is crucial to provide a proper psychological environment.

In that respect a number of intelligent software agents should be included in the design of habitat systems, which will determine the quality of life and work during the planetary stay. It should make the operation of systems possible without continued Earth-based monitoring and support. Also, for Mars sojourns, it is important to provide design conceptions based on ergonomics research and human-centered design with the integration of an anthropomorphic metric (Volkova T., 2017) which respect the specific activities of the residents. As a consequence of such design the safety (see "**1 - Safety and Health**") of the residents can be increased.

To create such a design, the best practice and feedback during/after human living experiences in extreme environments is being considered. Based on such experience from the medium-duration orbiting facilities including Skylab, Spacelab, Salyut 7, Mir, the ISS, the Shuttle, polar research stations in the Antarctica and Arctic and from Earth-based human space mission simulators, we can make assumptions about the necessary level of comfort and safety for a Colony on Mars (Volkova T. Bannova O., 2017). However, we still need to extend our knowledge about the adaptation of human locomotion to the reduced gravity environment of Mars. This experience can provide new insights into our understanding of the physiological and psychological effects of reduced gravity on residents, as well as the reduced gravity impact on the architecture itself.

As a principal style for this settlement, we propose the Bauhaus style (1919-1933), which is famous for its simplicity, functionality, and practicality. At the same time the expedient use of space with some aesthetically significant elements such as nature scenes, particularly sea-shapes and scenes involving wide-open spaces or forest views for amusement, or sweeping views from an elevated perspective, especially for views involving bodies of water (Figure 1 above) is recommended. With the help of this style, we can combine the human factor and the technical needs of the inside of the habitats. It should also help to achieve harmony between the internal and external environment. As an example, the horizontal and vertical elements should be in balance, and at the same time, asymmetrical, which creates a sense of dynamics inside, where maximum use of space is respected (Figure 2 above). Also, the effect of colors and texture, and illumination in the interior design should be based on scientific experiments. It can have a significant impact on human performance and psychology and for inventory management. For example, yellow, brown, red and orange can be useful in laboratories, workstations and other task-oriented areas, while, blue, green and violet might be appropriate in the sleeping quarters and recreation areas (Stuster, 1996). The display of personal items and the personalization of decor should not be discouraged in this settlement, in the sleeping chambers or the private quarters. But the posting of individually selected images in common areas is strongly discouraged.

## REFERENCES

- Martian Habitats: Molehills or Glass Houses? R. Heidmann, 2016 http://planete-mars.com/martian-habitats-molehills-or-glass-houses/
-Safety design for space systems, G.E. Musgrave, A. (Skip) M. Larsen, T. Sgobba, IAASS, 2009
- The Case for Mars, R. Zubrin, with R. Wagner, Touchstone Editions
- Documents from the Mars Foundation: http://www.marsfoundation.org/docs/
- Mars, Prospective Energy and Material Resources Viorel Badescu, Springer, 2009, ISBN 978-3-642-03628-6
- Resources Utilization and Site Selection for a Self-Sufficient Martian Outpost G. James, G. Chamitoff, D. Baker, 1998 NASA/TM-98-206538
- In Situ Resource-Based Lunar and Martian Habitat Structures Development at NASA/MSFC M. P. Bodifor et al. , 2005 AIAA 2005-2704
- The Mars surface environment and solar array performance, Paul M. Stella and Jennifer A. Herman, JPL
- Towards a closed life support system loop:
http://www.esa.int/Our_Activities/Space_Engineering_Technology/Melissa
- Microbial control: http://planetaryprotection.nasa.gov/file_download/97/MIDASS-ESA.pdf
- Agricultural production in a closed environment, artificialized and isolated. J. DUNGLAS, French Academy of Agriculture, 2018 https://www.academie-agriculture.fr/sites/default/files/agenda/jdunglasmilieuxfermes.pdf
- Vertical farming Skygreens (Singapore) : http://www.skygreens.com/
- Testing closed loop life support technologies Oikosmos :
https://www.letemps.ch/sciences/2014/10/23/aventure-un-voyage-vers-mars-commencait-suisse-romande
- The Economic Viability of Mars Colonization Robert Zubrin (then Lockheed Martin)
- The Emerging Inner Solar System Economy, 4Frontiers Corporation, 2008
- An Economic Model for a Martian Colony of a Thousand People. R.Heidmann, 2016
http://planete-mars.com/an-economic-model-for-a-martian-colony-of-a-thousand-people/
- A.Russian psychiatrist tries to make sure Russian cosmonaut stays up, Ignatius, Wall Street Journal, 1992.
- Interpersonal relations during simulated space missions. Aviation Space and Environmental Medicine, Sandal, G.M., R. Værnes, and H.Ursin, 66: 617- 1995.
- Behavioral, Psychiatric, and Sociological Problems of long-Duration Space Missions, Kanas, N. And W. E. Federson, Houston, Tex.: Johnson Space Center. NASA TM X-5867. 1971.
- Bold Endeavors: lessons from polar and space exploration ,Stuster Jack,, Naval Institute Press, 1996;
- - Diploma work "The new generation orbital station",Volkova T, Markhi/ENSAPLV, Moscow, Paris, 2017;
- Safety and comfort for Moon and Mars habitats: key design considerations, Volkova T, Bannova O. Volkova T.V., Bannova O.K,51st ESLAB Symposium: "Extreme Habitable Worlds", ESA ESTEC December 2017

# 2: IDEA CITY MARS COLONY

**Justyna Pelc, Beata Suścicka, Magdalena Łabowska, Piotr Torchała, Andrzej Reinke**
innspace.team@gmail.com

Figure 1 From the left side: Magdalena Łabowska, Piotr Torchała, Justyna Pelc, Beata Suścicka and Andrzej Reinke (in the rocket).

## PRELIMINARY ANALYSIS

After long hours of debate, we finalize the requirements for 1000 people for the village. The city is self-sustainable to the maximum extent possible. We produce all essential bulk materials and fabricate them into useful structures, together with food, clothing, shelter, power, common consumer products, vehicles, and machines with the only exception to the most important components needed from the Earth. To make life on Mars more fruitful than on the Earth, the crucial elements of the city are political and organizational aspects and aesthetic. Our priorities in the design of the city are a high level of automation, robotization, the system of systems, data-driven development, together with AI algorithms, and harmless effect on the Mars environment. To consider the strengths and weaknesses of the colony project, we conduct a SWOT analysis in table 1. As we may see there are a lot of Strengths and Opportunities, but there are lurking behind a corner a lot of weaknesses and threats, therefore, we strongly believe that a long-term attitude is the best idea to create a self-sustainable colony on Mars.

*Table 1 SWOT analysis*

| Strengths | Weaknesses |
|---|---|
| Development of cutting-edge technologies, new discoveries | Failure in assumption of self-sustainability |
| Making humans interplanetary spices | High cost |
| "Just-in-case" home for humanity | Hard to forecast all the scenarios in the Colony |
| Human desire to explore | Exchanging international wars to interplanetary wars |
| Inspiration, new possibilities | |
| Possibility to unite humanity | |
| Opportunities | Threats |
| Increase the level of education and awareness in people | Funding stop from countries or institutions (political issues, costs) |
| New natural resources on earth | Lack of public acceptance in the face of problems on Earth |
| Overpopulation of land and searching for new opportunities | Insufficient development of technologies for further exploration |

**Our research**

To have a more broad debate and make a significant contribution to the colony design we interviewed 167 people, with the various backgrounds. We asked about political issues, goods without they can't imagine life on Mars and social aspects. We used the results in our project to reflect the beliefs of a wider group of people. We highly believe that these answers brought a unique look to the project and helped us to improve our concept. One the conclusion is that technology will not completely replace a human being (fig. 2). Some conclusions can be found in the sub-chapters of this report. The whole comparison we hope to present at The Mars Society Convention 2019.

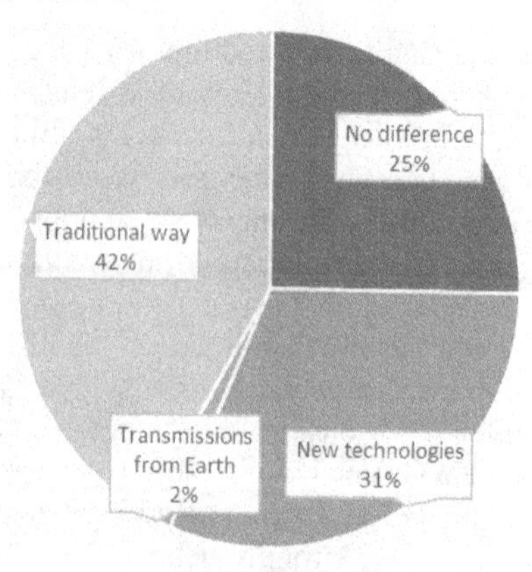


*Figure 2 "How services should be provided?"*

## Risk analysis

We also carry out risk analysis to gather all the issues we need to include or solve in our project. In a table 2 there is a shorter version with only main groups of risks. The most crucial part is the beginning of our exploration due to the possibility of lack of necessary resources like ores or water, so we should launch some exploration mission to make sure that our plans are possible.

*Table 2 Risk analysis*

| Type | Severity [S] (1-10) | Cause | Probability [P] (1-10) | Detection [D] (1-10) | Prevention | Risk level (S*P*D) |
|---|---|---|---|---|---|---|
| The first stages | | | | | | |
| Funding stops | 10 | political issues, bigger cost than expected | 8 | 3 | internationals treaties, accurate valuation | 240 |
| Law issues | 5 | no international agreement about basic aspects of mars base's functioning | 6 | 2 | preparation of plans and proposals of treaties, talks of leading institutions | 72 |
| Technological bareers | 7 | no proper rocket | 5 | 1 | additional funds, cooperation with private companies | 35 |
| Lack of expected raw materials | 8 | no ore deposits | 5 | 5 | exploration missions | 400 |
| Problem with water extraction | 9 | hard rocks, large depth of deposits | 7 | 5 | exploration missions | 315 |
| No adaptation of crops | 8 | radiation, other environmental conditions | 5 | 3 | exploration missions | 120 |
| The self-sufficient base | | | | | | |
| No electricity | 9 | malfunction of nuclear reactor | 6 . | 1 | solar panels and batteries | 54 |
| End of ore deposits | 8 | overexploitation | 3 | 3 | searching for new deposits, responsible management of resources | 72 |
| Mutations or new diseases (of people and plants) | 9 | radiation, other environmental conditions | 6 | 2 | continuous monitoring, previous research | 108 |
| No water | 7 | malfunction of pipelines due to environmental conditions or mechanical damage | 4 | 1 | continuous monitoring, maintenance | 28 |
| Interplanetary conflicts | 7 | attempt to enforce unfavorable solutions | 2 | 3 | international treaties | 35 |
| Buildings degradation | 7 | radiation, other environmental conditions | 4 | 2 | continuous monitoring, maintenance | 52 |
| No resources from Earth | 5 | rocket accident | 5 | 5 | stocks of resources | 75 |

# THE FOUNDATION OF THE MARS COLONY

## Localization

As a place of destiny for the Mars base, we chose Arcadia Planitia. It was shown in the Figure 3. The flatness, smoothness and the land-form without rocks allow making the base easier. It is mainly due to subsoil soft terrain [1]. Moreover, there is an excess of ice [2], which is a very important resource of the colony. Because of this place, we can achieve a compromise between distance to deposits of ice around the Pole (which could support human settlements into water) and the Equator (which is the best place to land a rocket). It is also at a low elevation so it's good for better thermal conditions and solar power [3].

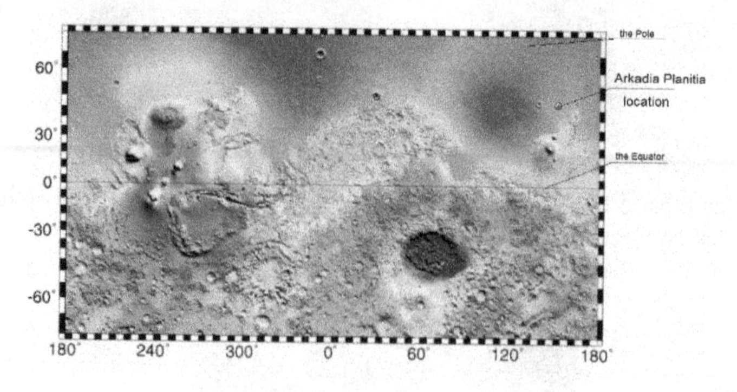


*Figure 3 Localization of our base [1]*

**Idea(of)city**

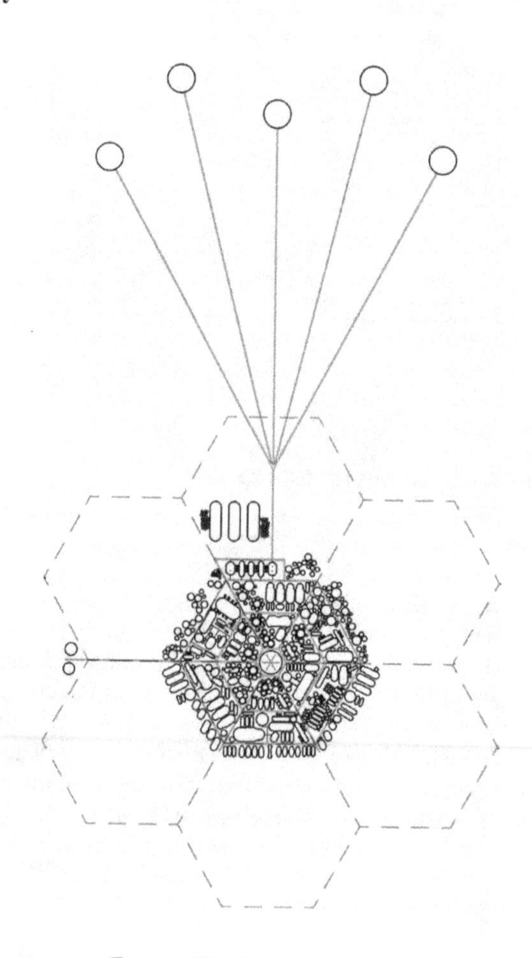

*Figure 4 2D plan of our Mars base*

We merged our knowledge about living on the planet Mars and the idea of the ideal city, as a trend of the new urbanization process in the 21st century. An ideal city is the concept of a plan for a city that has been conceived in accordance with a, particularly rational or moral objective. The "ideal" nature of such a city may encompass the moral, spiritual and juridical qualities of citizenship as well as the ways in which these are realized through urban structures including buildings, street layout, etc. The ground plans of ideal cities are often based on grids (in imitation of Roman town planning) or other geometrical patterns. [4] One of the examples of the ideal city is Leonardo da Vinci's project. Leonardo wanted to design a city that would be more united, with greater communications, services, and sanitation. His ideal city integrated a series of connected canals, which would be used for commercial purposes and as a sewage system. The city would feature lower and upper areas. The roads were designed to be very broad. Defining elements of new urbanism:

- The neighbourhood das a discernible center. This is often a square or a green and sometimes a busy or memorable street corner. A transit stop would be located at this center.
- Most of the dwellings are within a five-minute walk of the center and services, an average of roughly 0.25 miles (0.40 km) [5].
- Streets within the neighborhood form a connected network.
- Certain prominent sites at the termination of street vistas or in the neighborhood center are reserved for civic buildings. These provide sites for community meetings, education, and religious or cultural activities.

The city we propose is built on a grid of hexagons. In the city center is the main point of the city, where all other functions go. In the later stages, we assume the expansion of the base radially in relation to our center. Due to the radiation on the planet, the city is partially located underground.

**System Breakdown Structure**

We prepared System Breakdown Structure with all systems of our base and outputs of our report to make sure that we remembered about all functionalities in our project. We believe these are aspects that need to be considered in such a colony. Due to limited space, we presented only some subsystems more precisely in this report.

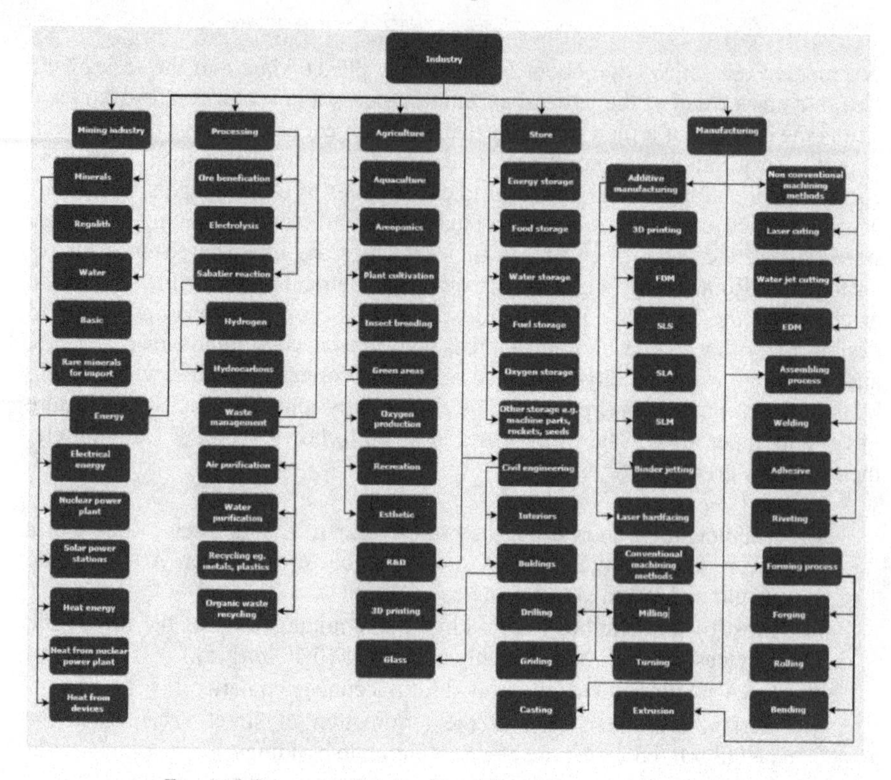


*Figure 5 One part of System Breakdown Structure (Industry)*

## CITY PLANNING

### Evolution

To facilitate the process of designing the base and determine the priorities, we decided to trace its evolution from the first settlers to the final version of the base. This allowed us to determine the most important functionalities, appropriate distribution of key buildings and layout of the remaining space. This simulation has also helped us in economic calculations. Such a project requires a far-reaching plan. We should build our base from the very beginning according to the final plan, not randomly. Thanks to this project implementation will be cheaper, faster and easier in the long run. Figure 6 shows only selected stages of colony development. We assume that the first base will mostly consist of ready habitats brought from Earth. In the next steps, we need to adapt the majority of terrestrial solutions (related to e.g., agriculture, fuel production or resource processing) to Martian conditions and ensure access to water, food and air. An important milestone is to find raw materials and develop a way to extract them. The further development of the colony is related to the delivery of all systems with reduced functionality and their gradual addition with the growth of the base.

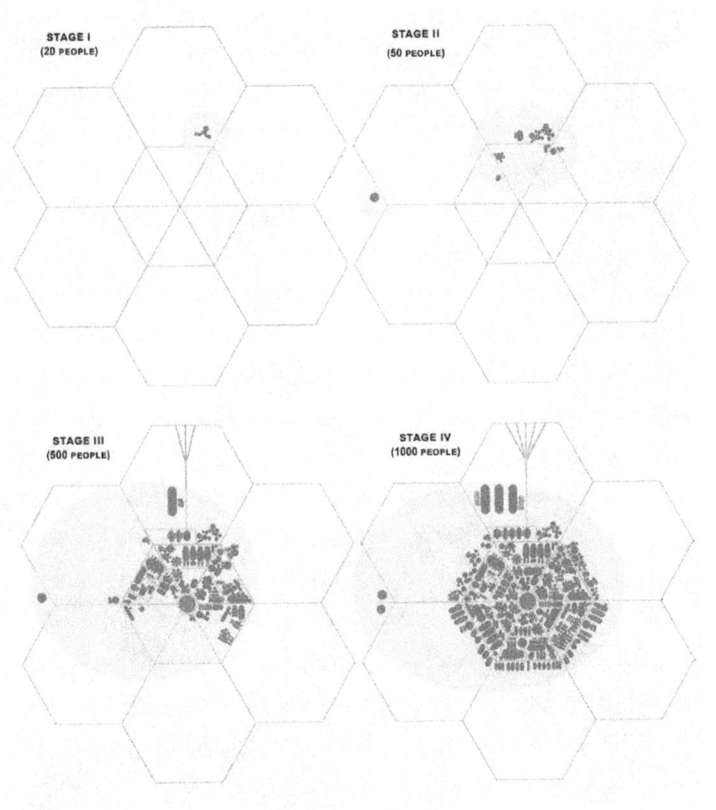


*Figure 5 Evolution of base*

**Paths**

To create the optimal architecture of our base, we conducted a path analysis. We included the flow of materials, electricity, water, and sewage to optimize the paths and simplify the flow of everything. This helped us to create more ergonomic and optimized mars base. We can save resources and time this way and facilitate the use and development of the base in the future. Our analysis is in Figure 7.

Electricity is generated by two sources – nuclear reactor and photovoltaic panels. A nuclear reactor generates energy using turbines that are set in motion by heat. Turbine rotational motion generates electricity. Turbines generate electricity with a voltage of over a dozen *kV*, which is transmitted to the power station. The power station distributes electricity and converts the medium voltage into a low voltage. Taking into consideration the need to ensure continuous access to electricity for all buildings, the station is connected to the receivers with three branches, which provides a redundant connection. In addition, each building is connected to at least two power lines.

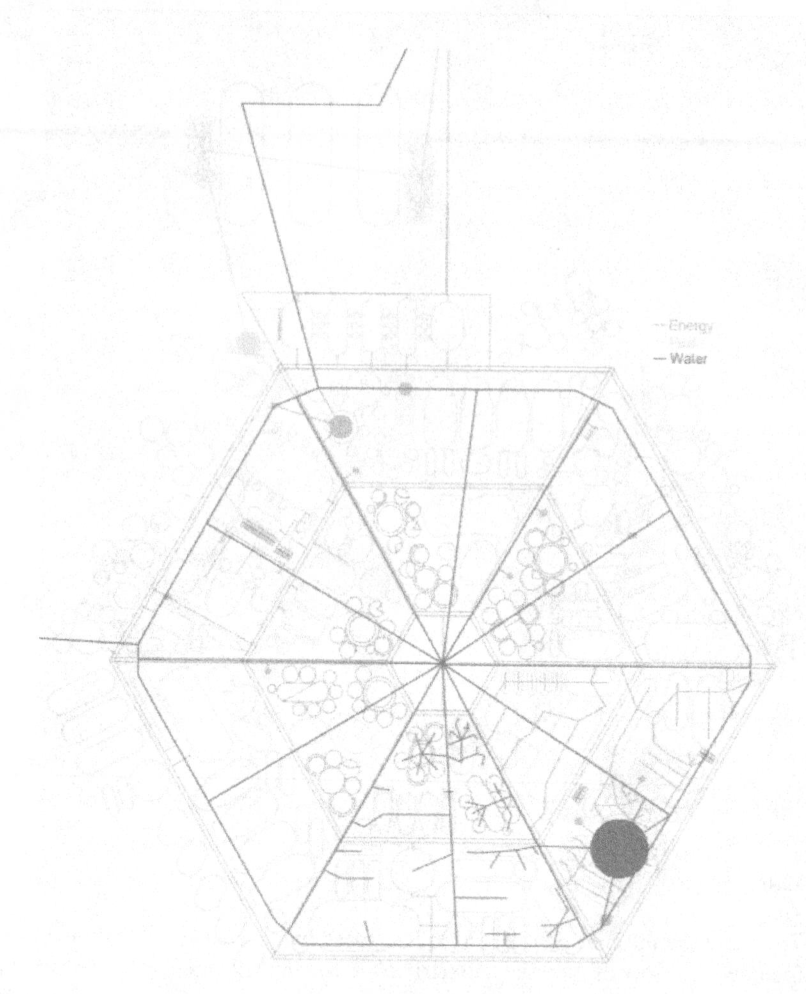


*Figure 6 Path analysis (fragment)*

We obtain water from ice deposits located under the surface of Mars. This is done by drilling into the deposit and then collecting the water in the solid or liquid state and transporting it via pipelines (under appropriate conditions keeping it in a liquid state) to the treatment plant. It is important to take care of the system tightness, because the contact of water with the atmosphere may cause its immediate evaporation. The water in the base circulates in a closed cycle thanks to the sewage treatment plant and the pipeline system. The largest part of the water is used for crops and the flotation process, so there we used the infrastructure for increased bandwidth. Despite the recovery of significant water, a continuous inflow is necessary. Due to the fact that the break in water supply could be a big threat and water consumption depending on the season may vary, we store it in appropriate tanks. The biggest change resulting from the analysis was a decision about reducing the sewage infrastructure by designing bathrooms and kitchens in the center of every building (e.g., a residential one). The air circulates also in the closed circle. We have an air purifier, but installed sensors detect the concentration of elements or substances in the air in a real-time and the air purification system works in a real-time thanks to filters like HEPA. Oxygen is mainly obtained from algae and produced by electrolysis and it is one of the elements contained in the air for humans. Conditions on the mars strongly depend on the season. The temperature ranges from $-125^{\circ}C$ to $20^{\circ}C$. Housing heating is therefore heavily dependent on the amount of heat supplied by the external environment. Depending on the season, the flat will need to be cooled or heated. Apartment heating consists of transferring excess heat from the active reactor to the flats. As the pressure decreases, the thermal conductivity of the air also decreases. This means that the heat passes through the dilution air 4 times more slowly, which makes it necessary to increase the surface area of the outdoor air conditioners.

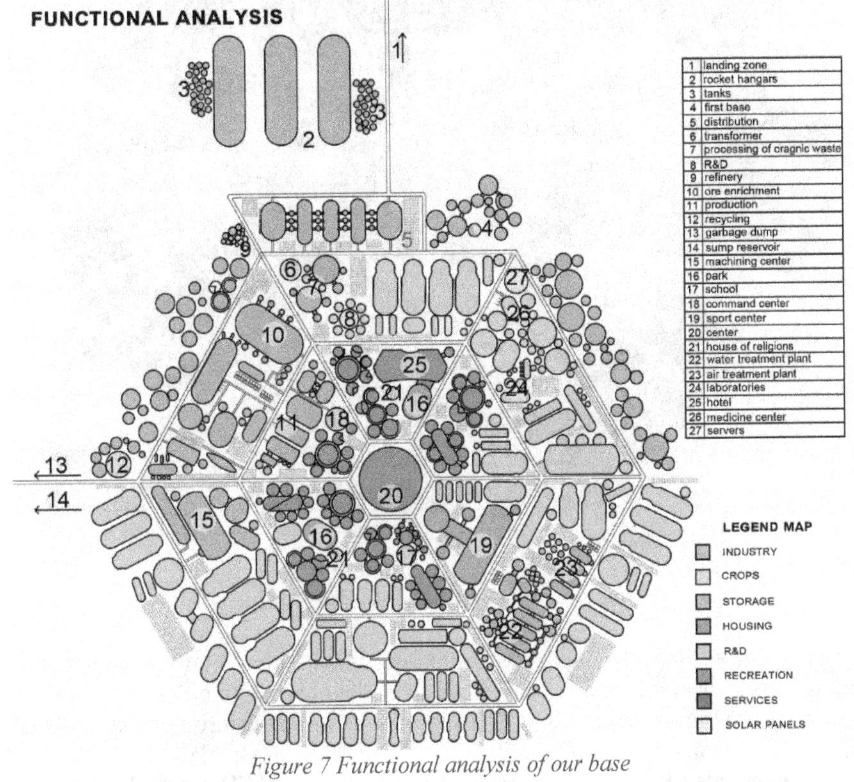


*Figure 7 Functional analysis of our base*

**Optimal architecture of the base**

After all analysis and optimization of our concept, our final version of Mars base looks as follows in Figure 4. We have carried the most important elements on a 2D plan and explained them in the legend (Figure 8). It will help to understand our project and have a better view of our concept. The basis for preparing the plan of our city was an idea of an ideal city and morphological analysis. We took into account the values and the visual experience, such as architectural quality of the building, form and management of public spaces, composition, mutual relations, proportions and specific objects, mutual relations and contrasts in the building or in the landscape, diversity of the city form and places, dynamics/activity, flows and readability and accessibility.

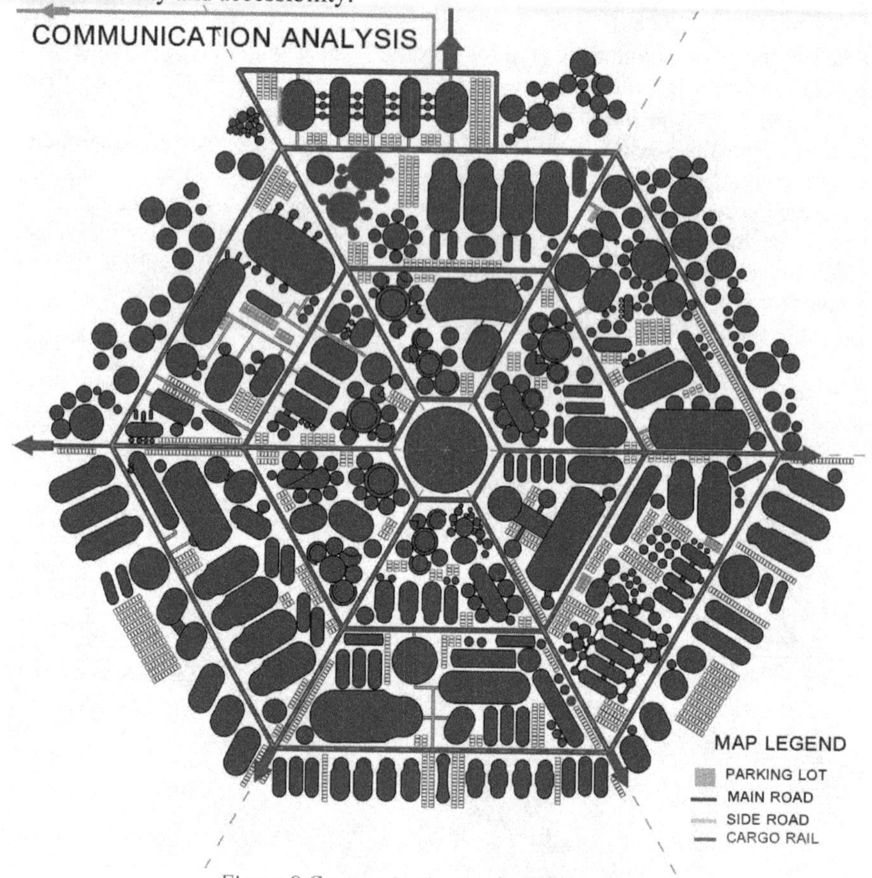


*Figure 8 Communication analysis of our base*

The structure of the urban space is composed of individual building complexes with different functions and forms, public space layouts, communication systems, complexes, and green spaces complexes as well as infrastructure systems. That is why our city is a very complex organism whose material (spatial) structure consists of elements above ground and underground (fig. 8). The paved roads are arranged on the outline of the hexagon and radiate to its interior, and their width is 5 *m* (the optimal width in which two Martian vehicles can pass). They connect buildings with a continuous flow of materials. For other places, transport takes place on a normal surface, using rovers, and reaching every building is possible by keeping the minimum distances (at least 3 *m*) between them. Also, the distances between the designed buildings and solar panels are proper for the rover to travel (min. 2.5 *m*) to allow maintenance. Communication for people is available in underground corridors connecting all buildings. For this reason, moving around the base does not cause additional difficulties, and thanks to short distances and vegetation, which grows on the walls of the corridors, additionally positively affect the condition and the psyche. The transport of resources, e.g., between rocket landing sites and distribution warehouses and between the base and mines, takes place by rail (fig. 9). Some resources, such as water or air, are transported via pipelines.

The key part of our base is located underground, which provides natural protection against adverse conditions prevailing on Mars. We are aware of the difficulties this solution will bring, but we consider developing a method of digging in Martian ground as a key aspect of ensuring security in such a project.

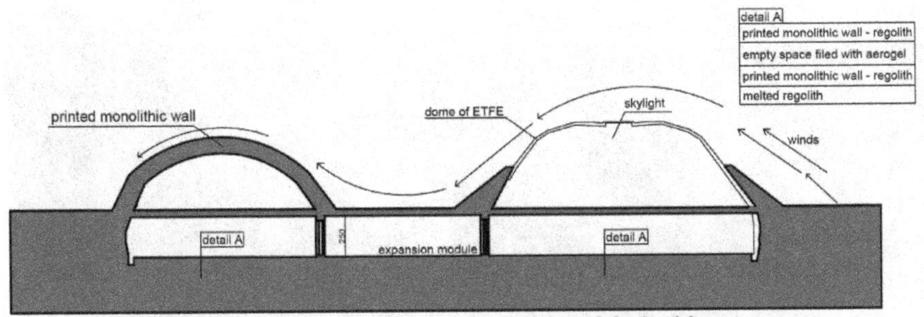


*Figure 9 Diagram of the upper part of the building*

Buildings have from two (in the case of industry) up to four floors (in the case of crops). The highest parties are places of everyday people staying above the surface. The lower floors provide security and implementation of key functionalities. Buildings are built on a circle plan, ellipses or related shapes. The roof surface is rounded in order to reduce sand deposition and thus avoid extra loading conditions. Key constructions are made by 3D regolith printing using printers placed on rovers, machines or robotic arms. Martian soil is the best raw material for building a base, as we have it in abundance, and the current development of 3D printing from the soil allows us to be calm about the possibility of using this technology on Mars. The roof surface is rounded in order to reduce sand deposition and thus avoid extra loading conditions. Some roofs (mostly above the common areas) are made of transparent material - ETFE [6], which ensures natural light of habitat, improving people's life quality and enables planting and vegetable cultivation in common spaces. We have carried out a strength analysis to make sure that our construction will withstand and analyze CFG external buildings to adjust their shape to diffuse air over buildings.

*Table 2 Controlling enviromental condition*

| Factor | Value | Sensor | Actuator |
|---|---|---|---|
| Temperature | 18-27 °*C* | thermometer | radiator |
| Pressure | 20,7-117,2 *kPa* | barometer | compressor |
| Humidity | 40 – 70 % | hygrometers and psychrometers | air moisturizer |
| Air purity | max concentra- | e.g., TCCA, | filters and ab- |

| | tion of indicated elements | MCA | sorbers (e.g., HEPA filters) |
|---|---|---|---|

In order to ensure optimal environmental condition, it must be created an environment management system, which will be controlled background parameters and regulate their value. The example of environmental factors with their optimal valuations, sensors to detect changes and actuators to regulate it was shown in table 3 [7]. It is important to control air purity, metabolic products produced by humans and carbon dioxide content due to the harmfulness of their excesses. All the above steps allowed us to obtain the final 3D model of the base, shown in a figure 11. The model was made in a 3D modelling program Solidworks.

*Figure 10 3D model of our base*

## INFRASTRUCTURE

### Industry

To be gradually self-supporting and developing city, the colony needs to utilize and process local raw Materials, for example for construction and machinery. However, there are also needs for Earth resources, especially that can't be made and replaced by Martian substitutes. First industry area processes the Martian raw materials, as well the Earth delivers. Its location is determined to minimize the distance to the missile launch area (1 in Figure 8) and warehouses (orange fields in Figure 8). Around the missile launch area, there are autonomous robots that unpack spaceships and place packages and materials on the conveyor belts. Those belts have their own destiny and most of them go directly to desire processing areas (10a in Figure 12) or the warehouses.

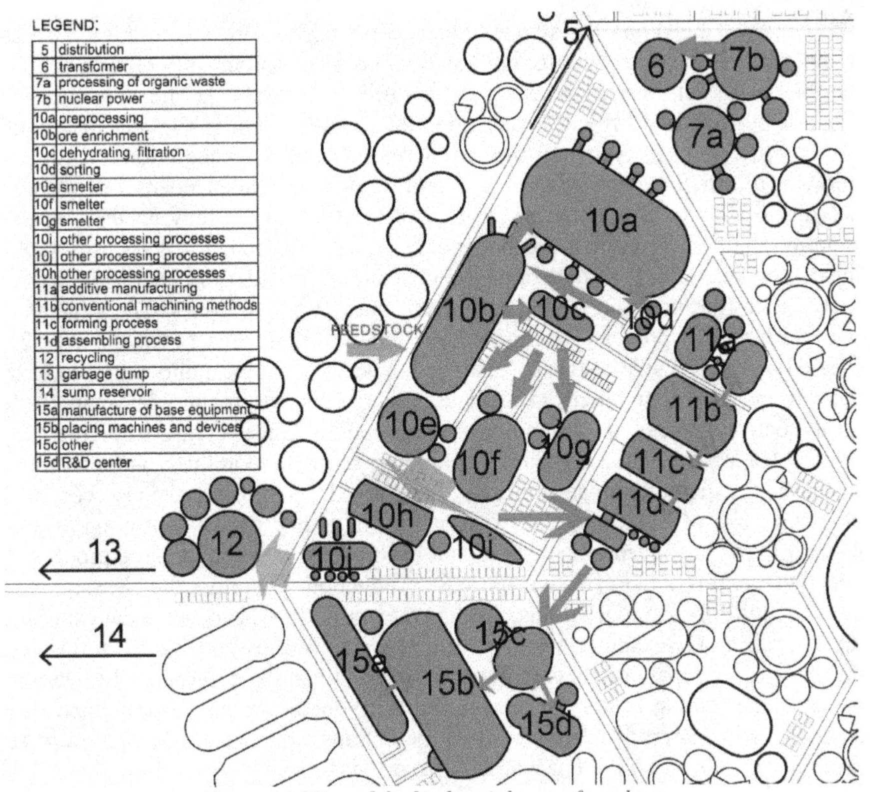


*Figure 11 2D model of industrial part of our base*

The Martian raw materials are commonly occurring ores of aluminum, chromium, iron, magnesium, and titanium in the Martian soil. This has been identified by The Mars landers Viking I, Viking II, Pathfinde, Opportunity Rover [8][9], and Spirit Rover [10]. In addition trace amounts of elements such as cobalt, copper, europium, gold, lanthanum, lithium, molybdenum, nickel, niobium, tungsten and zinc have been found [11] what gives important elements for future base development. With these elements we can produce plastics such as Polyethylene, Polypropylene, Polycarbonate, Polyethylene, Terephthalate, Poly(methyl methacrylate) or Polysiloxanes from the Sabatier reaction, so we can make i.e., pipes for complex constructions or links for rovers. The colony can create also advanced alloys as Inocel, a nickel-chromium alloy with the additions used heavily in the space industry. The one significant problem is with advanced electronics. Due to it is an advanced manufacturing process, there is a need for trading it with the Earth. If the rare element is found in the Martian soil, a kind of mines are built. They are partially automated, they still need human supervisor skills, but the transport from the mine to the warehouses is fully autonomous, it is possible mainly due to GNSS system that is composed of multiple orbiters around Mars, therefore Kalman filter is sufficient for robust and reliable operation.

The conveyor belts and the autonomous rovers with the ores reach the industrial complexes. There, the pre-crushed materials go to pre-processing (10a in Figure 12) before the entire enrichment process (10b in Figure 12). The materials are crushed by mills until the desired degree of fragmentation. In the meantime, the ore is sifted to classify it into the appropriate group. AI recognition techniques allow us to detect the size of the grains and create a few independent lines of processing to simultaneously have a few types of alloys. Then they go through the flotation plant. The result is a concentrate with a high content of the desired element. The next step is to dehydrate the concentrate in order to concentrate it. Finally, after filtration and drying the output goes to the smelters. The generated waste during this process goes to the postflotation waste tank (14 in Figure 12), located on the left side of our base, at a considerable distance, allowing free base growth and providing safety to residents. Semi-finished products leave smelters on the conveyor belts and go to the machine tooling center (11a-d in Figure 12). Even if we assume that 3D printing technology is matured at this point, and most of the things the colony can print out, especially from regolith, we also consider classical ways of machining. The reason for this is that even 3D printing allows using a lot of unconventional shapes and even use it for buildings, the process is very long, and it is not effective for mass production. Therefore the classical machining (milling, lathing, cutting, drilling), even if they have some dangers and drawbacks, as for example dustiness and shavings, are essential and useful. For this reason, we separated these in buildings 11b, 11c and 11d in Figure 12. Nearby the machine tooling center and processing areas there are other warehouses (orange fields in Figure 8) where conveyor belts transport final goods there to have products for trade with the Earth and utilize on Mars.

Another important aspect, especially in case of trading with the Earth is to develop fabrication of rocket fuel on the surface of Mars. Without this, the colony is not able to come back to the Earth and trade goods or exchange residents. The simplest scenario is to use rocket thrusters supplied by a mixture of methane and oxygen. The oxygen is generated due to plants, algae and electrolysis reactions. It is also extracted in Sabatier reaction from $CO_2$ and $H_2O$ to have $CH_4$ and $O_2$. Water needed to fuel production we obtain from underground deposits of ice and carbon dioxide from the atmosphere. The benefit of methane usage over hydrogen as rocket fuel gives some benefits, especially in easy of storing, because of high density and bigger size of particles. The methane storing should be below the temperature of liquefaction so $-162\,°C$. The temperature of hydrogen liquefaction is almost 100 degrees lower and is $-253\,°C$. To create one ton of methane there is a need for 17 *MWh* of energy [12]. The mass of methane for BFR is 240 *t*. Therefore to fill two rockets in two years it needs to be produced at least 1,3 *t* of rocket fuel. Energetic demand for methane production, having ready ingredients is 0,93 *MW*.

The last valuable opportunity for colony economy lies in other celestial bodies in the outer space, such as asteroids. Mars may be a good precursor for the cosmic mining development and our colony may lead in this enterprise. The first argument is that the orbit of Mars is easily accessible due to low gravity and thin atmosphere. Moreover, to the Asteroid belt, it is far closed than from the Earth. From the asteroids, the village extracts and mines palladium (44 000 *USD/kg*), platinum (27 000 *USD/kg*), gold (40 000 *USD/kg*), iridium (47 000 *USD/kg*). So for example, 100 tons of palladium gives the village 4 bln *USD* for trade what in comparison to GDP of Poland 520 bln *USD* looks quite remarkable.

**Energy and waste management**

The second aspect of the industry is electricity production. It is a crucial aspect for the continuous and safe operation of the village. We acquire electricity from two sources. The main source of energy is a nuclear reactor Kilopower. It uses uranium 235 and as medium sodium that is safer than water due to its properties. The energy we partially use for the current use and for accumulation. Water, without a proper cooling system, can rapidly change to the steam that may cause an explosion of the reactor and contamination of the environment. To change heat energy to an electrical one, the colony uses Sterling motors together with synchronous motors on the same shaft. Due to current usage of the uranium and its resources on the Earth, it should be exchanged in fusion rector in 120 years. The wastes from the nuclear reactor are placed 300 *m* under the Martian surface. It is possible to store them also in space disposal but there is a potential risk during taking off as a potentially catastrophic scenario and contamination of the whole planet. Assuming that for 4-6 person crew there is a need for 40 *kWh* [13], we estimate that for the village utility, for 1000 people, around 7 *MWh* is necessarily for the safe operation. The additional source of energy is photovoltaic panels located in free spaces in the colony. Their size is 100 x 100 *m* and provide around 250 *kWh*. Maybe it is not the most efficient ways of energy acquisition but it is ecological, and we treat it as a back up for dangerous scenarios, for example when something goes wrong with nuclear reactor, and there is a need for people survival with the low energy consumption until the reactor is fixed. Similar to Australian approach [14], the batteries are placed nearby windmills that ensure uninterruptedly energy storage. The photovoltaic energy production is strongly related to the season. During summer periods, the maximum energy generated is around $120 W/m^2$, however, during storm sols decreases only to $20 W/m^2$ [15]. To counteract such scenarios, there is a need for the nuclear reactor to increase its usage or reduce the amount of demand (i.e., stop carrying out some research experiments).

In order to manage waste disposal, the village should obey the balanced waste economy rule. The foundation of this is to reduce to the maximum productions of garbage. It is not so difficult since we do not have many resources on Mars to utilize, so we need to manage them carefully. Because of this, we try to fix devices that are broken and do not exchange for the new one unless there is a strong justification. The devices that are not qualified to repair, they go through the recycling process (12 in Figure 8). Bio-waste is converted to fertilizer and is used for agriculture. The metal elements are converted to semi-products i.e., plastic materials are milled and used as a stuffing material. Such principle of operation reduces the waste disposal, and the village maximizes the utility of resources. The same correspond to communal economy, where water (acquired from the ground as well sewage) and air are filtered and cleared in 21, 22 (Figure 8) and is reused in a closed system. Water is also treated to boost quality and to use it directly from taps. All the steps allow reducing the cost of the functioning of the city significantly.

**Communication and electronics**

Communication Earth-Mars is conducted via communication channels that orbits Mars (these are orbiters that have helped to settle the colony). For cases when the Sun eclipses the straight communication the kind of antennas may be placed in Lagrangian points of the Solar system. It provides robust communication channels with the Earth. Communication within the colony is arranged based on the G5 network with Ethernet and MQTT protocol that is used for IoT and smart cities. This ensures that all the devices are connected to the web and they provide real-time information about its state. It causes that if one device is about to be a danger for the colony information will be present immediately. IoT also gives a chance to more robustly utilize resources that the colony has by gathering essential data of various kinds and use AI algorithms to optimize the intake of water, food and reduce the other important factors. Cameras on the pavements and routs around the colony may also give our feedback about the humor of the people around (similar as HMI in the current systems) to asses quality of life and find very quickly the symptoms of depression or other illness that may influence the performance of the colony or even avoid the disaster. For better GNSS navigation kind of communication, posts are printed from the 3D printers in order to improve navigation for Mars exploration. For Power Electronics, the village has GaN-based power converters for quick clocking that influence the amount of space for these converters and reduce the heat loss. In the case of short circuits in the electrical grid, the village has two redundant cables to provide a power supply to each section of the colony all the time, without any chance of failure. In case of local small short circuits a normal fuse and over-current circuit breaker for each circuit.

**Research and development**

In our colony, we domesticate area around 9300 $m^2$ meters for research and development labs. They are located in the upper part of the base, near the missile launch facility and warehouses because it guarantees the well-established principle of cooperation and trust flow of information between industry and research. The labs are crucial for Mars colony because people there, need to adapt earthly technologies for the local harsh environment or even devise the new equipment for problems that we couldn't have expected or predicted. The necessity to work on new solutions with the given resources creates an opportunity for creating huge Intellectual Property portfolio. Lots of those solutions can be used on the Earth (or other celestial bodies) therefore IP is an element of economic exchange. Considering the importance of critical aspects for the base, we distinguish the following departments:

- Biological Chemistry Lab - research on compounds and particles found on Mars to utilize them in agriculture, especially genetically modified organisms (e.g., plants that react positively for radiation), new kinds of fertilizers and means to cultivate plants; fabrication of medicines, drugs and other dietary supplement; research on new sources of storing and generating energy.
- Electronics Lab - research on transistors made from Martian geological resources for CPU, power electronics manufacturing; creation of new kind of components for electrical machines and devices (e.g., a cooling unit for power machines); collaboration with Biological Chemistry Lab to develop new sources of storing and generating energy.
- Material Science Lab - research on properties of new materials and composites made to ensure long-lasting and safe base; manufacturing of the new kinds of Smart Materials as well development of the state-of-the-art materials in a cheaper way.
- Artificial Intelligence and Robotics Lab - work on an adaptation of robotics to the Martian environment and conditions; research on Virtual Reality technology for the distant rover teleoperation (i.e., quadruped exploring caves or other dangerous environment); development of automation of the base powered by data-driven approach to optimize resources utilization; AI algorithms for medical diagnosis.
- Technician lab - repairing and maintenance of the devices in the base.
- Geological lab - planning ventures in search of new resources of mineral deposits, water or place for new colonies; research on materials found during those ventures.
- Social lab - research on a behavioral aspect of people on Mars, especially people state of the mind in such specific environment; work on the better interpersonal relationship between people in the colony; analysis of data acquired from permanent monitoring people behavior and data-driven approach for Human Machine Interface to detect early signs of depression and solitude.

**Diet and agriculture**

The Martian environment does not provide conditions for food cultivation due to lack of basic factors such as oxygen and nitrogen supply, optimal temperature, hydration, isolation, and non-toxic soil. In addition, regolith that covers the surface of Mars has an arid material that contains perchlorate chemicals that are toxic to humans. To enable food production, our base needs to create suitable conditions in a glasshouse (fig. 13) and plan food management. The project provides a build-up of glasshouse (total 20 *ha*) and food storages (total 7 *ha*). The size of glasshouses and food storage allows to provide daily food, but also survive while an accident has happened and rescue operation is necessary. To reduce the prevalence rate risk of diseases we recommend for citizens a plant-based diet. This type of diet reduces total and LDL cholesterol and lower the risk of cardiovascular diseases. A diet high in fruits and vegetables may decrease the risk of high blood pressure, stroke, and colorectal cancer [16] and help with weight loss

[17]. To fulfill all the health requirements the diet is supplied also by proteins from insects, dietary fiber, vitamins and minerals from algae. The rest of human demands is supplemented by processed food and food additives. According to research [18] The Basal Metabolic Rate (BMR) is the amount of energy expended while at rest in a neutrally temperate environment and in a post-absorptive state we need to calculate average calories per day and divide them into proteins, carbohydrates, and fats. In table 4, the example was shown for 25 age person with height 180 *cm* and weight 72 *kg* we show a sample diet for one-day bases for him is 1717 *kcal*, so less than on Earth due to less gravity. The coefficient and the fat rate, the weight will be monitored and change according to keep the weight of the people within the accepted range. With the diversification of the BWT 50 percent, 30 percent, 20 percent, according to the acceptable macronutrient distribution ranges.

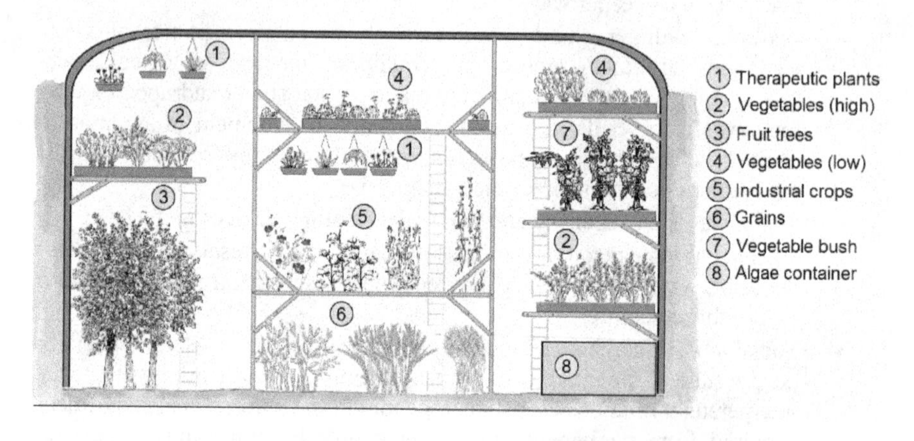


*Figure 12 Example of glasshouse*

*Table 3 Exemplary one-day meal for a man who is 180 cm and weights 72 kg - 1717 kcal*

| Breakfast | Lunch | Dinner | Supper |
|---|---|---|---|
| Garden cress + processed food | Soup: Tomato cream | Algae pasta + insects | Protein bar + fruit |

After reviewing research done by NASA [19] and the average nutrition requirements for humans we decided that, based on mostly the size-to-pace coefficient that the following plants are suitable for people in our colony: Tomato (*Solanum lycopersicum*), Carrot (*Daucus carota s. sativus*), Stinging nettle (*Urtica dioica*), Garden cress (*Lepidium sativum*), Potato (*Solanum tuberosum*), Soya (*Glycine max*), romaine lettuce (*Lactuca sativa L. var. longifolia Lam.*) and other, but also it will be insect breeding (*Dubia Cockroach*) and seaweeds bioreactors (spirulina (*Spirulina platensis*), chlorella (*Chlorella vulgaris*)). Plants will be cropped also for medical applications. Moreover, the hydroponic system allows to cultivation almost every plant, which is not obtainable in Martian soil. Production plants such as flax and cotton enable to manufacture own clothes. Cultivation of Martian plants intended for food will take place in a glasshouse with purified Martian soil – Regolith with the addition of nutrients, clay or peat in order to the maintenance of soil moisture and the addition of cyanobacteria. Cyanobacteria are a phylum of bacteria able to produce oxygen, they are applied for better plants development. Moreover, they are suitable to live in the harsh environment on Mars and they are able to extract and adapt to the assimilable form from Martian resources. [20] The conditions in the glasshouse must be similar to earthly air conditions (i.e., temperature, humidity, air mixture (an appropriate ratio of oxygen and nitrogen) under adequate pressure) with proper availability of light (in the project was applied artificial lighting systems). As our city is developing, there will be more data obtained from plants quality, which are farmed on Mars. This allows us to apply the optimization of AI to crops cultivation and harvesting. Moreover, we also want to migrate to artificial meat after the development of clean and sufficient method.

**Residential buildings**

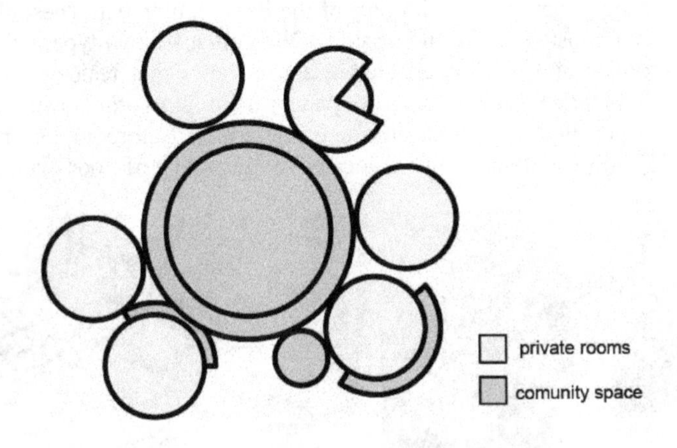


*Figure 13 Functional plane*

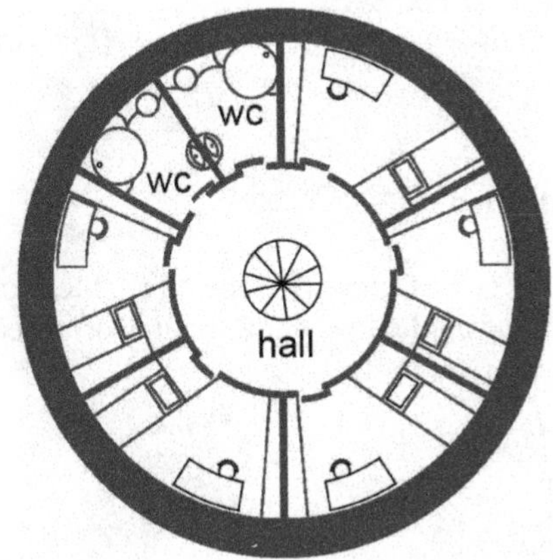


*Figure 14 Typical floor plan of a residential building*

Our idea for a living is cohousing. Cohousing is an intentional space where people know each other and look after one another. People have their own place, but you also share significant spaces, both indoors and out. We designed a scheme of residential buildings. In the middle is a common house, on the sides rooms with private or common bathrooms. In our research over 50 percent of checked answered that common bathroom is not a problem for them. Common house is the place where the social interaction and community life begin, and from there, it radiates out through the rest of the community. Inside common house, we have a large dining room and kitchen. In support of those meals, we have a large kitchen so that we can take turns cooking for each other in teams. Psychology has proven that people who interact with other people have a better frame of mind. In addition, we also included natural plants in this space. Plants provide oxygen, have a calming effect on humans and are familiar associations. The room dimensions are optimally adapted to the needs of the resident. There is the possibility of moving internal walls and connecting rooms. Inhabitants have their own space with a bed, desk, chair, wardrobe, computer and music player, as our respondent decided to. In Figure 18 we see private room with part of the basic equipment. These things turned out to be the most essential in our study. We anticipate two types of room decorations: futuristic and traditional. Our research shows that residents would like to have such a choice. Figure 16 shows part of bathroom with shower which has adjustable water flow and hydromassage and toilet. Bathroom has round shape. Figure 17 shows stairway which is located in the middle of floor plan.

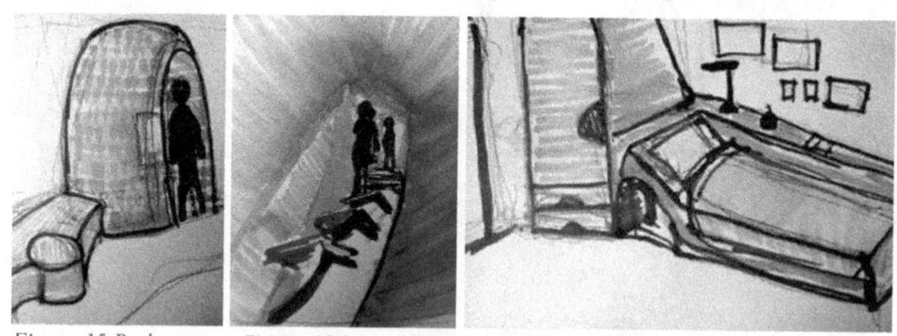

*Figure 15 Bathroom* *Figure 16 Stairway* *Figure 17 Part of private room*

## SOCIETY

### Government and organization

With managing the colony comes great responsibility, therefore, to decide what type of governance should be applied we carried out short research shown in Figure 18.

*Figure 18 Answer for questions in our research about "How should the colony govern itself?"*

The first answered the question "How the base should be managed" and it shows that people want to have a high-skilled-and-educated decision body. The only 30% of people chose the system different than with the technocratic approach. In our opinion, the difficult conditions on Mars require to make wise decisions based on true data. It does not mean that settlers do not have anything to say. Our chosen model assume that democratic decisions are made for daily problems and ways of functioning in the colony and decisions taken by the council only in the most tactical aspect for the colony. The council schedules and determines the long-term goals such as how to explore Mars, security of the base and direction of research and development, but also foreign policy, economy, especially communication and trade with the Earth and the law. The, however, choose about family policy, working system or building issues. The habitants should choose the council based on their qualifications for each sector independently. However, according to our research (Figure 18, chart 2), in response to the question who should choose the council, the people are aware that these decisions are crucial for the base development and 39% of them support the answer that some representatives from the

Earth should also choose the council. We incline to those respondents and we reckon this way of choosing as the optimal model, for mutually beneficial and long-term good relationship development. It allows having the habitants shape their own style of living while keeping the fundamental direction of development as an objective judge. At the same time, considering the long-term development of the colony, in the future the colony should aspire for being totally independent of the Earth. In regards to question 3 (Figure 18, chart 3) whether who should own the base, interviewees weren't consentaneous. In our opinion, the colony should be at the beginning of international territory and only founders should be pointing out the direction of development. In the future, when the colony is found on the proper level of development (i.e., GPD will have some value) it should be converted to an independent country.

**Economy**

We asked people how should the ideal situation on Mars look like. They could choose max. 5 answers. Most of the respondents think that everyone should be equal on Mars. It is also not surprising to indicate the need to provide basic needs for every settler, taking into account the lack of the possibility of, for example, the problem of homelessness due to external conditions. This can cause a complete lack of necessity to enter a circulation of money.

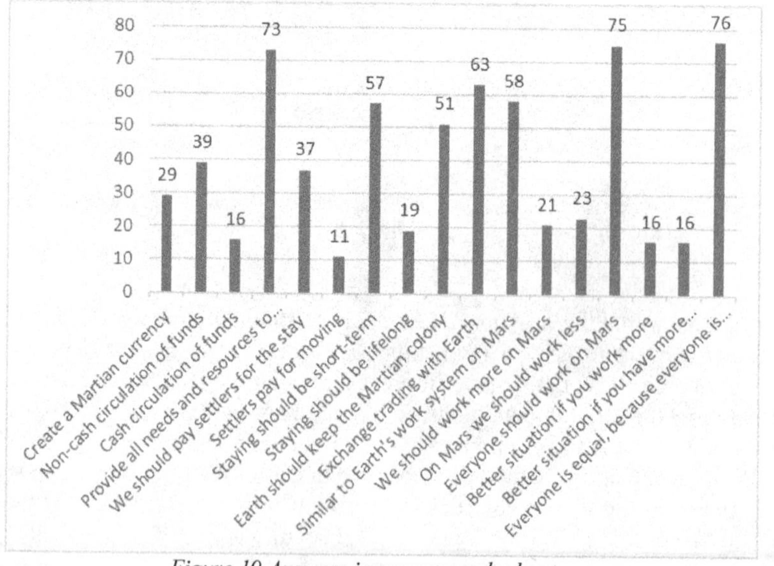


*Figure 19 Answers in our research about economy*

On the other hand, they say that in return work should be obligatory. As far as the work system is concerned, most of the respondents do not see the need for a longer or shorter working day than on Earth and indicate a similar dimension of hours. On these assumptions, we base the economic system of our colony. There is no need to enter a cash flow. All settlers are provided with both basic needs and those of a higher level (higher according to Maslow's definition). Of course, in certain limits related to the limited resource abundance and the need to use them in a sustainable way. The respondents also believe that at this stage the base should trade at the partner level with the Earth, and not be maintained, which confirms the concept of a self-sufficient colony. We can be expected that the cost of shipping goods from Earth to Mars will be $500/*kg* and the cost of shipping goods from Mars to Earth will be $200/*kg*. The exported and imported goods and other ways to obtain the necessary resources to be self-sufficient are presented in the tables 6 and 5. With expenditures equal to $372.95 million and revenues of $464.8 million the colony can not only maintain itself, but even make profits. The largest amounts are costs and profits from technology, which is why we believe that our base should work like a high tech research center selling its own technology as well as conducting commissioned research. Also, in case of using the SpaceX rocket for transport, we can assume we need a maximum of one flight a year to provide the needed resources.

*Table 4 The outcomes*

| Outcome | Amount [t] | Price with delivery/per year [mln $] |
|---|---|---|
| Rare resources | 4 | 80,8 |
| Research placement for scientists | - | 4 |
| Exceptional infrastructure for services (e.g. the biggest clean room) | 5 | 12 |
| Implementation of projects for space agencies | 15 | 69 |
| New (or better) methods of production | 2 | 5 |
| Commercialization of scientific achievements | - | 5 |
| Scientific research commissioned | - | 80 |
| Academics exchanges | 10 | 2 |
| Plant strains with specific properties or better method of growing | 0,5 | 3 |
| Tourism | 10 | 180 |
| Marketing industry (gadget stores, advertisements) | 0,5 | 6 |
| Entertainment industry (movies, exhibitions, streams) | 1 | 8 |
| Valuable items (pieces of space items, art, spices) | 2 | 10 |

*Table 5 The incomes*

| Income | Amount [t] | Price with delivery/per year [mln $] |
|---|---|---|
| Uranium 235 | 3,5 | 8,75 |
| Advanced electronics | 1,2 | 150,6 |
| Software licences e.g. Matlab | - | 14 |
| Spare parts | 2 | 12 |
| Maintenance of the machine park, e.g. welding gases, laser crystals | 5 | 47 |
| Prepared food and supplements | 20 | 11 |
| Seedlings and seeds | 10 | 5,6 |
| Solid additives | 2 | 1,5 |
| Licences for technology | - | 24 |
| Ready projects (e.g. machines) | - | 94 |
| Everyday use items (e.g. energy-saving light bulbs) | 1 | 3 |
| 3d printing materials | 1 | 1,5 |

## Social and cultural aspects

In such a closed environment, social aspects will play an extremely important role. Our goal is to integrate the base residents on many levels. For this reason, the rooms are grouped into modules connected by a common space, in which there is a kitchen, dining room, relaxation zone with home cinema or basic fitness equipment (like treadmills) or living plants. The implementation of other activities, such as education or hobby also requires integration with other residents. We have chosen a few buildings for it. A week of work looks on Mars just like on Earth, that's why residents have a lot of free time to develop. Most of the services are carried out in the center (20 in Figure 8). There is a gastronomic area with a canteen (caring for the basic catering of all residents), as well as a restaurant and bar area and entertainment area. The entertainment part includes cinemas, a bowling alley and a concert hall (to carry out your own events). In the center you can also find a part related to cosmetic services, such as SPA. An important place is played by the cultural part, which includes a theater hall or an artistic studio to pursue your own passions. We believe that culture will play an important role in the life of the inhabitants of the base, in view of the fact that their basic needs will be assured. The necessity of increased traffic (through reduced gravity) to maintain proper health and condition makes the extended facility a sports center (19 in Figure 8). There is a multifunctional playground for the implementation of many sports, a gym with training equipment and classrooms for fitness classes (both individual and group).

The education of minors and the continuous development of adults is taken care of by the school (17 in Figure 8). Our education puts a strong emphasis on practice, on the one hand it allows individualization of teaching, and thus better results, on the other, reduction of teachers. Traditional classes would be related to general knowledge and implemented using both 16 new technologies (online) and in the form of traditional exercises with the teacher. Taking into account the results of the survey (Figure 2) we do not want digital education completely, but we want to minimize the role of teachers to they would learn advanced knowledge during internships in individual departments and laboratories of the base. Learning from practitioners (i.e., specialists working in a colony) would allow to have up-to-date knowledge, individual approach and involvement of all residents in the upbringing of the younger generation. The resources (books, films, etc.) allow for a relatively small space to contain a huge amount of knowledge. We have access to huge resources of earthly literature, music and art, which is invaluable in our own development. We also care about continuous improvement throughout our lives, hence we have a huge number of online (individual) courses as well as we organize group classes on selected topics. An interesting solution is the hotel (25 in Figure 8). On the one hand, he accepts visitors from the ground - tourists or researchers, allowing for mutual knowledge of cultures, but also for holidays for Martian residents. In order to preserve the mental health of residents (rest from work) and high comfort of visitors (not used to the Marian conditions) our hotel is the only place in the database with such a high standard. Windows simulate various places, swimming pool with beach, large vegetation and non-standard entertainment, like water attractions, a paper wall extending over several floors, having a start in the pool, entertainment providing adrenaline (laser-tag, go-carts). It is also the starting point for all tourist trips around Mars. We have also designed several parks (16 in Figure 8), which allow you to calm down and commune with nature, and near them there are relaxation buildings, in which there are prayer rooms and rooms adapted for relaxation (yoga, meditation, sounds, lights).

The respondents in our research were to indicate a maximum of 5 activities that would be most important to them in the Martian base (figure 21). The majority of votes pointed catering services (pubs, bars, restaurants) and physical activities (gym, pitches, swimming pool). A lot of votes has gathered also personal development (libraries, additional classes) and traditional entertainment like cinemas, bowling, etc. They are standard forms of spending free time, popular on Earth. The indicated activities occupy an important place in our project. However, considering the psychological aspects, we decided to provide the base with a wide spectrum of leisure activities to minimize the feeling of being closed and the lack of perspectives.

**Design and aesthetics**

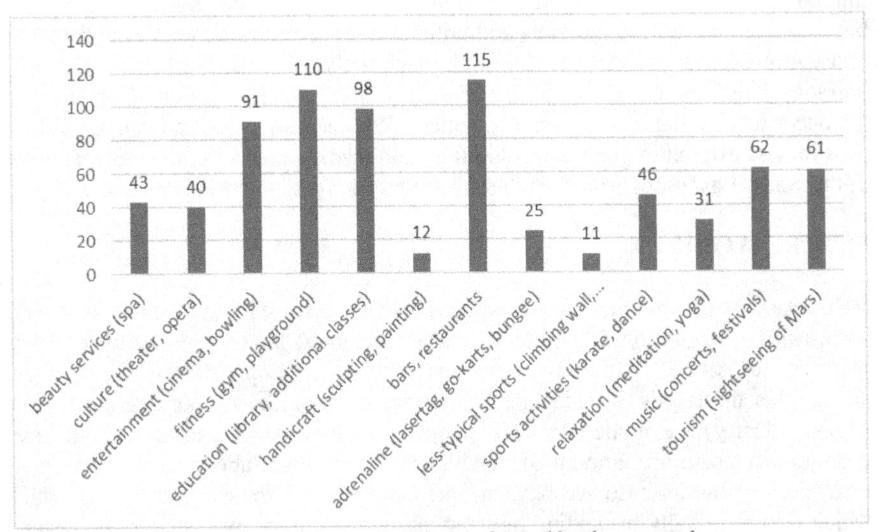


*Figure 20 Activities on the Mars indicated by the respondents*

We adopted bionic and parametric architecture to designed our habitat. Using the relationship between digital tools and architectural design. Computational design encompasses both advanced fabrication techniques for non-standard geometries and new dynamic and adaptive modes of spatial operation. Structure system is demonstrated by working with models from mathematics, biomimicry, human physiology, and psychology. A synthesis of these domains via computational tools and techniques with the aim to derive performative habitat formations is what drives our design process. What's missing on the Mars? In our research people most often answer classic interior design, known fragrances, colors, and living plants. Firstly we decided to print the flower garden. Like real flowers and plants, their 3D printed equivalents would be aromatic, to freshen the air, and perform the work of an air recycling unit changing carbon dioxide to oxygen. They would also 'grow' as each plant would be made up of smaller components that could be rearranged or added to over time as if growing in nature. We also included natural plants in every possible place. They not only provide oxygen but also have a calming effect on people. We put them in the corridors, common spaces, living rooms, and we have even chosen some buildings for parks. Secondly, we used the psychology of colors to make the interiors more friendly to everyday living. Thirdly, it is ergonomics, that is the process of designing and arranging workplaces, products, and systems. It refers to the work of people - working spaces, sports and leisure, health and safety. Uses anthropometry, biomechanics, environmental physics, body systems, applied psychology, social psychology. During the design, we have taken into account natural light, which is extremely important. In all places, we took care of other environmental conditions, like the intensity and color of artificial light. Other factors were less important for our respondents, but we included them in our project. We designed various spaces to avoid the impression of monotony and created designer interior furnishings (3D printing gives us unlimited possibilities in this matter). Taking care of the balance between responsible management of raw materials and unlimited space on Mars, we have provided much more space to live than the minimum for the well-being of residents. Important for us was using various textures (e.g., materials like cotton, flax, silk) in the decor to give the impression of coziness. We decided to take advantage of the possibilities offered by the new technology and display views in the place of some of the walls. Last aspects were including pieces of arts to decorate the space.

**FUTURE WORK**

1000 residents in a base is a good moment for becoming self-supporting as a base, but it is just a beginning. It is obvious that if we can create a safe place on another planet which satisfies all people's need we should develop it further for another and another thousands of residents. Some systems in our base (like a refinery or a whole industry) are made for 1000 people, but they can be used without any changes also for more amount of inhabitants. We thought about further development during this base so we have adapted our project for easy further growth, which occurs radially by adding housing and greenhouses. We believe, however, we should look for another place on the surface of Mars for the second colony and every next one to fully colonize this planet. In this case, we should also start to work on Mars terraforming to make this planet more friendly to people. We should take care to not destroy Mars like Earth, so we should develop our colony in a sustainable way. Our goal is to create a better place and it won't be possible if we will not take lessons from our earthly behavior. It is also important to remember that creating a Mars base as isn't a "just-in-case" home, but we should do this to have the opportunity to transfer the perfect solutions from Mars to Earth and make both planets a better place.

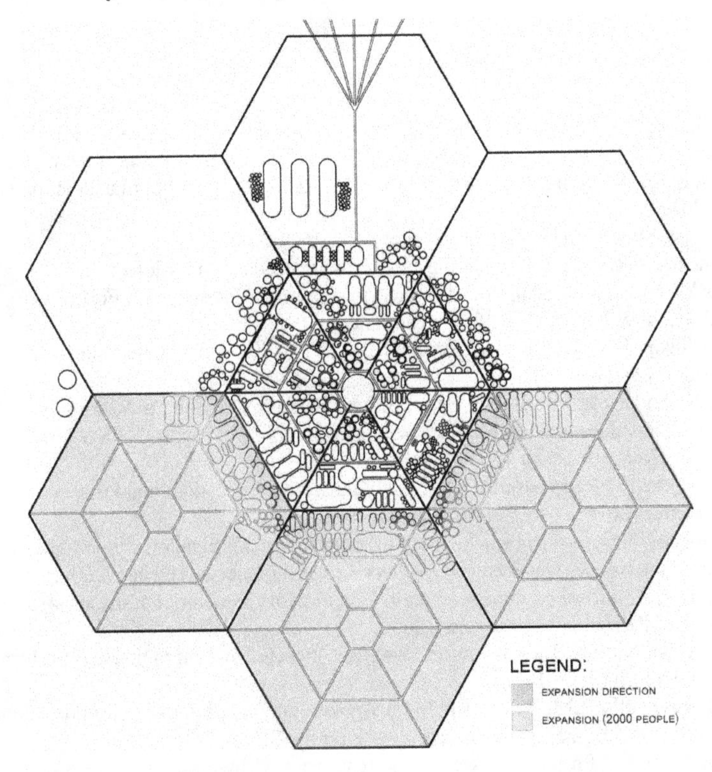


*Figure 21 Evolution of base*

## SUMMARY

We highly believe that humanity should create a colony on other planets. Exploring is inscribed in our nature. The best idea is to create not a temporary base for a few astronauts but a colony for residents. We should take care that Mars is even a better place to live than Earth and develop our base in a sustainable way. It is necessary for such a colony to produce all essential materials like food, power, shelters, etc. on Mars, not bring them from Earth. This way we can create a new place for living, not only a holiday destination. That's why we include all systems in our base. We think also, that creating a colony on another planet isn't only a technical problem, but a big social issue. We tried to find a perfect solution for all aspects of people's functioning as a society on a Mars base. We believe that our small research (survey) has helped us to create a solution that satisfies a larger group of people. It was quite a challenge to present our work on only 20 pages. We tried to explain the main concept and our approach to the problem. We've touched every aspect of the functioning of our base with a closer look at selected issues. We will be happy to present the rest of the project during the final of this competition.

## REFERENCES

[1] Arkadia planitia. https://en.wikipedia.org/wiki/Arcadia_Planitia. [access 30.03.2019].
[2] Nathaniel E. Putzig Sarah Sutton Jeffrey J. Plaut T. Charles Brothers John W. Holt Ali M. Bramson, Shane Byrne. Widespread excess ice in arcadia planitia, mars. Geophys. Res. Lett., 42, 2015.
[3] Jeff Foust. Spacex studying landing sites for mars mission, 2017. https://spacenews.com/spacex-studying-landing-sites-for-mars-missions/.
[4] E.Szpakowska. Architecture of ideal city. introduction. Przestrze´n i Forma, (16), 2011.
[5] Transit capacity and quality of service manual, 2013.
[6] Amy Wilson. Etfe foil: A guide to design, 2013. http://www.architen.com/articles/etfe-foil-a-guide-to-design/.
[7] Molly S. Anderson, Michael K. Ewert, John F Keener, and Sandra A. Wagner. Life support baseline values and assumptions document. Technical report, NASA Johnson Space Center Houston, Texas 77058s, 2015.
[8] et al. Squyres. The opportunity rover's athena science investigation at meridiani planum. Science, 2004.
[9] D. et al. Rodionov. An iron-nickel meteorite on meridiani planum: observations by mer opportunity's mossbauer spectrometer. Meteoritics and planetary science, 2007.
[10] et al. Yen, A. Nickel on mars: constraints on meteoritic material at the surface. Journal of Geophysical Research Atmospheres, 2006.
[11] Iulian Ivan. Nimish Biloria, Andrei Dan Mus, etescu. Mission to mars: A performance driven design approach. 2014.
[12] A.C. Muscatello R.M. Zubrin and M. Berggren. Integrated mars in situ propellant production system. Journal of Aerospace Engineering, 2012.
[13] Karol Z˙ ebrun´. Projekt kilopower - jak reaktory ja˛drowe zapewnia˛ przyszłos´c´ misji kosmicznych. http://www.benchmark.pl/aktualnosci/kilopower-reaktor-jadrowy-dla-misji-kosmicznych-mars-sondy.html. [access 30.03.2019].
[14] Thuy Ong. Elon musk's giant battery is now delivering power to south australia, 2017. https://www.theverge.com/2017/12/1/16723186/elon-musk-battery-launched-south-australia.
[15] Gianmario Broccia. Energy production in martian environment powering a mars direct – based habitat, 2018.
[16] A. Bub S. Ellinger H. Boeing, A. Bechthold et al. Critical review: vegetables and fruit in the prevention of chronic diseases. European Journal of Nutrition, 51(6):637–663, Sep 2012.
[17] Susan E Steck, Mark Guinter, Jiali Zheng, and Cynthia A Thomson. Index-based dietary patterns and colorectal cancer risk: A systematic review. E Adv Nutr. 2015 Nov; 6(6): 763–773, 2015.
[18] G.J.Anderson S.K.Auer, K. Salin and N.B. Metcalfe. Individuals exhibit consistent differences in their metabolic rates across changing thermal conditions. Comparative Biochemistry and Physiology Part A: Molecular & Integrative Physiology, vol 217, 1-6, 2018.
[19] Martina Heer Scott M. Smith, Sara R. Zwart. Human Adaptation to Spaceflight: The Role of Nutrition. National Aeronautics and Space Administration (NASA), 2014.
[20] Rochelle M.Soo, James Hemp, and Philip Hugenholtz. The evolution of photosynthesis and aerobic respiration in the cyanobacteria. Free Radical Biology and Medicine, 2019.

# 3: AENEAS COMPLEX
# A PLAN FOR A SUSTAINABLE, PERMANENT 1000 PERSON SETTLEMENT ON MARS

**James D. Little**
Chief Designer, Anumá Aerospace
james.little@anumaaerospace.com

## INTRODUCTION

A permanent human settlement on Mars of a thousand people does not simply pop into existence, but necessarily evolves from a smaller settlement, and if successful, continues to grow beyond a thousand inhabitants. Certainly, to be successful as a remote, isolated community in the harsh environment of Mars, such a settlement would need to become as self-sufficient as possible, as rapidly as possible, and would need to grow well beyond the one thousand inhabitant mark to fully achieve this goal.

Therefore, it is essential that the community immediately establish a productive and sustainable economy. The initial focus must be to create a solid economic foundation, and the foundation of any successful, sustainable economy is primary productivity, which on Mars would be defined as energy production, mining, and farming. Reliance on in situ resource utilization (ISRU) without long distance travel, and the use of only existing technologies, are key aspects of this plan.

Siting of the settlement would be on, or very near, the equator. This location provides the warmest surface temperatures, reducing heating energy requirements; the highest levels of solar insolation, reducing photovoltaic array and solar thermal array size requirements; and provides the highest level of surface velocity for boosting rocket launches, reducing rocket fuel requirements, at least in the long term. It should be noted that the largest, and perhaps only, drawback to an equatorial location is that this latitude represents regolith water content as low as 2% by weight[1].

There are several interesting sites. There is one just west of the southern end of Echus Chasma right on the equator. This location, with a relatively neutral altitude, provides a centralized location to several astounding areological formations. It's on the edge of the Tharsis bulge with Pavonis Mons due west. Fortuna Fossae, Ascraeus Mons, and Tharsis Tholus to the northwest, Echus Chasma just to the northeast. Hebes Chasma to the east, Perrotin crater, Ophir Chasma, and Candor Chasma to the southeast, and Tithonium Chasma to the south (Figure 1).

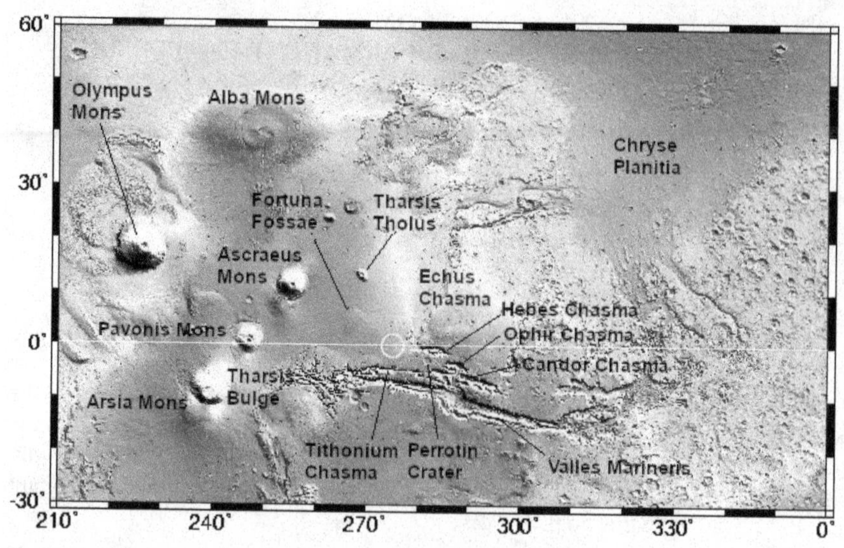


Figure 1 - Proposed site circled on equator. Map based on NASA/JPL/GSFC Mars Orbiter Laser Altimeter data.

Other potential sites with interesting areological formations are uplands overlooking Baetis Chaos, Orson Welles Crater, Oxia Chaos, Da Vinci Crater, Hydraotes Chaos, Nicholson Crater, Medusae Fossae[2], or Amazonis Mensa.

## ECONOMIC ASPECTS OF THE AENEAS SETTLEMENT

The Aeneas (/ɪˈniːəs/) settlement would first and foremost establish a mining, industrial manufacturing, and farming community for the purposes of establishing a sustainable, permanent settlement on Mars. Mining and industrial manufacturing enable the construction of settlement infrastructure, while farming provides food and clothing. Success in laying out the settlement infrastructure will enable the later establishment of the Mars Science and Technical Innovation Institute, which, beyond working to solve technical issues for the Aeneas community itself, would be working to solve issues for the general human settlement of Mars. It would also contract to house visiting scientists from Earth-based universities and research institutions such as NASA and ESA, providing them room and board, laboratory space and equipment, and a base from which to operate on Mars, with extensively more safety and comfort than could be provided with temporary, remote shelters, and at greatly reduced mission cost. Commercialization royalties from patents and technologies developed at the Institute can provide another source of return on investment (ROI) income. Subscriptions to Mars climate and geological data libraries, geological samples export, precious metals and minerals export, value-added product export (e.g., jewelry), and pleasure tourism are all potential sources of Earth-originating cash flow, or ROI. The Aeneas settlement would also, necessarily, operate as a spaceport with spacecraft refueling, maintenance, and repair capabilities, as well as providing crews with rest and relaxation opportunities. Further, due to Mars proximity to the asteroid belt, an established Martian spaceport stands to profit from providing spaceport services to asteroid mining ventures.

Aeneas is envisioned as incorporated, with shareholders, and with community-members being awarded stock grants. This provides residents with a sense of ownership and a vested interest in the future success of the settlement. Political management would be accomplished with a democratically elected governance board, who would in turn recruit and appoint qualified experts (technocrats) to manage specific sectors, projects, and activities within the community and economy. While management hierarchies have some benefits, it has been demonstrated that flatter work hierarchies provide individuals with a stronger sense of empowerment, encouraging creativity and problem solving, all of which benefit the community[3]. It is, therefore, suggested that the community be organized and led in this way.

This would most certainly be a planned economy, and it can be argued that a hybrid economic model makes the most sense. All products and services required by all the community's inhabitants would be provided by the community to its permanent members. Centralized control of infrastructure development and management of life sustaining activities is simply more efficient and secure. A basic income would be paid to each permanent resident, which would allow them to purchase their food and clothing. Life support, water, shelter, all levels of education, and basic medical care would be managed as the public sector of the economy. Those working in the public sector would be paid salaries above the basic income by the community. There would be no private real property ownership, but private ownership of personal possessions, objects of value, and personal services would be owned and traded as the private sector of the economy. Commercial space would be rented from the community to allow for small business entrepreneurs, such as restaurateurs, baristas, craftsman, artists, and personal services vendors to produce, sell wares, and provide services. Those working in the private sector would earn their salaries above the basic income by charging for their goods and services. Once the community has met its investment obligations, surplus earnings can be directed toward settlement expansions and improvements, or even in providing shareholders (which include community members) with a dividend.

**Primary Production – Energy, Mining, Farming**

The first technical challenge, beyond basic life support on the surface of Mars is energy production. An abundant energy supply is the key to unlocking Mars for humankind, because energy is the limiting factor for industrialization. The industrialization of Mars is essential to making it habitable, and with adequate energy, industrial processes can be driven to solve nearly any technical problem.

Energy can initially be provided with imported Radioisotope Thermoelectric Generators (RTGs), solar photovoltaic, and solar thermal dynamic power generation equipment. These sources would be followed later with the development of in situ geothermal energy production using imported drilling and imported or repurposed power generation equipment.

In the earliest stages of development, electric power will be provided with RTGs. This will provide the electricity for lighting, heating, and life support for the initial teams, while the solar photovoltaic system, including a relatively small battery storage system is assembled. Initially, the RTGs can be stationed in crudely excavated pits, covered with insulated inflatable dome tents, utilizing their waste heat to sublimate water ice in the soil subsurface. Water vapor released from the soil in this way can be collected with Pressure Swing Adsorption (PSA) equipment stored within the tent, the liquid water shunted to insulated, pressurized storage tanks for use as potable water, energy storage, and heat distribution to temporary living quarters.

A very large solar array can be constructed. Imported thin-film panels would be used initially until PV manufacturing, such as inkjet-printed, thin-film perovskite (calcium titanate) solar cells[4], is underway. The panels are aligned on long guide rails upon which rides a robotic cleaner. The robotic cleaner uses an air compressor to blow dust off the panels on an ongoing scheduled basis to keep them clean and operating efficiently. The robot's control system uses a wind direction sensor to stage the beginning of a cleaning cycle at the most upwind location and moves progressively downwind to optimize the cleaning process.

The next phase would be construction of a solar thermal dynamic power generation[5] facility utilizing compact linear Fresnel reflectors and multiple absorbers to heat water stored in insulated, pressurized tanks. Pressurized hot water can be flashed for steam turbine electric generation, with the waste heat being circulated through inflatable hydroponic greenhouses and temporary living quarters. Solar thermal has the distinct advantage of being able to store thermal energy using in situ resources, such as water or salts, allowing for electric generation around the clock. Once the initial settlement has been secured with enough available electricity, heat, and water, work on boring geothermal[6] wells can commence. Once geothermal power is secured, the settlement can continue expansion and development of its industry.

The second technical challenge is water. It would be enormously beneficial if a site can be identified that has significant quantities of subsurface water ice at, or near, the surface, and ongoing Mars reconnaissance seems to indicate that water ice and mineral hydrates are relatively abundant and widespread[7]. This plan calls for the eventual extraction of approximately 21 million metric tons of regolith, which at a conservative estimate of 2% by weight water, would yield 420 thousand metric tons of water. Closed-loop recycling of water used for life support is essential. PSA equipment can be used to regulate atmospheric gas and water content concentrations within the complex. Ongoing production of water will be necessary to replace water lost due to electrolysis and other industrial uses.

Water, while obviously necessary for human consumption such as drinking, cooking, and hygiene, is also necessary for heat distribution into greenhouses and living areas, for agricultural and aquacultural endeavors, exercise, and entertainment, as well as energy production (heat storage, steam turbine electric generation, geothermal circulation). Water is also crucial for chemical and manufacturing industrial uses. Water will be used for electrolysis to produce hydrogen and oxygen,

(a) $2H_2O \rightarrow 2H_2 + O_2$

forming the basis of industrial chemical production.

Initial mining activity will focus on regolith extraction and processing, particularly heating the regolith to chemically reduce metal oxides, perchlorates, sulfates, carbonates, and nitrates. This processing includes gas capture, such as water vapor, oxygen, nitrogen oxides, hydrogen sulfide[8], sulfur dioxide, etc., utilizing PSA equipment.

By developing the techniques to process generalized Martian regolith into a series of useful products, the settlement puts itself in a position to industrialize Mars far more rapidly than could be done through the typical mining processes of exploration and identification of individual mining sites rich in single minerals. By creating this economic stability, later identification of mineral rich sites merely represents a windfall, rather than a prerequisite to building a sustainable economy.

Combining mining operations directly with habitat construction is the most efficient use of energy and labor. Vertical shafts bored to extract regolith are to be lined with 3D printed sulfur concrete cylinders, topped with 3D printed sulfur concrete domes for utilization as subsurface habitats. Horizontal mining shafts lined with 3D printed sulfur concrete shoring provide passageways between the vertical sections. These subsurface habitats will provide far more protection from extreme surface temperatures, meteorites, and radiation than would surface structures.

Vertical Shaft Sinking Machines (VSM)[9] are used to bore vertical shafts in the regolith (at least the smaller diameter shafts). The regolith is processed by grinding it and feeding it through an electric rotary kiln. Gases from the kiln are collected and separated via pressure swing adsorption (PSA). The VSM 3D prints a continuous shaft wall as it sinks it into the excavation. Once a vertical shaft has been completed, the boring machine 3D prints a sulfur concrete[10] floor. The VSM then moves to another shaft location and passageways between boreholes are horizontally bored with tunnel boring machines, the walls shored with 3D printed sulfur concrete. A dome printing machine then moves in over the completed cylinder and prints a dome on top of the subterranean cylindrical structure[11]. Boreholes of various diameters and depths are created in this fashion.

Generalized Martian soil constituents, in order of content, based on data from Spirit, Opportunity, and Curiosity rovers, are silicon dioxide ($SiO_2$), iron oxides ($FeO^*$), aluminum oxide ($Al_2O_3$), magnesium oxide ($MgO$), calcium oxide ($CaO$), sulfur trioxide ($SO_3$), sodium oxide ($Na_2O$), titanium dioxide ($TiO_2$), phosphorus pentoxide ($P_2O_5$), chlorine ($Cl$), potassium oxide ($K_2O$), chromium oxide ($Cr_2O_3$), manganese oxide ($MnO$), nickel ($Ni$), zinc ($Zn$), and bromine ($Br$)[12,13] (Figure 2).

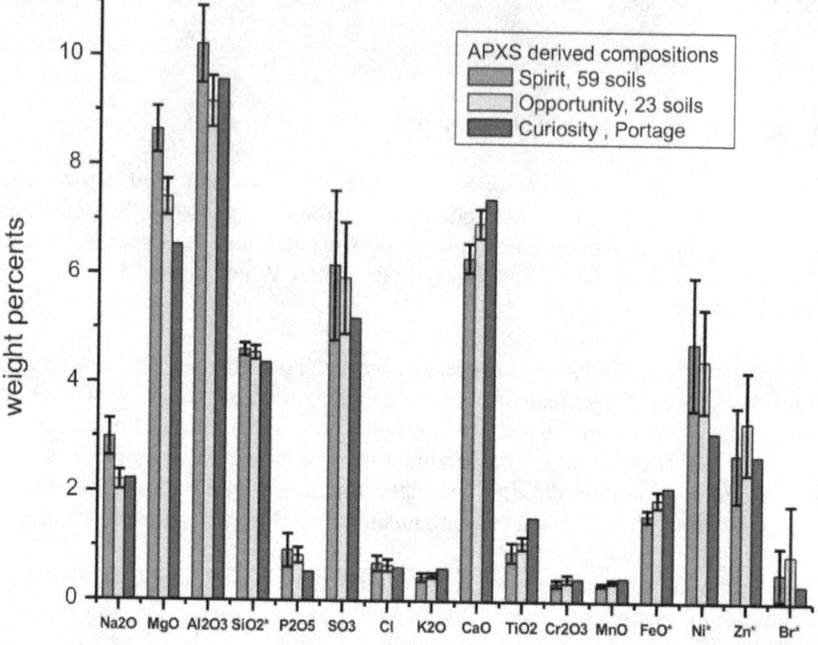


Figure 2 - Martian Soil Composition - NASA/JPL-Caltech/University of Guelph

Note that concentrations of silicon dioxide and iron oxide were divided by 10, and nickel, zinc and bromine levels were multiplied by 100. Additionally, there have been reports of sulfates[14], nitrates[15] and borates[16] in the regolith.

Excavated regolith will be processed by crushing, dry-grinding, and running it through a continuous feed, electrically heated rotary kiln, heating it to temperatures above 500 °C to decompose perchlorates[17,18],

(a) $[Ca(H_2O)_4](ClO_4)_2 \rightarrow Ca(ClO_4)_2 + 4H_2O$
(b) $Ca(ClO_4)_2 \rightarrow Ca(ClO_4)_2$ (melt)
(c) $Ca(ClO_4)_2 \leftrightarrow CaCl_2 + 4O_2$

oxides, sulfates, carbonates, and nitrates driving off gases and water vapor. The gases are processed through a series of PSA equipment. The remaining regolith can be sifted and sorted, using various ore processing techniques, such as magnetic separation for iron rich minerals, and sifting to produce aggregate mixtures suitable for producing 3D printed sulfur concrete.

Covering vertical shaft regolith excavations with inflatable tents will allow work to progress regardless of surface conditions, protecting personnel and equipment from dust and extreme cold, allowing work to continue even during dust storms, and facilitating the capture of water vapor, and other gases released during mining with PSA equipment.

Significant levels of oxygen ($O_2$) will also be released from oxides, perchlorates, carbonates, nitrates and borates present in the regolith. This is also easily captured with PSA equipment.

Hydrogen sulfide ($H_2S$) released from the regolith during kilning would be collected with PSA equipment and processed with the Claus process,

(a) $2H_2S + 3O_2 \rightarrow 2SO_2 + 2H_2O$
(b) $4H_2S + 2SO_2 \rightarrow 3S_2 + 4H_2O$

to produce elemental sulfur ($S_8$). This sulfur is to be used to produce sulfur concrete for 3D printed construction, particularly vertical and horizontal shaft shoring walls and covering domes (35% sulfur, 65% aggregate)[19].

Large amounts of sulfur dioxide ($S_2$) can be expected to be generated through thermal decomposition of sulfur compounds in the regolith. Catalytic reduction of sulfur dioxide with hydrogen[20,21],

(a) $3SO_2 + 7H_2 \rightarrow 6H_2O + S_2 + H_2S$

can also be used to produce elemental sulfur for use in sulfur concrete production with the hydrogen sulfide shunted to Claus process.

Additionally, some sulfur dioxide can be processed into sulfuric acid via the contact process,

(a) $2SO_2 + O_2 \rightarrow 2SO_3$
(b) $H_2SO_4 + SO_3 \rightarrow H_2S_2O_7$
(c) $H_2S_2O_7 + H_2O \rightarrow 2H_2SO_4$

for industrial uses.

Nitrogen oxides ($NO_X$) are also likely to be released from the regolith during heating. PSA equipment can be used to capture these gases for later conversion into nitric acid for chemical use, nitrates for fertilizers, or Selective Catalytic Reduction (SCR) (Tungsten oxide catalyst),

(a) $4NO + 4NH_3 + O_2 \rightarrow 4N_2 + 6H_2O$
(b) $2NO_2 + 4NH_3 + O_2 \rightarrow 3N_2 + 6H_2O$
(c) $NO + NO_2 + 2NH_3 \rightarrow 2N_2 + 3H_2O$

can be used, which injects ammonia into the nitrogen oxide gas at relatively low temperatures of 300 to 400 °C, converting the nitrous oxides ($NO_X$) and ammonia ($NH_3$) to nitrogen ($N_2$) and water ($H_2O$). In this process, 80 to 90% of the nitrogen oxide is converted to nitrogen, which can be used for the complex's atmosphere.

| Regolith Average Density | 1.52 *g/cm³* | | | | |
|---|---|---|---|---|---|
| | Composition % | Molar Fraction | Mass *kg* | Density *kg/m³* | Volume *m³* |
| Total Excavated | 100.00% | | 21,011,740,478 | 1520 | 13,823,513 |
| Silicon Dioxide | 43.30% | | 9,098,083,627 | 2650 | 3,433,239 |
| Iron Oxide | 17.60% | | 3,698,066,324 | | |
| Iron | | 0.70 | 2,586,541,345 | 7874 | 328,491 |
| Alumina | 7.30% | | 1,533,857,055 | | |
| Aluminum | | 0.53 | 811,540,223 | 2700 | 300,570 |
| Sulfur Trioxide | 6.50% | | 1,365,763,131 | | |
| Sulfur | | 0.40 | 546,939,880 | 2000 | 273,470 |
| Magnesium Oxide | 6.30% | | 1,323,739,650 | | |
| Magnesium | | 0.60 | 798,262,527 | 1738 | 459,299 |
| Water | 2.00% | | 420,234,810 | 997 | 421,499 |

While hydroponic farming, commonly used for growing crops like lettuces, spinach, strawberries, tomatoes, peppers, and herbs will be important initially, a wider variety of crops can be grown in soil. One important consequence of regolith processing will be the thermal decomposition of perchlorates, chemical reduction of oxides, and removal of metal ores and salts, leaving behind mineral content suitable for mixing with composted organic material to create soil suitable for edaphon, and consequently farming. Enriched soil produced in this way can be used for farming a wider variety of crops, as well as those that aren't suited for hydroponic farming. The following table provides a sample methodology for agricultural production with the goals of providing suitable crop rotation, production of animal feed, adequate caloric intake and balanced nutrition, while providing enough variety to keep it interesting and delicious.

| Crop | Area *hectares* | Avg Annual Yield *metric tons/hectare* | Total Annual Yield *metric tons* | *kg/person* (1250 People) | Energy Density *kcal/kg* | Areal Energy Density *kcal/m²* | *kcal/person* | *kcal/person-day* |
|---|---|---|---|---|---|---|---|---|
| Maize[a] | 3 | 25 | 75 | 60 | 860 | 2150 | 0 | 0 |
| Rice | 1 | 9 | 9 | 7.2 | 1300 | 1170 | 9360 | 26 |
| Wheat | 5 | 8 | 40 | 32 | 3270 | 2616 | 104640 | 287 |
| Barley | 7 | 11 | 77 | 61.6 | 3520 | 3872 | 216832 | 594 |
| Potatoes | 12 | 45 | 540 | 432 | 770 | 3465 | 332640 | 911 |
| Sweet Potatoes | 1.9 | 33 | 61.3 | 49.1 | 900 | 2970 | 44170 | 121 |
| Kale | 1 | 17 | 17 | 13.6 | 490 | 833 | 6664 | 18 |
| Bananas | 5 | 45 | 225 | 180 | 890 | 4005 | 160200 | 439 |
| Avocados | 5 | 30 | 150 | 120 | 1600 | 4800 | 192000 | 526 |
| Lentils | 1 | 2 | 2 | 1.6 | 3530 | 706 | 5648 | 15 |
| Peanuts | 1 | 5 | 5 | 4 | 5700 | 2850 | 22800 | 62 |
| Bamboo[b] | 1 | 80 | 80 | 64 | 0 | 0 | 0 | 0 |
| Crickets[a] | 0.5 | 150 | 75 | 60 | 1210 | 18150 | 0 | 0 |
| Eggs/Hens | 2 | 1.7 | 3.4 | 2.7 | 1550 | 264 | 4216 | 12 |
| Tilapia | 1 | 5 | 5 | 4 | 1290 | 645 | 5160 | 14 |
| Catfish | 1 | 4 | 4 | 3.2 | 1440 | 576 | 4608 | 13 |
| Rainbow Trout | 1 | 4 | 4 | 3.2 | 1410 | 564 | 4512 | 12 |
| **Total** | **49.4** | | | | | | **1113450** | **3051** |

Annual numbers based on 365-day Earth year
[a]Feed for animal production
[b]Cellulose/lignin for rayon/clothing/industrial use

Algae, particularly Chlorella, can be grown in bubble column photobioreactors, fertilized with human and animal waste and Mars atmospheric carbon dioxide ($CO_2$) to produce oxygen ($O_2$). Excess Chlorella can be used as feedstock for tilapia and/or shunted to biodigesters to produce methane and compost.

Bamboo, a rapidly growing, renewable crop, can be farmed to produce cellulose fibers and lignin. The cellulose fibers can be used to produce rayon, which can subsequently be used to produce various textiles as needed. The lignin, a phenylpropanoid-based biopolymer, can be used as a substitute source for phenol in most of phenol's industrial applications such as phenolic resins, surfactants, epoxy resins, adhesives, or polyester. It can be processed into a variety of products, not the least of which are activated charcoal and carbon fiber[22].

High productivity animal production such as vermiculture, cricket farming, and brine shrimp can produce high quality proteins, which can be converted to more palatable forms of proteins for humans when fed to chickens (both for egg and meat production), or aquacultural varieties such as catfish and rainbow trout.

Apiary in greenhouses will ensure successful plant pollination and provide honey.

Note that first generations of chickens, bees, crickets, and fish can be transported to Mars as cryopreserved embryos, which can then be thawed, incubated, and hatched.

Human, animal, plant, and food waste will be composted in biodigesters, producing both methane and compost. The compost is mixed with processed regolith to create enriched soil for farming, while the methane is compressed for fuel or converted to ethylene.

In these ways, biotechnologies, biogeneration, and biorefinery operations allow recycling of biomass and its use as source material to produce food, clothing, and a myriad of valuable organic products.

**Secondary Production – Synthesis, Materials, and Manufacturing**

While it is beyond the scope of this paper to define the entire structure of an industrialized Mars, it is important to note that the basis of secondary production is the chemical industry. Recommended guiding precepts for the development of an economically viable and sustainable chemical industry on Mars can be borrowed from the Green or Sustainable Chemistry movement, and would be: a preference for the prevention of waste, over treatment or cleanup; a preference for processes favoring atom economy over yield; a preference for lower toxicity methods and substances; a preference for energy efficiency and methodologies for heat recovery and cogeneration; a preference for renewable feedstocks when practical; a preference for the reduction of derivatives; a preference for catalysis over stoichiometric reagents; a preference for the production of easily recycled materials with inert decompositional products; continual analysis, in-process monitoring, and control for the prevention of the formation of hazardous substances, waste production, or energy loss; and, a preference for substances and processes which minimize risks of accidents, such as unintended releases, fires, or explosions.

As previously discussed, water electrolysis will produce hydrogen and oxygen,

(a) $2H_2O \rightarrow 2H_2 + O_2$

forming the basis for industrial chemical production.

The Sabatier reaction can then be used to catalyze ($Al_2O_3$ catalyst) hydrogen and atmospheric carbon dioxide into methane and water,

(a) $CO_2 + 4H_2 \rightarrow CH_4 + 2H_2O$

Methane production is primarily for fueling Earth return vehicles[23], but excess methane from rocket fuel production and biogas digesters can be oxidized to produce supplemental heat for the steam turbine electric generators in instances of low solar availability, such as during dust storms, until geothermal energy production is achieved.

Additionally, oxidative coupling of methane (OCM) (LiMgO catalyst) can be used to produce ethylene,

(a) $2CH_4 + O_2 \rightarrow C_2H_4 + 2H_2O$

Reaction inefficiencies lead to some production of carbon dioxide and carbon monoxide. The carbon dioxide can be shunted back to the Sabatier process and the carbon monoxide can be used in the Boudouard reaction,

(a) $2CO \rightarrow CO_2 + C$

to produce elemental carbon and carbon dioxide. The carbon dioxide is again shunted to the Sabatier process with the elemental carbon being used to produce carbon electrodes for the Hall-Héroult process to produce aluminum.

Ethylene gas can then be polymerized into polyethylene with the Ziegler-Natta catalyst titanium trichloride ($TiCl_3$). Low- and high-density polyethylene (LDPE, HDPE) can be used for the manufacture of piping, water and chemical storage tanks, electrical conduit and wire insulation, and radiation shielding. Boron isolated during regolith processing can be added to HDPE at 5% by weight to enhance its radiation shielding capabilities (5% Boron Enhanced HDPE)[24].

The Martian atmosphere, while primarily composed (95.32%) of carbon dioxide ($CO_2$), is about 2.7% Nitrogen ($N_2$). PSA equipment can be used to concentrate pure nitrogen gas from the atmosphere. Nitrogen and hydrogen can be fed into the Haber-Bosch process ($Fe_3O_4$ catalyst),

(a) $N_2 + 3H_2 \rightarrow 2NH_3$

to produce ammonia ($NH_3$).

As previously discussed, some ammonia is used in Selective Catalytic Reduction (SCR) to convert nitrogen oxides to nitrogen. Additional ammonia would be used in the modified Solvay (Hou's) process to produce sodium bicarbonate,

(a) $NaCl + CO_2 + NH_3 + H_2O \rightarrow NaHCO_3 + NH_4Cl$

The sodium bicarbonate is heated to produce sodium carbonate,

(a) $NaHCO_3 \rightarrow Na_2CO_3 + H_2O + CO_2$

Sodium carbonate ($Na_2CO_3$) can be used for glass making, water treatment, and aluminum processing.

Sodium carbonate ($Na_2CO_3$), silicon dioxide ($SiO_2$), aluminum oxide ($Al_2O_3$), and boron trioxide ($B_2O_3$) can be combined to manufacture borosilicate glass, which has very low coefficients of thermal expansion, high resistance to thermal shock, high chemical durability, high light transmission, pristine surface quality, and blocks neutrons due to its boron content. Borosilicate glass can be used for bottles, windows, cookware, solar panels, mirrors, evacuated tube collectors, electrical insulators, lighting fixtures, laboratory equipment, medical equipment, telescope mirrors and optics, and fiber optics.

Twenty-six percent sodium chloride (NaCl) brine can be processed using chloralkali membrane cells to produce sodium hydroxide (NaOH), chlorine gas ($Cl_2$), and hydrogen ($H_2$),

(a) $2NaCl + 2H_2O \rightarrow Cl_2 + H_2 + 2NaOH$

The sodium hydroxide (NaOH) can be used in the Bayer process to purify alumina from crushed regolith. The chlorine and hydrogen can be reacted with UV light,

(a) $Cl_2 + H_2 \rightarrow 2HCl$

and dissolved in water to produce hydrochloric acid (HCl).

In Bayer process 1,

(a) $Al_2O_3 + 2NaOH \rightarrow 2NaAlO_2 + H_2O$

crushed regolith (which appears to contain around 9 to 10% alumina by weight) is hydrated and mixed with sodium hydroxide then heated in a pressure vessel to around 200 °C. Calcium oxide, which is already present in the regolith induces silica to precipitate as calcium silicate. This mixture is filtered to remove undissolved compounds and concentrate the sodium aluminate. Initially the sodium aluminate has carbon dioxide bubbled through it (Bayer process 2),

(b) $2NaAlO_2 + 3H_2O + CO_2 \rightarrow 2Al(OH)_3 + Na_2CO_3$

to create aluminum hydroxide crystals to seed slurry for ongoing Bayer process 2,

(c) $2H_2O + NaAlO_2 \rightarrow Al(OH)_3 + NaOH$

Then the aluminum hydroxide is desiccated with heat,

(d) $2Al(OH)_3 \rightarrow Al_2O_3 + 3H_2O$

to produce pure alumina. This alumina can then be processed with the Hall-Héroult process,

(a) $Al_2O_3 + 3C \rightarrow 2Al + 3CO$

to produce pure aluminum. The carbon monoxide can be shunted back to the Boudouard reaction for carbon production, or to the Fischer-Tropsch process (Fe catalyst),

(a) $3H_2 + CO \rightarrow CH_4 + H_2O$

to create additional methane. The aluminum can be used to produce castings, extrusions, wires, and powders, all of which can be used for construction of structures, vehicles, machines, electrical wiring, and for 3D printing.

The Alumina Refinery Residues (ARR), or red mud, which is highly caustic due to sodium hydroxide (NaOH) content can be neutralized using hydrochloric acid (HCl)[25],

(a) $NaOH + HCl \rightarrow NaCl + H_2O$

This will allow it to be safely handled for further processing. It can be processed into pavers and bricks for interior construction[26], or further refined into purified silicon dioxide ($SiO_2$) for glassmaking.

A thermal reduction process with silicon can be employed to convert magnesium oxide to magnesium vapor,

(a) $2MgO + Si \rightarrow SiO_2 + 2Mg$

which can be condensed into pure magnesium. The silica is combined with calcium oxide to form calcium silicate,

(a) $SiO_2 + CaO \rightarrow CaSiO_3$

Magnesium can be used in alloying aluminum and steel, and in the production of titanium.

Steel can be produced using iron compounds separated from the regolith using magnetic separation and processed using a Direct Reduced Iron (DRI) methodology, such as the HIsarna process Cyclone Converter Furnace (CCF) and Smelting Reduction Vessel (SRV) to produce sponge iron, where hematite ($Fe_2O_3$) becomes magnetite ($Fe_3O_4$) by reduction with carbon monoxide (CO) and hydrogen ($H_2$),

(a) $3Fe_2O_3 + CO \rightarrow 2Fe_3O_4 + CO_2$

$3Fe_2O_3 + H_2 \rightarrow 2Fe_3O_4 + H_2O$

magnetite ($Fe_3O_4$) becomes ferrous oxide (FeO) by reduction with carbon monoxide (CO) and hydrogen ($H_2$),

(b) $Fe_3O_4 + CO \rightarrow 3FeO + CO_2$

$Fe_3O_4 + H_2 \rightarrow 3FeO + H_2O$

ferrous oxide (FeO) becomes sponge iron (Fe) by reduction with carbon monoxide (CO) and hydrogen ($H_2$),

(c) $FeO + CO \rightarrow Fe + CO_2$

$FeO + H_2 \rightarrow Fe + H_2O$

for conversion into steel in an Electric Arc Furnace (EAF). Steel can be used for framing, construction, piping, machines, and wire or powder for 3D printing.

The examples above, such as the production of methane, plastics, aluminum, magnesium, steel, and glass, are merely examples of the chemical synthesis and secondary production capabilities that will be necessary in creating a sustainable settlement on Mars. Equipment shipped from Earth to set up these capabilities would only be for small scale production. Once initial production is established, the produced material can be used to scale up machinery and capability, creating a positive feedback loop of increasing industrial scale. The Aeneas settlement is designed to provide the energy and industrial spaces necessary to set up chemical and industrial processing, as well as, manufacturing.

## ENVIRONMENTAL ASPECTS OF THE AENEAS SETTLEMENT

Aeneas is intended to be a walkable community for a thousand permanent residents in a climate-controlled, underground system of tunnel-connected, cylindrical living structures of varying size containing an atmosphere at a pressure of 81.2 kPa (approximating an Earth altitude of 1600 m above mean sea level), composed of a mixture of 21.00% oxygen, 78.75% nitrogen, and 0.25% water vapor at a steady background temperature of 20 °C, all of which is intended for maximum health, safety, and comfort of its residents.

Below ground construction provides living areas with far greater radiation shielding and safety from small meteorites than is possible with surface construction without extensive and expensive remediation. Additionally, it provides far greater thermal efficiency by reducing heat loss to the frigid Martian atmosphere. The goal would be to develop the living areas such that they would provide an average radiation dosage limit of <=2.0 mSv/y. As normal Earth background is ~2.4 mSv/y, this will help compensate for higher radiation doses experienced during planetary transfer and surface excursions while living on Mars.

There would be few if any windows or skylights in the settlement complex. Most of the complex is below grade, and windows are both expensive and dangerous, as they are difficult to seal, and provide potential points of decompression failure and increased levels of radiation exposure. Additionally, Martian natural light is dim enough to compare with significant overcast on Earth, meaning naturally lit spaces would be dark and dreary. To that end, all interior spaces would be brightly lit with energy conservative LEDs, using broad-spectrum LEDs suitable for plant growth in public spaces and controlled to mimic a natural day/night cycle. This will help promote plant growth, and to prevent psychological problems, such as Seasonal Affective Disorder (SAD), or problems associated with the disruption of natural circadian rhythm.

Views outside the complex can be provided on high-definition screens (virtual windows) which broadcast from stationary or robotic, mobile cameras on the surface. The actual night sky over the complex could be projected onto the undersurface of the Campus Martius (/ˈkampəs/ /ˈmär-sh(ē-)əs-/) central dome to provide a place to observe the night sky and reinforce the feeling of being outdoors.

Given the harsh and dangerous surface conditions, actual surface excursions would be rare for most inhabitants, as it would generally be limited to individuals involved in necessary construction and maintenance tasks, scientific and exploratory expeditions, and tourism excursions.

The hexagonal-tiled, fractal-like pattern of the complex's layout keeps it compact, simplifies construction procedures, and minimizes the required equipment. As the same pattern repeats itself, it makes it easier to seal off sections both during construction or in disaster events and provides redundancies. If each sector has the technical capability of supporting twice its normal population, then it is easy to safely accommodate displaced individuals in the event any section in the complex needs to be evacuated. Any technical system can fail, so redundancies are required. Passive systems are preferred when possible, as they tend to have greater stability and are easier to manage with less required intervention. Having multiple sources of resources is also useful. An example would be water electrolysis and algal photobioreactors both being capable of producing breathable oxygen and both systems being able to be run with electricity produced from one of several different production options (radioisotope thermoelectric generator, solar photovoltaics, solar thermal, methane combustion, geothermal).

**Construction and Layout**

An apron ring trench is excavated surrounding the outside edge of the intended shaft and a sulfur concrete ring is 3D printed with the 3D dome printer into the trench to support the vertical shaft sinking machine (VSM) (Figure 3).

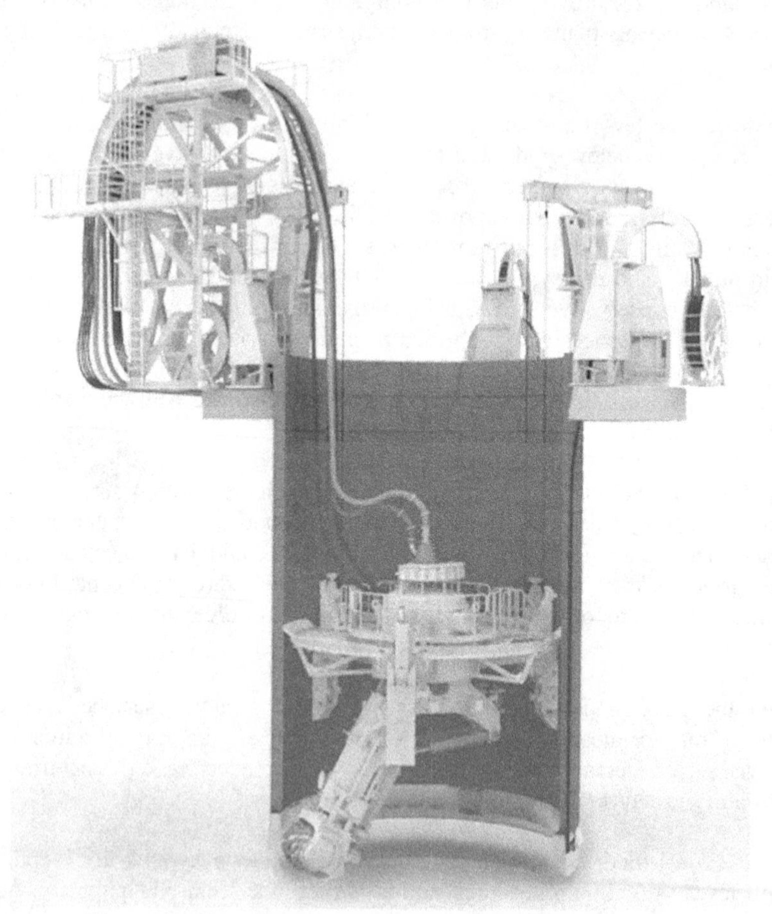

Figure 3 - Herrenknecht Vertical Shaft Sinking Machine

Once the VSM is moved into place, the VSM cutting tool begins excavating regolith material from the shaft. Once the minimum starting depth has been excavated, the machine prints the first section of the cylindrical shaft support wall, the first meter of depth being solid, but the remainder of the wall is printed with internal voids which serve to insulate the cylinder, and with channels on the interior wall side for fitting radiant heat tubing. With a final shaft wall thickness of 1 meter, the sulfur concrete, with internal voids, is printed at 0.8 m thick. The cylinder shell is lowered into the shaft as the shaft is excavated, with the cutting tool undermining the shell as the print head continues printing along the top, continuously elongating the shell as it is gradually lowered into the excavation. Horizontal tunnel openings are printed into the shell at the appropriate elevations. Once the cylinder is lowered to its finished depth, the print head is lowered to the bottom of the pit where it prints a 0.8 m thick floor with internal insulation voids and radiant heat tubing channels in its top surface. The VSM machine is then raised and moved to another excavation site and set in motion excavating and printing a new vertical shaft.

The horizontal tunneling machine (Figure 4) can be lowered into the shaft and set to work boring the connecting tunnels. The horizontal tunneling machine 3D prints a similarly constructed 1 m sulfur concrete shoring wall behind it as it moves down the tunnel toward the adjoining vertical shaft.

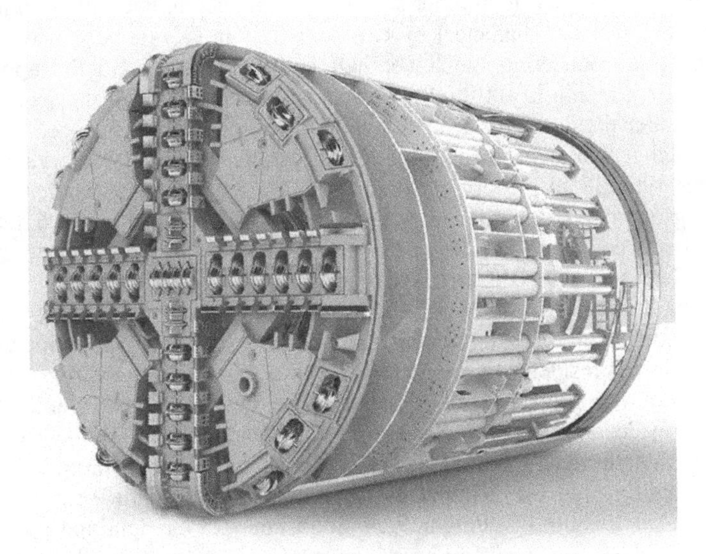

Figure 4 – Herrenknecht Horizontal Tunneling Machine

Once the shaft and connecting tunnels have been bored, and any equipment or material has been removed, the dome printer is used to cap the cylinder with a 3D printed sulfur concrete dome, resulting in what we will refer to as a silo, designating them by their inner diameters.

Airtightness of a finished silo or tunnel is then accomplished through a process of over pressurizing the silo tunnel with atmospheric carbon dioxide and spraying a fine mist of water-soluble acrylic polymer emulsion. The mist is pulled into leaks by the escaping air where it dries in place, building upon itself until the void or leak is plugged. Performing this process up to significant overpressure (121.59

kPa) ensures sealing at nominal pressures (81.2 kPa ideal to a not to exceed pressure of 101.3 kPa).

The interior surfaces can then be lined with a radiant barrier, such as aluminized biaxially-oriented polyethylene terephthalate (BoPET) film. Polyethylene tubing is then installed in the channels for warm water radiant heating circulation. The floor can be covered with 0.2 m thick sintered flat pavers, and the walls can be lined with 0.2 m thick interlocking sintered bricks made from neutralized AAR (red mud)[27], machine made to match the wall radius. These will provide internal thermal mass to assist in environmental thermal management, while also serving to protect the airtight barrier from damage, and providing a warm, beautiful interior.

The first constructed shaft would be bored to a diameter of 10 m to a depth of 13 m. This would leave an interior cylindrical space of 12 m in depth and 8 m in diameter covered by a hemispherical dome. Two aluminum truss framed floors are installed, one midway up the cylinder and one at the top of the cylinder. This 8 m silo would be connected to another silo of the same depth, but with a 20 m outer diameter (18 m inner diameter) by two 4 m ID/5 m OD tunnels, one over the other, the first connecting the lower half cylinder, the second the upper half cylinder. The 8 m silo is outfitted to serve as 2 apartments, with the space on the top floor under the dome used for life support equipment and storage. The 18 m silo has 4 other 8 m silos surrounding it, all 5 of them set at 60° intervals around the center. The 18 m silo is outfitted as 2 shared living spaces for the five connecting apartments on the respective levels. The 18 m silo has 6 sets of 2 tunnels, 5 sets connecting to the five 8 m apartment silos, the 6th connecting to another 18 m silo, which is outfitted with a midlevel catwalk and stairs going down to the lower level. Additionally, this silo has no upper floor, and is intended to be planted with an orchard tree and other plants, perhaps with a “living wall”. It will allow the ten residents to access their level of shared living quarters and serve as a transition from private to public space. These transition domes would not only transition residents from their residence levels, but as these domes would be planted with trees and other plants and have automatically controlled natural spectrum lighting (as would most of the public spaces in the settlement) that would simulate a natural day/night cycle of 15 hours of daylight and 9 hours, 37 minutes, 35.244 seconds of darkness, would transition them to an environment that feels like the “outdoors” and assist residents in maintaining a normal and healthy circadian rhythm. The additional 37 minutes, 35.244 seconds of darkness beyond the Earth normal 24-hour day allows the interior “day” to maintain alignment with the actual Martian sol.

The five 8 m apartment silos and one 18 m transition silo surround one 18 m shared living silo. This grouping forms a 10-resident subsector (Figure 5).

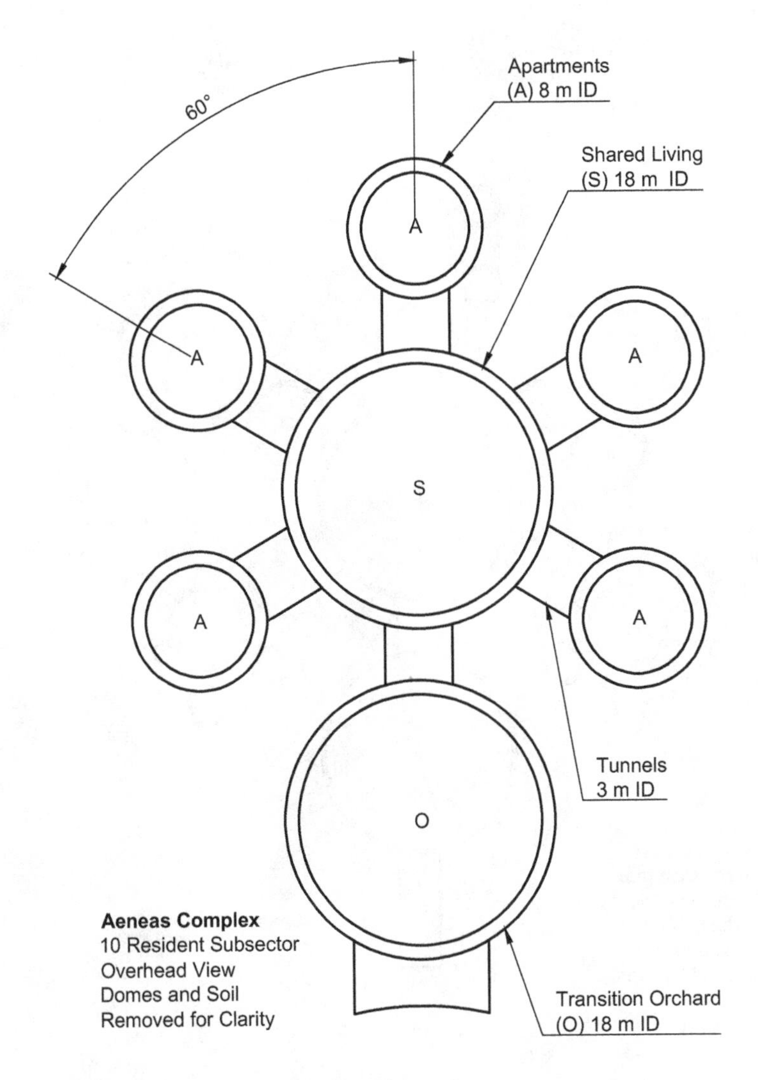


Figure 5

The 10-resident subsector is repeated 5 times at 60° intervals around a 46 m silo, which serves as a single level park. The sixth connecting tunnel connects to another 46 m silo, which is outfitted for cafes and market stalls. These five 10-resident subsectors and the one market silo surrounding the 46 m park silo at 60° increments form a 50-resident subsector (Figure 6).

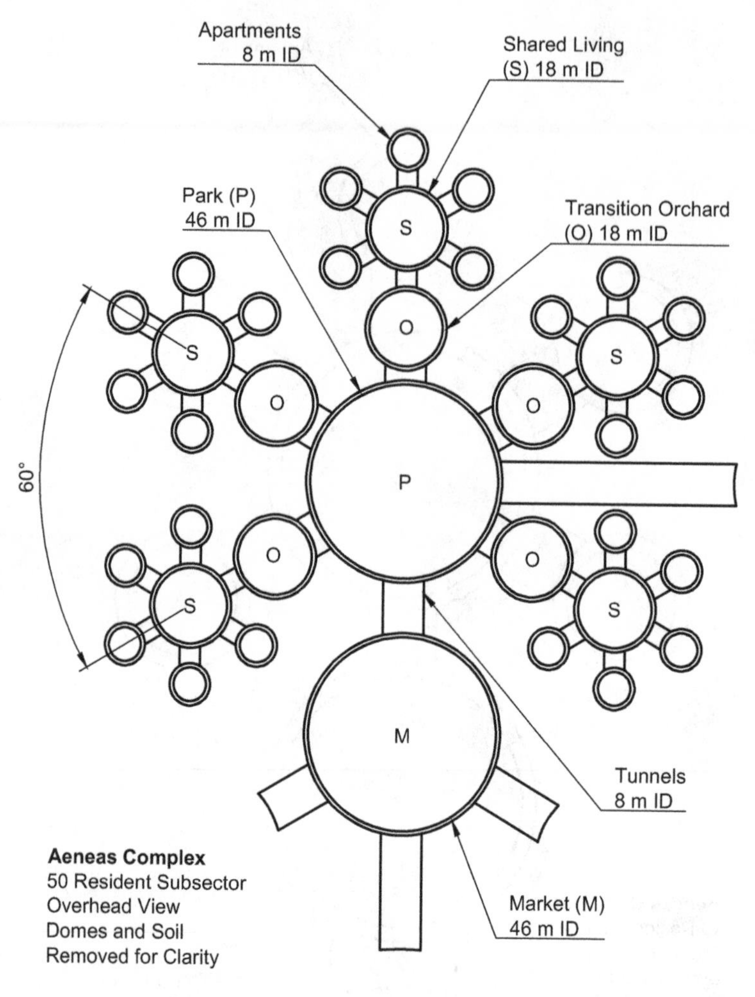


Figure 6

Five 50-resident subsectors and a 138 m farm silo surround a 138 m central park silo to form a 250-resident sector (Figure 7). Also included in the 250-resident sector, are 6 more 46 m market silos surrounding the 138 m park silo at the 30° increments between the 46 m market silos which are part of the 50-resident subsectors. Connected to those six 46 m market silos by longer tunnels radiating outward from the center are six more 46 m silos intended for equipment and tankage.

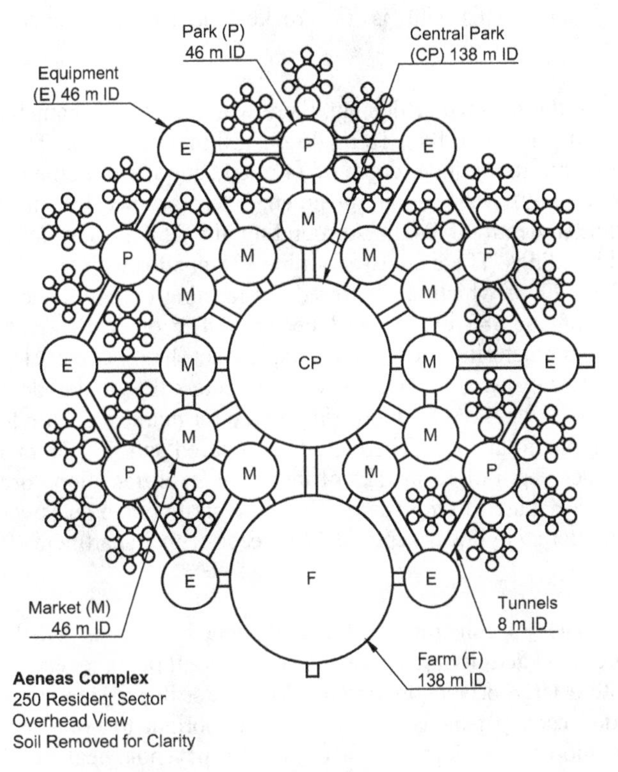


Figure 7

The culmination of construction and forming the center of the Aeneas settlement, is a 258 m diameter silo, with a 24 m interior depth and covered with a saucer dome (23 m high above top of cylinder (interior)) for a total interior height at the center of 47 m (24 m + 23 m). This is the Campus Martius green, forming a 5.2 ha (52280 m$^2$) park of botanical gardens, lawn, trees, and freshwater pond, complete with aquatic plants, aquatic insects, and fish. This central green, is surrounded by six 138 m silos (interior 24 m depth and divided into 8 floors) at 60° intervals to be used for institutional needs; hospital, classrooms, laboratories, offices, etc. Five of these six institutional silos connect to the five farm silos of the five 250-resident sectors surrounding the Campus Martius/Institutional complex at 60° increments. This provides the settlement with 1250 single resident apartments, 1000 for the permanent residents and 250 rental units for visitors, such as scientists, pleasure tourists, and astronauts. The sixth 138 m institutional silo connects to a 138 m industrial silo which is one of six 138 m industrial silos forming a ring around a central 138 m industrial silo at 60° intervals. Radiating outward from the central 138 m industrial silos are pairs of 46 m industrial silos at the 30° increments between the other 138 m silos on the 60° increments. All but one of these industrial silos have 12 m interior depths and are divided into 3 levels. The outermost 138 m silo in the industrial sector has a 24 m depth and an 18 m tunnel extending eastward and up out of the ground toward the rocket launch and landing pads. This silo is the rocket hanger for rocket storage and maintenance.

In summary, in the finished settlement, there would be 5 silo diameters, identified by their inner diameters of 8 m, 18 m, 46 m, 138 m, and 258 m. There is one 258 m silo with an interior cylinder depth of 24 m and an interior dome height of 23 m forming the Campus Martius central green. There are six 138 m silos with an internal cylinder depth of 24 m surrounding the 258 m silo forming 8 story institutional buildings. There are twenty-nine 138 m silos with an internal cylinder depth of 12 m, 11 of which are outfitted for farming (3 floors each), for a total farming area of 493,585 $m^2$, or 49.4 hectares. The other 18 are for industrial processing, manufacturing, rover hangars, rocket hangar, etc. There are one hundred and twenty-two 46 m silos with an internal cylinder depth of 12 m. Twelve of these are marked for industrial, 55 are for markets. 25 are for parks, and 30 are for equipment/tankage. There are two hundred and fifty 18 m silos with an internal cylinder depth of 12 m. 125 of these are level transition “orchards”. The other 125 are for shared living space. There are six hundred and twenty-five 8 m silos with an internal cylinder depth of 12 m, each with 2 apartments (Figures 8, 9, and 10).

In the early stages of construction, the settlement inhabitants will have to share tighter, more crowded living spaces. Eventually, though, everyone would be provided with a large private apartment. While first impressions may be that the space afforded each person is extreme, it is important to note that there is no habitable “outdoors” on Mars, so it is critical for psychological health to provide habitable spaces that don’t feel confined or claustrophobic. Additionally, while residents would certainly have the freedom to share their space with partners, each adult individual having their own private space available to them would help to eliminate interpersonal conflict. The architecture allows the flexibility that some 5 resident floors could be configured to accommodate families with children.

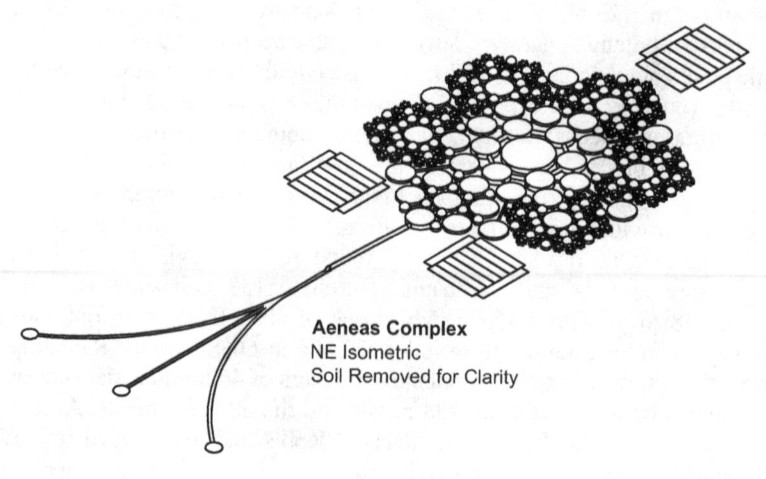


Figure 8

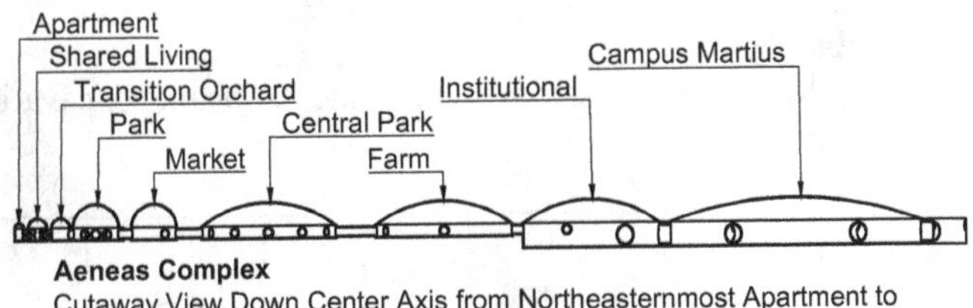


Figure 9

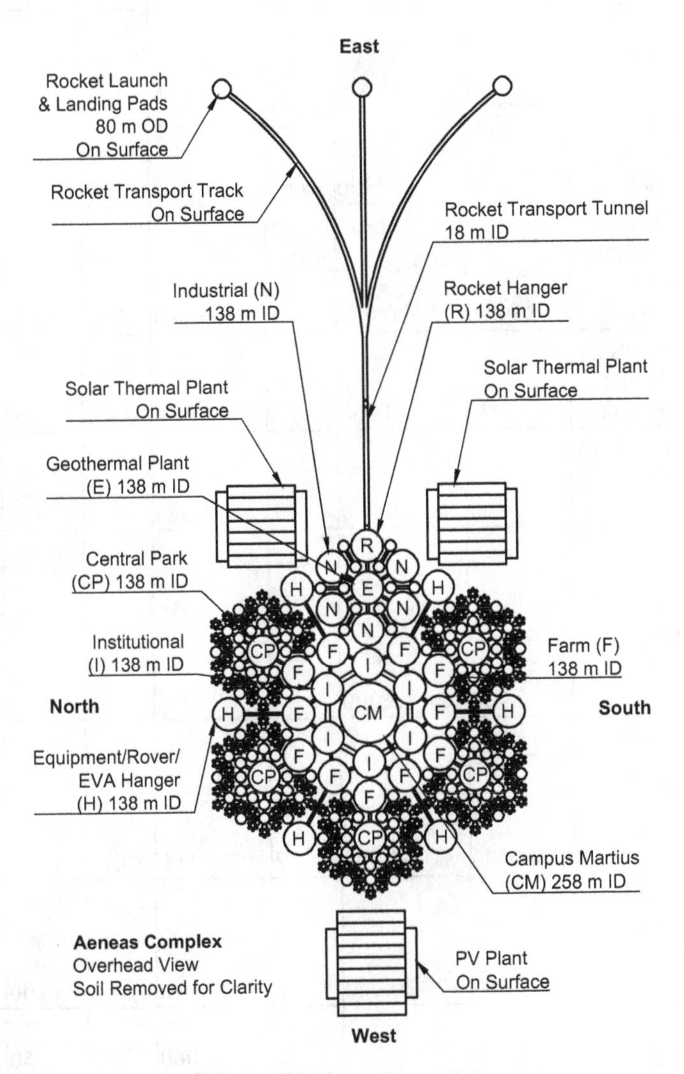


Figure 10

**SUMMARY PLAN**

**Phase 1 - 6 Months**

Major goals are to survey, assemble equipment, begin excavation, water production, sulfur production

| Assumptions & Resources | Mass (kg) | Unit Cost ($US) | No. Req | Total Mass (kg) | Total Mtrl Cost | Ship. Cost* ($M US) | Total Cost ($M US) |
|---|---|---|---|---|---|---|---|
| RTGs 10 kWe - $400/W @ 5 W/kg | 2,000 | 4,000,000 | 1 | 2,000 | 4.00 | 1.00 | 5.00 |
| Thin-film Solar Panels, 5 MWe @ 100 W/panel | 3 | 300 | 50000 | 150,000 | 15.00 | 75.00 | 90.00 |
| Vertical Shaft Sinking Machine bore 5 m/day autonomously | 185,000 | 2,500,000 | 1 | 185,000 | 2.50 | 92.50 | 95.00 |
| Backhoe/Front Loader, Mining Larger Boreholes | 5,000 | 20,000 | 2 | 10,000 | 0.04 | 5.00 | 5.04 |
| Construction Tent 1.36 kg/m^2 - 1000 m^2 | 1,360 | 10,000 | 1 | 1,360 | 0.01 | 0.68 | 0.69 |
| Cone Rock Crusher - Consistent Particle Sizes, Increased Surface Area | 6,100 | 5,000 | 2 | 12,200 | 0.01 | 6.10 | 6.11 |
| Electric Rotary Kiln Regolith Proc. Thermal Decomposition | 47,500 | 54,000 | 2 | 95,000 | 0.11 | 47.50 | 47.61 |
| Water Storage 4 m D, 12 m L FRP tank winding machine | 12,000 | 20,000 | 1 | 12,000 | 0.02 | 6.00 | 6.02 |
| Water Storage FRP Yarn | 1,000 | 1,000 | 5 | 5,000 | 0.01 | 2.50 | 2.51 |

| Assumptions & Resources | Mass (kg) | Unit Cost ($US) | No. Req | Total Mass (kg) | Total Mtrl Cost | Ship. Cost* ($M US) | Total Cost ($M US) |
|---|---|---|---|---|---|---|---|
| Water Storage FRP Resin Polyester Resin | 220 | 330 | 10 | 2,200 | 0.00 | 1.10 | 1.10 |
| Gas Mining from regolith kilning PSA - H2O, SO2, H2S, O2, NOx | 500 | 15,000 | 10 | 5,000 | 0.15 | 2.50 | 2.65 |
| Gas Mining For PSA Zeolite Adsorbent | 500 | 1,000 | 5 | 2,500 | 0.01 | 1.25 | 1.26 |
| Gas Storage Gas Compressor | 500 | 2,000 | 10 | 5,000 | 0.02 | 2.50 | 2.52 |
| Gas Storage Cryo Tanks SO2, H2S, O2, NOx, CH4 15000L | 6,500 | 10,000 | 10 | 65,000 | 0.10 | 32.50 | 32.60 |
| Sewage Proc. CH4/Compost Biodigester | 500 | 4,000 | 2 | 1,000 | 0.01 | 0.50 | 0.51 |
| Magnetic Separator Separation Iron Minerals | 1,500 | 12,000 | 2 | 3,000 | 0.02 | 1.50 | 1.52 |
| Vibrating Sifter Sifting Aggregate | 580 | 1,000 | 2 | 1,160 | 0.00 | 0.58 | 0.58 |
| Sulfur Recovery Unit Elemental Sulfur Product. | 50,000 | 2,000,000 | 2 | 100,000 | 4.00 | 50.00 | 54.00 |
| Crew of 50 Construction, Mining, Industrial, Support | 62 | 0 | 50 | 3,100 | 0.00 | 1.55 | 1.55 |
| Temporary Housing and Life Support - 25 crew/unit with 50 crew capability | 100,000 | 10,000,000 | 2 | 200,000 | 20.00 | 100.00 | 120.00 |

| **Assumptions & Resources** | **Mass (kg)** | **Unit Cost ($US)** | **No. Req** | **Total Mass (kg)** | **Total Mtrl Cost** | **Ship. Cost* ($M US)** | **Total Cost ($M US)** |
|---|---|---|---|---|---|---|---|
| Freeze-dried Food for 50 for 3 years | 15,000 | 13,500 | 50 | 750,000 | 0.68 | 375.00 | 375.68 |
| Mobile Hospital | 50,000 | 2,000,000 | 1 | 50,000 | 2.00 | 25.00 | 27.00 |
| **Total** | | | | **1,660,520** | **48.68** | **830.26** | **878.94** |

**Phase 2 – 6 months**

Major goals are to finish out initial subsectors, move in personnel, begin methane production

| **Assumptions & Resources** | **Mass (kg)** | **Unit Cost ($US)** | **No. Req** | **Total Mass (kg)** | **Total Mtrl Cost** | **Ship. Cost* ($M US)** | **Total Cost ($M US)** |
|---|---|---|---|---|---|---|---|
| Compact Linear Fresnel Reflector - 1 m^2 for Solar Thermal 5 MWe @ 7000 m^2 per MWe | 8 | 300 | 35000 | 280,000 | 10.50 | 140.00 | 150.50 |
| Solar Thermal Steam Turbine - 745 kw 1000 HP | 818 | 40,000 | 10 | 8,180 | 0.40 | 4.09 | 4.49 |
| Solar Thermal Generation Equipment 600 kw | 1,500 | 4,000 | 10 | 15,000 | 0.04 | 7.50 | 7.54 |
| Solar Thermal High Pressure Hot H2O 15000L Storage Tanks | 6,500 | 10,000 | 2 | 13,000 | 0.02 | 6.50 | 6.52 |
| Electrolysis Plant H2 /O2 Production Water Electrolysis | 4,000 | 120,000 | 2 | 8,000 | 0.24 | 4.00 | 4.24 |
| Gas Storage H2, O2 Tanks 15000L Cryo Tank | 6,500 | 10,000 | 4 | 26,000 | 0.04 | 13.00 | 13.04 |
| Gas Mining Atmospheric Mining PSA - CO2, N2 | 500 | 15,000 | 4 | 2,000 | 0.06 | 1.00 | 1.06 |
| Gas Mining PSA Zeolite Adsorbent | 500 | 1,000 | 2 | 1,000 | 0.00 | 0.50 | 0.50 |

| Assumptions & Resources | Mass (kg) | Unit Cost ($US) | No. Req | Total Mass (kg) | Total Mtrl Cost | Ship. Cost* ($M US) | Total Cost ($M US) |
|---|---|---|---|---|---|---|---|
| Gas Storage<br>Gas Compressor | 500 | 2,000 | 6 | 3,000 | 0.01 | 1.50 | 1.51 |
| Gas Storage<br>CO2, N2, CH4 Tanks<br>15000L Cryo Tank | 6,500 | 10,000 | 6 | 39,000 | 0.06 | 19.50 | 19.56 |
| Methane Plant<br>Sabatier Reactor | 6,500 | 10,000 | 2 | 13,000 | 0.02 | 6.50 | 6.52 |
| Tunnel Boring Machine<br>Connecting Tunnels | 6,000 | 300,000 | 2 | 12,000 | 0.60 | 6.00 | 6.60 |
| Silo Lining<br>Spray Acrylic Polymer Emulsion | 1,000 | 2,000 | 2 | 2,000 | 0.00 | 1.00 | 1.00 |
| Silo Lining<br>Hydraulic Brick Press | 2,500 | 10,000 | 1 | 2,500 | 0.01 | 1.25 | 1.26 |
| Silo Lining Brick Sintering<br>Electric Tunnel Kiln | 3,500 | 10,000 | 1 | 3,500 | 0.01 | 1.75 | 1.76 |
| Silo Lining<br>12 Micron<br>Aluminized BoPET Film | 1,000 | 1,700 | 1 | 1,000 | 0.00 | 0.50 | 0.50 |
| Silo Lining<br>PEX Radiant Heat Tubing<br>Radiant Heat Tubing | 35 | 200 | 10 | 350 | 0.00 | 0.18 | 0.18 |
| Dome Printer<br>Prints 3D S Concrete Domes | 3,500 | 150,000 | 1 | 3,500 | 0.15 | 1.75 | 1.90 |
| **Total** | | | | **433,030** | **12.17** | **216.52** | **228.69** |

**Phase 3 - 48 Months**

Major goals are to complete construction, geothermal drilling, set up farming and basic industrialization

| Assumptions & Resources | Mass (kg) | Unit Cost ($US) | No. Req | Total Mass (kg) | Total Mtrl Cost | Ship. Cost* ($M US) | Total Cost ($M US) |
|---|---|---|---|---|---|---|---|
| Geothermal<br>Drilling depth to 2600 m with 127 mm drill pipe<br>Rotary Drilling Rig | 9,960 | 70,000 | 1 | 9,960 | 0.07 | 4.98 | 5.05 |
| Geothermal<br>130 mm<br>130 mm Drill Bit | 8 | 1,000 | 10 | 80 | 0.01 | 0.04 | 0.05 |
| Geothermal<br>9.5 m length<br>127 mm Drill Pipe | 500 | 220 | 300 | 150,000 | 0.07 | 75.00 | 75.07 |
| 50 MW of<br>Steam Turbine - 745 kw<br>1000 HP | 818 | 40,000 | 50 | 40,900 | 2.00 | 20.45 | 22.45 |
| Power Generators<br>600 kw | 1,500 | 4,000 | 50 | 75,000 | 0.20 | 37.50 | 37.70 |
| Regolith Proc.<br>Sawtooth Wave Jig Separator | 610 | 1,500 | 1 | 610 | 0.00 | 0.31 | 0.31 |
| NaOH Production<br>Chloralkali Membrane Cell | 1,000 | 2,000 | 2 | 2,000 | 0.00 | 1.00 | 1.00 |
| Alumina Refining<br>Tank Reactor | 1,000 | 2,000 | 2 | 2,000 | 0.00 | 1.00 | 1.00 |
| Alumina Refining<br>Classifier | 700 | 1,000 | 2 | 1,400 | 0.00 | 0.70 | 0.70 |
| Al Production<br>Hall-Héroult | 1,500 | 20,000 | 1 | 1,500 | 0.02 | 0.75 | 0.77 |
| Glass Furnace | 2,000 | 80,000 | 1 | 2,000 | 0.08 | 1.00 | 1.08 |
| Ethylene Production<br>Oxidative Coupling of Methane Reactor | 1,800 | 20,000 | 1 | 1,800 | 0.02 | 0.90 | 0.92 |
| Polyethylene Prod.<br>Ethylene Polymerization Reactor | 1,800 | 20,000 | 1 | 1,800 | 0.02 | 0.90 | 0.92 |
| Iron Production Direct Reduction of Iron Cyclone Converter Furnace | 1,000 | 25,000 | 1 | 1,000 | 0.03 | 0.50 | 0.53 |

| Assumptions & Resources | Mass (kg) | Unit Cost ($US) | No. Req | Total Mass (kg) | Total Mtrl Cost | Ship. Cost* ($M US) | Total Cost ($M US) |
|---|---|---|---|---|---|---|---|
| Iron Production<br>Direct Reduction of Iron<br>Smelting Reduction Vessel | 1,000 | 25,000 | 1 | 1,000 | 0.03 | 0.50 | 0.53 |
| Steel Production<br>Electric Arc Furnace | 1,000 | 10,000 | 1 | 1,000 | 0.01 | 0.50 | 0.51 |
| Steel/Aluminum<br>Metal Powder 3D Printer | 900 | 250,000 | 2 | 1,800 | 0.50 | 0.90 | 1.40 |
| Plastic Filament Printing<br>Plastic Filament 3D Printer | 10 | 300 | 5 | 50 | 0.00 | 0.03 | 0.03 |
| Inkjet-printing thin-film perovskite solar cells<br>Digital Flatbed Inkjet Printer | 15 | 24,000 | 2 | 30 | 0.05 | 0.02 | 0.06 |
| Bamboo Processing<br>Fiber Machine | 1,000 | 5,000 | 1 | 1,000 | 0.01 | 0.50 | 0.51 |
| Bamboo Processing<br>Pulping Machine | 1,000 | 2,000 | 1 | 1,000 | 0.00 | 0.50 | 0.50 |
| Viscose Spinning<br>Viscose Filament Spinning Machine | 800 | 3,000 | 1 | 800 | 0.00 | 0.40 | 0.40 |
| Rayon Fabric<br>Rayon Weaving Loom | 2,000 | 6,000 | 1 | 2,000 | 0.01 | 1.00 | 1.01 |
| Clothing Production<br>Sewing Machines | 50 | 200 | 4 | 200 | 0.00 | 0.10 | 0.10 |
| Algal Production<br>Photobioreactors | 110 | 1,000 | 10 | 1,100 | 0.01 | 0.55 | 0.56 |
| Seeds/Seedlings/Eggs<br>Seeds/Embryos | 1,000 | 20,000 | 1 | 1,000 | 0.02 | 0.50 | 0.52 |
| Chicken Production<br>Egg Incubator | 60 | 300 | 1 | 60 | 0.00 | 0.03 | 0.03 |
| Aquaculture<br>Aerator | 14 | 500 | 2 | 28 | 0.00 | 0.01 | 0.02 |
| Grow Lights<br>LED Lights | 5 | 90 | 1000 | 5,000 | 0.09 | 2.50 | 2.59 |
| Vertical Tower<br>Growing Pots | 1 | 1 | 5000 | 2,500 | 0.01 | 1.25 | 1.26 |
| **Total** | | | | **308,618** | **3.25** | **154.31** | **157.56** |

**Phases 1, 2, & 3 - Total 60 Months**

| | Total Mass (kg) | Total Material Cost ($M US) | Ship. Cost* ($M US) | Total Cost ($M US) |
|---|---|---|---|---|
| **Grand Total** | **2,402,168** | **64.10** | **1,201.08** | **1,265.19** |

* Shipping cost assumed at $500/kg

Important Note: Equipment weights are based on off-the-shelf equipment largely constructed without regard to weight limitations. While building lighter-weight custom machines would increase the unit costs, there is likely a positive gain in reduced transport costs.

The above summary includes just enough to get the infrastructure in place to feed, clothe, and support 1000 plus people and bootstrap the mining, construction, industrial production, and farming industries within five years. It is reasonable to assume this will cost significantly more than $1.3 billion, but, even at an order of magnitude above this at $10.3 billion, it's clear that this is a cost-effective path toward permanent human settlement.

As for return on investment, Moon rock samples are currently valued at approximately $50,000 per gram. Even if Mars rocks were valued at half that, it would only take about 412 kg of samples returned to Earth to offset the entire $10.3 billion cost. Meteorites found on Earth sell for between $5000 to $1 million per kilogram. One should expect Mars meteorites, which can be collected all over the surface to have significantly more value. Then, there are gold and platinum group metals such as ruthenium, rhodium, palladium, rhenium, osmium, iridium, and platinum that will likely be isolated during regolith processing. At current Earth prices, and an assumed Earth return cost of $200/kg, all of these represent significant profit margins in ranges from $2600/kg to $87,000/kg.

It costs the International Space Station program about $1.3 million each day for each astronaut aboard. If the Aeneas settlement charged only $200,000/day for room and board at the settlement for visiting astronauts, scientists, and tourists, with a capacity for 250, this alone would represent an income of $18.25 billion per Earth year at full capacity.

## SOCIAL ASPECTS OF THE AENEAS SETTLEMENT

To produce a thriving community and culture on Mars will require a comprehensive approach to providing for the needs of its human inhabitants. While an imperfect model, Abraham Maslow's 'Hierarchy of Needs' does provide a basic framework for discussing the physical, emotional, and social needs of human beings. They are, as follows: physiological needs, safety needs, social belonging, esteem, self-actualization, and self-transcendence.

The settlement complex is foremost designed to provide for the immediate physiological needs of the inhabitants; providing safe shelter, with a warm, breathable atmosphere, drinking water, food, and clothing.

The settlement's social, legal, political, and economic organization are designed to provide its residents with personal, emotional, and financial security, as well as healthcare. Work schedules should be organized to provide all residents with ample amounts of leisure time. The marketplaces, with cafes, bars, restaurants, shops, gyms, theatres, etc. are intended to provide residents with opportunities for entertainment, socializing, pursuing hobbies, exercising, and playing sports (it isn't difficult to imagine people having a lot of fun adapting Earth sports to Martian gravity).

Settlement recruitment would have to consider numerous factors. Moving to Mars is a long, difficult, and potentially dangerous trip. The orbital mechanics of the Earth-Mars Hohmann transfer orbit necessitate that work contracts would be a minimum of 26 Earth months. The community will require experts in mining, construction, industrial processes, robotics, energy production, chemical engineering, metallurgy, manufacturing, horticulture, life support systems, doctors, dentists, massage therapists, physical trainers, chefs, waiters, brew masters, bartenders, musicians, artisans, etc. Potential community members would be chosen from applicants based on their experience, expertise, and emotional intelligence, as well as mental health and stability, Those chosen to become members of the settlement, would have the option of renewing their contracts, or returning to Earth, avoiding the possibility that anyone feel permanently trapped.

At the small community size of 1000 people, and as intimacy and sex are as vital to human emotional health as water, food, and sleep are to physical health, preference should be given to partnered couples[28,29].

It's critical to the success of the settlement that a culture of mutual respect be established and maintained. To this end, it would be recommended that founding leadership be composed of experienced leaders, who have demonstrated a history of mentorship and an adherence to the philosophy of servant leadership[30].

**CONCLUSION**

This paper started by saying that a settlement of a thousand people doesn't pop into existence; that it would grow from a smaller settlement into a thousand people and if successful would grow beyond that thousand. So far, it has been discussed how to establish it and grow it to that moment in time where the population reaches that one thousand, but what of its future? What lies ahead for Aeneas?

One can envision a similar pattern of development continuing around the original Aeneas complex. The boreholes, tunnels, and subsurface excavations would become wider, deeper, longer, and more sophisticated as the settlement grows, continues industrializing, and develops new tools and construction techniques. Eventually, it would grow into a large city with enormously large diameter boreholes a kilometer deep ringed with apartments and balconies, like inside-out skyscrapers plunged into the ground, each housing thousands; open space caverns dwarfing the original Campus Martius central green. One can imagine a future city of millions, with underground open spaces of forests and lakes, people walking, riding bicycles, and using a tube transport system to get around.

As this city grows, the original Aeneas complex at the historic center of the city would be transformed into the first Martian University; the original apartments becoming dormitories; the original Campus Martius now, quite literally, a campus green.

In May 2003, Dr. Richard E. Smalley presented his 'Humanity's Top Ten Problems for the Next 50 Years'. He prioritized them in a top down hierarchy: Energy, Water, Food, Environment, Poverty, Terrorism and War, Disease, Education, Democracy, and Population. The idea being that as the problems at the top of the list, starting with energy, are solved, the issues below become easier to resolve. These are the challenges for humanity on Earth.

The Aeneas settlement won't be a utopia. It will have significant challenges, but by providing the freedom, security, opportunity, and lifestyles that people aspire to achieve; by focusing on creating a settlement with abundant energy, water, and food; a pleasing living environment; opportunities for education, art, and entertainment; an economy where each individual is provided the opportunity to not only contribute to the advancement of their community, but to the establishment and advancement of human civilization beyond Earth; the political and economic instability, inequality, terrorism, war, disease, and overpopulation of Earth will provide the impetus for human migration to Mars.

**POSTSCRIPT**

A note about the names, Aeneas and Campus Martius. Aeneas (/ɪˈniːəs/) is a Trojan hero of Greco-Roman mythology, and is featured in Virgil's epic poem, the *Aeneid*, where he is portrayed as having fled to Italy after the fall of Troy to become a progenitor of the Romans. Aeneas is believed to have founded the settlement that would later become Rome, several generations later, when Romulus laid out its fortifications and became its first king. The Romans, in turn, revered Mars, the god of military power (viewed as a way of securing the peace, the Pax Romana) and the guardian of agricultural. The Campus Martius (/ˈkampəs/ /ˈmär-sh(ē-)əs-/), or Field of Mars, was about a 2 square km publicly owned area of ancient Rome.

## REFERENCES

[1]Audouard, J., Poulet, F., Vincendon, M., Milliken, R., Jouglet, D., Bibring, J., Gondet, B., Langevin, Y., Water in the Martian Regolith from OMEGA/Mars Express, Journal of Geophysical Research: Planets, Aug. 2014, Vol. 119, Iss. 8, pp. 1969-1989.
[2]Watters, T., Campbell, B., Carter, L., Leuschen, C., Plaut, J., Picardi, G., Orosei, R., Safaeinili, A., Clifford, S., Farrell, W., Ivanov, A., Phillips, R., Stofan, E., Radar Sounding of the Medusae Fossae Formation Mars: Equatorial Ice or Dry, Low-Density Deposits?, Science, Nov. 2007, Vol. 318, Iss. 5853, pp. 1125-1128.
[3]Zwick, T., Employee Participation and Productivity, Labour Economics, Dec. 2004, Vol. 11, Iss. 6, pp. 715-740.
[4]Tan, H., Jain, A., Voznyy, O., Lan, X., García de Arquer, F., Fan, J., Quintero-Bermudez, R., Yuan, M., Zhang, B., Zhao, Y., Fan, F., Li, P., Quan, L., Zhao, Y., Lu, Z., Yang, Z., Hoogland, S., Sargent, E., Efficient and Stable Solution-Processed Planar Perovskite Solar Cells via Contact Passivation, Science, Feb. 2017, Vol. 355, Iss. 6326, pp. 722-726.
[5]Badescu, V., Solar Thermal Power Generation on Mars, International Journal of Energy, Environment and Economics, Jan. 2007, Vol. 15, pp. 1-50.
[6]Fogg, M., The Utility of Geothermal Energy on Mars, From Imagination to Reality: Mars Exploration Studies of the Journal of the British Interplanetary Society (Part 2, Base Building, Colonization and Terraformation), 1997, Vol. 92, p. 187.
[7]Dundas, C., Bramson, A., Ojha, L., Wray, J., Mellon, M., Byrne, S., McEwen, A., Putzig, N., Viola, D., Sutton, S., Clark, E., Holt, J., Exposed Subsurface Ice Sheets in the Martian Mid-Latitudes, Science, 2018, Vol. 359, Iss. 6372, pp. 199-201.
[8]Izumi, J., Morimoto, T., Tsutaya, H., Araki, K., Hydrogen Sulfide Removal with Pressure Swing Adsorption from Process Off-Gas, Studies in Surface Science and Catalysis, 1993, Vol. 80, pp. 293-299.
[9]Schmäh, P., Vertical Shaft Machines. State of the Art and Vision, Acta Montanistica Slovaca, 2007, Vol. 12, Spec. Iss. 1, pp. 208-216.
[10]Loov, R., Vroom, A., Ward, M., Sulfur Concrete - A New Construction Material. Prestressed Concrete Institute, 1974, Vol. 19, Iss. 1, pp. 86–95.
[11]Khoshnevis, B., Yuan, X., Zahiri, B., Xia, B., Construction by Contour Crafting using Sulfur Concrete with Planetary Applications, Department of Industrial and Systems Engineering, University of Southern California, Rapid Prototyping Journal, 2016, Vol. 22, Iss. 5, pp.848-856.
[12]https://photojournal.jpl.nasa.gov/jpeg/PIA16572.jpg
[13]Allen, C., Morris, R., Jager, K., Golden, D., Lindstrom, D., Lindstrom, M., Lockwood, J., Martian Regolith Simulant JSC Mars-1, 29th Annual Lunar and Planetary Science Conference, Houston, TX, Mar. 1998, Abstr. No. 1690.
[14]Wang, A., Zhou, Y., Rates of Al-, Fe-, Mg-, Ca- Sulfates Dehydration Under Mars Relevant Conditions, 45th Lunar and Planetary Science Conference, 2014, Abstr. No. 2614.
[15]Tirsch D., Airo A., Nitrates on Mars, Encyclopedia of Astrobiology, 2014, Springer-Verlag, Berlin, Heidelberg.
[16]Gasda, P., Haldeman, E., Wiens, R., Rapin, W., Frydenvang, J., Maurice, S., Clegg, S., Delapp, D., Sanford, V., McInroy, R., First Observations of Boron on Mars and Implications for Gale Crater Geochemistry, American Geophysical Union, Fall General Assembly, 2016, Abstr. ID P21D-04.
[17]Marvin, G., Woolaver, L., Thermal Decomposition of Perchlorates, Industrial & Engineering Chemistry Analytical Edition, 1945, Vol. 17, Iss. 8, pp. 474–476.
[18]Bruck, A., Sutter, B., Ming, D., Mahaffy, P., Thermal Decomposition of Calcium Perchlorate/Iron-mineral Mixtures: Implications of the Evolved Oxygen from the Rocknest Eolian Deposit in Gale Crater, Mars., 45th Lunar and Planetary Science Conference, Mar. 2014.

[19]Omar, H., Production of Lunar Concrete Using Molten Sulfur, University of South Alabama, Department of Civil Engineering, Mobile, AL, 1993.
[20]Paik, S., Chung, J., Selective Catalytic Reduction of Sulfur Dioxide with Hydrogen to Elemental Sulfur over Co-Mo/$Al_2O_3$, Applied Catalysis B: Environmental, Feb. 1995, Vol. 5, Iss. 3, pp. 233-243.
[21]Habashi, F., Mikhail, S., Vo Van, K., Reduction of Sulfates by Hydrogen, Canadian Journal of Chemistry, 1976, Vol. 54, Iss. 23, pp. 3646-3650.
[22]Agrawal, A., Kaushik, N., Biswas, S., Derivatives and Applications of Lignin - An Insight, The Scitech Journal, Jul. 2014, Vol. 1, Iss. 7, pp. 30-36.
[23]Zubrin, R., Baker, D., Gwynne, O., Mars Direct: A Simple, Robust, and Cost Effective Architecture for the Space Exploration Initiative, 29th Aerospace Sciences Meeting, Jan. 1991, Reno, NV.
[24]Mortazavi, S., Kardan, M., Sina, S., Baharvand, H., Sharafi, N., Design and Fabrication of High Density Borated Polyethylene Nanocomposites as a Neutron Shield, International Journal of Radiation Research, Oct. 2016, Vol. 14, Iss. 4, pp. 379-383.
[25]Rai, S., Wasewar, K., Mishra, R., Puttewar, S., Neutralization of Red Mud using Inorganic Acids, Research Journal of Chemistry and Environment, Jul. 2013, Vol. 17, Iss. 7, pp. 10-17.
[26]Bhaskar, M., Akhtar, S., Batham, G., Development of the Bricks from Red Mud by Industrial Waste (Red Mud), International Journal of Emerging Science and Engineering, Feb. 2014, Vol. 2, Iss. 4, pp. 7-12.
[27]Garg, A., Yadav, H., Study of Red Mud as an Alternative Building Material for Interlocking Block Manufacturing in Construction Industry, International Journal of Materials Science and Engineering, Dec. 2015, Vol. 3, Iss. 4, pp. 295-300.
[28]Satcher, D., The Surgeon General's Call to Action to Promote Sexual Health and Responsible Sexual Behavior. American Journal of Health Education. 2001, Vol. 32, Iss. 6, pp. 356–368.
[29]Umberson, D., Montez, J., Social Relationships and Health: A Flashpoint for Health Policy, Journal of Health and Social Behavior, Oct. 2010, Vol. 51, Iss. 1, Sup.1, pp. 54-66.
[30]Rivkin, W., Diestel, S., Schmidt, K., The Positive Relationship Between Servant Leadership and Employees' Psychological Health: A Multi-Method Approach, German Journal of Human Resource Management: Zeitschrift für Personalforschung, Feb. 2014, Vol. 28, Iss. 1-2, pp. 52-72.

# 4: THE NERGAL MARS SETTLEMENT

**Cassandra Plevyak**
Brightworks, Stanford Online (Class of 2020)
casplev@gmail.com
**Audrey Douglas**
Urban School of San Francisco (Class of 2022)
audrey.maya.douglas@gmail.com

## INTRODUCTION

In ancient Sumeria there was a god called Nergal. He was a god of war and a bringer of death but also called the "king of the sunset". He is representative of violence and destruction but is the king of the sunset, which is one of the most reliable and beautiful events we experience. We believe this is a worthy representation of what will be experienced on Mars. The climate is incredibly harsh, likely doing its best to kill those who disrupt it, but if we find our way past what it throws at us, we will be lead to something truly incredible. Our city will be almost entirely self-sufficient, using the natural resources found on Mars to build to build a place that will be a home to many of the bravest explorers of our time. We propose a city not only about survival but about culture and beauty. We will develop a warm and supportive community where voices are heard and citizens are happy and able to express themselves; a city that will develop high end intellectual services and keep contributing to the interplanetary world through art, science, and engineering.

## AREA

The area we have chosen for our Mars base is Milankovic crater on Mars. This is not to be confused with Milankovic Crater on the moon. Milankovic Crater on Mars is located just north of Olympus Mons on Vastitas Borealis. It is at a high enough altitude to have water extremely close to the surface, so much so that some of it is sublimating off, but not at such a high altitude that we would lose too much sunlight. Milankovic crater is especially interesting because NASA has confirmed there is water there. In fact, we have confirmed that there is enough water there to be seen from satellites, trapped in sheets in the cliff walls. Another good thing about this location is that we know a lot about the soil in this part of Mars. Vastitas Borealis is also where the Phoenix lander touched down, and the soil analysis from that mission applies to the area around Milankovic.

## STRUCTURE

Martian soil contains a lot of elements and molecules that would be great to build with, including silicas and iron deposits. It primarily consists of Silica Dioxide(43%), the same molecule glass is made of. This means that we can create fiberglass. Fiberglass is a very useful material, and can be used as both insulation and building materials, although creating building materials from fiberglass would require resins which are not easily produced on Mars. The soil also contains large amounts of Iron III oxide(18%), the ore used in commercial iron production. This is what gives Martian soil its color. Aluminum oxide is also present(8.6%), but nowhere near the amount found in Earth basalts of similar origin. However, there is still enough present to use as a construction material. There are recipes for creating concrete out of sulfur and martian soil, and early tests of this materials show that it is as strong as concrete found on Earth.

The Mars settlement will be made out of Martian concrete and steel, as these are easily manufactured from the raw materials available on Mars. The internal structures and machinery will be made out of Aluminium, as this is non corrosive and lightweight. Glass can also be used in the interior, although it will be primarily used for the solar farm. Other materials will have to be mined from veins while on mining expeditions, grown on the farm, or imported.

The main structure of the base, including the floors, will be made of Martian concrete and steel to provide structural support. The floors of each level will be attached straight to the super structure and also be made of concrete and steel. There will be 16 by 16 foot pillars every 33 feet to provide structural stability. The base will then be insulated with fiberglass.

**LAYOUT**

The majority of the Mars settlement will be inside the cliff wall of Milankovic crater. Since it is cut into the wall, there will be over 2 meters of martian regolith and ice sheets above it, providing radiation protection for the whole base. The excavated area will provide all of the building materials, and water. The total square footage of floor space will be 3 square miles, consisting of six half mile squared levels . Each level will be 16 feet tall from floor to ceiling. This is a total of 836,352,000 cubic feet of evacuated mars rock.

The settlement will have six levels. The first two will be for the majority of the stations human support: housing, work, recreation, healthcare, etc., while the next four will be primarily for growing food.

Level one, which is level with the floor of the crater, will be for critical systems, communication heavy labs, manufacturing, and airlocks. Since the solar panels are nearby on the surface, the main power station will be located here. The algae farms where the majority of the stations oxygen is produced will be here, as will the processing plant for the Lime filters. So that the plumbing system can take advantage of gravity, the clean water tank will also be on level one. This level will also house both maintenance and manufacturing. While maintenance personnel for the stations various systems will have workrooms throughout the station, the maintenance headquarters with large or expensive equipment will be on this level. Manufacturing will require raw materials found on mining expeditions, so there will be both manufacturing and ore processing. In a similar vein, the staging and mission control areas for the mining expeditions will be on level one.

Level one is also going to contain all the communication support. This includes the main communications dish, and the people who talk with Earth support. Servers housed here will download heavily used websites, like Wikipedia or Netflix. We will also have a remote Mission Control, which will be used for long-distance missions for NASA and other space agencies. Because Mars is usually in a different position relative to earth, this mission control can take over if it has a better connection or less time delay. Attached to this will be an astronomy lab. Immediately outside of Level one, next to the solar panels, will be a launch pad. For safety, incoming ships will land outside of the solar panel area and people and materials will be driven in..

Level two will provide places for people to live, recreation spaces, schools, along with the laboratories. The majority of the population will live in apartments, which would be on either the first or second floor. These could range from 1000 - 3000 square feet. There will also be a movie theater, a gym, and a large library. We will also have several parks, with dwarf trees and grass. We will also have schools for anyone who has children or moves to Mars without the proper skill sets, and a hospital. Level two will have laboratories for research into Geology, Chemistry, Engineering, and Biology. Mars has many interesting features that are as of now unresearched, from the unstudied rock formations to the effects of low gravity on life and evolutionary development.

Level three through five will contain farms divided into zones according to optimal growing temperatures. Grow zones 4 and 3, as mentioned below, will be on level three. Grow zone 3 contains many important crops, so it will take up all of level four. Grow zones 2 and 1 will be on level five.

Level six will contain parts of grow zone 1 and the sewage plant. The sewage plant is located on the lowest level so that the pipes can work primarily on gravity. When the water is purified, it will be pumped to the holding tank at the top of the base. The remaining waste will be processed into fertilizer and hauled to where it is needed.

**MAIN SYSTEMS**

**Air**

The air system will be almost completely decentralized. This means that we don't have to use extensive duct systems to move air around, and there is no single point of failure for this extremely important system.

A single vent will cross each room, and when there are large open spaces there will be a series of cross cutting vents covering the room. Intake vents will take

$CO_2$ level readings. If the levels are within an acceptable range (under 1 mmHG), then the air will be sent through a very fine filter to remove any particulates, followed by a dehumidifier. This dehumidifier will just cool down the air to a temperature at which the water precipitates out before heating it up again with a heat exchanger.

If the $CO_2$ levels are over 1 mmHG the air will move through the Carbon Dioxide retrieval unit. The $CO_2$ filter will be made of Calcium Oxide, also known as Lime, which is 6% of the Martian soil. A tank will then replace the filtered out $CO_2$ with Oxygen. When the $CO_2$ filters are full they will be taken up to the main air system station, where they will be processed. That is also where the 2000 square feet of algae that produces the Oxygen will be located. Air tanks and filters will be replaced when needed.

The many plants of the space settlement are also involved in the air system. While we can not make all of our oxygen from farms and parks, some of it will be produced there naturally. The amount of oxygen will vary greatly depending on the season and plant growth stages so the more stable and responsive algae system will be used to produce the required balance..

**Water**

The water treatment plant will use conventional processes to recycle the water and create fertilizer out of waste. The only difference is that this will be on a smaller scale than an Earth city sewage treatment plant. The waste will flow from the houses on level 2 and base dehumidifiers to a treatment plant on the 6th level. After processing the water will be pumped back up to a tank on the first level, so that it can flow according to gravity.

**Power**

Mars has a very different power supply than earth. While on Earth there are a number of options to choose from when generating power, on Mars there are only really two: solar and nuclear. A nuclear plant would create a large amount of energy per square foot, and the power plant should run smoothly for quite a long time. Nuclear power plants are used on a lot of deep space missions, including the Curiosity mission. The hindering factor here is fabrication. The elements we would need, such as Uranium, are hard to find on the unexplored Martian surface, which means that any nuclear power source would have to be built on Earth. There are also safety concerns. Radiation is dangerous, and a meltdown could destroy not only the settlement but any hope of making another one. Funding would disappear overnight.

Solar panels are the better option in the long run. First of all, certain models are made almost completely of silicon, so we can make as many as needed using processed Martian soil with minimal imported materials. They are also completely safe. The downside is that we will need a lot of them to do anything. The latitude we are at is not very great for sunlight, and Mars is farther from the sun than Earth. There are also issues with dust storms. Mars is pretty famous for its dust storms, and they have a tendency to kill solar powered machinery like our dear departed rover Opportunity. We will have to correct for that.

In order to build and run our base, we will have to use both fission power and solar power. For the initial building of the base we will have to use a nuclear fission reactor, which will also run the spacecraft that takes the crew and cargo there. Once on the surface, the nuclear reactor will be used to power the machinery required to mine the base out of the cliff wall. This is because these tools need a lot of power now, and setting up solar panels would be too much work for a crew that has nowhere to live. The fission reactor will be used further on mining missions, which also need power immediately and cannot afford to set up solar arrays.

Once built the base will move to solar power. Our base has the entire area around the floor of Milankovic crater to put solar arrays, stretching out as far as is needed for the base at that time. These solar panels will be maintained by rovers, accessible from the base's main entrance in the cliff face. The solar panels will also have wipers attached to them, like windshield wipers on a car, which would allow them to remain generating power after a dust storm.

**Food**

The settlements food will be grown on the floors as stated above, which will be heated or cooled to the correct temperature, as noted below. Total amount grown will depend on other factors, like what people like and what has been grown in the soil before (crop rotation). (1st number = priority level, 2nd number = temp zone: zone 1: 65° zone 2: 68° zone 3: 73° zone 4: 80°)

- Potatoes 2-1 (45-80°)
- Common wheat 1-3 (70-75°)
- Coffee beans (species: Coffea Arabica) 3-2 (64-70°)
- Soybeans 1++(extra important)-3 (68-77°)
- Peanuts 1-2 (65-70°)
- Mushrooms 2-1 (55-68°)
- Cacao beans 3-4 (65-85°)
- Beets (species: Sugar beets) 3-1 (50-85°)
- Lemons (species: Meyer lemons) 3-1 (50-80°)
- Hot peppers 3-3 (65-80°)
- Lettuce 2-1 (60-70°)
- Carrots 2-1 (50-85°)
- Bamboo (species: Bambusa) 1-4 (70-100°)
- Linen 2-2 (50-80°)
- Bananas 1-4 (77-102°)

## VEHICLES

The base will be equipped with two main types of rovers. The first are short-range rovers which are primarily for doing maintenance on the solar panel arrays. While comfortable, these rovers are only equipped to run for 2-3 days, enough to do maintenance or replace solar sells. The second type of rover would be used on mining missions. It would be more like a very large transportation truck, and would contain the necessary mining equipment. These rovers would also be the ones used for the initial excavation of the area of the mars base.

## POLITICAL SYSTEM

### Overview

The political system we believe will be most effective and fair given the number of people we have is a democratic socialist government in which all citizens are provided with the materials to fulfill their basic human needs and a job. However, there will not be a money system within the colony. We believe that money will only lead to class lines and tension within the colony, which will not benefit a colony the size of ours. However, a month before the end of the year, the money that the Nergal council does not use for the colony as a whole will be split evenly among the citizens can be spent either on Earth or to ship something from Earth to Nergal. We believe that all citizens should have their needs fulfilled because no one should be left to fend for themselves on Mars. If someone in the colony doesn't have the money for food they would likely resort to stealing as they will not have another option, and we do not want that to be a part of the culture of Nergal. Being left to fend for yourself on Mars is essentially death and we will not condemn our citizens to that.

The Nergal Council will handle diplomacy and trade with earth organizations, and with that Nergal's collective money, for the benefit of all. The Council stewards the hard currency of the settlement for trade and external purchases, because there is no internal money. The Council will be empowered to negotiate all things trade and foreign affairs, but they must bring bigger actions before the citizens in advance of finalization. They will handle relations with multiple Earth governments and companies as well as other Mars settlements as they develop.

### Mechanics of Governing

The Nergal Council will be a legislature of 75 people consisting of an elected city council of 74 and 1 appointed advisor. The council members will be elected by the different groups of jobs (see section on jobs). It is a representative democracy with 1 representative per every 25 people. Each member of each job group votes on the amount of city council members their group is alloted. The city council will discuss and vote on most major decisions. Their discussions are open and citizens are welcome to listen in.

Internally, Nergal's socialist system will be governed by direct democracy. Citizens can choose to vote on as many or as few decisions as they like, however the largest decisions are mandatory. Citizens are allowed to call a town meeting if they would like to hold further discussing before a decision is made and be part of the voting process. We want all Nergalites to be informed about government happenings if they would like to be. Transparency will foster trust in the council since citizens will know that they are not hiding anything. All proceedings are published by journalists to make sure all citizens are informed. If over ⅔ of the population disagrees with a decision, it can be vetoed. Anyone who commits a major crime or a experiences serious longlasting medical emergency will be sent back to Earth.

**JOBS**

Everyone in the Nergal Mars settlement will be provided food, water, and a home but working is also a requirement. Each job category (see below) will have a head of emergencies and volunteers who support them. They are in charge when emergencies occur. The jobs are split into categories that operate as one when voting on representatives. The full list, including personel distribution is: (mechanics count as build team when they are not working as mechanics):

- Gardeners - 200
    - Greenspace manager
    - Garden manager
    - Garden life support
    - General gardeners
    - Chef/food prep
- Mechanics - 100
    - Solar panel and power cell upkeep
    - Heating sysops
    - Life support mechanics
    - General mechanics (fix many basic systems)
    - Rover driver and upkeep
    - Air lock monitor
    - Waste treatment operators
    - Server operator
        - Communications optimizers
        - Server Mechanics
        - Server room manager
        - Communications Dish upkeep
- Doctors/nurses/medical professionals - 20
- Manufacturers/build team - 300
- Miners - 100

- Government (city council) - 75
  - Foreign relations advisor
  - Specialized positions for city council members
    - Emergency coordinator (once the city council is elected people will vote again to decide this)
    - City planner - multiple plans will be proposed whenever expansion is need and people will vote on plan they want (once the city council is elected people will vote again to decide this)
    - Event planner (self selecting)
    - Econ. team (once the city council is elected people will vote again to decide this)
    - Any other position the community feels is needed as time goes on
- Scientists - 200
  - Robotics engineers
  - Astronomers
  - Mission control
- Journalists/communications - 5
- Any other positions that need arises for

Our inhabitants will launch to Nergal in stages. We will prelaunch rovers, mining equipment, a temporary habitat, and other essential building materials. The build team, five of the doctors, half of the miners, and five gardeners including the garden manager will be the first people sent up along with the original foodstock and water supply and they will construct the most essential rooms and systems of the settlement (such as water collection, air regulation, etc.) After the next group arrives the build team will continue to expand the settlement until the blueprints have been completed and after as needed. When work on the settlement is not being done they will take part in the Nergal manufacturing service (more on that in the exports/imports section). After the build team the rest of the gardeners and miners will be sent next along with the major seedstock and gardening tools and any other building materials needed for this stage of the plan. The third launch will carry mechanics and scientists along with scientific instruments and any building materials needed. The fourth launch will carry everyone else as well as art supplies (musical instruments and such), personal items and any other building materials needed.

**EXPORTS/IMPORTS**

Nergal will aim to be self-sufficient in basic life support as soon as possible however it will need to import the original foodstock and water supply as nwell as building materials for the construction of the site. We will also need continuous supplies of seeds and chemicals for fertilizer that will not be available to us otherwise. We will also import mechanical parts and scientific instruments as needed, rare Earth elements, and online entertainment.

Nergal will achieve long term sustainability through developing a high end service based economy on top of the internal economy designed to sustain the necessities of life and culture within the city. We will offer a variety of different services to Earth, each costed $1,000.00 per hour for the purposes of estimation:

- If we are the first colony on Mars, more will likely want to follow. We will offer an architectural and scientific consultant service that will advise on how, where, with what, etc. to build on Mars. Our scientists and some of our mechanics will serve as these consultants trading off days. Our consulting service will work hand in hand with our manufacturing service.
- Since Mars has one third of Earth's gravity and ready access to a near vacuum, it will be much easier to construct large and heavy objects on Mars than it is on Earth. We will use this to our advantage by selling a service that will build items easier to manufacture on Mars near Nergal and ship them back to elsewhere in the solar system, or if the client would like to build and launch something from Mars they can send Nergal their materials and we will build it and launch it, the price will depend on size, time, and complication of each design (shipping of materials must be paid for by the client). We are willing to build anything from small robots to whole ships or stations (not everything has to be for space travel though) but we will likely not help with the manufacturing of space weapons for any specific government or company because aiding one specific party would be likely to put Nergal and its citizens in danger.
- We will run a remote mission control so countries on Earth may pay us to launch from Mars and to monitor various things more visible from Mars or any spacecraft they launch in our general direction. Because of the way our services connect, people will be able to pay use to design, build, launch, and monitor a project for them on Mars.
- We will work with scientists, governments, private companies, etc. and conduct scientific experiments for them. We will work with media (video games, movies, etc.) creators to do motion capture and filming for space physics. For certain science work Nergal will provide results to Earth for all of humankind. Some will be private and we require payment for them.
- We will create an infrastructure for tourism and virtual tourism for the sake of education.

| YEAR 1 | Staff | Hours/year | Dollars/hour | Income | Lift capacity (E-M) | Lift capacity (M-E) |
|---|---|---|---|---|---|---|
| Scientists | 3 | 900 | $1,000.00 | $2,700,000.00 | 5,400 | 13,500 |
| Creatives | 1 | | $1,000.00 | $0.00 | 0 | 0 |
| Builders | 3 | 900 | $1,000.00 | $2,700,000.00 | 5,400 | 13,500 |
| **Totals** | **7** | **1800** | **$3,000.00** | **$5,400,000.00** | **10,800** | **27,000** |

The first year our services will not be extremely profitable because most of our staff will be focused on getting the colony up and running, but we will have a few people devote their time to getting the services up and running. We will not use all of the money for lift capacity, but the chart shows how much we could lift if we did (in kg).

| YEAR 3 | Staff | Hours/year | Dollars/hour | Income | Lift capacity (E-M) | Lift capacity (M-E) |
|---|---|---|---|---|---|---|
| Consulting | 30 | 1200 | $1,000.00 | $36,000,000.00 | 72,000 | 180,000 |
| Build projects | 50 | 1200 | $1,000.00 | $60,000,000.00 | 120,000 | 300,000 |
| Mission services | 50 | 1200 | $1,000.00 | $60,000,000.00 | 120,000 | 300,000 |
| Contract science | 15 | 1200 | $1,000.00 | $18,000,000.00 | 36,000 | 90,000 |
| **Totals** | **145** | **4800** | **$4,000.00** | **$174,000,000.00** | **348,000** | **870,000** |

In our third year we will be able to spare a few more people. This is when we start to need less support from Earth.

| YEAR 5 | Staff | Hours/year | Dollars/hour | Income | Lift capacity (E-M) | Lift capacity (M-E) |
|---|---|---|---|---|---|---|
| Consulting | 30 | 1800 | $1,000 | $54,000,000 | 108,000 | 270,000 |
| Build projects | 100 | 1800 | $1,000 | $180,000,000 | 360,000 | 900,000 |
| Mission services | 50 | 1800 | $1,000 | $90,000,000 | 180,000 | 450,000 |
| Contract science | 30 | 1800 | $1,000 | $54,000,000 | 108,000 | 270,000 |
| **Totals** | **210** | **7200** | **$4,000** | **$378,000,000** | **756,000** | **1,890,000** |

At five years we will put even more people towards our service program, which will be fully established by this point.

| YEAR 10 | Staff | Hours/year | Dollars/hour | Income | Lift capacity (E-M) | Lift capacity (M-E) |
|---|---|---|---|---|---|---|
| Consulting | 50 | 1,800 | $1,000 | $90,000,000 | 180,000 | 450,000 |
| Build projects | 200 | 1,800 | $1,000 | $360,000,000 | 720,000 | 1,800,000 |
| Mission services | 50 | 1,800 | $1,000 | $90,000,000 | 180,000 | 450,000 |
| Contract science | 100 | 1,800 | $1,000 | $180,000,000 | 360,000 | 900,000 |
| **Totals** | **400** | **7,200** | **$4,000** | **$720,000,000** | **1,440,000** | **3,600,000** |

When Nergal has been around for a decade, we believe we will be able to make enough money to lift everything we need from Earth and have extra to spare. We will spend some of the extra on person items requested by citizens each year.

**GREEN SPACES**

To be sustainable long-term, the city must also fulfill the human need for beauty, inspiration, and mental health. We also want to grow a steady and supportive community so psychologists will be included in the selection of doctors we bring so if someone is having a mental health issue they may reach out to them but if some only wants a person to talk to we will have volunteer therapists who will be trained by our psychologists. One main garden will be the location for the growth of the plants that will be used for food. (See plant list.)
In addition to the large garden there will be smaller gardens, green spaces if you will, around Nergal. They will be managed and mainly taken care of by a small team of about ten people from the garden staff with one of them being the main coordinator. Citizens may plant what they like or volunteer to help take care of the existing plants. There will be digital art equipment in the green spaces where citizens may draw/paint/photograph the garden. The equipment can be removed from the green spaces but one of the green space caretakers must be notified and the citizen must specify when the equipment will be returned. The green spaces will be hooked into the air system as well. We would also like to send musical instruments to Mars. They will be kept in common spaces but they can be taken back to a private area if one of the greenspace coordinators is told.

**ENTERTAINMENT**

As mentioned above, one form of entertainment will be art and movies and online content. Another form, probably the most common, will be VR. We want to set up sensory deprivation tanks and use them game simulations/video games, virtual tours of other places, 360° movies. The smaller gardens mentioned above will also serve as common spaces offering people places to sit and talk and children a place to play "outside".

**SCHOOL**

In every job department there will be a head teacher who will organize as many people as are needed to train new people for whatever job they are going to do. There will be written manuals about everything you need to know for each job that will be looked over and updated yearly unless the job is not currently being done. When a new position is created, the first person or people to do it will write the manual at the end of the first three months they do it for, unless the duration is shorter in which case they will write it at the end of their time doing it. In addition, we will have people from different job departments who only work part time and in their spare time they will work as teachers (when there are kids) each teacher will teach a different subject and will teach for about an hour a day. As the colony grows there will start to be more teachers and eventually there will be several full time teachers. As an adult if there is a skill you would like to learn you may contact the city council who has a list of volunteers who are willing to spend an hour a week teaching (one person at a time) and they will connect you to that person. If there is no person than you may talk to the communications and media regulation group about downloading a textbook or some other resource that may teach you.

**SPORTS**

Since gravity is low and muscle mass must be kept up if we want it to even be remotely possible for people to return to Earth, there will be a gym and citizens will be advised by doctors how much they need to exercise and of what kind. There will also be recreational activities provided for fun. Because of the low gravity, gymnastics will be much less dangerous and a way to teach agility especially to kids. It also does not need as much equipment as some things, which is optimal because of shipping costs between planets. A trampoline will also be provided for recreation. There will be nightly playground games, after most people are off work, for anyone interested (things like capture the flag and and don't touch the ground tag if people are willing to move furniture). These games will be held in the common space and there will be a volunteer supervisor. We think that projectile sports such as archery or ultimate frisbee could serve as an interesting experiment if we obtained the space and materials for them.

**CONCLUSION**

What makes this colony is community. Many of our systems rely on volunteers and trust that our inhabitants want what is best for the colony. We believe that in our system the power does not belong to a small group of people and that decisions truly are up to the community as a whole. There are many moving parts within our colony and we rely on our people being competent and willing to work. We will select our original citizens carefully and foster a culture that supports these traits. While it will be hard work to live in these conditions, especially in the early years, we want to create a community where people support each other and where our citizens can feel comfortable and safe.

Nergal will slowly grow horizontally, around the walls of Milankovic crater. It will be almost completely self-sufficient, and become even more self-sufficient as it makes its mark on Mars. It will be an Oasis of human life in a place where nothing grows, and it will support future Mars colonization.

**BIBLIOGRAPHY**

(APA)

Albert, S. (1970, May 03). How to Grow Hot Peppers. Retrieved from https://harvesttotable.com/how_to_grow_hot_peppers/

Albert, S. (2019). *How to Plant and Grow Potatoes*. [online] Harvest to Table. Retrieved from https://harvesttotable.com/how_to_grow_potatoes/

Bradford, J. (2012, January 13). One Acre Feeds a Person. Retrieved March 31, 2019, from http://www.farmlandlp.com/2012/01/one-acre-feeds-a-person/

Chow, D. (2012, July 20). New Mars Rover Could Far Outlive Its Lifespan. Retrieved from https://www.space.com/16679-mars-rover-curiosity-nuclear-power-lifespan.html

Climate.gov. (2019). *Climate & Coffee | NOAA Climate.gov.* [online] Retrieved from https://www.climate.gov/news-features/climate-and/climate-coffee

Domenghini, C. (2016, October 07). What Kind of Climate Do Beets Grow In? Retrieved from https://homeguides.sfgate.com/kind-climate-beets-grow-in-71927.html

Efficient Earth-Sheltered Homes. (n.d.). Retrieved March 31, 19, from https://www.energy.gov/energysaver/types-homes/efficient-earth-sheltered-homes

Esterhuizen, G. S., Dolinar, D. R., Ellenberger, J. L., & Prosser, L. J. (2011). Pillar and roof span design guidelines for underground stone mines. doi:10.26616/nioshpub2011171

Farm Progress. (2019). *High Temperature Effects on Corn, Soybeans.* [online] Retrieved from https://www.farmprogress.com/corn/high-temperature-effects-corn-soybeans

Growing Guide: Carrots. (2006). Retrieved March 31, 2019, from http://www.gardening.cornell.edu/homegardening/scenea765.html

How to Grow the Most Delicious Mushrooms Right at Home. (n.d.). Retrieved from https://www.bhg.com/gardening/vegetable/vegetables/how-to-grow-mushrooms/

India, I., Us, C. and Yadav, S. (2019). *Climatic conditions for Growing Wheat - ImportantIndia.com.* [online] ImportantIndia.com. Retrieved from https://www.importantindia.com/12612/climatic-conditions-for-growing-wheat/

LaLiberte, K. (n.d.). How To Grow Salad Greens All Year | Gardener's Supply. Retrieved from https://www.gardeners.com/how-to/how-to-grow-salad-greens-all-year/7272.html

Leschmann, T. (2011, May 03). What is a Good Climate for Growing Peanuts? Retrieved from https://www.hunker.com/13428124/what-is-a-good-climate-for-growing-peanuts

Los Alamos National Laboratory. (2002, April 15). Imagine No Restrictions On Fossil-Fuel Usage And No Global Warming. *ScienceDaily*. Retrieved March 27, 2019 from www.sciencedaily.com/releases/2002/04/020412080812.htm

Mackenzie, A. (2016, October 07). Temperature Requirements for Bamboo. Retrieved from https://homeguides.sfgate.com/tempature-requirements-bamboo-69804.html

Malik, T. (2018, June 21). Mars Dust Storm 2018: How It Grew & What It Means for the Opportunity Rover. Retrieved from https://www.space.com/40888-mars-dust-storm-2018-and-opportunity-rover-images.html

Mars Colony Prize - Design the First Human Settlement on Mars. (2018, October 24). Retrieved from http://www.marssociety.org/mars-colony-prize-design-the-first-human-settlement-on-mars/

Materials Scientists Make Martian Concrete. (2016, January 05). Retrieved from https://www.technologyreview.com/s/545216/materials-scientists-make-martian-concrete/

Michaels, K. (2019, March 24). How to Grow a Beautiful Meyer Lemon Tree in a Pot. Retrieved from https://www.thespruce.com/growing-meyer-lemon-trees-in-pots-848166

Mulvaney, D. (2014, November 13). Solar Energy Isn't Always as Green as You Think. Retrieved from https://spectrum.ieee.org/green-tech/solar/solar-energy-isnt-always-as-green-as-you-think

Oritz, A. R., Rygalov, V. Y., & De León, P. (2015). Radiation Protection Strategy Development for Mars Surface Exploration. Retrieved March 31, 2019.

Peters, Greg & Abbey, William & Bearman, Greg & Mungas, Greg & Anthony Smith, J & Anderson, Robert & Douglas, Susanne & Beegle, Luther. (2008). Mojave Mars simulant—Characterization of a new geologic Mars analog. Icarus. 197. 470-479. 10.1016/j.icarus.2008.05.004.

Richter, C. (2017, April 22). Fibre Flax Planting and Processing Instructions. Retrieved March 31, 2019, from https://www.richters.com/show.cgi?page=InfoSheets/d2701.html

Shuler, R. L., & Affens, W. A. (1970). Effect of Light Intensity and Thickness of Culture Solution on Oxygen Production by Algae. *Applied Microbiology*. Retrieved March 31, 2019, from https://aem.asm.org/content/aem/19/1/76.full.pdf.

Stallsmith, A. (2010, May 12). How to Grow Cocoa Beans. Retrieved from https://www.hunker.com/12001728/how-to-grow-cocoa-beans

Voosen, P. (2018, January 26). Ice cliffs spotted on Mars. Retrieved from https://www.sciencemag.org/news/2018/01/ice-cliffs-spotted-mars

## APPENDIX

**Room list (13/20)**

- **Main Water System**
  - Sewage treatment
  - Water tanks (33 by 48 feet)
  - Dehumidifiers (Decentralized, add lots to all areas)
- **Air locks** (Try to get a complete vacuum before opening to outside.) (10 square feet)
- **Centralized power**
  - Power station
  - Solar arrays
  - Radiation arrays
  - Extra Batteries
  - Initial RTG
- **Air processing**
  - Carbon Dioxide processing / monitoring (2ft^3 of algae per person, each in a room, 200 square foot control room,)
  - Air Filtration (In ducts)

  - Vents (All around, in ceiling. Equipped with fans) This is a balanced air filtration system.
- **Gardens**
  - Main Growing Garden (4.356e+7, 1000 acres min)
  - Main people garden (5 acres, 217800 square feet)
  - Mini Community Gardens (500-1000 square feet)
  - Gardener work station (500 square feet)
  - Bike rack (10 square feet)
  - Garden Storage (500 square feet)
- **Mission Control room**
- **Maintenance/Janitorial stuff**
- **Food preparation**
  - Kitchen (1300 square feet)
  - Food Storage (30000 square feet)
  - Cleaning room
  - Fertilizer production (Algae and compost)
- **Science Lab**
  - Engineering
  - Chemistry
  - Bio lab
  - Geology
  - Astronomy
- **Emergencies**
  - Emergency Shelters (10 square feet per person, in 100 person blocks. 1000 square feet, have own air system, Food stuff)
- **Communications**
  - Server Room
  - Dish operation/Maintenance/the actual dish
  - Communications optimization workstation
- **Hospital** (300,000 square feet)
  - All the hospital stuff
- **City Council**
  - Council offices (200 square feet for each, 14400 total)
  - Council meeting chamber (40,000 square feet)
- **Apartments**
  - Smaller, one-two person apartments (1000 square feet)
  - Larger, family apartments (2500 square feet)
  - "My elderly parents have to move in help" apartments (4000 square feet)
- **Entertainment**
  - Movie theaters (40,000 square feet)
  - Library/bookstore (3,000 Square feet)
  - Workout area (1000 square feet)
- **Rovers**
  - Elevator (22 square feet)
  - Rover Maintenance

  - Rover storage
  - Really awesome rovers (~300 square feet? Figure that out.)
- **Mining stuff**
  - Mining mission Control (500 square feet)
  - Mining area for more rooms (N/A)
  - Building room (750 square feet)
- **School**
  - CAN GROW
  - Job training
- **News room** (750 square feet)
- **Suit-up rooms** @ airlocks (200 square feet)
- **Manufacturing**
  - Ore Processing (Steel, Fiberglass, ect)
  - Export Manufacturing
  - Interior stuff-ness manufacturing (Appliances, tools, ect)

# 5: THE "TEAM ENSC" VIZZAVONA DESIGN

**Laetitia Calice, Caroline Cavel, Adrien Leduque, Mateo Mahaut**
Ecole Nationale Supérieure de Cognitique, Bordeaux INP, France
**Jean-Marc Salotti**
Ecole Nationale Supérieure de Cognitique, Bordeaux INP, France
Univ. Bordeaux, CNRS, Bordeaux INP, IMS, UMR 5218, France
INRIA, IMS, UMR 5218, France
Association Planète Mars, France
Jean-marc.salotti@ensc.fr

## 1. INTRODUCTION

**Fig.1**. Vizzavona is a forest of mountain pines in Corsica.

The settlement of Mars represents such difficult technical and human challenges that its feasibility has been questioned. One of the most challenging problems is to be able to satisfy the needs of the settlers during the long period required for developing the settlement. How many rockets and how much is it going to cost to support such a settlement? Before presenting our project, let us define the context and clarify the statements that explain our choices. According to several authors, the cheapest way to send a cargo payload to Mars is to use an interplanetary re-usable vehicle that exploits solar electric propulsion (SEPV) [1,2]. The trip would be longer but there would be important mass savings for the propellant and the SEPV could be re-used several times. It is assumed in this proposal that this kind of SEPV will be developed and used in the future for sending payloads to Mars. In order to send some payload to LEO, a heavy launcher (at least 100 tonnes capability) will also be available (same capability as Saturn V, SLS etc.). If electric propulsion is used, up to 50% of the 100 tonnes LEO capability will represent the cargo payload to Mars. Obviously, another space vehicle will be used to send astronauts. All in all, let us assume with careful estimates that, at most, two heavy launchers will be built and launched each year (or 4 launches every 2 years to take the planetary configuration into account) in order to send:

- 1st rocket: 5 to 10 settlers to the surface of Mars (the mass is allocated to propulsion systems, life support systems and the landing module).
- 2nd rocket: 50 tonnes of payload to Mars (100 tonnes to LEO, but some mass is allocated to propellant for SEPV re-fueling and another part of the mass is allocated to the structure and EDL (Entry, Descent and Landing) systems of the landing craft).

Remark: In his plan to settle Mars, Elon Musk proposed a very powerful re-usable rocket with very high transportation capability. Such an interplanetary vehicle could be a game-changing technology. However, the testing and qualification of such a vehicle (especially Mars entry, descent and landing systems, which are supposed to be re-usable) might be very difficult and uncertain. In order to make sure that the constraints of the settlement problem are not underestimated, it is assumed in the proposed study that such a vehicle will not be available for the settlement process.

In order to better understand the difficulty of the settlement, we present a mathematical model in section 2, showing that the required mass of things that have to be sent to Mars each year to sustain the lives of the settlers during the growing phase is incredibly high. This model defines the economic constraints of the settlement and helps us in determining a global strategy for the settlement process. Section 3 presents the technical choices. Section 4 discusses social and cultural issues. Section 5 is dedicated to economic aspects of the settlement. Political issues are presented in section 6. Finally, well-being and aesthetic aspects are addressed in section 7. The name of our project is Vizzavona, a beautiful forest of mountain pines in the island of Corsica. These pines are able to grow on rocky terrain and survive in harsh environmental conditions.

## 2. MATHEMATICAL MODEL OF THE SETTLEMENT

### 2.1 Settlement equation

In order to better understand the difficulties of the settlement process, a mathematical model is proposed [3]. The idea is to determine the annual payload mass required to satisfy the needs as a function of the number of astronauts. See equation (1).

$$p(n)= k*n*(r/s(n) - w) \quad (1)$$

Where:

- $p(n)$ is the required annual payload mass that has to be sent to the planet to sustain the lives of the settlers.

- w (in hours) is the average individual working time capacity per year. This parameter may vary according to the type of work, the organization of the society, habits, etc. In modern societies, a person works approximately 2000 hours per year.
- r (in hours) is the minimum individual working time requirement per year and on average (considering that the settlement could survive forever) to produce on the planet all objects and consumables that are necessary to sustain the life of one person during one year. It includes agricultural time to grow plants, industrial time to extract chemical elements from the atmosphere and ores from the soil in order to produce metals, plastics and then tools and complex objects, as well as medical time, teaching time, administration time, etc. As all objects have a limited lifetime, the working time also includes the time to create or build each object divided by the lifetime of the object. The working time therefore includes a percentage of the construction time of buildings, cars, and all complex objects that are assumed to be required for living a decent life.
- s(n) is a function of n and is called the sharing factor. For instance, if for 4 astronauts, on average, each object is shared between 2 persons, s(4) = 2. It is expected that the number of shared objects grows with the number of persons. For instance, a kitchen may be shared by four or five persons of the same family and an electric power plant may be shared by a thousand people. Importantly, for a high enough value of n called $n_t$, $r/s(n_t)$ equals w, which means that a threshold is reached and that there will be enough people in the settlement to produce everything. In other words, the settlement would achieve full autonomy as soon as the number of persons is greater than $n_t$.
- Given w, r and s(n), the right part of the expression computes the time that is missing per person and per year to build all required objects to sustain the life of one person. It is multiplied by n to obtain the total missing time for all persons and by a coefficient called k to transform the missing time into tonnes.
- k is a mass conversion coefficient. If the missing time is greater than zero, it is necessary to compensate by sending objects from the Earth. A difficulty is to convert a time into mass. This is the role of the constant k. It is obviously not required to send all missing objects. It is possible to send only tools or to provide parts of the missing objects, with the objective of minimizing the total payload mass and therefore k. A method is proposed in section 3 to determine k.

It is difficult to estimate the values of the parameters. In a first trial, considering the needs of a modern society, the following values are proposed:

- w=2000 hours
- r = $10^6$ hours. This means that 1 million hours of work per year would be required to produce all objects and consumables needed each year for a single person.
- $s(n) = n^{\alpha}$
  This is an easy way to consider the sharing factor. This function is interesting

because it starts as expected with n equal to 1 whatever the value of α and it regularly increases with n but in less proportion, which is also expected. Intuitively, as for 2 persons more than 50% of all objects would be shared (house, car, life support system, facilities, etc.), it is assumed that α is higher than 0.5. For the sake of simplicity, 3 examples are taken to illustrate the model with α equal to 0.52, 0.54 and 0.6, respectively. See Fig. 2.

- k = 0.00001

  It is proposed to determine k by looking at p(1), which is the expected payload for a single person. For n=1, the sharing factor is equal to 1, which simplifies the equation. A single person would not be able to produce large amounts of objects or consumables on the planet. As most objects would have to be sent from Earth, an estimate of the payload can be made by looking at the payload mass that is expected for the first manned missions to Mars, which is in the order of 30 tonnes for 3 astronauts and a stay of 1.5 years on the planet. It is proposed here to determine k such that p(1) would be equal to 10 tonnes given the set of parameters already fixed. In an initial approximation, k can therefore be set to 0.00001.

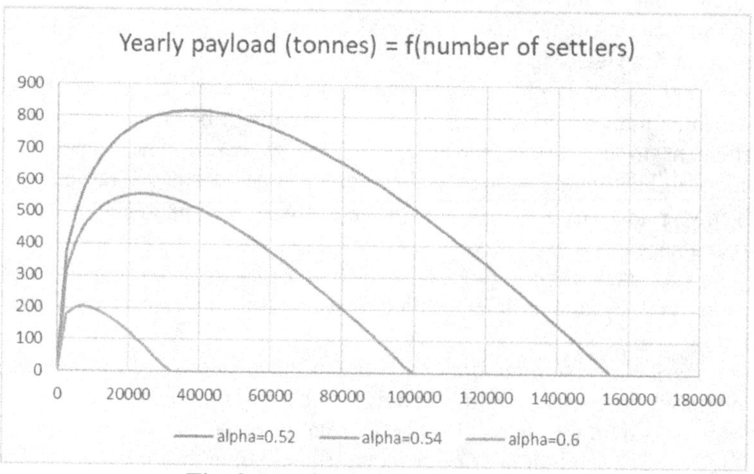


**Fig. 2.** Yearly payload for scenario 1.

The result is presented Fig. 2. According to our model, for α=0.54, approximately 100,000 people are required in the settlement to achieve full autonomy. Another important result is the peak payload requirement that is close to 550 tonnes and occurs when there are 20,000 people in the settlement. This number is 10 times as high as the annual payload capability that has been proposed in the introduction. This high payload requirement, which has to be maintained over a long period of time, is clearly a tremendous constraint that could discourage any public or private organization from investing in the settlement process. If important commercial exchanges are possible (tourism?), a settlement could become economically viable. However, according to us, it is doubtful that the ticket to Mars could become affordable anytime soon without significant technological revolutions and, therefore, that tourism could become economically viable (preliminary cost estimates from Elon Musk for one way tickets are assumed to be too optimistic). Based on these considerations, do we have to give up the idea of a Mars settlement? Not at all, let us see in the following chapters how difficulties can be overcome.

## 2.2 Appropriate parameters in the search space

***Reduction of the needs***

In order to make the settlement possible, a key idea is to reduce the needs. In our equation, reducing the needs means reducing r, which is the working time to produce all required objects. Let us consider two important examples.

- If robots are used in the settlement, robots will have to be imported from Earth as long as the settlers are unable to build their own robots. If only a small number of robots are needed, this would not be a big issue, but for a settlement of several thousand people, the mass of all robots might be very high and since the lifetime of robots is usually not greater than ten years or so, the annual importation of new robots might quickly become a problem. In addition, robots are typically complex objects that require many different industries and factories for their construction (metallurgy, manufacturing electronic components, plastics, computers, etc.), which means that the impact on parameter r is certainly high. On the other hand, if it is accepted to live on Mars without robots, the productivity would be lower, but all in all there would be a notable reduction of r.
- Surface vehicles will undoubtedly be used on Mars for the transportation of ores, goods, consumables, or astronauts. However, what sort of technologies will be involved in these vehicles? In modern cars, computers are everywhere, but fifty years ago, there weren't any computers at all and cars were almost as efficient as they are now. In other words, producing cars on Mars will be easier if it is decided to build simple old-fashioned cars rather than modern cars. In addition, in order to maximize the ability to repair or maintain old cars on Mars, it would be preferable to use simple technologies and simple materials.

Similar considerations can be made for other technologies. In order to reduce parameter r (and therefore the list of needed objects that have to be (re)sent to Mars), a kea idea is to rely as far as possible on simple objects that can be easily manufactured on Mars. It is important to notice that these considerations are only valid for the first stages of the settlement process. It is obviously expected that high-tech tools will also be produced on Mars one day, but probably not before human resources are numerous enough to support the industrial organization that would be needed to produce them.

***Sharing factor***

Another way to reduce the needs is to increase the sharing among the settlers. For example, instead of allowing individual belongings, all vehicles can be shared and used according to an appropriate policy. If it is desired and accepted to maximize the sharing, important reductions of needs can be made. Washing machines, ovens, showers, toilets, clothes, tools, etc., the list of possibly shared objects is potentially very long. The acceptability, however, is questionable. Such a society would clearly be closed to a kind of communism, which is considered by many people as inefficient and not adapted to human nature. It is not proposed here to maximize the sharing and to try a kind of full communism on Mars, but to look at what can be reasonably accepted to minimize the needs. For instance, it is probably acceptable to share the production of consumables, especially water and food, the vehicles and the main tools among all settlers without asking them to pay for them. In fact, another important issue is to minimize human resources for administration, education, health, justice, trade management, finance, entertainment, etc., which are not directly concerned with the production of consumables and goods for the support of the settlers. These services should be shared among the settlers with the strict minimum requirements. Once again, it does not mean that a local currency would not be created one day, as well as a justice department, a police department, etc., but as long as they can be reduced to the strict minimum and the resolution of problems can be shared by the community of settlers, it will help considerably to increase the sharing factor.

***Time to mass conversion***

In our model, the time required to produce consumables, goods or tools, is converted into payload mass according to a conversion coefficient. In order to reduce that coefficient, various strategies can be chosen. The first idea is to send missing objects. However, this is not an optimal strategy because it is often possible to send only an efficient tool that could make the production of that object possible. For instance, a 3D printer can help to produce complex plastic objects. Building a 3D printer from scratch on Mars might be very difficult. According to the recommendation made in previous paragraphs, we should try to avoid the construction of such complex objects to reduce industrial needs. However, there exist tools that allow important time savings in production processes and 3D printers certainly belong to that category. A trade-off has therefore to be found. Complex tools can be sent to Mars in the early stages of the development of the settlement if they are not too heavy and if they enable the production of numerous objects with little effort. In fact, 3D printers might help in reducing the mass conversion coefficient of our model, and therefore the peak payload requirement. When the number of astronauts becomes higher, as numerous industries will be available, it will be easier to avoid the use of 3D printers or at least to avoid the use of thousands of 3D printers.

***Smoothing the curve***

In order to reduce the peak payload of our model, another important idea is to split the settlement process into several steps. In the first step, the objective would be to accumulate assets and resources when the number of astronauts is kept low (5 astronauts for instance). In the second step, the number of astronauts could rapidly grow, but as the resources would be abundant, the peak payload would be more easily passed. In other words, it means that the variables of our model can be dynamically adjusted to cope with the 50 tonnes payload per year constraint. Then, as it is desired in the long term to build more complex objects, a new industry can be started but only when there are some margins for the payload.

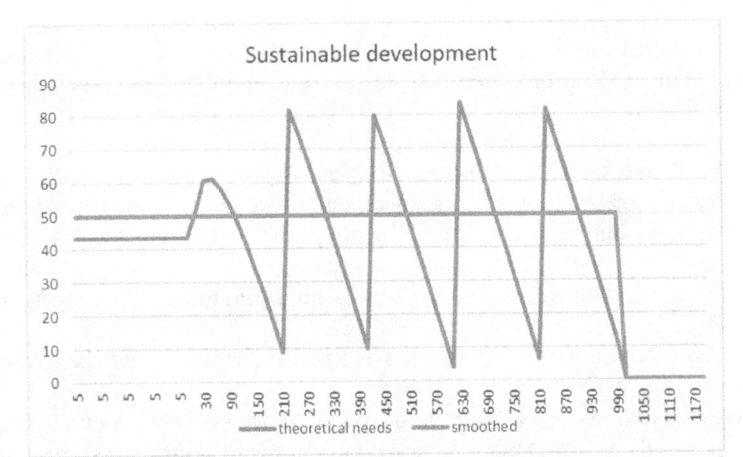


**Fig. 3:** Smoothing the payload requirement curve. Horizontal axis: number of astronauts. Vertical axis: mass imported from Earth in tonnes per year.

For example, concretely, at the beginning of the settlement, many greenhouses can be built, as well as many houses, facilities, etc., which could sustain a large settlement. Then, many settlers come and benefit from the numerous infrastructures. When the peak payload is expected, the needs are reduced because more facilities and resources are available (see Fig. 2). Later on, the number of settlers continues to increase and a new autonomy phase is reached (low importation requirement). This would be the right moment to start a new industry and to release the constraints on complex object requirements and also on the sharing factor. Once again, the new peak payload can be passed thanks to the accumulation of resources before it is reached. And so on, and so forth. This process is illustrated in Fig. 3.

All in all, providing that significant efforts are made to reduce the needs (very few modern complex objects), to increase the sharing factor and to optimize productivity, it might be possible to achieve the first autonomous phase on Mars with a very low number of astronauts, perhaps a few hundreds. By doing so, there are two advantages:

- First, obviously, we cope with astronautics constraints (50 tonnes of importation per year) and the project remains feasible.
- Secondly, if support from Earth is sudden abandoned, temporarily or definitively, the impact is low and the settlement process is only delayed, since the settlement is close to autonomy.

# 3. TECHNICAL DESIGN

## 3.1 Main principles

The mathematical model of the settlement process provides important cues for technical choices. It is proposed here to derive 4 important guidelines from the model, as follows:

a) **Make it simple**: The settlement can use complex objects, but must not rely on them in order to reach an autonomous phase as quickly as possible with the smallest number of settlers. In addition, whenever possible, the simplest objects or tools have to be chosen so that they can be produced/maintained/repaired on Mars with a minimum of energy/industrial effort.
b) **Develop step by step:** New habitats and new industries have to be introduced step by step, as illustrated in Fig. 3.
c) **Accumulate resources**: In order to pass a "payload peak", the best strategy is to accumulate resources and assets when the settlement is close to an autonomous phase and before the development of any new industrial process.
d) **Maximize sharing:** The sharing factor has a large impact on the organization of the settlement and therefore also on technical choices. For instance, as energy distribution has to be shared, it is preferable to have a large power plant rather than multiple small ones.

## 3.2 Energy

### *3.2.1 General considerations*

Energy production capability is probably the most important parameter driving the settlement process. Large amounts of energy are indeed required for growing plants, for mining, for industrial production, for life support, etc. As there is no oil on Mars, energy must primarily come from the sun. It might also be possible to use thermal energy from underground, but at the moment the feasibility of that option is not clear. Nuclear energy might also be used with nuclear power plants and uranium imported from Earth. Nevertheless, the "make it simple" principle suggests that the settlement should not rely on that capability. According to the "accumulate resources" principle, it should be possible to increase energy production or storage with little effort, especially when the settlement is close to an autonomous phase and before the start of any new industry.

### *3.2.2 Solar arrays*

Solar arrays are chosen because very thin and light solar arrays already exist (easy to import from Earth), and they can be easily deployed and maintained [4]. There are, however, two difficulties. The first is to be able to build solar panels from local resources. As silicates are already required for the production of glass (see next sections), we propose to try to implement a factory for photovoltaic cells to produce solar arrays. Other resources like copper may not be available and producing pure silicon might be very difficult. Nevertheless, even if the efficiency of locally produced solar panels is below 5%, it is worth doing because, in case of an abandon from Earth, it would be difficult to find alternatives. The second problem is to provide energy at night and during dust storms.

### *3.2.3 Energy storage*

Energy can be stored in many different ways (e.g., thermal energy, mechanical energy, gravitational energy, chemical energy, biomass, etc.). The "make it simple" principle suggests the use of biomass. It is indeed very easy to grow plants (energy storage) and to burn plants to obtain energy. However, though it can be recommended to use plants as a backup energy source, it is not possible to convert electricity into plants for energy storage. Fuel cells or batteries are typically proposed in combination with solar arrays but they are complex systems that will be difficult to produce on Mars using local resources [4]. Another important element of the problem is the way vehicles are powered. The simplest solution is to use chemical energy, for instance methane or ethanol, which is also suggested by Zubrin [5]. If it is agreed to use methane, then methane might also be chosen as energy storage to be used in combination with photovoltaic arrays. This is what is proposed here - see Section 3.3 on chemistry.

Remark: In case of dust storms, some energy would still be collected, in the order of 10% of the array capacity. The simplest solution is therefore to multiply the size of solar arrays by a factor of 10.

### *3.2.4 The case for plants*

It is important to include plants in the energy category. The photosynthesis process is indeed an energetic process allowing plants to grow. Then, plants are the primary source of food (energy for humans), and can also be used as biomass for energy accumulation. More importantly, provided that they are integrated in an appropriate ecosystem, plants grow on their own with little human intervention, and the growing process does not depend on the use of complex objects. In addition, plants are also very interesting for the production of oxygen, trees are interesting for the construction of objects made of wood (tables, chairs, etc.) and can also be used for other functions (glue, paper, etc.). For these reasons, it is proposed in this project to increase the role of plants and trees. The idea is to create a huge number of greenhouses and to use plants as energy and resource accumulators that will help attaining the first autonomous phases. This idea suggested the name of the project called "**Vizzavona**" (see Fig.1). Such trees can grow almost everywhere on rough rocky terrain with very limited organic soil.
**Greenhouses:** Ideally, plants will be able to grow in greenhouses without human intervention. However, there are several important technical challenges:

- Brightness. On the surface of Mars, the light of the sun is weaker (Mars is further than the Earth from the sun). How to increase the illumination? The first idea is to settle not too far from the equator, for instance 10° north. As a complement, it is proposed to add mirrors behind the greenhouse to reflect more sunlight towards the plants. It is expected that this strategy will avoid artificial lighting.
- Transparency. Greenhouses have to be as transparent as possible to maximize illumination from the sun. It is proposed to use glass supported by an iron structure. Glass and iron can be produced using local resources - see industrial section.
- Plants need an atmosphere. It is proposed to maintain in all greenhouses a pressurized atmosphere in the range 350 to 400 mb with approximately 25% $N_2$, 70% $O_2$, other elements being mainly $CO_2$ and $H_2O$. Such an atmosphere is breathable by humans and animals and the pressure is relatively low in order to minimize structural constraints (glass thickness reduced).
- Plants need water. There are huge amounts of water ice on Mars and sometimes very close to the surface. However, extracting ice, transporting ice and converting ice into water will be quite difficult and will require lots of energy. As it is desired to "make it simple" and to maximize autonomy, it is proposed to create an ecosystem, in which water is recycled. The idea is to grow plants on an impermeable terrain. Two options are possible to recover water. On a flat terrain, after watering, water can be collected on the side and pumped back to the reservoir. On the slope of a mountain, water drops will penetrate the first centimeters of the soil but will then follow the slope. At some point, all water is concentrated in the same place and is filtered by a sandy/rocky conglomerate. Below that point, water is collected and pumped back to the highest point of the greenhouse. The pumps will be alimented by solar panels placed outside the greenhouse.
- Plants need carbon dioxide and nutriments. There are two main options to satisfy these needs. The first option is intensive hydroponic cultures, and the second would be creating acceptable soils and growing plants on them. As there are advantages and disadvantages in each case, the choice is not easy. Hydroponic cultures are often suggested because they enable an accurate monitoring of plant feeding. However, the difficulty is to be able to produce the nutriments using *in situ* resources. As the number of plants will be huge (the requirement is to feed 1000 people and to carry on increasing the number of settlers), the mass of nutriments, pumps, sensors, and other hydroponic accessories would quickly represent a non-negligible part of the payloads sent from Earth. More importantly, it would be very difficult to build all facilities allowing the production of local nutriments using *in situ* resources because it would require mining different ores, extracting very specific minerals, and controlling mineral concentrations accurately. Creating soils for plants would also be difficult. However, numerous tests have been conducted using simulated Martian soil and it has been demonstrated that it is possible [7]. It is therefore proposed here to avoid hydroponic cultures and to grow plants on Martian soil, with the addition of fertilizers and decomposers that can be imported from Earth at the beginning and, later on, produced on Mars.

Carbon dioxide can be extracted from the Martian atmosphere (compressor) and pumped into the greenhouse. Worms, insects, bees and other animals also have to be introduced to help recycling organic waste and stabilizing ecosystems. Different ecosystems will have to be implemented, as has been suggested for Biosphere II experiments that were conducted in Arizona a few years ago.

- Dust storms. Thanks to the greenhouse effect, it is expected that temperatures will easily increase above 20°C. At night, as external temperatures will drop below -50°C, inside temperatures might drop below 0°C. In order to avoid freezing, big black rocks can be inserted into the greenhouse to collect the heat during the day and release it at night. The main problem, however, is to be resilient to dust storms, which would be equivalent to winters on Earth, with very weak sunlight and very low temperatures. The proposed strategy is twofold. On the one hand, it is possible to accumulate food resources in normal sunlight periods. During dust storms, plants would simply wait for the sun to come back as they do on Earth. Though some plants do resist low temperatures (pines for instance), in most greenhouses, it will be preferable to keep temperatures above 0°C. To make sure that this constraint is respected, it is proposed to heat up greenhouses using solar panels and heaters. During dust storms, solar panels would not be very efficient, but the idea is to dimension them so that 10% efficiency would be sufficient to warm the greenhouse to acceptable temperatures.
-

Remark: It is important to make sure that the proposed greenhouses will work properly. Tests have therefore to be carried out in the first phase of the settlement process, with very few people on Mars and the growing phase will only take place in a second phase, when numerous greenhouses have been built and checked.

**3.3 Chemistry**

There are two good reasons to develop a chemical industry on Mars. First, it will be fundamental to control the composition of the atmosphere, keeping $O_2$, $N_2$, $CO_2$ and $H_2O$ at acceptable levels. Secondly, it will be necessary to produce methane ($CH_4$) for vehicles and other engines. Let us present here the main chemical reactions that will help the settlers:

- Water electrolysis:

  $$2\,H2O \rightarrow 2\,H2 + O2 \qquad \text{(180 MJ/kg)}$$

  Water will be an important resource. Water electrolysis will be used to produce $O_2$, which is required in habitats, spacesuits, pressurized vehicles and also in numerous industrial processes. It will also produce $H_2$, which is a key element of the Sabatier reaction.
- Sabatier reaction:

  $$CO2 + 4H2 \xrightarrow[400°,pressure]{} CH4 + 2H2O$$

  $CO_2$ can easily be extracted from the Martian atmosphere and $H_2$ can be obtained with water electrolysis. Methane will be used as fuel. The Sabatier reaction has been proposed by several authors to produce methane for Martian vehicles [5,6]. It can be used in combination with the reverse water-gas shift reaction to simplify the production of methane. Remark: There is no oxygen in the Martian atmosphere. In order to use methane as a fuel for surface vehicles, it will be necessary to bring also oxygen.

- Oxidative coupling of methane

$$2CH4 + O2 \underset{900°}{\longrightarrow} C2H4 + 2H2O$$

Methane and oxygen can be obtained thanks to the previous chemical reactions. The result of the proposed equation is ethylene, which is a key molecule for plastics and organic chemistry, including polymerization. The transformation rate of the proposed equation is not optimal - some $CO_2$ may also result - but the result is very promising for the future of the settlement. Very complex objects can be manufactured if the production of ethylene is mastered.

### 3.4 Industry

As suggested by Zubrin, several industrial processes can be implemented on Mars [5]. It is proposed here to manufacture in priority ceramics, glass, iron and silicon.

Glass can be manufactured using Martian sand. Sand is abundant on Mars. A selection of the best sandy terrains will be required but not sufficient to avoid a filtering stage to obtain appropriate silicate proportions. Specific chemical elements also have to be added to obtain transparent glasses. An important issue is to build ovens to melt silicates. A possible solution is to bring ovens from Earth, but as it is required to reach autonomy as quickly as possible and to minimize Earth importations, an interesting option is to build a solar furnace. See Fig. 4 for an illustration. This is what is proposed here for ceramics and glass production.

Iron can be manufactured on Mars from hematite, which is abundant on Mars. The process is not simple, because ore quality will probably be poor. However, even if iron production is difficult and not optimized and if the product is not as robust as it is expected on Earth, it will be sufficient to support many of the metallic structures that are needed, especially for greenhouses and domes. The idea is strengthening the structures wherever it is necessary to make sure that it will be resistant to the large difference in pressure between inside and outside.

Some sand can also be used to produce silicon that can be exploited for solar array production. Simplified photovoltaic cells can be manufactured, as already proposed by several authors for lunar settlements.

**Fig. 4.** Solar furnace with double reflexion in Mont-Louis, France.

## 4. ECONOMIC ISSUES

### 4.1 Main principles

In the long term, it is expected that there will be a rocket industry on Mars, interplanetary trade and several launches per year to export goods from Mars to Earth. However, in the context of the proposed study, with no more than 1000 settlers, there will be very limited industrial capacity and it is quite unrealistic to believe that there will be a rocket industry on Mars. For that reason, it is assumed here that the rocket going back to Earth must first arrive from Earth. If it is a kind of shuttle, bringing some goods back to Earth might be possible and not so expensive. Before examining that option, let us consider first the economic issue of shipping goods from Earth to Mars in the first stages of the settlement.

### 4.2 Shipping goods from Earth to Mars

As already highlighted in earlier sections, there are two important organizational challenges:

a) To be able to maintain interplanetary transportation (astronauts and cargo) on a regular basis over a long period of time.
b) To pass the peak payload requirements (Figs. 2 and 3).

In the astronautics industry, it is well known that the annual rate of rocket construction must be as stable as possible. It is very difficult to increase that number because it takes a long time to design and integrate all parts. If it is required to double the production rate, new facilities have to be built and new teams have to be hired to achieve that goal with a tremendous cost impact. Ideally, as illustrated in Fig. 3, the best way to minimize the cost is to impose that the payload mass is always the same during all the settlement process, as long as there is no paradigm change in interplanetary transportation. This is what is assumed here.

In 2017, the cost of launching 1 kg into orbit with a Falcon 9 was $1,891. In 2020, it is expected that the cost will drop to $951 using a Falcon Heavy. Further reductions might be achieved with another rocket, but it is not yet proven. For the sake of simplicity, let us assume that the cost of sending 1kg to a low Earth orbit is $1,000.

The objective is to send the payload to the surface of Mars, not to a low Earth orbit. As suggested in the first chapter, in terms of energy, the most efficient way to send a payload to Mars is to use solar electric propulsion. If that vehicle can be re-used multiple times, the main cost is associated with the cost of sending the propellant to a low Earth orbit. In a first approximation, considering $\Delta V$ calculations, let us assume that the mass of that propellant is equivalent to the mass of the payload (see NASA Design Reference Architecture 5.0 [1]). The cost of 1kg to a Mars orbit is therefore $2,000. Then, the payload has to be sent to the surface by means of entry, descent and landing systems. Such systems generally are heavy. In the last NASA study as well as in Salotti's architecture [8], the entry mass is on the order of the double of the payload mass. A kind of re-usable shuttle might be used to minimize the cost but the qualification and the re-qualification for the next flights might be very difficult and impractical in the early stages of the settlement process. In order to avoid possible criticism of such an estimate, it is assumed here that the cost of sending 1kg to the surface of Mars is once again doubled at $4,000.

Finally, the payload itself has a cost. A pressurized rover, for instance, might cost several million dollars which, as a first order approximation, is as expensive as the cost of its payload mass. As it is necessary to send very robust objects and tools with strict qualification procedures, such costs will not be negligible. All in all, it is assumed that shipping 1 kg to the surface of Mars will cost around $5,000.
Remark: In the document describing the contest, a cost of $500 per kilogram is suggested. At the same time, it is asked to cost it out. Considering that the goal is to propose a realistic settlement project, it is proposed here to take our own estimate into account at $5,000 per kilogram.

Finally, as 50 tonnes of payload are expected each year, the annual cargo cost is about $250,000,000. Since it is also desired to send people to Mars, the recurrent annual cost will be at least twice as expensive, thus about $500,000,000.

Such a cost is high for private investors but it is easily affordable for nations such as the United States, Russia, China, etc., even if these efforts have to be maintained a long time. More importantly, as illustrated in Fig. 3, it is proposed here to develop the settlement step by step, trying to reach an intermediate autonomous phase as soon as possible. The key idea is to make sure that the settlement does not decline in case of a sudden abandon of investment. It must be resilient. Ideally, if resiliency is high enough, even in the case of a definitive stop of interplanetary travels, the Martian settlement should be able to slowly develop itself on its own.

### 4.3 Shipping goods from Mars to Earth

If it is required to send objects from Mars to Earth, the SEPV can be used but there will be an impact on the mass of propellant at departure from Earth. The problem is that it will be important to maximize the payload sent from Earth to Mars. As a consequence, there will be no margin for additional propellant. A possible option is to bring propellant from Mars, but it is highly impractical because ion thrusters generally use gases that cannot be produced on Mars.

Another idea is to send other vehicles to Mars to bring some objects back to Earth. Nevertheless, as mentioned in the previous section, it will be very difficult to increase the number of annual launches.

All in all, it would be very expensive to send goods from Mars to Earth. Occasionally, when the settlement is close to an autonomous phase, some people and some objects might be sent back to Earth, but it will be difficult to generalize and to make substantial benefits arising from such opportunities.

Remark: In the document describing the contest, shipping goods from Mars to Earth is considered possible at low cost ($200/kg). Though it might become affordable in a distant future, it is assumed here that the cost will be much higher and impractical for interplanetary trade in the early stages of the settlement process. The same conclusion applies for interplanetary tourism. With a ticket to Mars in the order of $100,000,000 (assuming 10 tonnes per person), tourism seems unaffordable.

### 4.4 Return on investment (ROI)

There is no direct ROI. However, there is considerable indirect ROI. As highlighted by several authors [9], space exploration brings huge ROI in many different domains:

- Technology: It is difficult to know *a priori* what technologies will be developed for the settlement process. It might be for instance in the domain of energy. As there is no oil on Mars, renewable energy systems have to be chosen from the start. It might also be in the domain of ecology with new air control systems, new paints that do not pollute the air or new water purification systems, etc.
- Arts: Many science fiction books and movies have been inspired by the early stages of the space conquest. No doubt that a Martian settlement will continue to stimulate imagination and creation in all countries.
- Education: Many children are curious and love space because space is full of incredible things like microgravity, planets, moons, stars, etc. Many of them want to work in the space domain and work hard for that.

- Psychology: Many people are depressed because they do not have a project or they feel that the world is dull and the future is not interesting. A settlement project will stimulate many people. It might help mitigating psychological depressions and violent reactions, which have a tremendous cost on productivity, health and security.
- Other ROI: There will be also books, games, toys, clothes and other objects derived from the settlement process as there was for the Apollo program.

All in all, the ROI is difficult to estimate, but it is probably huge. To end this section, we propose the famous quote from Walter Peeters, when he was teaching economics courses at ISU (International Space University): "The best ROI ever has been made by Christopher Columbus in 1492" [10].

## 5. SOCIAL AND CULTURAL ISSUES

### 5.1 Main principles

There are several important and difficult social and cultural issues:

- Will it be an international settlement or a national one?
- Do we have to select the settlers? And if yes, what criteria will be used? Do we have to impose constraints on the age and sex of the new settlers? Do we also have to send kids? Do we have to favor families or singles?
- Do we have to foster having children on Mars? And if so, how to care for them and how to educate them?
- After some decades on Mars, many settlers will be old. How to care for the elderly?
- It is not possible to avoid accidents. Some people might die and others might become disabled. How to face the death and how to care for handicapped people?
- How to take religion into account?

It is not possible to address all questions in this document. We propose to focus on the age pyramid and the role of families and children. Importantly, over time, adults grow old and can no longer work or, at least, not with the same effectiveness. If many settlers are getting old at the same time, the labor force rate would drop and the settlement might collapse due to insufficient human resources. In addition, if there are too few persons in the same age group, there is a risk of not finding a compatible sexual partner. In order to avoid these difficulties, it is important to select the settlers according to their age group and sex such that the age pyramid is globally homogeneous.

As the objective is to settle on Mars without compromising well-being, people will like to live on Mars like people on Earth, looking for a partner and having children. If the age pyramid is balanced at the beginning of the settlement, new children are expected at a reasonable rate and will contribute to the expansion of the settlement.

## 5.2 Habitats

One of the most important parameters of the settlement is the “sharing factor”. As already suggested, the idea is to maximize the number of shared objects without compromising with comfort. The family level will be emphasized. Habitats will be sized for families or groups of 2 to 10 persons, possibly including grandparents a few decades later to avoid leaving old persons alone. They will be built in giant greenhouses (domes) and will be the property of the family/group living inside. All furniture (tables, chairs, beds, etc.) and all personal belongings will also be the property of the family/group. However, in order to maximize the sharing factor, it is proposed to share the management of life support systems (energy, air, food, water, including showers and toilets) and the management of household waste among all habitats of a village (see next section).

## 5.3 Villages

A small village will be a group of 50 to 200 persons living in 10 to 20 habitats located inside a giant dome (50 meters wide). If possible, the dome will be located inside a small Martian crater. Each village will be in charge of several important life support systems:

- Energy production and supply (mainly electricity production using solar panels and methane production for vehicles).
- Exploitation of greenhouses for agriculture (vegetables, fruit), farming (mainly chickens and rabbits) and other needs (wood, etc.).
- Air production, supply and revitalization.
- Water extraction from ice (assumed to exist underground close to the base), transportation and recycling.
- Spacesuits, pressurized vehicles, quads, as well as 3D printers, general tools (hammer, saw, etc.), will also be shared at the village level.

In addition, some villages will be in charge of a specialization, which will be implemented at some distance from the main greenhouse of the village.

- Ore mining.
- Iron industry.
- Glass industry.
- Chemical industry.
- Wood industry.
- Maintenance and repair, mechanics, electronics, etc.

If possible, all places will be accessible from the center of the village thanks to pressurized tunnels and some villages can also be interconnected if they are not too far from each other.

### 5.4 Radiation risks and acceptability

High radiation levels are expected on the surface of Mars. According to scientific studies on the subject, such radiations are not strong enough to create immediately observable injuries but, in the long term (several years), the cancer risk is increased by a few percentage points. The risk can be decreased if the settlers hide deep underground during the most important part of their life. Obviously, all settlers will know the risks before the trip to Mars and the trip itself is very risky. If they nevertheless decide to go, they already have a high risk acceptability level. Are they willing to live underground in small habitats all the time or will they rather live in large spaces, exploring Mars as they want, even at the expense of a few years of life expectancy? It is assumed in this proposal that most settlers will have the psychological profile of adventurers and will accept the risks. They will go to Mars to be part of a great project and will prefer the second option. Eventually, a tradeoff can be chosen: At night, the settlers can sleep at the lowest level of their habitat with important shielding. During the day, however, depending on their activity, they might have to stay under the dome or in a greenhouse with relatively low shielding. But they will certainly prefer these conditions rather than staying underground all the time, never seeing the sky and the sun.

### 5.5 Behavioral competency

On Earth, the selection process will take psychological profiles into account. In addition, as for NASA astronauts [11], there will be behavioral training activities to improve communication, problem solving and team performance. These behavioral competencies will help a lot to get the settlers collaborative and constructive in their everyday activity, finding solutions to unexpected problems and avoiding interpersonal conflicts. It will certainly also impact the culture of the settlers, favoring the development of new social activities.

## 6. POLITICAL ISSUES

### 6.1 On Earth

It is assumed in this proposal that the settlement will be implemented by a state or multiple settlements by multiple states. Private companies might eventually propose launchers and interplanetary vehicles but a commitment to a long and expensive investment is required and can be supported only by nation states. Then, once a state is committed to the settlement project, there needs to be a group of persons - we can call it the Mars Settlement Organization Committee (MSOC) - in charge of the organization of the settlement:

- Technologies to be developed.
- Organization of the settlement process.

- Selection of settlers. For political, economic and pragmatic reasons, all settlers will probably belong to the same state. The advantage is the sharing of the same language and culture. Different nations might eventually invest in their own Martian settlement.
- Planning of launches.

Once the first base is built, the settlers will be in contact with the MSOC in order to adjust the payload according to the needs of the settlement.

**6.2 On Mars**

During the first development phase, it will not be possible to implement a complex political organization with different groups of persons in charge of executive tasks, legislative tasks, justice issues, etc. As the objective is to maximize the sharing factor and to optimize human resource allocation, it is rather proposed to manage the settlement using simple management principles:

- At the level of villages, a Village Direction Council (VDC) of 5 to 10 persons will be elected every 4 years and will be in charge of all decisions linked to life support systems and organization of the village. One person will be the Village Chief, one person the assistant and other persons will represent different activity domains. The exact number and the exact role of each VDC member will be determined by the VDC itself. The idea is to maximize the flexibility of the system in order to adapt the decision process to the difficulties that are going to be encountered and to allow quick re-organizations.
- At the settlement level, as long as the number of settlers is lower than 1 thousand (10 villages), each village chief will participate to a Settlement Direction Council (SDC). The SDC will be in charge of the development of the settlement, assigning industrial development tasks and production objectives to each village and organizing the sharing of all resources.

The VDC and the SDC will function as most organizations function in western societies. Each important decision will require the vote of each member of the committee and a decision is taken only if the number of positive votes exceeds the number of negative ones (or abstentions).

**6.3 Conflict management**

It will be important to minimize the risks of interpersonal conflicts on Mars. Behavioral competencies (see Section 5.5) will help a lot, but it is not possible to avoid all conflicts and, inevitably, there will be a person injured or even killed by another person. If the conflict involves several persons of the same village, and providing that they are not members of the VDC, it is the role of the VDC to address the problem and to find a solution, for instance a work punishment. If the conflict involves several persons of different villages, it is the role of the SDC to find a solution.

## 7. WELL-BEING AND AESTHETIC ISSUES

Although there exists a wide variety of landscapes, Mars is a desert with no vegetation, no ocean or lakes and no rivers. In addition, going out for a walk or for exploration will require complex procedures with spacesuits, pressurized vehicles and safety concerns. As a consequence, if nothing else is proposed, many settlers will have to live in the same small place during long periods of time with very few changes. Important efforts have therefore to be made to increase the well-being of settlers and to avoid cases of depression:

- First, a village should be as beautiful as possible. Ideally, we should have large spaces and numerous gardens. It is proposed here to build a village under a transparent dome made of glass with an iron structure. The dome will be 10 to 20 meters high and 30 to 50 meters wide, allowing the presence of numerous houses and gardens inside. This concept has several advantages:
    - The width and transparency of the dome allows the feeling of freedom and wide-open spaces.
    - The presence of gardens brings nature into the village.
    - The sharing of the dome increases social relations and avoids isolation.

A drawing of our concept is presented in the appendix to the document. The dome is deeply incrusted in the rocks of the crater to resist external pressure forces.

- Secondly, it is proposed to build numerous refuges far from villages in order to provide safe backup habitations and to facilitate exploration and excursions. These refuges will provide minimum resources for settlers: air, water, food and a place to rest. However, as resources have to be saved, refuges will not be a place in which to live for long. All resources will be stored in containers. Such refuges will allow long distance trips on the surface of Mars.
- Thirdly, as already suggested, it might be possible to build greenhouses with robust ecosystems and very limited use of artificial systems to maintain living conditions. Such greenhouses will provide abundant plant and wood resources and will mobilize low human resources. Furthermore, they will also play the role of refuges and gardens for all settlers.

All in all, everything is done to provide comfort, well-being, exploration capabilities and development opportunities on the surface of Mars.

## 8. CONCLUSION

The settlement of Mars will not be easy. However, it is deemed feasible, providing that an adequate organization is implemented and appropriate efforts are made to achieve that goal. As resiliency is probably the key capability required of the settlement project and, as the step by step approach towards a modern Martian society will take a long time, we propose to end this document with a quote from Sheryl Sandberg, which could become the motto of the settlers:

"To fight for change tomorrow we need to build resilience today."

**REFERENCES**

[1] G. Drake ed., Mars Architecture Steering Group, 2nd Addendum of the Human Exploration of Mars, Design Reference Architecture 5.0, NASA Johnson Space Center, 2014.

[2] G. Genta and J.M. Salotti (ed.), Global Human Mars System Missions Exploration, Goals, Requirements and Technologies, Cosmic Study of the International Academy of Astronautics, January 2016.

[3] J.M. Salotti, Parametric equation for the settlement of Mars, proceedings of the 70th Internatioanal Astronautical Congress, IAC-19,A5,2,7,x49674, Washington D.C., 21-25 October 2019.

[4] C. Cooper, W. Hofstetter, J. A. Hoffman, E.F. Crawley., Assessment of architectural options for surface power generation and energy storage on human Mars missions, Acta Astronautica, vol. 7-8, pages 1106-112, 2010.

[5] R. Zubrin and R. Wagner, The Case for Mars, The Plan to Settle the Red Planet and Why We Must, Free Press, Touchstone Ed 1996.

[6] R. Zubrin, A. Muscatello and M. Berggren, Integrated Mars In Situ Propellant Production System, Journal of Aerospace Engineering. 26: 43–56, 2012.

[7] Cartier, K. M. S. (2018), Tests indicate which edible plants could thrive on Mars, Eos, 99, https://doi.org/10.1029/2018EO090749

[8] J.M. Salotti, Robust, affordable, semi-direct Mars mission, Acta Astronautica, vol. 127, pp.235-248, 2016.

[9] N. deGrasse Tyson, Space Chronicles: Facing the Ultimate Frontier, W.W. Norton, 2012.

[10] W. Peeters, International Space University, Economic course of the Master of Space Studies, 2002.

[11] NASA (2008), International Space Station Human Behavior and Performance Competency Model, NASA/TM–2008–214775, Vol. 2, Langley, USA.

**APPENDIX**

Illustrations of the settlement. Two domes are built above Martian craters. Under each dome, there is a village of ten houses and numerous trees. From each dome, there are corridors to reach greenhouses and industrial facilities. The next village can also be reached using a pressurized underground corridor. Large arrays of photovoltaic cells provide energy to villages.

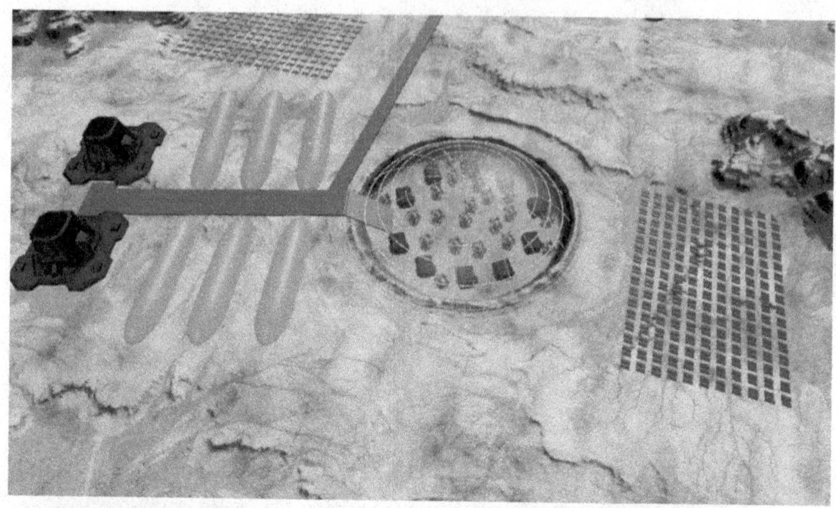

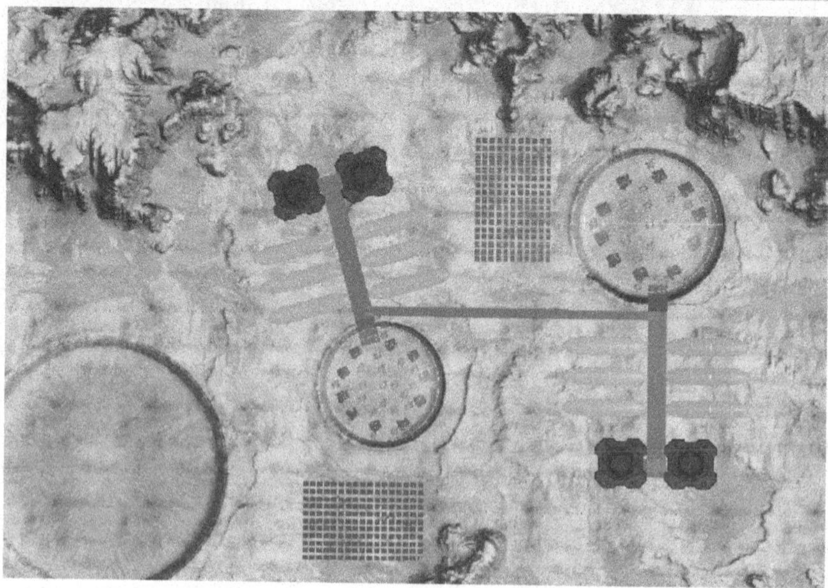

# 6: THE EUREKA DESIGN

**Kent Nebergall**
MacroInvent.com
kent@macroinvent.com

The Eureka concept is specifically created to surmount the technological and economic arguments opposing or deferring Mars settlement. This design assumes that every debate point opposing space settlement is valid. It then proposes the most practical engineering and financial solution set against that highwater baseline. For grand projects, it is often wiser to start with an overbuilt concept, then scale it back based on future evidence to more efficient solutions. If something is designed without this wholistic approach for an unknown environment, there is too much sponsor risk to justify the near-term expense. Difficult, remote environments require simple, rugged, affordable solutions to be maintained on-site.[1] Conversely, systems that are overbuilt and overpriced due to risk aversion will become a fiscal and technological dead end, or result in systems too fragile for practical development.

Eureka's construction methods are highly modular and adaptable to local conditions, unexpected assembly difficulties, and later repurposing. The central area contains elongated, LED-lit greenhouse domes, a 100-meter diameter recreational permaculture central dome, research labs, and manufacturing facilities. The foundational sizing units are the limits of the SpaceX Starship payload bay for hauling large equipment (9-meter full-diameter cargo, 4.5-meter conventional payload), and the dimensions of a functioning habitat centrifuge. Large structures are assembled from interlocked, framed blocks of reinforced ice (Pykrete) or concrete, augmented with magnetic shielding against Galactic Cosmic Rays (GCR) and solar flares. The ice is frozen in bags flexible enough to adapt to a guide track that contains tension structures and utilities. Indoor conditions within the centrifuge rings are nearly indistinguishable from life on Earth, giving a null starting point for expanding knowledge and adaptation.

## EUREKA'S PURPOSE

With off-world life sciences, the most logical and ethical roadmap starts with something as close to Earth as possible and works out experimentally. Eureka is focused on epigenetic life sciences, working from known to unknown, across a broad range of plant, animal, microbiome, and human tests. It also develops mining, construction and industrialization methods for surface settlement. Mars is the ideal environment for a space settlement research facility due to access to ice shielding, higher surface gravity, and a broad spectrum of available raw materials. Life sciences are notoriously "noisy" in terms of experimental versus environmental and control variables. Having a GCR-shielded centrifuge provides a large volume for experimentation with as close to a single variable test environment as possible. The large box at the top represents the range of facilities available at Eureka. With similar bases in a lunar lava tube and Phobos, the remaining range of gravity/GCR stresses can be explored (below). With these labs, every inner solar system environment can be explored. The practical limits and novel variants of each industrial process, animal adaptation range, and plant epigenetic expression are covered in this full-spectrum research network.

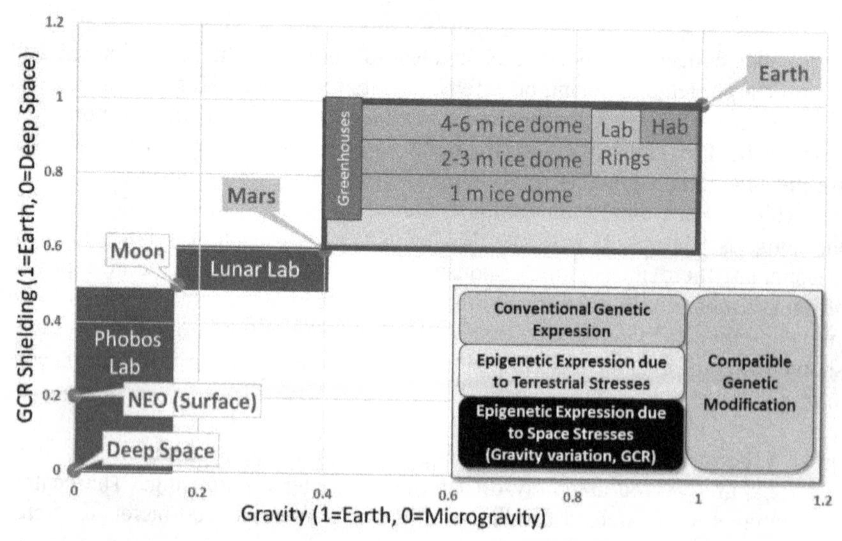


The inset box (lower right) represents the range of biological research possible on earth (left top and middle). The black box represents off-world research facilities. With understanding the epigenetic expressions in these three environments, GMO elements (right) may be added to expand the range further or take advantage of traits discovered in the silent gene research. Since many plants and microbes have far more base pairs than humans, this opens a broad new window in understanding and extending the genomic expression of terrestrial life. These silent genes have potential use in improved strains for disease resistance, adapting to challenging environments, and other productive agricultural and industrial uses. Crops can expand into broader hardiness zone maps to reliably feed and fund populations. Further, epigenetic expressive work rather than pure GMO work is closer to the natural state of heirloom genomes for multigenerational study.

## IDENTIFYING AND REMEDIATING TECHNICAL CHALLENGES

The table below identifies and classifies the grand challenges of space settlement. Areas currently addressed by SpaceX Starship are shaded. Areas addressed by Eureka are shown in white. Black boxes are out of scope for this concept.

| Launch/LEO | Deep Space | Moon/Mars | Settlement |
|---|---|---|---|
| Affordable Launch | Solar Flares | Moon Landing | Air/Water |
| Large Vehicle Launch | GCR: Cell Damage | Mars EDL | Power and Propellant |
| Orbital Refueling/ Mass Fraction beyond Earth Orbit | Medication/ Food Expiration | Spacesuit Lifespan | Base Construction |
| Space Junk | Life Support Closed Loop | Dust Issues | Food Growth |
| Microgravity (health issues) | Medical Entropy | Basic Propellant Production | Surface Mining and Extraction |
| ***Grand Challenges of Space Settlement*** | Psychology | Return Flight to Earth (speed, mass, etc.) | Hybrid Manufacturing |
| | Mechanical Entropy | Planetary Protection | Reproduction |

For each challenge, the section below describes the *State* (planned remediation method for the Eureka settlement when operational) and *Goal* (the focus of Eureka's research to extend life throughout the solar system).

**Solar Flares and Galactic Cosmic Rays (GCR)**
***State:*** Permanent inhabited structures are covered in 2-5 meters of ice blocks reinforced with a fiber matrix (advanced Pykrete). Some structures are augmented with magnets in Holbach arrays that bend the paths of incoming particles to maximize the GCR flight path through the ice, effectively making it a thicker shield. ***Goal:*** Find the optimum protective combinations of ice thickness and magnetic field arrangements and flux for remediation in environments (drier areas of Mars, the moon, NEOs, deep space vehicles, etc.) where less shielding material is available.

**Closed Loop Life Support**
***State:*** All main structures are at least triple-sealed and have multilayered

construction with active air and moisture recovery. Greenhouses and permaculture parks form a basic natural ecosystem, supplemented with Sabatier systems. Additional plants assist with trace contaminant removal in living and work areas. Systems are modular to allow adaptation and design optimization over time. Air is actively circulated through ducts and passively via the centrifuges and thermal convection patterns. ***Goal:*** Optimize and advance the suite of agricultural and mechanical systems for future use in larger and remote facilities.

**Medication Expiration and Nutrient Reduction in Deep Space**
*Food and medication shelf life in space is reduced due to radiation exposure.* ***State:*** Import dehydrated/preserved drugs from Earth and transport/store in deep radiation protection. On-site agriculture and drug production with testing to reduce the risk of epigenetic issues in food, drugs, or conversely in the consumers. ***Goal:*** Develop nutraceutical, chemical, and GMO-produced replacements in local conditions based on in-house research for intellectual property (IP) export. Individualize plans for nutrition, exercise, and radiation protection to reduce the need for medications. Identify unaltered heirloom strains for settlement use.

**Increased Pathogen Virulence in Reduced Gravity**
*This is an issue with some bacteria (C-diff, salmonella, etc.) in microgravity. This is a known issue in microgravity transit and a possible risk on the surface.* ***State:*** Determine the risk gradient in the full range of reduced gravity. Facilities and equipment (restrooms, kitchens, etc.) are built for automated cleaning and periodic decontamination. ***Goal:*** Epigenetic virulence vectors allow A:B isolation of the disease-causing genes to accelerate the development of vaccines and other treatments. This is already being done with ISS research. This process would be expanded at Eureka testing at all gravity levels from 1 to 0.4 G.

**Environmental Psychology**
*Avoiding "cabin fever", homesickness, and other stresses of being far from Earth.* ***State:*** A spinning living area has rooftop gardens with a projected ceiling that follows the habitats on the ring to give a sense of distance. The central dome has open permaculture gardens with trees, bodies of water, and a projected landscape and sky on the dome ceiling and walls. Settlers feel a reduced dread of cosmic radiation or other space risks due to thick shielding. Rounded ceilings with wrap-around sightlines avoid the monotony of seeing everything from one vantage point. For direct Mars surface views, there are shielded observation decks, periscopes, windows or 8K+ monitors to show the view outside. ***Goal:*** Test additional methods for use in other space settlements. Determine the scale of gardens, etc. and any risk factors with dual-gravity living over time.

**Mechanical Entropy and Design for Unknowns**
***State:*** Both imported and site-built systems are highly modular to allow swapping, rearrangement, upgrades, and expansion to adapt to field use. Engineers design modular systems for a range of unknowns rather than over-engineering a single, expensive configuration for anticipated risks. Mission-critical automation controls can be swapped out or bypassed with manual controls

in emergencies. This reduces catastrophic risks from software or hardware failure, hacking, or supply chain isolation. Systems are designed to avoid rapid cascade failure across modules. ***Goal:*** All key systems are manufactured locally to allow the option of independence from both imports and system failure. As construction is localized or imported from other non-terrestrial sources, then space settlement is no longer limited by launch capacity and may expand exponentially across the solar system.

**Planetary Protection (Forward Contamination)**
*ISS studies indicate bacterial spores buried more than a meter under clay could survive over 70 million years in space[2]. Conversely, surface Mars dust is sterilized due to dust-storm turnover and solar UV exposure.* **State:** Initially, material inputs to the construction base are surface dust and rock, atmosphere, and subsurface ice. Any subsurface mining uses fully sterilized equipment, robotic controls, and an autoclave airlock for replacing worn parts. ***Goal:*** If no native life is found, some effort to keep human contamination reversible remains in place to protect future settlers from possible pathogens. *If native or historically-transplanted life is found, see below.*

**Planetary Protection of Crew from Possible Indigenous or Isolated Life**
*Given the ISS study mentioned above, lithopanspermia from Earth to Mars over the history of the solar system is a near certainty. Native life is also possible.* ***State:*** Surface and subsurface areas are explored with sterile robotic instruments. Core samples are examined for biomarkers prior to resource extraction. Mining sites are verified safe prior to actual mining. Samples are examined in telerobotic labs. The settlement is built in an older crater to ensure a GCR-sterilized surface. Returning spacecraft will not be exposed to the sub-surface or other potential active habitats (caves, etc.). See also, Water Production. ***Goal:*** If native life is found of terrestrial origin (lithopanspermia), or indigenous life is found, genomic compatibility and risks are assessed, and any relevant protection methods remain in place. Experimentation on risk and compatibility would take place in deep space labs to avoid back contamination.

**Air Production**
***State:*** Air components are extracted at the propellant plant (oxygen, nitrogen, argon, etc.) and by baking off from surface materials in the dust extraction plant (additional nitrogen). Carbon dioxide and contaminants are removed by plants, filters, and systems such as Sabatier reactors as appropriate. ***Goal:*** Optimize the reliability and mix of these systems, while maintaining enough backup systems to compensate for crop failure, partial hardware shutdown, or other contingencies. Refine system designs for other worlds and deep space.

**Water Production**
***State:*** The main source is relatively clean subsurface ice (a nearby covered glacier), mined via hot $CO^2$ injection and steam extraction with sterilized equipment. This is transported to Eureka by pipeline. For ecosystem use, water is further purified to the needed levels. Waste is recycled via anaerobic digesters

and other conventional methods for use with crops. ***Goal:*** Expanded methods are tested for extracting water from lunar and NEO sources and dry Martian soils.

**Power Production**
***State:*** Low enriched uranium (LEU) modular heat sources such as those in the Kilipower and Megapower[3] designs are low risk and can provide useful heat for over 200 years. They are coupled with modular heat-differential generators (Stirling cycle, turbine, etc.). Turbines can also be used for direct mechanical work, such as driving compressors for atmospheric mining. To extract the large volume of water needed from the subsurface glacier, compressed atmosphere carrying reactor waste heat is channeled down the drill well to both remove the heat and extract water. ***Goal:*** Build the heat exchangers and generators on-site using imported LEU cores. Eventually, use local and other space-sourced resources for power, such as orbital solar platforms built in the asteroid belt, when they become available. A transition to fusion is also possible.

**Food Production**
***State:*** With a backstop of dry goods and supplements from Earth, food production begins with stacked hydroponic beds and LED light sources in agricultural domes and epigenetic testing is done for possible health risks. Plants that require more earth-like gravity would be grown in the centrifugal rings. Different fruit and vegetable species would be rotated into the system for dietary variety. ***Goal:*** With a fully characterized suite of basic species for a closed ecosystem, space settlement can be done wherever energy and resources are available. Research into genetic expression also aids food production for Earth that can grow in harsh conditions or with higher nutritional content.

**Mining**
***State:*** Initial work is focused on ice, surface dust and rocks in the top meter for planetary protection and accessibility. Dust is separated magnetically, electrostatically, and chemically into useful materials. Igneous deposits with little chance of harboring biology would be next, followed by metamorphic and sedimentary. Any mined area is scanned (radar, x-ray, etc.) and core samples are stored for ongoing geological research. ***Goal:*** Expand the ability to mine any environment in the solar system using a mix of localized and universal equipment with experience gained in construction and extraction techniques at various settlements.

**Manufacturing**
***State:*** The initial landings set up a propellant/water manufacturing base using Mars Direct principles and an LEU reactor power plant. This is expanded to include basic steel production (carbon monoxide and surface iron oxide feedstock), which is then formed into bulk structures using additive/subtractive manufacturing and imported parts. The first goal is local production of foundational construction base plates, pilings, and brackets. Experiments with different gravity levels, mixes, and purities will take local production in novel directions. Chemical industries use a mix of local components and feedstock from

Earth, with an emphasis on creating more feedstock from local materials over time. Industrial equipment is modular and can be reconfigured for different production runs. ***Goal:*** All major systems for ongoing habitation, industrialization, and ecology expansion are built on site.

**Reproduction – Plant/Animal**
***State:*** The normal range and epigenetic localized expressions are mapped for each species and strain according to gravity and limited radiation. As the epigenetic expression of each species is mapped, the range of species available for different settlement environments is defined for agricultural and experimental biosphere applications. ***Goal:*** Epigenetic and GMO research to expand ecosystems well into the solar system and beyond. Understand the full genome of all primary species, and gain knowledge of genetic and biological risks in reproduction across various gravity and radiation environments. Eventually engineer entire ecosystems for different artificial habitats, expanding options for future terraforming and exoplanet settlement.

**Human Reproduction**
***State:*** After animal testing in the ring (most earthlike) environment, couples could start families within the ring. Pregnant women and young children remain in the earth-like environment and are monitored for stresses. Limited exposure to Mars gravity is allowed as children age with monitoring in the early generations. ***Goal:*** A practical set of guidelines on minimum gravity and shielding levels for multigenerational space settlements is defined to allow the human ecosystem to expand off-world. Gravity rings on Mars may prove unnecessary except as labs.

**Shelter Construction**
***State:*** After site prep, structures begin with a set of steel plates anchored to the ground using sterilized steel pilings. Each interlocked base plate is 1-meter square and allows for an insulation/sealing layer above. From here, a light framework of flexible T-beams and stringers are assembled by robots into various structures and bolted together using a system that allows tracks for mounting of ice blocks, magnets, and utilities. Frames are typically two to five layers of gridwork, with each open cell within the beam structure sized to roughly one cubic meter. Seals and insulation layers are clipped to the inner and outer surfaces. The ice block bags are clipped in place within those framed cells as construction progresses, with a middle utility gap. The frame is assembled first and a dust barrier seal installed on the outermost skin. Then the contents of the remaining cells are filled in. The ice bags are filled with a fiberglass-like "fluff" of high strength composite filaments to add strength and heat capacity (thermal mass) to the blocks. Once robotically filled with water and frozen, they are stronger than concrete and mitigate cosmic rays. The bags/blocks have dovetail arrangements to lock them together with the rails and each other while allowing for some shifts. This construction method allows flexibility and settling during construction while still giving precision in the final build. The blocks are filled from the bottom row up and allowed to freeze and settle by layers until the inner wall is completed and partial pressure kept in the dome. Sections that are continually above freezing

(internal foundation pillars, etc.) can fill the bags with cement instead of water. Once available, base foundations plates are built of locally-made steel.

Pressurized to one atmosphere, structures must contain ten metric tons of internal pressure per square meter. This would require an ice dome 25 meters thick to offset this pressure by mass alone in Mars gravity. A 12-meter thick ice dome would offset cosmic ray exposure approximately as much as Earth's atmosphere. A thinner dome partially offsets the pressure load by hanging components on cables from the ceiling, rather than placing them on the floor. A greenhouse, for example, would hang the plant tracks from the ceiling and use a thinner roof, because plants require less cosmic ray protection. Such a system can be tuned to different pressure loads by lowering the bottom-most hanging components to the floor or bolting the anchor cables to the floor. Outside stem wall hoop stress is offset with strong steel footings and a berm of more ice blocks outside the wall. Other walls within the city are backed up to other walls with similar berms and trusses. Typical wall construction for the operational base would be two, 1-meter blocks of ice, a 1-meter service gap with the magnetic shield elements at 500 millibars, and another two meters of ice. If necessary, more layers can be added to adjust pressure differentials. This system allows for structures of any size, shape, or thickness (in roughly cubic meter units) to be built and adapted to local conditions. Inhabitants can disassemble and reconfigure structural blocks and tracks into new buildings. The inner surface contains fiber optic lighting arrays and (where appropriate) sprinkler heads for watering plants and fire suppression. The ice walls would be insulated inside (from the heat of the habitat), and have an active temperature management system to prevent the inner surfaces from melting and reducing strength. Having the hardest, coldest ice on the outside and warmer, softer ice on the inside would mimic the structure of a quality knife blade or medieval castle wall and thus boost protection from meteor splash damage or launch accidents. Magnetic Holbach arrays (where the magnetic field is much stronger on one side) not only limit radiation but can help augment stress loads in the framework where needed. Block dovetailing can increase seal tightness under stress. ***Goal:*** The base plates, bags, track structures, and conduits are eventually manufactured from local materials. Methods for building them with materials from other worlds (lunar beta cloth, etc.) would expand the use of the technique.

## SITE CONSTRUCTION AND BUILD-OUT SEQUENCE

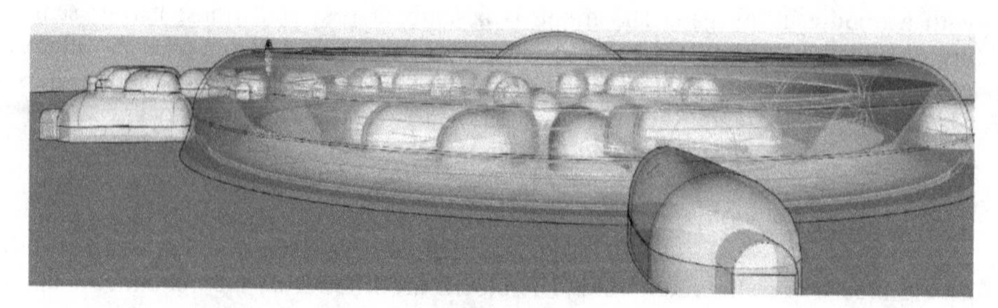

**Site Location**

The main city is in an 800+ meter diameter older, dust-filled crater at or near a large deposit of subsurface ice. At a safe distance, a second crater 300+ meters in diameter hosts the spaceport and initial manufacturing domes. In both cases, crater walls would provide a natural barrier from nearby meteor strikes or spaceport accidents. A subsurface radar and acoustic survey would find and remediate issues prior to site preparation. A roadway is cleared between the two, with an eventual pressurized surface corridor and pipelines built between facilities. The ideal site location would also have useful metal and mineral deposits nearby.

**Initial Landing Site**

The first landing would be on the nearest, most accessible, solid, flat area in proximity to a subsurface glacier. On-site landings could be done with a "blanket" landing pad anchored down by a crew to prevent engine erosion of the landing site, with a simple frame underneath to keep the ship level during fueling and takeoff. As such, the first mission may consist of landing, a "hop" to the proto-spaceport, then fueling and return. Conversely, a robotic starship hosting the fuel production plant may intentionally bury its footpads to help anchor it from the blast of being in proximity to future vehicles landing and launching nearby.

**Initial Settlement Site Preparation**

Machines gather all loose regolith from the surface to a reasonable depth. The subsurface scans are repeated and archival cores drilled.

**Equipment Boxes (Construction Camp Phase)**

Equipment boxes are standardized modular architecture to enclose power, living, production, and related equipment. They are the space settlement equivalent to the intermodal shipping container standard. Modules vary in length, with a core 3.5 meters tall by 3 meters wide. They are designed to be loaded horizontally through a cargo bay door that could accommodate a standard satellite payload (4.5-meter diameter cylinder). The boxes themselves may be part of a cylindrical module (below). Equipment boxes provide basic services while building the spaceport and accommodate settlement equipment for both surface and in-dome use. The sides of some units can unfold and be pressurized like a Bigelow habitat, and the mid-layer gap may be filled with water or ice to minimize radiation exposure. This provides a dense block of hardware in the center section with shirtsleeve work access from both sides. Boxes are positioned on insulated, level skids and interconnected as appropriate.

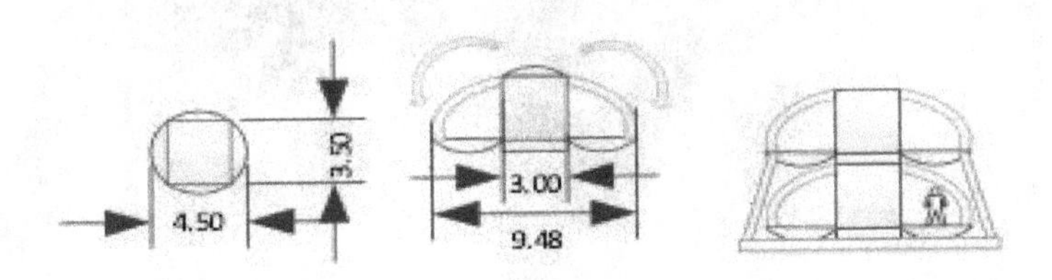

Some modules can be stacked, with support frameworks for the upper pressurized sections mounted to a wider ground pallet. Airlocks, ladders, and passageways are incorporated into the ends of some unit cores. Modules build in this way include power plants, basic surface habitats, fuel production, life support, vehicle servicing, dust separation and mineral/metal extraction systems, basic foundries, manufacturing systems, water extraction, and waste treatment. Elements are connected to pass power, crew, and industrial feedstock directly between boxes.

**Long Dome Structures**

Long domes provide ample workspace and are the principle static structures for manufacturing, hangars, and LED greenhouses. They are scaled to provide unloading and service hangars for Starship. They have either flat or tapered rooflines depending on the purpose. The 18-meter diameter Starship 2 would still fit in the "tall dome" version of the structure if the hangar door were expanded and put on the larger dome end.

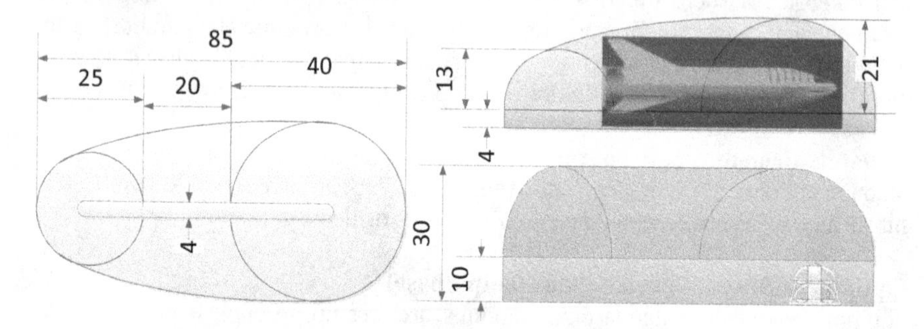


Interior dimensions are shown in meters, with an equipment box at the lower right for scale. Outside these dimensions, walls are 1-5 meters thick. Stem walls of larger structures are 10 meters tall to permit full Starship cargo-scale hardware to fit through airlock doors and passages within the main base. The shorter domes with 4-meter stem walls (to reduce hoop stress) are for remote installations like hangars and industrial space.

**Spaceports**

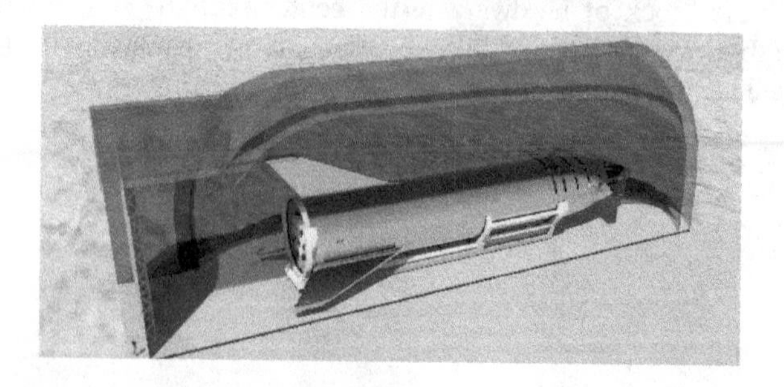

Each port would have domes for a propellant plant storage tanks, with oxygen and methane storage separated in separate domes for safety. A third dome would house propellant, power, and other gas mining systems, and additional domes would serve as hangars. The domes have the smaller end toward the pad to minimize the effect of shock waves, with blast-out doors at the back to direct any tank ruptures away from other facilities. The tower on the landing pad is the crew gantry, cargo crane, and refueling arm. Additional hangars would be built as needed for vehicle maintenance and storage of locally-kept Starships for use in point-to-point exploration and mining of Mars, as well as asteroid missions with Mars as the home base.

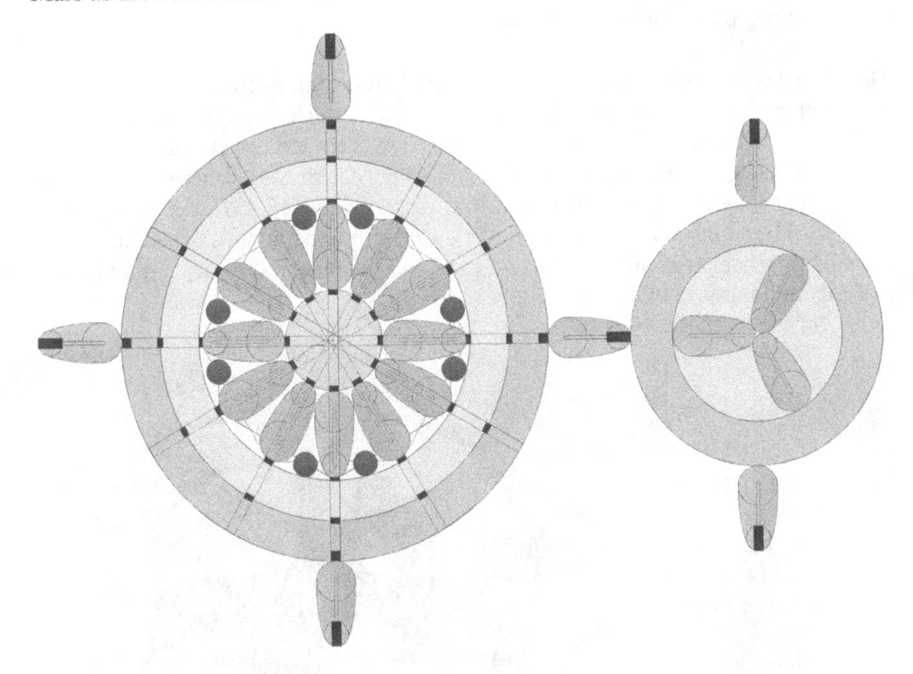

**Demo Ring and Main Facility Layout**
The main settlement consists of two concentric rings, various long domes for main airlocks, work areas, and greenhouses, and a central, 100 m diameter dome. The entire structure has an outer 10 m stem wall backed by a berm outside the rings. This schematic is the main settlement adjoined to the smaller prototype settlement. For the main settlement, the outer ring is 440 meters in diameter. The dark circles are storage silos with dome ceilings, 25 meters in diameter. The light interior domes are agricultural spaces, and the darker ones are factories.

The smaller prototype ring on the right is 260 meters in diameter and is the minimum size to comfortably generate 1 G equivalent on Mars. It also contains three work/lab long domes in the center and has a 10-meter stem wall. If arranged as apartments, the prototype ring could house 1300 people. The mini-ring resolves any engineering issues prior to building the main settlement and would remain active as a lab facility. Additional facilities are built using these common elements (long domes, storage towers, rings, central domes, etc.) and are interconnected with the main facility. More mini-rings with radial exterior domes would be built where appropriate as mining and manufacturing outposts.

**The Ring Tracks**

The larger the ring, the greater the track speed needed to reach 1G. For this reason, it is better to build multiple rings in this smaller size range than to attempt a still-larger ring. Having three ring sizes here gives a range of experimental options for future replication as needed and allows different gravity levels to be used simultaneously for long term experiments.

The cross-section below shows the central park dome, an internal long dome, and the inner and outer ellipsoid rings with a banked track. For 1G of centrifugal force, a track angle of 57 degrees is needed for Mars gravity (lunar version, 74 degrees; deep space, 90 degrees.). The mini-ring (not shown), inner hab ring, and outer ring move at 70, 110, and 125 kph respectively (45, 66, and 80 mph) to maintain 1G. Rings may operate at lower speeds for experiments. There are three planned ring habitat arrangements, and each will be done in turn depending on the maturity and purpose of a given ring.

***Train Habitats:*** Initial outposts have rail cars that merge onto tracks above and below a main flat plate and can be interconnected both across and down the tracks. This allows habitat and lab cars to move at different speeds on different tracks in the same ring, and for passenger cars to clip onto and off from the back of the trains to exit and enter the ring from the stationary facility. Labs with multiple experiment durations and gravity levels may use this system indefinitely. But living facilities will advance to the block system shown above.

***Cubic Habitats and Service Cars:*** For a habitat with more living and lab space, rail cars are replaced by 2 by 10 by 10-meter interconnected gondola service cars. The cars contain water and waste treatment systems, propulsion, and power utilities. Waste and sewage would be anaerobically digested, pasteurized, dried and compressed into blocks within the service cars themselves. The resulting blocks would be removed via a service system under the tracks. Water input to the ring would be a separate track of containers "uphill" on the ring from the waste tracks. A third system handles mail, commuters, and supplies. The "sky" (interior top surface of the stationary ring that appears above the inhabitants due to spin) of the ring would remain open and have a projected sky. The system is built so that any module, car, or rail can be replaced while the track stays in motion.

On top of each service car is a cubic habitat, 10 by 10 by (up to) 9 meters (three stories). These habitats include homes, restaurants, work areas, labs, and agricultural spaces. The ring arrangements are up to 112 cubic module slots for the mini-ring (2 tracks of 56 modules each), 260 for the outer ring (2 by 130), and 300 for the inner ring (3 by 100). If all habitats were three levels, Eureka would have 200,000 square meters of gravity-boosted space, plus 67,200 square meters of rooftops, across 672 cubic habitats. Since airflow in the ring would be accelerated to roughly match the habitats, rooftop spaces would be open for small gardens under the illuminated curved surface of the track pressure vessel. Sky and horizon visuals on the curved ceiling are projected to either match the speed or remain stationary, depending on which induces the least vertigo. Earth or other landscapes are projected at the lower edge, and animated like a rail journey through the mountains, ocean voyage, or whatever suited the population. Given the steep angle, furniture would be built in, with interior paths and storage arrangements designed to allow for an emergency ring stop. An exit track on the inside of each lower wall would allow a safe exit if there were an evacuation.

***Nested Habitats:*** In the third phase, the entire interior of the ring would be a solid toroid shape that fits within the stationary torus. Architecture can be even more integrated as a single structure, and every cubic meter is available for use. The bottom of the ring is open to onramps and off-ramps below the spinning torus. The open area that represented the sky to the inhabitants in the block phase would remain visually but would be part of the same rotating inner structure. The ring would have to be able to de-spin for maintenance, so this option may only be used for specific situations where it is judged superior to the other two options.

**Aesthetics**

Eureka is designed to fill the senses with as many psychological prompts of comfort and inspiration as possible. Pastoral environments give a feeling of being outside in natural smells, sights, and daylight. The central dome provides a set of fruit trees, berry bushes, and other natural elements in a permaculture "food forest" that is both a park and a food source for the settlement, possibly with some wildlife. It also allows for epigenetic studies of trees and a mini-ecosystem at Mars gravity. Since trees at Biosphere 2 became embrittled by a lack of wind, these domes have systems that would provide artificial winds from various directions, rain, and seasonal temperature changes to maximize fidelity to the Earth-outdoor experience while isolating the variables for plant growth to gravity. The dome may include a lazy river pool and a central tower that overlooks the dome interior. Overlooks on the edges and in the core pillar would give a beautiful escapist set of garden spaces and plaza design opportunities. With the park domes and rings, an artificial "sky" is projected on the ceiling, with mountain or seascapes at the horizon edge changed regularly. Artificial sun LED skylights[4] throughout structures would bring simulated daylight into interior spaces, and some locations may use a sun tracker and mirror to project sunlight into the facility. Eventually, an additional dome would be built outside the rings for an ocean wave park with a beach and reef for diving and ecosystem experiments.

With the high stem wall, Eureka has the feel of interconnected vaulted cathedrals or train stations in the dome interiors. The internal architecture is a combination of right angle (Frank Lloyd Wright) elements in the factories and hydroponic farm spaces, flowing curved organic (Roger Dean) architecture in the ice domes, and angular yet flowing elements (Syd Mead) in various equipment and mechanical systems. People also feel comforted with spacious city plazas and a feeling of permanence beyond a human lifespan[5]. The vaulted ceilings with 10-meter base walls give this timeless feel along with a general openness. Observation decks on towers and the rings overlook the Mars environs while still being protected from GCR. Surface images on 8K monitors feel more like windows than televisions in indoor spaces, and show live feeds from the surface. Light shows on the ceilings and exterior, fountains, and 3D sculpted public works provide a unique art gallery to display local creativity.

Residences, restaurants, and offices would have an allotment of material that is then 3D-printed and CNC carved into whatever furniture, art, and fixtures matched the tastes and needs of the space. Homes or public modules could have a décor theme of steampunk, sci-fi, Victorian, Parisian, medieval, the age of sail, garden, contemporary, retro-future, or whatever suited the owners. Restaurants and businesses would have a similar flair to break the monotony with endless variety and creativity. If tastes or residents change, the interior can be traded or recycled into an entirely new space. Entire blocks of the ring, including the projected sky and horizon, would mimic world cities at different periods in time, and fictional cities in nearly any setting that can be built. A visit with friends or to restaurants would be like a microvacation to fully counter the sense of isolation.

Good architecture avoids as many imperceptible annoyances as possible to avoid a form of subliminal, though cumulative, irritation. This stress dampening quality would be important to the health and happiness of the settlers. Since everything is built from a clean sheet to be maintained robotically, robots have a much lower AI threshold to do cleaning and organizing tasks. Power, light, and data systems would be universal and hacker-proof with quantum-resistant communications protocols and simplified software/hardware systems. Restrooms, labs, workshops, and kitchens use steam-cleaning robots. Manufactured goods are stamped with unique IDs that would also designate ownership in the blockchain. In short, the worries of information security, theft, cleaning house, or even having the right plug for a given socket are largely remediated based on this clean-sheet approach.

## ECONOMIC MODELS

The early space settlement revolution is unlikely to support many large enterprises until the economic benefits are democratized. Items in the second and third column of the Space Settlement Grand Challenge table above have no practical pay-back model in the early stages of development. Successful technology revolutions typically begin with one or two lead efforts, followed by a wave of follow-on products from a broad spectrum of manufacturers once the barriers to entry are removed. The first column is currently undergoing this transition.

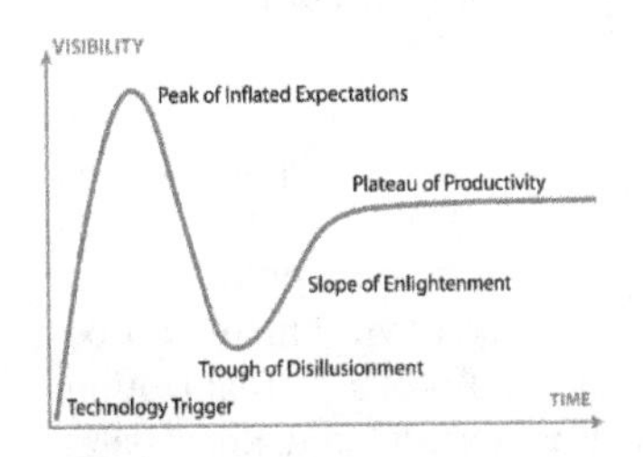


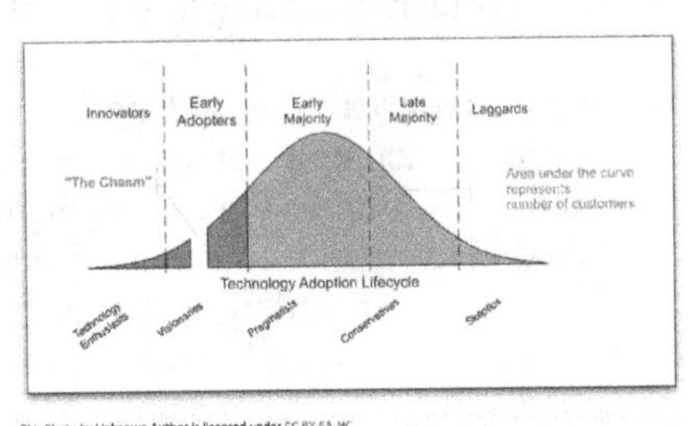




Any start-up enterprise faces a phase called "the chasm" (above right), where early expectations and seed funding wains, and mainstream markets have not yet taken root. Similarly, new technologies face the Gartner Hype Cycle (above left) where early promises collapse and investment fades prior to a slow normalization and plateau of that technology as a "new normal". Eureka's economic model must be strong enough to survive both collapses simultaneously, and do so for an entire technology revolution, all while competing with simultaneous revolutions in other opportunities for technology investment. The good news is that like a conventional periodic table of elements, the introduction of the first few elements provides a foundation for creating all the others. As with the engineering challenges described in the first section, the resolution is a matter of finding the investment path of least resistance, then follow that path with the greatest force.

### The Limits of NewSpace

Modern technology revolutions use agile methodologies. These methods enable rapid innovation. But they do so at the expense of being nearsighted in terms of strategic goals. Planning voyages of millions of kilometers over months cannot be done tactically, so agile methods must converge with long term thinking.

The NewSpace revolution enables the Space Settlement revolution, but it is not the same wave. While the market increases as launch costs drop, and the high-speed satellite business model becomes viable, the launch market will have a mini-collapse of minor players, much as the early PC market created and dropped many smaller enterprises. Without a transitional economic model, the NewSpace revolution may undergo a launch service hype cycle collapse just as the space settlement revolution begins to need investment.

The hope by SpaceX, Blue Origin, et al is that the new affordable launch market will finance ongoing efforts to develop crewed deep space technologies. This entire interdependent matrix of Grand Challenge technologies (new surface suits, life support, etc.) must become affordable as a complete system. There is a limit to the democratization of settlement technology demand for these mid-table systems because the demand for spacesuits et al will not be digitized or otherwise flattened to commodity levels. Until settlement technology (column four) is democratized, the technologies of all four columns will remain niche products. This is much as business jets remain a small market in the decades since their invention, rather than devolving to the level of cars or even prop aircraft.

**Harnessing Concurrent Technology Revolutions**

That said, this concurrent set of technology revolutions has the potential to combine into a powerful economic engine of factorial-scale growth in inventions. This is called convergence, punctuated equilibrium, or multifactor productivity. Further, the stresses of settling and industrializing the space environment can be turned into advantages in biological and material sciences. Eureka's organization combines the novel environment on Mars, affordable sensors to gather data, artificial intelligence to analyze multi-variable data, and quantum computing as a back-end calculator to model difficult elements such as material chemistry and genetic expressions. This mutually-supportive engine, along with Earth labs built on the same model to further expand the knowledge base, would drive fast iterations in experimentation, discovery, and invention.

**Getting Started**

***Assumptions:*** The foundational infrastructure of Mars missions will be created by SpaceX and joined by Blue Origin, with more competitors and systems added later. SpaceX plans a Martian outpost for refueling and power near a buried glacier for ease of water extraction, much as outlined earlier. Eureka then provides the foundation for ongoing work past this foothold.

The cost of getting city-scale shipments to Mars will be prohibitive if done without a democratized sponsorship model. Mars-to-Earth exports (discussed later) need a ten-fold mark-up to offset the cost of imported equipment and supplies for the first large settlement. A shrewder approach is needed to bridge the fiscal gap from exploration to settlement- one that monetizes and democratizes the historicity of the moment.

The supplies for early expeditions have unique historical and promotional value. First, the company that builds early equipment judged premium and reliable enough for use on Mars would be motivated to sponsor that equipment for promotional purposes. Second, if that equipment could be shipped back, in whole or in part, then it becomes a historic collector's item far more valuable than a terrestrial copy. Third, the equipment is then tested in Mars conditions and may be marketed for commercial off-world use. These are three key motivations for manufacturers to sponsor or discount equipment.

***The Eureka Foundation (TEF):*** Eureka's foundational organization sets standards, performs testing, and routes requests for proposals (RFPs) for equipment needed for space settlement to competing organizations. NASA could have some involvement in this effort using their knowledge and test facilities. Qualified participating manufacturers would then submit designs and products for certification and flight evaluation. Companies entering competitions would retain partial ownership of their flight articles as part of the financial incentive for participating in these historic missions at reduced organizational cost. TEF also coordinates the activities and below.

***MarsSpec:*** TEF sets standards for testing and certification of hardware or software exported to an off-world space settlement. These systems must be maintainable with site-manufactured parts where possible. They must be highly reliable and secure, have electronics encapsulated into replaceable units, use a new universal set of ports, common bolt sizes, and so on. Right to Repair must be expanded to include "Right to Bypass" (replace existing machine controls with simplified systems) for emergencies or independent on-site support. As with fire engines and other mission-critical technologies currently, manual bypasses must be available in emergencies to prevent hackers, misinformed AI, or software errors from putting key infrastructure at risk. Crews would be trained in manual systems and drilled to avoid skill erosion. This kinesthetic practice mentally grounds them in the principles of how the systems work, enhancing the depth and integrity of future site-created inventions.

Components must be modular, interconnect with other systems on the same standard, and be "stackable" to create larger capacity systems. This also makes equipment adaptable to unexpected situations. By making standardized systems work together, there is a built-in terrestrial market for ruggedized or luxury equipment. For example, if a luxury car or home builder can do MarsSpec product lines, that differentiates the product (and manufacturer) from the competition. This same halo effect and research model is why automotive manufacturers invest heavily in racing and concept cars. Conversely, field-maintainable equipment on the low-end of the cost spectrum would be useful in developing economies and remote locations. The specifications are compartmentalized in such a way that these spin-offs could benefit both luxury and developing markets on Earth.

Since the qualified systems are simple, maintainable, and reliable, they will often be affordable to fabricate due to the prevalence of additive and subtractive manufacturing techniques. The organization would also help avoid intellectual property issues such as led to the "sewing machine wars" and other conflicts in the mid-twentieth century. Membership in the foundation would entitle the company to common knowledge and intellectual property applicable to any system within the Foundation ecosystem where standards and components are concerned. This opens catalog marketing opportunities within the ecosystem for component manufacturers, which further democratizes design, manufacturing, and marketing.

We are seeing this now to some degree with cube-sats, just as we have with computer parts for the past few decades.

***LaunchFest:*** Each year, the operational authorities of planned missions, bases, and settlements would put out RFPs for equipment needed for the next few missions, including lunar and Mars missions. Finalists would attend an exhibition and stress test competition to select winning products and equipment. This space settlement exhibition would have a similar dynamic to entrepreneurial pitch competitions, technology exhibitions, and University Rover Challenge combined. Judging would be held early, or as a prerequisite to entering the show, so that the competitors could show off their awards. A second festival cycle would invite the manufacturers to the appropriate spaceports to see their handiwork launched into history during the next launch window. A similar ritual would celebrate the return of ships with used hardware for analysis, collectible sale and promotional display.

***Certification Seals:*** Since the equipment is designed to be maintainable on Mars with 3D printing, forgeries of these historic items would be quite simple without a certification seal. Seals are attached to equipment and may also be used like mission patches for crew and sorties. Certification seals would prevent forgeries of space-flown equipment from undermining the sponsorship model above.

Seals are small blocks that identify each flown item and contain unique attributes that make them easy to identify and impractical to counterfeit. Originally envisioned as passive cosmic ray trace recorders to make the space environment itself part of the seal, the more stringent requirement is that the seal is readily identifiable and able to be validated as authentic using a specialized reader linked to a blockchain. A similar method is used to validate diamonds as mined versus synthetic or conflict, by mapping the occlusions in the diamond and placing this data in a blockchain. As a blockchain element, ownership of the item can also be linked to the record, sold, and exchanged. Via smart contracts, ownership of a seal also entitles the owner and investors to dividends from profits related to the item's use on Mars. This convergence would unify stores and transfers of economic value the way the iPod/iPhone unified portable digital media access. It is analogous to how Isaac Newton's invention of the gold standard enabled the economic expansion of the British Empire by providing a trusted, standardized economic foundation for value transfer and storage. Trusted value stores are a key enabler to any commerce epoch in human history, particularly when traditional value stores from the previous epoch are too highly leveraged to be trusted.

There is a potential conflict between the demand for the return of sponsored artifacts and the desire for settlers to keep hold of refined materials. A win-win option would put all the unique elements in a front-plate, including the multiple certification seals, to allow the returned identification plate to be split as a "limited-edition" series of artifacts (conceptual illustration, below). The equipment behind the plate could be reused on-site or recycled into new equipment, with some parts returned with the ID plates for wear analysis and design improvements. Since the plates and parts would take less space/mass on return flights, more collectibles would be returned on each flight. They could also be returned while equipment is still in use, so sponsorship provides a return on investment from both Earth and Mars simultaneously. Below is a notional diagram of the concept. The control components would be modules throughout the machine structure rather than simply in a box on the side.

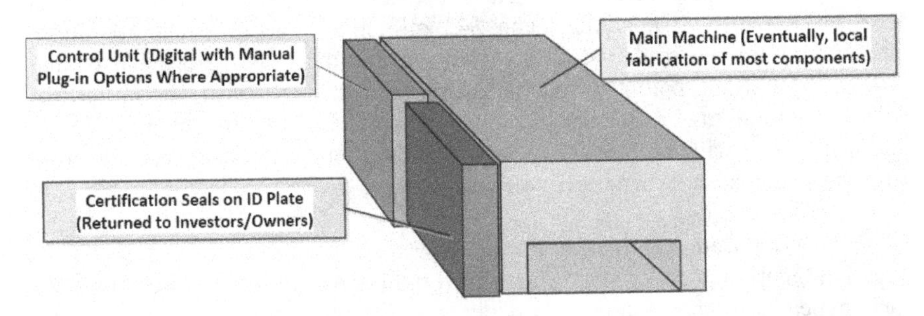


***Investment Forum:*** The certificate/sponsorship systems above provide the financial foundation for getting the settlements profitable and democratizing production, investment, and base ownership far more quickly and dramatically than would otherwise develop. These contracts avoid ownership conflicts between settlers and remote institutions. Investing in settlement equipment would be like owning non-voting stock in Eureka's day-to-day operations. Once Eureka becomes profitable, twenty percent of that profit level is reserved for these pioneering suppliers above and beyond the other benefits. In addition to suppliers and equipment collectors, anyone could purchase a certification seal in sponsorship of specific missions or equipment. These investor payments would help fund for non-returnable items such as reactor cores and foodstuffs. Many sponsors would keep returning to invest in blocks of missions to complete collections of seals and equipment related to some topic of interest. For example, a collector may display seals for helping sponsor coffee shipments for Mars, with an additional seal for sponsoring early coffee growth experiments included.

Since the end goal is space independence rather than a return to mercantilism, the certificate of ownership would be non-voting and would have an expiration date in terms of ownership and dividend payments. This date provides an investment analogous to options trading. The permanent value of the investment would be the certification seal itself, which would increase in historical value over time. Eventually, the original Eureka Foundation would become a subsidiary of the Mars-based organization.

***Minor Markets.*** Eureka could also sell sliced Mars rocks and surface meteors that would have the double rarity of being meteorites from the surface of Mars. An industry of making unique teas, coffees, and liquors on Mars for the Earth luxury market is also possible since they are easy to preserve and compact.

**Import/Export Trade Balance**

Our baseline assumption is that Earth to Mars transport costs \$500/kg and exports to Earth are \$200/kg. Raw imported material for onsite manufacturing such as chromium powder average $30/kg. Exports may average $100/kg (opportunity cost). Under a flat model, this would mean that exports would have to be marked up 830 percent to break even with imports. This model is unsustainable because the return on investment would take too long. The LaunchFest model would help sponsor each import flight and raise the profit closer to the full $210 million per sortie. If twenty percent of the profit is a dividend for the sponsors, that gives $168 million to help operations and $42 million for the investors per flight. Investors in Eureka itself would also receive a comparable percentage on the intellectual and informational exports and research arrangements. The billions of dollars required to get Eureka built and expanded would be returnable in a reasonable time with far less effort, and a funding model is established for other outposts as humanity expands across space.

**The Profitable Democratization Phase**

This combination of investment methods would be enough to get Eureka past the early hype/chasm gaps and on to producing intellectual property, inventions, and commercialized and scientific research. The early enthusiasm and public engagement at this moment in history are embraced and democratized so that it may become distributed, normalized and permanent. Technology revolutions only fully mature when everyone has some mundane ownership of them. It may seem demeaning to break the grand challenges down in this way, but the alternative is to never advance at all. By the time settlement hardware becomes too common for certificates to have high value, conventional investment in research will be self-supporting. Once the city becomes profitable, the overall profit breakdown becomes twenty percent for the investors proportional to investment and the value of equipment sent, then twenty percent for onsite entrepreneurial start-ups, ten percent as a fund for earth-based entrepreneurial projects carried out on-site, twenty percent for infrastructure capability expansion (things with no direct financial return but necessary, like better sewage disposal or mining techniques), and twenty percent base expansion. Ten percent would be for resilience improvements such as on-site systems to displace imported products, starting with the most massive and easiest ones. The twenty percent for onsite projects can be expanded by local settlers investing their own funds, and gaining commensurate increased returns. Research done for clients is sponsored by those clients (pharmaceutical and agricultural organizations, universities, engineering firms, companies, governments, or individuals on both Earth and Mars). This provides ample investment for growing trade and the domestic economy. A three-phase simulation shows the city expanding over four decades to $450 billion in annualized profits, and supporting a population of 36,000. Eureka provides 500 percent ROI in phase 1 (rock and collectible exports), peaking at 12-fold profit relative to import costs in phase 3 (pharma exports). This provides a strong margin to fund experimental work and cover GCR-related depreciation in exported pharmaceutical and agricultural products.

**Mars-based Consumer Products**

Common necessities such as basic foodstuffs, toiletries, and so on are produced in fixed batches in adaptable workspaces. Production of limited-run items such as furniture would be done in common maker spaces with contract labor or hobbyists. That leaves a middle zone that can be filled with Kickstarter-like operations that design new products, take pre-orders, and commit to building the items, along with spare parts. This avoids inventory and marketing issues. As in the age of sail, settlers will be used to pre-ordering complex items from their civilization of origin two years in advance of needing them. There are spares on site for replacements, but inventories are limited. Items designed on Mars can also be licensed and sold back on Earth with local production and royalty distribution.

**Social/Cultural Structure**

Settlers must have individual independence with responsibility and property rights. The framework for this, in the near term, takes the form of employment and liability agreements.

***Organization Scales:*** A good breakdown point for an informal organization is Dunbar's Number, where organizations have 150 or fewer members. Breaking groups into organizations with 100-120 members reduce the risk of infractions and social isolation. Organizational units of this size are large enough for ambitious projects, particularly if grouped as departments within larger projects.

***Guilds:*** Eureka has a guild style system based on the tasks required to keep the base operational, and additional guilds for research, engineering, and entrepreneurial projects. Guild membership would typically be seven years, with people having the freedom to switch guilds. This helps with knowledge transfer across professions such as biomimicry in engineering. Guilds would sponsor and educate new residents and provide a social safety net in terms of health and accident insurance. Guild groups would be limited to Dunbar numbers, but contribute to common credit unions and insurance organizations.

All organizations are structured like a credit union for transparency and profit-sharing and follow Generally Accepted Accounting Principles (GAAP). Over time, there would be multiple competitive guilds in the same field. They would sponsor infrastructure and quality of life improvements for the entire settlement. They provide the talent pool for institutional knowledge, and they would be represented like corporate departments within the settlement leadership.
While the Eureka organization itself would focus on building a framework for life support and research, ad hoc teams and startups within the base would self-assemble and bid for investment for specific projects and research. This avoids institutional sclerosis when teams and systems outlive their original purpose. It also provides a dynamic engine for inventive start-up work. Thus the Innovator's Dilemma is defused before it can start. Guilds would be responsible for maintaining and expanding key systems. This prevents the population from neglecting core infrastructure when allocating resources for projects.

***Political Structure:*** Leadership based on geographic representation is a small council selected by lottery (like jury duty) to serve a single, six-month term. No one guild would provide too many necessary services, and the geographic council would prevent regional neglect in services. This provides a check on guilds holding excessive political or economic power. Guild members would not congregate in specific living areas to avoid synching guild and council power structures. This would also prevent a localized disaster from wiping out all talent in a necessary field.

***Property Rights:*** Employees may launch new businesses in their off-hours. To simplify this dynamic, rental offices, labs and maker spaces give shared resources for freelance or start-up operations until those enterprises can afford their own facilities. They may also start new manufacturing spaces and be the primary client, then rent out excess capacity when not needed. Those living on Mars must have the freedom to experiment with local materials. All businesses must respect the safety and property of the settlement and environs, and submit locally-extracted materials for scientific cataloging and analysis. While the materials of a space settlement may be privately owned, the base scientific knowledge of planetology is shared as a "common heritage of mankind" so long as mining rights and security are respected (knowledge can be copied and shared; atoms cannot).

***Education and Careers:*** On-site education would be a mix of first principles, classical, and practical work, and be continued online and in study groups on a lifelong basis. Jobs would typically be done in seven-year contracts with a six-month transitional sabbatical to re-tool the mind for another career or return to the same roles, return to Earth or a tour on another outpost.

***Competitions:*** Sports would be designed with rings and open domes in mind, with smaller courts and allowances for gravity or centripetal effects. Robotics competitions are a local mirror image of LaunchFest, particularly where mining and manufacturing are concerned.

***Permanent and Guest Residents:*** There will be a complex mix of permanent residents and visiting specialists. Developing permanent tribal knowledge versus providing opportunities for shorter stays is another cultural complication. Visitors may over-prioritize their own research or business goals while sidelining local standards or the long-term well-being of the settlement. New residents would be required to have extensive training and stress testing prior to arrival. An ideal split between permanent and visiting populations is dependent on the culture and conditions, but there should be bias towards those who must live with the situation permanently rather than simply passing through in the name of career advancement.

## CONCLUSION

Elon Musk said of Starship/Raptor design, "The answer flowed once the question could be framed with precision. But framing [it] with precision was very difficult."[6] Eureka is focused on asking the hardest foundational questions, then giving the best possible answers from first and financial principles, to provide solid cornerstones for ongoing work. Any engineering or economic model must allow for the realities of physics, finance, technology, and culture to disrupt those plans in the coming decades. Since Eureka is planned from a worst-case start to a full-solution finish, any advances or setbacks can be absorbed by this design.

**REFERENCES**

For details and updates on Eureka, visit https://macroinvent.com/mars-1000/.

---

[1] Most of this paper is written in the present tense to save space. Eureka is named for the 2006-2012 US TV series about a hidden small town devoted to advanced science. 3D illustrations by Michel Lamontagne.

[2] Natural Transfer of Viable Microbes in Space from Planets in Extra-Solar Systems to a Planet in Our Solar System and Vice Versa. The Astrophysical Journal, Dec 1, 2009. https://iopscience.iop.org/article/10.1088/0004-637X/690/1/210/meta

[3] Kilopower (Official NASA website). https://www.nasa.gov/directorates/spacetech/kilopower

[4] CoLux is the inventor/leader in this field with very bright simulated sunlight, though there are other brands that are more like televisions. https://www.lightology.com/index.php?module=vend&vend_id=774

[5] This is a common architectural concept. https://www.architecturenow.co.nz/articles/longevity/

[6] YouTube Dear Moon announcement. https://youtu.be/zu7WJD8vpAQ?t=2806

# 7: K-TOWN: A THOUSAND-PERSON COLONY

**Jeffery Greenblatt**
Chief Scientist, Emerging Futures, LLC, Berkeley, CA, USA
jeff@emerging-futures.com
**Akhil Rao**
Assistant Professor, Middlebury College, Middlebury, Vermont, USA
akhilr@middlebury.edu

## 1. Overview

Welcome to K-Town, the largest human settlement on Mars! It is located at 47°N, 274°E in a flat region of western Tempe Terra adjacent to a number of beautiful and scientifically interesting natural features. The town itself occupies 0.6 $km^2$, not including its solar array or various surface mining operations. It is also sufficiently far north to access large quantities of subsurface water that are available there. In the future, Mars may be terraformed, so K-Town was located at sufficiently high altitude to avoid being flooded if Mars' ice caps were to melt. It is also adjacent to many interesting geologic features including numerous outflow channels from an ancient Mars ocean. About 800 km to the southwest lies Alba Mons, one of Mars' great volcanoes. Further south off the map is the Tharsis Rise, containing the three Tharsis Montes volcanoes and Valles Marineris, the "Grand Canyon" of Mars. See Fig. 1.

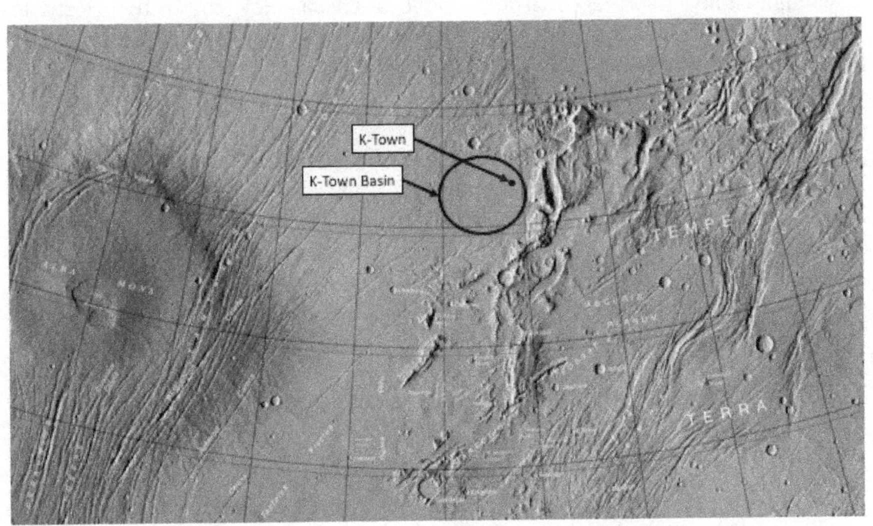


*Fig. 1. Location of K-Town at 47°N, 274°E within the Tempe Terra region.* Source: [1]

K-Town is comprised of people with a wide variety of skills and backgrounds (see section 2), and herald from various corners of Earth; however, a few recent arrivals herald from K-Town itself—we will discuss the safety of giving birth on Mars below. In addition to a significant flux of tourists, young adults seeking work, and permanent retirees, several major corporations on Earth have recently started sending representatives to K-Town, often along with their families, to scout out new business opportunities and broker deals with other businesses already there. Companies in the Forbes Interglobal 2000 including Amgen (biotech), ArcelorMittal (metals), AT&T (communications), Bank of China (finance), BASF (chemicals), BHP Billiton (mining), Caterpillar (heavy equipment), Heidelberg Cement, Hilton Hotels, LabCorp (medical), LG (appliances), Microsoft (software), Mitsubishi Electric, Nestle (food processing) and Waste Management have recently opened small offices in the commercial sector of K-Town.

So named because it recently passed the 1,000-person population threshold, K-Town may have to change its name again in the future, as its population continues to grow at about 7.1% per Earth year ("E-year"), or 13.7% per Mars year ("M-year"). K-Town is also not the only settlement on Mars, but these other pockets of humanity can hardly be called "towns" at this point, as they only consist of a handful of people living in the most basic life support accommodations, and heavily dependent on both K-Town as well as Earth for their continued existence. Scientific outposts, mining operations, and a few tourist attractions are scattered across the globe, with some accessible by Hyperloop spurs that have been built at private expense to allow for efficient transport of people and goods. Other locations, such as Hellas Planitia Outpost or South Pole Station, are still too far away and can only be reached via suborbital rocket "hop," and residents of K-Town almost never see people from those locations, as they tend to fly directly to or from Earth. Nonetheless, a nearly-constant flux of ~100 people flow through K-Town each M-week (equal to eight Martian days or "sols"; see Timekeeping section for more information).

So while K-Town is certainly the largest settlement on Mars, it is not its only spaceport, and in principle any location with a spaceport and access to sufficient raw materials could grow into a K-Town or larger, given sufficient interest and investment. So why has K-Town flourished while others remain quiet backcountry outposts? Certainly K-Town has age on its side: it is one of the oldest settlements on Mars, celebrating its 10th M-year (19 E-years) this spring. But beyond longevity, its location, proximity to resources, aesthetic beauty (for the town IS beautiful; see below), wealth, government organization, investment opportunities, and culture all contribute to its success.

## 2. Demographics and physical organization

### 2.1. Demographics

People in K-Town come from all over Earth (and now, also, Mars), and as a result enjoy a high degree of ethnic and cultural diversity. While the official languages are English and Chinese, it is common to hear Spanish, French, Luxembourgish, Arabic, Russian, Japanese and Korean spoken as well. People who have lived in

K-Town a long time tend to be conversant in several languages, and children in school must study at least two besides their native tongue.

In terms of population breakdown, K-Town is currently composed of 607 permanent residents and 393 visitors, with the latter consisting of 204 professionals (business, medical, scientific, etc.), 96 tourists, 44 spacecraft crew on furlough (see spaceport section for more details), 26 guests from other Martian settlements, and 23 visiting college students. In addition to the temporary visitors who arrive during every Hohmann transfer window (also known as a synodic cycle, equal to about 2.14 E-years or 1.14 M-years), K-Town recently welcomed 209 immigrants, 31 of which were spacecraft crew members planning to make Mars their permanent home. On the other end, roughly 5% of K-Town's population or ~50 people per synodic cycle opt to return to Earth due to financial, health, marital or other reasons, bringing the net immigration rate down to 159 per synodic cycle. Among colonists, there were five healthy births in the past M-year but, sadly, also seven deaths, making the net population growth rate slightly negative, though immigration far outweighs this trend.

In terms of composition by age, 50 are under 18 E-years, 874 are adults younger than 65 E-years, and 76 are seniors. Of working-age adults, 46 are college students between 18 and 24 E-years, though 50% leave to study on Earth, and are replaced by a similar number who come to K-Town to study in its unique environment. There is a small school for children in K-Town (supported by 10 teachers and staff), but due to the small size of the college-aged population, there is not a traditional college or university available. Instead, those who opt to spend their college years here learn all about the workings of its small society and many technological innovations, in a uniquely self-directed, independent study program, supported by five professors and three staff.

For the 96 tourists currently here, half stay in the single hotel in town, *The Cangwu* (named for the nearby crater and city in China), along with up to 52 other guests, and tended by 50 hotel staff. Everyone else, including remaining tourists, stay in either apartments or, if they're very wealthy, standalone homes. Other major types of employment in K-Town include medical care (90), food processing and restaurants (80), spaceport (55, though seasonally it grows to 88), government (48, including 18 in foreign embassies and 13 in the K-Town legislature), agriculture (40), manufacturing (39), tourism (27), arts & entertainment (25), maintenance (20), banking (17), scientific research (16), propellant plant operations (12), and astronauts/explorers (8). Another 86 are involved in other business activities of various kinds. Remaining people (124) include non-working parents, furloughed crew, wealthy individuals, retirees and unemployed.

### 2.2. Physical layout

K-Town is arranged as a set of semi-buried airtight structures providing radiation and micrometeorite protection for residents, while allowing ample interior space for multi-story buildings and open-air parks, gardens and fields. While built mainly of opaque materials, the curved roofs contain many large glass panels to allow for ample natural light and views of the breathtaking surrounding topography that is part of K-Town's unique appeal.

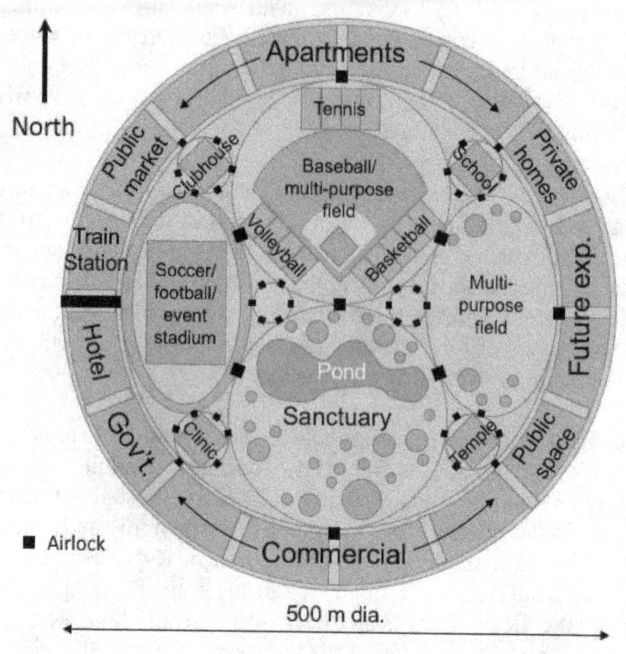


*Fig. 2. Detail of K-Town central region showing parks, fields and buildings. Outer diameter is 500 m.*

Fig. 2 shows the structures within the heart of K-Town. The centermost region contains several interconnected open spaces including a sanctuary, event stadium, multi-purpose fields, gardens, K-12 school, sports clubhouse, medical clinic, and multifaith temple. Two spaces (sanctuary and the larger multipurpose field) are enclosed by 200 m dia. domes, the largest pressurized structures currently in K-Town, representing the state-of-the-art in design. These domes are mainly composed of thick "cold" glass supported by structural steel members; the glass is coated to reject all infrared and ultraviolet light, reducing heat loads and sun damage. Other structures are enclosed by smaller circular or oval domes, and are interconnected via airlocks that are normally open but can be closed in emergencies. Within the central region are several unenclosed natural areas (indicated by orange in the diagram) in which artists have placed a number of rock, metal or glass sculptures that can be viewed from inside the domes.

Apartments are located in the northern quadrant, with a few private homes adjacent to the east. To the west of the apartments is a public market, which sits just adjacent to the train station on the other side. Just south of the train station is the *Cangwu* hotel, flanked on the opposite side by government buildings. These public facilities are also adjacent to the sports stadium just inside the central region to allow for easy access during events. Continuing around the ring to the south are four blocks of commercial buildings, housing offices and light industrial activities such as electronics repair, 3D printing of small parts, textile manufacturing, etc. The eastern quadrant houses a multi-purpose public space and two empty blocks reserved for future expansion. A rendering of the sanctuary dome is shown in Fig. 3.

*Fig. 3. Interior view of Central Park.* Source: [2] (used with permission).

Fig. 4 shows the full extent of K-Town, revealing large agricultural areas surrounding the urban ring. Seven access roads radiate from the central region providing access; the eighth direction (west) is occupied by the spaceport train line. The physical structures are similar to those in the urban area, but with smaller, 25 m dia. half-cylinder rings arranged concentrically out to a diameter of ~1,000 m. (Only the inner 700 m is currently in use, with the unused areas reserved for agricultural expansion.) Rings are periodically interconnected to allow free passage of people and equipment. Coverings are mainly cold glass in order to minimize the use of artificial lighting and efficiently utilize the available area for production; during dust storms, however, such lighting is essential to maintain crop production. Although they have less radiation protection as a result, all aspects of farming are highly automated, limiting human radiation exposure. The sole non-agricultural structure in this region is the main hospital, located just inside the town boundary at its western edge, adjacent to the spaceport train line. It is located here to provide rapid emergency access from the spaceport; see further discussion in the Medical facilities section. A dedicated airlock at the hospital also allows access from outside the town in case of emergency.

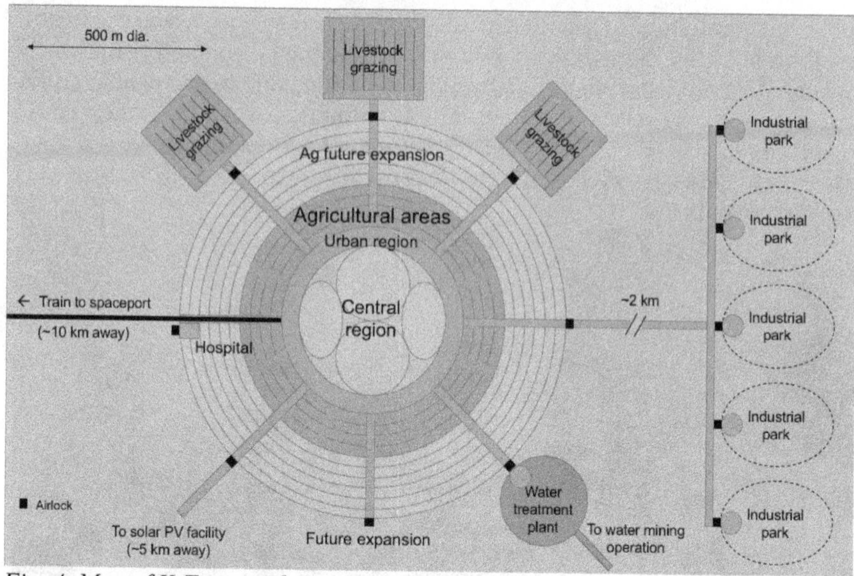


*Fig. 4. Map of K-Town and vicinity*

Large airlocks surround the town at the terminus of the seven access roads, separating the town from other facilities that are either too far away or present various hazards. Among those presenting hazards are livestock grazing areas, which must remain separated from the town air and water supply to prevent animal-to-human infection, as many human epidemics originate in animal populations [3]. For similar reasons, the town's water treatment plant is also isolated to prevent pathogens from entering the air supply. A different type of hazard presents itself in the form of industrial facilities, which are located ~2 km to the east through a connecting road to a series of individual airlocks at the entrance to each area. This isolation is mainly to prevent an explosion, toxic chemical release, or other mishap from affecting town infrastructure or its inhabitants. Similarly, the spaceport is located ~10 km away to the west, to provide ample area for wayward rockets to land without damaging K-Town or other critical facilities.

Finally, the central solar PV facility is accessed via a ~5 km connecting road, and is mainly isolated due to its size and anticipated future needs. A final spare airlock is present for future expansion.

### 2.3. Medical facilities

K-Town possesses two medical facilities: a world-class hospital staffed by 78 personnel (about double the typical number for a town of its size) located on the outer edge of K-Town nearest the spaceport train line, and a community health clinic staffed by 12 people that provides non-urgent care, mental health, acupuncture, chiropractic, massage and physical therapy, optical care, dentistry and veterinary services, and is located in the central region. Because the unique environment of Mars presents many first-of-its-kind research opportunities, about one-third of the hospital staff are engaged in full-time medical research, studying myriad effects ranging from the impacts of reduced gravity on bone-setting to the epigenetics of lactose intolerance among K-Town's population.

Fifteen staff, including three pilots, are on rotating shifts to operate up to three round-the-clock, rapid-response medical shuttles with the capability of reaching anywhere on the Martian surface within 100 min. via suborbital flight. Each shuttle is equipped with a surface rover to provide access in treacherous terrain, an inflatable field hospital if urgent care is required, and sufficient food, water, oxygen and medical supplies to support 10 people for up to four M-weeks if necessary. The shuttles operate out of the K-Town spaceport, with an emergency express train link (that overrides regular traffic) to transfer patients to the main hospital.

There is still some concern over the health of fetuses and young children in the reduced gravity environment on Mars. While research continues on rotating space stations with artificial gravity around Earth, medical authorities recommend that pregnancy, at least, take place in full Earth gravity to avoid developmental problems. Most parents thus avoid having children on Mars, but for the few pregnancies that do occur, there is a small centrifuge facility located within the main hospital that allows a woman to spend nearly her entire pregnancy at 1 g, though the experience can be isolating. As an alternative, the *Kubrick* research and tourist station currently under construction on Phobos' near side will provide full radiation shielding and artificial gravity ranging from micro-g to 1.5 g. Women will have the option of residing there with their families throughout their entire pregnancy, and for as long after birth as desired.

### 2.4. Plans for future expansion

K-Town's population is currently expanding by about 14% per M-year, so it needs to build 80 new dwelling units (assuming an average occupancy of two people) every synodic cycle, along with additional commercial, agricultural, manufacturing, power and other facilities. Within 10 M-years, K-Town expects to grow to ~4,000 people, at which point its planned urban area and overall footprint will reach capacity. Assuming growth continues at this rate, population will expand within 30 M-years to almost 50,000, and within 54 M-years (102 E-years) K-Town will be renamed M-City as it surpasses 1 million.

For these reasons, the design of K-Town is highly modular, so that multiple "copies" of the basic layout can be replicated across the K-Town Basin (an ellipse of approximately 225 km x 190 km, or 33,600 km$^2$, about 38% of the area of Greater Los Angeles [4]). While each K-Town "module" is designed to feature a large open space in its center, part of the M-City Master Plan is to build a huge, enclosed forested park in the center of K-Town Basin occupying ~1% of its total area or ~30,000 hectares—roughly the size of Arches National Park in Utah [5]. This park will serve as a wildlife preserve as well as "outdoor" recreation area for residents weary of living for many years in sealed and often claustrophobic spaces. However, the self-regulating ecosystem technology as well as the enormous required dome structures are still in development.

## 3. Major industrial activities

### 3.1. Overview

Physical flows of materials in K-Town are nearly closed-loop, as making things completely closed would require exorbitant amounts of time and/or energy to achieve. Designs strive for ≥95% reuse of materials, and reuse rates are expected to improve over time as technologies mature. K-Town and other space settlements will form an important proving ground for completely closed life support systems that will be required when the first interstellar starships are launched to Alpha Centauri in the next century.

While K-Town engineers understand the flows of energy and materials in their own systems very well, an overall picture requires a computer model called ASTER (Analysis of Space Technologies, Economics and Resources) [6] to provide estimates of the total flows of critical resources and energy throughout K-Town, its surrounding operations, and trading partners. In what follows here, ASTER has been used to generate these estimates, which help to set the scale of primary mining activities, water treatment, power plant, greenhouse operations, and other key activities. A summary of the major energy and material flows of interest, broken down by consumption type, is shown in Table 1.

*Table 1. Major flows of energy and materials within K-Town, broken down by consumption type*

| | | | Fraction of total consumed | | | | |
|---|---|---|---|---|---|---|---|
| Quantity | Total output | Units | Urban | Agricultural | Propellant | Other industrial | Exported |
| Water | 128 | kg/s | 5.53% | 14.58% | 14.11% | 65.73% | 0.05% |
| Food (dry basis) | 0.038 | kg/s | 30.56% | 35.83% | 0.00% | 3.36% | 30.25% |
| Fiber (dry basis) | 0.0032 | kg/s | 49.73% | 0.00% | 0.00% | 1.04% | 49.23% |
| $CO_2$ | 20.2 | kg/s | 0.00% | 0.51% | 97.92% | 1.57% | 0.00% |
| $N_2$ + Ar | 0.92 | kg/s | 4.24% | 0.00% | 0.00% | 53.76% | 42.00% |
| $CH_4$ | 7.23 | kg/s | 0.00% | 0.00% | 89.91% | 3.58% | 6.51% |
| $O_2$ | 28.9 | kg/s | 0.03% | 0.05% | 80.71% | 13.03% | 6.18% |
| Steel | 1.18 | kg/s | 7.28% | 0.34% | 1.72% | 82.65% | 8.00% |

| | | | | | | | |
|---|---|---|---|---|---|---|---|
| Concrete | 2.24 | kg/s | 30.39% | 35.41% | 0.10% | 34.09% | 0.00% |
| Glass | 2.50 | kg/s | 83.51% | 2.46% | 0.00% | 14.03% | 0.00% |
| Regolith | 275 | kg/s | 0.56% | 0.32% | 0.00% | 99.13% | 0.00% |
| Electricity | 1,400 | MW | 0.09% | 3.43% | 33.51% | 62.97% | 0.00% |

### 3.2. Regolith mining

Raw regolith is mined in a joint venture of K-Town Water Co., K-Town Concrete and Gravel Co., Mars Silicon Manufacturing Co. (who supplies silicon to Self-Reliant Photovoltaics, LLC), ClearView Glass, Inc., Red Planet Metals Ltd., and Interplanetary Resources, LLC, which primarily mines gold and platinum group metals (PGMs). All these companies share in the cost of mining, and then each extract their portion of beneficiated regolith to produce the materials they need. Nothing is discarded, as the tailings from the silicon, glass and metal processing operations are used as additives in concrete and gravel products. Interplanetary Resources operates a dedicated PGM mine about 50 km the southeast of K-Town where an enriched deposit was discovered two M-years ago, and now produces 15 t/M-yr. (see Exports section). Elements produced as byproducts from other processes, such as chromium, vanadium, nickel, cobalt and copper, are used as additives to make high-quality metal alloys, glasses and ceramics.

Primary regolith mining is accomplished through a combination of downscaled Earth surface mining shovels [7] and trucks [8], suitably modified to operate on $CH_4/O_2$ fuel cells, and chemical explosives. Because of the large quantities of material involved, the extremely small (50 kg) robotic mining rovers [9] used in K-Town's early years have been replaced with heavy-duty teleoperated or directly human-driven machinery. A fleet of two shovels (133 t each) and 13 trucks (50 t each) collectively move 275 kg/s (16 Mt/M-yr.) of regolith while consuming ~40,000 t/M-yr. of compressed $CH_4$ and $O_2$ propellant. Raw regolith is transported to the industrial area where it is processed through a series of interdependent steps, described below.

### 3.3. Water, oxygen and propellant

Each human requires 12 kg of water and 0.84 kg of oxygen daily for sustenance and hygiene in K-Town. In addition, these supplies are needed for livestock, plants and industrial processes. About 26 kg/s of dirty water is treated in a sophisticated chemical-biological hybrid reactor that produces potable water with a 95% overall recovery rate. The makeup water requirements would be modest, except that an additional 17.5 kg/s is needed (along with 18 kg/s of $CO_2$ from the Martian atmosphere) to produce propellant, mainly for spacecraft. Since combustion products from spacecraft propellant cannot be recovered, mining must occur to continually replace them.

K-Town is located at 47 °N where frozen ground water is more abundant than in equatorial regions. The K-Town Water Co. (a collective owned by all citizens) estimate that 8% of the surface regolith by mass is water. Initially, water was extracted merely by placing airtight domes over the surface and applying heat to the subsurface regolith [10], but as this surface water was depleted and more regolith was also required for other industrial processes, the approach shifted to wholesale surface mining, whereby deposits are first fractured into manageable chunks using explosives, and the regolith is then scooped up, crushed to ~1 cm diameter particles, heated mildly to extract volatiles like water, and then passed onto other processes to extract metals and other materials.

The resulting water vapor is filtered, treated via reverse osmosis to remove salts and other impurities, and its pH and hardness are adjusted to make it acceptable for human consumption. Water used for making propellant is further treated to ultrapure standards so it can be electrolyzed without destroying the equipment. Because spacecraft propellant does not need the full stoichiometric amount of oxygen produced through water electrolysis, plenty of leftover oxygen is available, which is more than enough to supply the colony with all its needs, and much more available for export.

Altogether, 275 kg/s of regolith is mined to produce water for the colony. The energy consumption of water production is 61 MW; water treatment, 0.3 MW; propellant production, 382 MW; and liquefaction, 87 MW—bringing total water-based operations to 530 MW, or 38% of total colony power.

### 3.4. Perchlorate and other salts

Martian regolith contains ~0.5% perchlorate ($ClO_4^-$) by mass in the vicinity of K-Town [11]. As it can impair thyroid function, perchlorate is an acute human health hazard, but fortunately, it is very soluble in water, along with other salts (NaCl, etc.) so can be easily removed from soils and dust-exposed equipment by spraying with water. Once isolated, it is biochemically degraded into $O_2$ and chloride ($Cl^-$), which are both non-hazardous and useful. In fact, people venturing onto the surface routinely carry a portable emergency $O_2$ system based on bacterial enzymes to provide 60 minutes of breathable $O_2$, activated when ~6 kg of raw regolith are placed in a bag and water is added [11]. Of the perchlorate processed by the colony, a small amount is also converted into hypochlorite ($ClO^-$), a powerful oxidizing agent otherwise known as bleach, through the use of ionizing (x-ray) radiation. This is the primary disinfectant, along with ozone ($O_3$), used by the colony. Chloride is also used to make industrial chemicals such as chlorine gas, hydrochloric acid and phosgene ($COCl_2$), polyvinyl chloride (see plastics section), and, of course, table salt.

### 3.5. Food and fiber

K-Town grows all of its own food, with enough excess capacity to export ~50% of it to other locations. A wide variety of foods, including grains, fresh fruits and vegetables, herbs, legumes, nuts, animal protein (see below) as well as some specialty crops (such as coffee and tea) are grown. Plants grown for human consumption constitute 55% of total output, with animals consuming the balance; 90% of human nutrition needs are satisfied by plants. Agriculture takes up a total of ~237,000 m$^2$, with livestock requiring 29% of this for grazing. Due to space constraints, livestock only receive 25% of their calories from grazed pasture grasses; the rest comes from intensively-cultivated plants grown along with human food in the main agricultural area. Fig. 5 shows an example of an automated greenhouse in K-Town.

*Fig. 5. Interior view of agricultural area.* Source: [12] (used with permission).

While a purely vegetarian diet would reduce agricultural space, water consumption and electricity needed for artificial illumination, the modest amounts of livestock present is due both to a desire for sustainable animal products (milk, eggs and wool fiber), and a preference to eat meat among both colonists and visitors. Moreover, animal products for export are in high demand, with 50% of these and other food products exported. While the majority of animal protein comes from fish (tilapia, 55%) and insects (crickets, 23%), a small number of land animals (cows, goats, sheep, pigs and chickens) are raised to produce milk (16%), eggs (3%) and meat (3%). On a whole food basis, K-Towners enjoy 644 kg of fresh animal protein per person per M-year.

A small portion of agricultural land, ~31,000 m$^2$ or 13%, is dedicated to growing fiber plants for textiles, wood and paper. Four species are cultivated, producing 188 t/M-yr. in total: hemp (70%), flax (25%), and bamboo and hybrid poplar (2.5% each). Hemp and flax, which reach mature heights of 5 m and 1.2 m, respectively, can be grown in multiple levels. The other plant types (bamboo and poplar) reach nearly the full height of the agricultural cylinders, so are grown in a single level. These woody species are mainly used to produce ornamental and craft wood for furniture and related items, plus some paper.

### 3.6. Organic chemicals and plastics

Organic chemicals (containing carbon, hydrogen, oxygen, nitrogen and/or sulfur) are synthesized from atmospheric $CO_2$ and $N_2$, regolith-mined $H_2O$, and sulfur minerals. Because there are no fossil fuels on Mars, even the simplest molecules such as methane, ethylene and benzene must be synthesized using energy-intensive processes. Once created, however, a wide variety of more complex molecules can be created. Through the use of sophisticated computer modeling and automated chemical synthesis, Terra Tempe Chemical Co. is able to create almost any desired specialty chemical from a limited number of starting materials.

Acheron Polymers, LLC produces ten basic plastics used in K-Town: polyethylene terephthalate, polyethylene, polypropylene, polyvinyl chloride, polystyrene, polycarbonate, polymethyl methacrylate, polyoxymethylene, acrylonitrile butadiene styrene (ABS) and nylon [13], which are synthesized from about three dozen reagents made from simple organic precursors. Biodegradable polylactic acid is produced using plant-based materials, and is used extensively in 3D printing along with ABS. Together, these polymers enable a wide array of uses and significantly reduce the colony's Earth dependence. Although most polymers are not biodegradable, K-Town collects used plastics and reuses, recycles or incinerates to avoid landfill. $CO_2$ and water from incineration are also captured and re-used. Acheron produces 85 t/M-yr. of plastics, and exports ~50%.

### 3.7. Iron and steel

Due to its relatively low melting temperature, after extracting water and other volatiles from regolith, iron is extracted next via crushing and heating to 1,800 °C [14]. The resulting iron oxide (FeO) is then converted to steel using the direct reduction iron process, with small amounts of methane added to provide the carbon necessary for strength, and lime (CaO) produced as part of the cement-making process (see below), added to remove impurities. Other elements, such as chromium, nickel or vanadium, are added to produce steel alloys.

Initially, a small (6.5 t) steel plant based on the Mars Aqueous Processing System developed by Pioneer Astronautics [15] was sent as part of the seed colony equipment, capable of producing 136 t/M-yr. of steel. This steel, along with concrete from the small concrete plant, and a number of specialized materials shipped from Earth, was then used to construct a full-scale steel plant requiring 164 t of steel, 49 t of concrete and 8 t of other materials [6]. This plant was capable of producing ~1,400 t/M-yr. of steel, and consumed 3.7 MW of energy. A second such steel plant was built five M-years ago, and a larger version is due to be built next year, that will once again double the total steel output capacity.

### 3.8. Cement, gravel and concrete

Cement is produced on Mars through the combination of lime ($CaO$), silica ($SiO_2$) and alumina ($Al_2O_3$), with smaller amounts of magnesia ($MgO$). All of these compounds are found in Martian regolith, but with much lower amounts of $CaO$ and $Al_2O_3$ than required. By first heating the regolith to separate the lower-melting $FeO$, and, to a lesser extent, $Al_2O_3$, $MgO$ and $SiO_2$, a mixture with composition of approximately 52% $CaO$, 31% $Al_2O_3$, 15% $SiO_2$ and 2% $MgO$ is obtained [14]. This cement is then mixed with coarsely crushed regolith (gravel) and water to produce concrete.

Similar to the initial steel plant, a small cement plant was initially shipped to Mars, with a mass of 5.5 t and an output capacity of 210 t/M-yr. of cement (or 1,900 t/M-yr. of concrete). This plant was used, along with steel from the initial steel plant and specialized parts shipped from Earth, to construct the full-scale cement plant in use today. This plant required 83 t of steel, 9 t of concrete and 4 t of other materials, produces ~14,500 t/M-yr. of cement, and consumes 21 MW (mostly in the form of $CH_4$ and $O_2$ combustion). Like the steel plant, a second, larger version of the concrete plant is currently under construction.

### 3.9. Glass

Glass is produced through a combination of $CaO$ and $SiO_2$, with small amounts of sodium oxide ($Na_2O$) other components [16]. The first pair of glass plants were built primarily from concrete and steel supplied by the initial plants, plus 3.6 t each of specialized materials imported from Earth. After this small initial capacity went online, a total of five larger plants were constructed, each with a mass of 400 t. The total capacity now produces ~150,000 t/M-yr. of glass and consumes 7 MW of electricity.

### 3.10. Aluminum, magnesium and other metals

In the process of making cement, large amounts of $Al_2O_3$ and $MgO$ are produced, which are the starting points for making aluminum and magnesium, two lightweight industrial metals used in smaller quantities in K-Town for construction, electrical equipment and other applications. Aluminum is produced via the FFC Cambridge process [17], which requires significant amounts of electricity similar to the Hall-Héroult process on Earth, but without the need for cryolite ($Na_3AlF_6$) that is difficult to obtain on Mars due to the relative scarcity of fluorine. Instead, calcium chloride ($CaCl_2$) is used to dissolve the alumina, and due to the large quantities of chlorine available from perchlorate, this material is readily produced. Magnesium metal is made via the Pidgeon Process [18], whereby $MgO$ is reacted with silicon metal (see PV manufacturing below) at high temperature to produce magnesium and $SiO_2$. The normally unfavorable thermodynamic equilibrium is driven toward completion by distilling away magnesium vapor as it forms.

### 3.11. Solar photovoltaic manufacturing, power plant and heat rejection

An early goal of K-Town was to establish its own solar photovoltaic (PV) manufacturing capability, so beginning seven M-years ago, Self-Reliant Photovoltaics, LLC began making their own PV panels from Martian regolith, based on earlier lunar systems [19]. While not as efficient as those supplied from Earth, the cells are much cheaper to manufacture, and were used to greatly accelerate K-Town's energy generating capability, since the plant can produce many more panels annually than are needed to replace those powering the plant as they wear out. Like the other industrial facilities described here, an initial PV manufacturing plant of mass 9 t was flown to Mars, whereupon it began producing 662 t/M-yr. (~30,000 $m^2$/M-yr.) of solar PV cells. This was eventually replaced by the current plant, which has a total mass of 266 t, consumes 237 MW of electricity, and produces ~40,000 t (1.8 million $m^2$) of solar PV cells per M-year.

Electricity is generated primarily through the large PV array located on the outskirts of K-Town. This plant produces 1.4 GW continuous output from a 3.8 GW peak power system, and takes up an area of 86 $km^2$ (including gaps between panels) located southwest of K-Town. The system stores 20% of the power it generates in a high-efficiency battery system for nighttime use, and an additional 5% in a reversible fuel cell system for long-term contingency storage during dust storms, etc. About one-quarter of the array are 40% efficient cells from Earth, and the rest are the 20% efficient, locally-made cells from Self-Reliant. All solar PV is now made locally.

A set of five 10 MW nuclear reactors left over from the earlier days of the colony are maintained as a secondary source of contingency power, along with some experimental wind turbines. It is expected that the use of the nuclear reactors will be phased out eventually, as ample supplies of $H_2/O_2$ fuel cells are viewed as providing all the security the colony would ever need, even if many systems fail.

Even on frigid Mars, the combination of waste heat from devices operated within K-Town, solar gain from numerous windows, and excellent insulation from the thick shielding, result in a net positive thermal balance; heat must be removed from the colony to prevent overheating. Air ducts throughout the colony pick up heat, and additional sources of high-temperature heat are injected into the air stream prior to exiting the colony. This air is then directed into a large underground heat exchanger where the energy is rejected to the –63°C Martian subsurface [20]. During this process, most of the moisture is condensed and removed, and is subsequently sent to the water processing plant for recovery and reuse.

### 3.12. Spaceport

The main function of the spaceport is to facilitate interplanetary transport between Earth and Mars. While the vast majority of the ~2,900 spacecraft per M-year (about 4.5 per sol) that arrive or depart carry cargo, K-Town also received 423 visitors from Earth during the last synodic cycle and returned 264. About 75 of these arrivals were spacecraft crew, and 44 returned to Earth. The crew-to-passenger ratio for these roughly 180-day journeys is about 20%. Crew perform many essential functions including command, navigation, communications, safety, maintenance, sanitation, exercise training, medical care, conflict resolution, food preparation, entertainment, and even haircuts, are cross-trained to provide redundancy across multiple shifts and in case of incapacity.

In addition to providing access to space, the spaceport serves as a transportation hub between K-Town and the rest of Mars, via both Hyperloop train lines and suborbital rocket "hops." Four main train lines emanate from the spaceport, connecting with K-Town to the east, and various destinations to the north, west and south. The spaceport also operates the emergency medical shuttle services described under medical facilities. Altogether, surface travel makes up about 50% of spaceport passengers.

When spacecraft flights are not arriving from or departing to Earth, roughly 100 people pass through the spaceport each M-week, and during Earth-Mars launch windows, this number skyrockets to nearly the population of K-Town itself over the ~4 M-week period. During these peak times, spaceport staff is increased from its normal level of 55 people to 88, with most of the extra personnel provided by spacecraft crew arriving from Earth, and some seasonal K-Town labor.

### 3.13. Mars internet

Due to significant time delays between Earth and Mars even at closest approach, a live internet linking the two planets is not possible. As a result, a "store and forward" protocol, in use since the early 2000s, was adopted, enabling efficient, secure and fault-tolerant interplanetary communications [21] for text, e-mail, ftp, video, and even financial transactions. Data transfer employs a trio of lasercom links: one located just outside K-Town (250 Gbit/s average, consuming 100 MW), with two in remote bases in the southern highlands of Noachis Terra (30°E) and Terra Cimmeria (150°E) (each 25 Gbit/s, 10 MW), connected to K-Town via buried fiber optic cable (using ZBLAN to reduce the number of repeaters needed). This arrangement provides redundancy and nearly continuous line-of-sight with Earth, while avoiding reliance on large, power-hungry, vulnerable satellite links.

Moreover, Martian citizens led by K-Town invested heavily in a 50 MW data center that, together with an enormous data store, runs local copies of Earth's most important websites, providing instant access. Non-time critical data transfer is provided by regular shipments of storage devices from Earth; together with other computer hardware, internet support equipment now makes up ~80% of imported materials.

## 4. Economic considerations

### 4.1. Overview

K-Town's economic system was designed with five interrelated goals in mind:

1. Provide citizens with economic freedom;
2. Minimize distortions and maximize allocative and productive efficiencies;
3. Minimize rent-seeking and economically unproductive activity;
4. Encourage savings, and therefore investment, for largely self-sustaining economic growth;
5. Prevent labor shortages or surpluses.

The first goal is achieved through a universal basic income (UBI)—see details below. By ensuring citizens access to a source of funds, they are free from anxiety over their ability to live (particularly important in an environment like Mars, where even air is economically scarce) and able to pursue activities which they believe are valuable, even if not financially lucrative. The second goal is achieved by using depreciating licenses and a land value tax as the main sources of government revenue, rather than other distortionary taxes, such as on income or labor. Depreciating licenses contribute to the third objective, as does the cooperative ownership structure of many K-Town enterprises. The fourth is achieved by the Central Bank of K-Town as part of their mandate, and is critical to the self-sustainability of K-Town's economy. The fifth is achieved through a combination of immigration policy and the use of UBI to offset the need for living wages.

At a high level, K-Town's economy was approximated using a Solow model of economic growth. The model describes a "bootstrapped" economy: an initial infusion of resources from outside its economy is provided, after which the economy grows using almost entirely its own resources. The key idea is that a substantial amount the output produced in each time period is reinvested in additional capital stock, with which to produce more output in the future. Similar models have been used to study space colonies at Sun-Earth L5 [22] and on the Moon [23], where an initial infusion of resources from Earth is used to springboard a mostly self-sustaining space colony. While K-Town does trade with Earth (and other locations), such trade is not a central driver of K-Town's economy.

K-Town's industrial base includes some heavy industries that on Earth are very greenhouse gas intensive, and now tend to be shunned due to their climate impacts. These include mining of PGMs and gold, which consume ~10,000 as much energy per kg as more common materials like steel. A number of other elements including some REEs, whose carbon footprints have recently grown nearly as large, will likely become important off-Earth activities in the near future. (See the section below for details.) Products from these industries tend to be traded with Earth, though they are also extremely useful to free-space colonies, other settlements on Mars, as well as K-Town's own development.

One of the central challenges facing the initial K-Town settlers was the question of legal systems. Under the Outer Space Treaty, citizens of any given country are held to account under their country's laws. On Mars, with settlers from multiple Earth nations, such a fragmented legal system creates barriers to economic efficiency. To provide predictability and consistency, K-Town established itself as a sovereign entity with its own legal system and jurisdiction. Anyone residing within the K-Town Basin boundary must adhere to its laws, while citizenship is granted over time to immigrants and, of course, children born there.

### 4.2. Monetary policy

K-Town's sovereignty also allows it economic flexibility to maintain its own currency (K-Town dollars, or K$) and monetary policy, which is conducted by the Central Bank of K-Town (CBK), a lender of last resort that is independent of the K-Town government. The CBK allows the exchange rate between K$ and other currencies to float freely within a band that depends on the needs of the moment and the foreseeable next five M-years. K-Town is generally open to free flows of capital in and out of the town, subject to the CBK's guidance on how capital flow policy should be set to support the monetary policy.

Similar to the U.S. Federal Reserve (Fed) system's dual mandate, the CBK has two objectives. Like the Fed, it is required to maintain price stability, and does this by targeting a low rate of inflation (typically around 2%). Unlike the Fed, the CBK does not aim to ensure full employment, but rather aims to target the savings rate (currently around 10%). Note that the "savings" targeted by the CBK is not the same as individual consumer saving. Such savings by an individual consumer, e.g., by putting money under a mattress, reduces the level of economic activity by reducing demand for goods and services. Savings in the aggregate, such as purchases of goods produced in prior periods or purchases of interest-bearing assets such as bonds, promotes economic activity generally and investment in particular. Such investment is critical for K-Town's economy to grow.

A major issue facing central banks on Earth in the post-2008 crisis period was the "zero lower bound" on interest rates: in general, central banks were unable to move nominal interest rates below zero, limiting their ability to stimulate the economy. To avoid this situation, K-Town uses "cashless" electronic currency, allowing the CBK to charge negative interest rates when necessary.

The CBK also serves the key function of providing general, unbiased, economic analysis for the region's needs. When a new project or policy (or a revision to an existing one) is being considered, the CBK provides rigorous analysis that is free of conflicts of interest. Toward that end, CBK employees are prohibited from owning shares in any specific K-Town asset. Instead, they are required to invest in a blind randomized portfolio (similar to a blind trust) of K-Town assets, where the randomization is conducted by a computer program stored in a secure facility running open-source (and therefore transparent) code. This incentivizes analysts to provide the best analysis possible, as they are equally likely to be a part of any disruptive new innovation as they are to be a part of incumbent institutions.

Generically, the health of the investment portfolio is correlated with K-Town's broad economic well-being.

### 4.3. Immigration policy

K-Town's immigration policy is best described as "flexible, but conservative." While tourism is encouraged, the town's stance on permanent immigration depends on the CBK's assessment of the current labor market and the projected labor market situation over the next two M-years. Where possible, temporary needs are filled through automation or by hiring visitors who are willing to work for a reasonable wage. Where necessary, longer-term needs are met by encouraging immigration through a combination of employment guarantees and land grants. In the most recent planning period, the CBK forecasted a need for 200 new workers (and their associated families) per synodic cycle. However, only 80% of this need was approved, resulting in 160 new immigration permits. But because 50 citizens chose to revoke citizenship and return to Earth during the last Hohmann transfer window, the total permits were increased again to 210 people. In fact, a total of 209 new immigrants arrived—very close to the CBK target.

So far, K-Town has not experienced a recession which has created a permanently-lower need for labor. However, given the political and practical difficulties of "de-immigrating" unnecessary labor, K-Town's immigration encouragement policy errs on the side of fewer rather than more immigrants. The incentives of these asymmetric adjustment costs may cause problems in the future, and will likely require further economic analysis and legislative action as K-Town grows.

### 4.4. Depreciating licenses

Given the importance of economic efficiency on Mars and the chance to start afresh, K-Town decided to implement a radical scheme of ownership for capital assets and real estate: depreciating licenses [24]. Such licenses solve issues of inefficient investment in asset development as well as inefficient production decisions, while traditional private property as practiced in most places on Earth only address the latter issue.

The idea is a simple tweak on private property: rather than granting the "owner" of a property a license to that asset in perpetuity, they are granted a potentially-indefinite license to the asset, which depreciates over time at an annual rate ($\tau$), with that share passing to the government. For example, in the first M-year after an asset's purchase, if $\tau = 5\%$, the government would receive a 5% ownership share in the asset. The licenses are available for auction at the valuation set by the current license holder once every fiscal year. If any prospective owner submits a bid greater than the reserve value, the current owner is required to sell to the prospective owner for the value bidded. To avoid incentives to assign arbitrarily high valuations, the current owner is required to repurchase the depreciated $\tau$ share of ownership from the government at their assessed price. The cost of repurchase is effectively a self-assessed license fee.

K-Town does not apply depreciating license ownership to all assets. Rather, the mechanism is concentrated on goods with public-resource qualities, like land and natural resource rights. Financial investments such as bonds and annuities, consumption goods such as food, and physical assets such as buildings and vehicles operate under a traditional private property model. To ensure people security in their residences, homes are treated as a combination of consumption and investment goods. An individual's primary place of residence is treated as an asset providing a stream of consumption services, giving them private property-like rights to reside without worrying about licenses. Any additional properties they own are treated as investment goods, and subject to the depreciating license structure. Renters in good standing are guaranteed the rights to completion of their occupancy agreement regardless of changes in building or land ownership.

### 4.5. Land value tax (LVT)

The highest-level principle of public finance is "tax the least-elastic decisions first." This minimizes deadweight loss from the tax discouraging economic activity while raising required revenues for public services. Economists on Earth have generally regarded the availability of land as inelastic with respect to taxes: no matter the tax rate on land, the available stock of land on a planet is unlikely to change (though terraforming Mars could change this dramatically). An LVT, most often associated with economist Henry George, capitalizes on this feature of land by taxing its unimproved value.

LVTs have several desirable properties. First, since the availability of land is essentially inelastic with respect to taxes, they do not discourage productive economic activity. Second, since they are levied only on the unimproved value of land, they provide land rights holders with incentives to maximize the land's productivity. Third, since the tax burden scales with the amount of land held, they discourage the type of asset concentration which leads to individual entities acquiring market power, and encourage a more equal distribution of resources throughout society.

Depreciating licenses applied to land implements an LVT. While LVTs on Earth suffer from incentive problems in valuation—the incentives of the assessor may not be aligned with society's interest in accurate valuations—in K-Town, the tax/auction structure of depreciating licenses encourages those with the most information about the value of unimproved land to value it as accurately as possible.

To allow people time to learn and incentives to start acquiring land, K-Town initially divided land into arc-shaped plots (25 m wide by ~22-25 m long) immediately around the 400 m-dia. central region—which itself was divided into ~30 irregularly-shaped parcels—with survey information attached, and auctioned the plots to potential colonists at low, but still revenue-raising, prices. Land further out is divided into one-hectare square plots. K-Town administers all land within the entire K-Town Basin—about 3.4 million parcels in total. The caveat, to ensure that potential colonists intended to use the land productively, was that the purchaser was required to claim the land in person on Mars within the arrival time from the next two launch windows. This system was phased out for existing properties, and the full depreciating license system for land phased in, at the end of 2045. New land grants to encourage immigration are given a 5 M-year grace period of depreciating license exemption.

### 4.6. Universal Basic Income (UBI)

First popularized by U.S. presidential candidate Andrew Yang in 2019, the UBI is a familiar concept that many Earth nations have embraced. All citizens of K-Town are entitled to a monthly stipend administered by the CBK. Note that tourists and other non-permanent residents (e.g., non-citizens) are specifically excluded from the UBI; if they run out of funds, they are offered to either become citizens and therefore eligible for the UBI, or are sent back to Earth. This UBI stipend is designed to provide a minimum subsistence standard of living; it is not glamorous by any means. Indexed annually to the cost of living in K-Town, it affords one a basic (one-room) residence; water, air, heat, electricity and internet; and adequate sanitation, nutrition, health insurance and pre-college education. Universal health insurance ensures that nobody is denied care for lack of funds, and college education is subsidized to ensure that everyone who works can afford it. The full value of the stipend is awarded to every citizen over 9.5 M-years (about 18 E-years) of age; parents of minors are provided with a per-child supplement that is always less than the full adult amount.

### 4.7. Exports and competition with other space resources

Earth has superior manufacturing and supply chain capabilities, but its deep gravity well makes it more expensive to ship anything into space if it can be made elsewhere. The lower gravity of the Moon and asteroids, however, present in some cases challenging economics for Mars to compete against. The main advantages Mars has over other space resources are twofold: 1). superior manufacturing capabilities in some areas, due to a larger human presence and more developed infrastructure, and 2). abundant carbon (C) and nitrogen (N), two elements very limited on the Moon and many asteroids.

In the first category, materials such as metal alloys, glasses and ceramics, as well as complex equipment such as rocket engines are either not possible to make, or can be made in only limited quantities, in locations other than Earth and Mars. K-Town has over the last 5 M-years begun to make such high-quality materials, and its export business to both Mars and Earth orbits is growing.

In the second category are many C- and/or N-containing materials including food, fiber, plastics, organic chemicals, fuels, fertilizers, and $N_2$ used in breathing air and inert blanket gas. (Argon, which is present in Mars' atmosphere at about half the concentration as $N_2$, is also used for this purpose.) K-Town's greenhouses grow twice the food and fiber required for its citizens, with the balance being exported to remote locations on Mars and in space. Non-biological materials containing C and N are also produced in excess of K-Town's need, with the balance exported. Moreover, water and oxygen, while not competitive with the Moon on cost, is exported to Mars orbital locations as well as some nearby asteroids.

There are also some materials that are now cheaper to make in space than on the Earth, such as PGMs and gold, that require enormous amounts of energy to purify and take a high toll on Earth's environment. Production of these materials on Earth is now strongly discouraged in favor of importing these materials from space. While many of these materials are also available from asteroid and lunar resources, K-Town's superior material independence has driven down operational costs, whereas the Moon and asteroids still tend to be heavily dependent on Earth for basic supplies such as food, labor and even building materials, which are very expensive to ship there. Energy is also now cheaper in K-Town that just about anywhere other than Earth, thanks to the completion of its 1.4 GW power plant built from indigenous solar PV. About 1% of Earth's PGM demand and 0.1% of gold demand are now supplied by K-Town mines. Jeff Bezos' vision of moving heavy industry into space [25] is starting to become reality.

Other elements, such as xenon, tantalum, indium and gallium are still a few years away from profitability, as are the REEs (particularly neodymium and praseodymium, used in wind turbine generators and maglev trains) [26]. While not yet economical, K-Town (and Mars in general) is beginning to invest in mining operations of these elements, as well as atmospheric separation to produce liquefied xenon gas. About the only energy-intensive activity that K-Town has not yet begun investing in is the fabrication of integrated circuits, where the Moon has had a considerable head start.

Finally, some materials made in K-Town are valuable simply because they're Martian. Examples include wine from Terra Tempe vineyards, wool sweaters from K-Town's sheep herd, and wooden objects produced from K-Town's small tree plantation, as well as ordinary Mars rocks that collectors on Earth can't seem to buy enough of. A handful of investors on Earth are also somewhat crazy about Mars gold, and are willing to pay a premium for it over Earth-mined gold. While exports of "Made on Mars" goods aren't as large as other types, it represents an important additional income source to the colony.

The vast majority of exports is in fact propellant: specifically, liquid methane and oxygen, which make up ~75% of exports by mass. Table 2 lists all of the currently exported commodities from K-Town.

*Table 2. Commodities exported to Earth and orbital destinations (2019 U.S. $)*

| Item | Destination(s) | Mass flow (t/M-yr.) | Gross revenue ($M/M-yr.) | Costs ($M/M-yr.)*** | Profit ($M/M-yr.) |
|---|---|---|---|---|---|
| Metals and metal parts | Orbital** | 5,586 | 729.0 | 656.1 | 72.9 |
| Water | Orbital** | 4,118 | 537.5 | 483.7 | 53.7 |
| Oxygen | Orbital** | 5,713 | 684.6 | 616.1 | 68.5 |
| Nitrogen/Argon | Orbital* | 22,851 | 3,488.3 | 3,139.5 | 348.8 |
| Food (dry) | Orbital* | 680 | 113.1 | 101.8 | 11.3 |
| Propellant - $LCH_4$ | Orbital* | 27,931 | 4,896 | 4,407 | 490 |
| Propellant - $LO_2$ | Orbital** | 100,086 | 12,824 | 11,542 | 1,282 |
| Plastics | Orbital* | 40.0 | 6.6 | 6.0 | 0.7 |
| Organic chemicals | Orbital* | 79.9 | 13.3 | 12.0 | 1.3 |
| Fiber | Orbital* | 93.1 | 15.5 | 13.9 | 1.5 |
| PGM | Orbital* | 9.13 | 373.0 | 335.7 | 37.3 |
| Mars gold | Orbital* | 5.93 | 382.3 | 191.6 | 190.7 |
| "Made on Mars" products | Orbital* | 18.8 | 17.3 | 4.5 | 12.8 |
| Total | | 167,212 | 24,081 | 21,509 | 2,572 |

*45% Earth orbits, 35% Mars orbits + Lagrange points, 10% nearby asteroids, 10% Mars surface

**Only non-Earth orbits: 70% Mars orbits + Lagrange points, 20% nearby asteroids, 10% Mars surface

***Includes 90% of launch costs + production costs

### 4.8. Economic growth overview

Export revenues make up ~10% of K-Town's economy, which is expected to be nearly $250 billion/M-year by the end of 2049 (note all costs are expressed in 2019 U.S. dollars). This value is close to what K-Town economists estimated back in 2030 when the colony was established, with an initial seed investment of $250 million. They used a Solow model with labor-augmenting technology to project

K-Town's long-run growth and the initial investment required in 2030 to get to the 2049 level. The model consists of three main mathematical features:

1. A production function which related capital and labor stocks to value of output
2. A law of motion for the capital stock
3. An exogenous savings rate (set by the CBK) and exogenously-assumed labor and labor productivity growth rates

The production function is of a Cobb-Douglas type with labor-augmenting technologies and constant total returns to scale in capital and labor. This implies that capital and labor are complements in producing valuable output, and that there are decreasing returns to scale in either input individually. That is, doubling only the amount of capital available while holding labor constant will result in less than double the output produced, and similarly for labor, while doubling both capital and labor will double the total amount of output produced. The long-run growth of this economy will be driven by growth in the productivity of labor-augmenting technologies.

The economists took a three step approach to calculating the initial investment level required to achieve the target size of K-Town's economy in 2049. First, they calibrated the parameters using calculations from ASTER (see section 3 above), stated assumptions about K-Town's population, and historical analogy to similar economies (primarily Japan and post-war "Asian Tiger" economies). Second, they used these calibrated values with the export numbers from ASTER to estimate K-Town's Gross Domestic Product (GDP) and the growth rate of GDP in 2049. Third, they used the estimated K-Town GDP and its growth rate to project backwards to 2030 to back out the initial investment of productive assets. This approach yielded an initial investment of $250 million in 2030, which was sufficient to cover initial seed equipment of habitat modules and conversion of Martian resources into construction materials.

It is important to note that the 2030 estimate from this process was only for productive assets set up on Mars. It did not include the cost of transporting those resources to Mars, nor one-time fixed costs such as acquiring intellectual property licenses, non-recurring engineering expenses, or imports of supplies before the colony became productive. These additional costs were estimated to be $2 billion in 2030, making the total initial investment $2.25 billion.

With an economy now >100x this size per M-year, this was a debt that has easily been paid and now returns substantial dividends to early investors. K-Town currently receives capital inflows valued at approximately $30 billion/M-year, and is one of the most attractive off-world investments available.

## 5. Political, organizational and social aspects

### 5.1. Governmental organization

As mentioned earlier, K-Town is a sovereign governmental entity separate from any Earth entity. Its jurisdiction applies to any person or equipment residing within the borders of K-Town Basin as depicted in Fig. 1. Its authority derives from its Constitution, which is now recognized by almost every nation on Earth, and is organized as a representative democracy with a unicameral legislature, similar to many smaller nations on Earth. These representatives are elected every M-year through district voting. All citizens who have reached 9.5 M-years of age (roughly 18 E-years), including those living abroad, have the right to vote. The legislature currently consists of 13 representative, or one per ~50 citizens, of which there are 656 in total. As K-Town grows, its Constitution will eventually cap the number of representatives at 101 (always an odd number to break ties).

A mayor is separately elected through direct vote of all citizens every two M-years. The mayor's office, plus associated governmental offices, employ a total of 12 people. The police department consists of three people (sheriff and two deputies) while the fire department employs two people, relying heavily on volunteers in case of emergency. Foreign embassies of eight Earth governments plus the United Nations constitute an additional 18 people.

### 5.2. Democracy 2.0: Quadratic voting system

The founders of K-Town wanted to ensure that all citizens had equal representation, but recognized that individuals would have different strengths of feeling on different issues and would want to be able to express those feelings. To achieve this, the founders enshrined Quadratic Voting (QV) into K-Town's Constitution.

Rather than a "one-person-one-vote" assignment, QV gives each voter a stock of non-transferrable, non-storable "vote credits" for each voting occasion. The voter is then free to allocate credits in favor of or against any issue on a ballot, with the cost of vote credits being quadratic in the number of votes allocated to a particular issue. For example, expressing one vote ("for" or "against") on a generic ballot issue would cost one vote credit, while two votes would cost four vote credits, three votes would cost nine vote credits, and so on. This system allows voters to express not only a "for" or "against" preference on issues, but also the strength of their preference. Given a fixed stock of vote credits, it forces voters to prioritize the issues that are most important to them. In general, this ensures that voters with the strongest opinions on any given issue can be heard even if they find themselves in the minority. Giving each voter the same initial allocation of vote credits ensures that all voters have an equal say in the process *ex ante*.

QV fixes a number of limitations with voting systems still in use on Earth, many of which fall under the heading of "problems of intense preferences" [27]. For example, one-person-one-vote systems tend toward tyrannies of the majority, with large numbers of voters with weak preferences able to drown out the voices of minorities with strong preferences. In situations where relevant domain knowledge is likely to be concentrated among a minority of experts, this can lead to significant welfare losses when uninformed majorities choose scientifically-poor options. Similar problems emerge in cases of minority rights, when the majority is either unaware of or indifferent to problems faced by a minority of the public. One way to think of the difference between QV and one-person-one-vote methods is that the latter approaches all *ration* votes, while QV *prices* them. At a high level, the efficiency gains from QV can be viewed as gains from removing an artificial supply constraint and moving to marginal cost pricing. Although QV is considered a radical concept on Earth [28], economic research in the 2010s established QV as an approximately optimal system of voting under general assumptions [29].

### 5.3. Timekeeping

Mars' day ("sol") is only slightly longer than Earth's. While several proposed systems were discussed and debated prior to the first human arrivals, early settlers quickly converged on a 24-hour clock identical to Earth's except for a "timeslip" of 39.587 min. once per sol [30] that K-Town, by dint of its seniority to other settlements, has chosen by overwhelming vote to take place at midnight. Settlements located in different time zones around Mars observe the timeslip at a different local hour to maintain synchronized hours across the planet, with a few exceptions. Generally, people use the timeslip to socialize, perform rituals or sleep; work during this period is considered taboo.

One M-year consists of 668.6 sols. Settlers wanted to preserve the familiar 12-period calendar from Earth, so they created M-months consisting of either 55 or 56 sols, and known by the same names as months on Earth. Eight M-months have 56 sols and three (May, August and November) have 55 sols. February has 55 sols for two M-years out of five, and 56 sols for the other three, in a pattern reminiscent of Earth's leap day. January 1 is designated as Mars' winter solstice ($L_s = 270°$).

After a period of experimentation, the citizens of K-Town and most other settlements on Mars enthusiastically chose an eight-sol M-week as being more conducive to recreation and recharge, offering an entire extra sol each M-week to do with as they please. As a result, M-months consist of (nearly) seven M-weeks each. Most businesses adhere to a five-sol work week as on Earth, with a three-sol weekend, but many self-employed people choose to work four sols followed by four sols of recreation (or work every other sol), claiming it results in better overall productivity. Many dedicate the "extra" sol to community service, and important civic functions such as elections always take place on that day.

### 5.4. Culture

Being dedicated to building a positive future on Mars, the culture of K-Town is one of intense discussion and debate about the best ways to accomplish things technically, economically and socially. Recognizing the many past and present failings of human society on Earth, K-Town citizens are committed to making things better by trial-and-error, self-criticism, and "failing fast." People have learned to critique the method or outcome and not the people, though exceptions still occur. Storytelling in particular is used as a means of developing intuitive models for visualizing the future, and its practice is widespread throughout all aspects of society. Science fiction as well as historic accounts of Earth-bound explorations are routinely drawn upon, and sophisticated mathematical models are frequently used for forecasting.

K-Town is an exciting place to live, and looks forward to growing in size and stature. Come join us!

## 6. Acknowledgments

The authors thank Jason Aspiotis, Andrew Granatstein, George Lordos, Brad Neuberg, Raluca Ostasz, Doug Plata, Kai Staats, Brian Unger, Bryan Versteeg and Eric Ward for contributing to the report.

## 7. References

1. *Wikipedia*, 2018. "Arcadia quadrangle." Last edited 13 Nov. https://en.wikipedia.org/wiki/Arcadia_quadrangle. Accessed 2 Feb. 2019.
2. Versteeg, B., "Crater dome cam 2," Space Habs. http://spacehabs.com/mars-gallery/#gallery-43873/19. Accessed 9 Feb. 2019.
3. Centers for Disease Control and Prevention, "Zoonotic Diseases," U.S. Department of Health & Human Services. Last reviewed July 14, 2017. https://www.cdc.gov/onehealth/basics/zoonotic-diseases.html. Accessed 20 Mar. 2019.
4. *Wikipedia*, 2019. "Greater Los Angeles." Last edited 19 Jan. https://en.wikipedia.org/wiki/Greater_Los_Angeles. Accessed 2 Feb. 2019.
5. *Wikipedia*, 2019. "Arches National Park." Last edited 16 Jan. https://en.wikipedia.org/wiki/Arches_National_Park. Accessed 2 Feb. 2019.
6. Greenblatt, J. B., 2018. "ASTER tool presentation to NASA," Emerging Futures, LLC. http://www.emerging-futures.com/single-post/2018/10/07/ASTER-tool-presentation-to-NASA. Accessed 9 Feb. 2019.
7. Caterpillar, 2019. "6015B," Specifications. https://www.cat.com/en_US/products/new/equipment/ hydraulic-mining-shovels/hydraulic-mining-shovels/1000012100.html. Accessed 30 Mar. 2019.
8. Caterpillar, 2019. "785D Mining Truck," Specifications. https://www.cat.com/en_US/products/new/ equipment/off-highway-trucks/mining-trucks/18089285.html. Accessed 30 Mar. 2019.

9. Mueller, R. P., R. E. Cox, T. Ebert, J. D. Smith, J. M. Schuler, A. J. Nick, 2013. "Regolith Advanced Surface Systems Operations Robot (RASSOR)," NASA Technical Reports Server, https://ntrs.nasa.gov/archive/nasa/casi.ntrs.nasa.gov/20130008972.pdf. Accessed 30 Mar. 2019.
10. Kornuta, D., et al., 2018. *Commercial Lunar Propellant Architecture: A Collaborative Study of Lunar Propellant Production.* https://www.philipmetzger.com/wp-content/uploads/2018/11/Commercial- Lunar-Propellant-Architecture.pdf. Accessed 9 Feb. 2019.
11. Davila, A. F., D. Willson, J. D. Coates, C. P. McKay, 2013. "Perchlorate on Mars: a chemical hazardand a resource for humans," *International Journal of Astrobiology* 12(4), DOI: 10.1017/S1473550413000189. https://www.researchgate.net/publication/242525435_Perchlorate_on_Mars_A_chemical_hazard_and_a_resource_for_humans. Accessed 8 Feb. 2019.
12. Versteeg, B., "Mars-farm-cam-22-Aa," Space Habs. http://spacehabs.com/mars-gallery/#gallery-43873/ 18. Accessed 9 Feb. 2019.
13. Creative Mechanisms, 2016. "The Eleven Most Important Types of Plastic," blog, July 21. https://www.creativemechanisms.com/blog/eleven-most-important-plastics. Accessed 5 Sept. 2018.
14. Hashimoto, A., 1983. "Evaporation metamorphism in the early solar nebula-evaporation experiments on the melt $FeO$-$MgO$-$SiO_2$-$CaO$-$Al_2O_3$ and chemical fractionations of primitive materials," *Geochemical Journal* 17:111-145. https://www.terrapub.co.jp/journals/GJ/pdf/1703/17030111.PDF. Accessed 29 Mar. 2019.
15. Berggren, M. R. Zubrin, C. Wilson, H. Rose, and S. Carrera, "Chapter 21: Mars Aqueous Processing System," In: *Mars: Prospective Energy and Material Resources*, V. Badescu, ed., Bucharest, Romania, ISBN 978-3-642-03628-6.
16. Johnson, R. D. and C. Holbrow, 1975. "Appendix J," in: *Space Settlements: A Design Study*, National Aeronautics and Space Administration Summer Study, Ames Research Center, NASA SP-413. https://settlement.arc.nasa.gov/75SummerStudy/4appendJ.html. Accessed 29 Mar. 2019.
17. Tripuraneni Kilby, K. C., L. Centeno, G. Doughty, S. Mucklejohn, and D. J. Fray, 2006, "The Electrochemical Production of Oxygen and Metal via the FFC-Cambridge Process," *Space Resources Roundtable VIII*, Lunar and Planetary Institute, 31 October–2 November. https://www.lpi.usra.edu/meetings/roundtable2006/pdf/tripuraneni.pdf. Accessed 29 Mar. 2019.
*18.* Wikipedia, 2018. "Pidgeon Process." Last modified 28 Sept. *https://en.wikipedia.org/wiki/Pidgeon_process. Accessed 30 Mar. 2019.*

19. *Ignatiev, A. and E. Caroll, 2018. "Lunar Vacuum Deposition Paver," Survive and Operate through the Lunar Night Workshop*, Lunar and Planetary Institute, Columbia, MD, 13 November. https://www.hou.usra.edu/meetings/survivethenight2018/eposter/7013.pdf. Accessed 30 Mar. 2019.
20. Williams, D. R., 2018. "Mars Fact Sheet," NASA Goddard Space Flight Center. Last edited 27 Sept. 2018. https://nssdc.gsfc.nasa.gov/planetary/factsheet/marsfact.html. Accessed 9 Feb. 2019.
21. Mann, A., 2013. "Google's Chief Internet Evangelist on Creating the Interplanetary Internet," *Wired*, 6 May. https://www.wired.com/2013/05/vint-cerf-interplanetary-internet/. Accessed 4 August 2019.
22. O'Neill, G.K., 1979. *The High Frontier: Human Colonies in Space*. Morrow.
23. Sandler, T. and Schulze, W., 1981. "The economics of outer space." *Natural Resources Journal*, *21*(2), pp.371-393.
24. Weyl, E.G. and A. Zhang, 2018. "Depreciating Licenses". *SSRN* https://papers.ssrn.com/sol3/papers.cfm?abstract_id=2744810. Accessed 30 Mar. 2019.
25. Stella, R., 2016. "To save Earth, Jeff Bezos wants to move heavy industry to outer space," *Digital Trends*, 1 June. https://www.digitaltrends.com/cool-tech/jeff-bezos-says-big-industry-factories-should- be-built-in-space/. Accessed 7 Feb. 2019.
26. *Critical Materials Institute*. "10 Things You Didn't Know About Critical Materials. Ames Laboratory," U.S. Department of Energy. https://cmi.ameslab.gov/materials/ten-things. Accessed 2 Feb. 2019.
27. Posner, E.A. and Weyl, E.G., 2015. "Voting squared: Quadratic voting in democratic politics." *Vand. L. Rev.*, *68*, p.441.
28. Lalley, Steven P., and E. Glen Weyl. 2018. "Quadratic Voting: How Mechanism Design Can Radicalize Democracy." *AEA Papers and Proceedings,* 108 : 33-37.
29. Weyl, E.G., 2017. "The robustness of quadratic voting." *Public choice*, 172(1-2), pp.75-107.
30. Robinson, K. S., 1992. *Red Mars*. Random House. ISBN 0-553-09204-9.

# 8: NEW JAMESTOWN
# A PROPOSAL FOR A 1000-PERSON COLONY ON MARS

**Alex Dworzanczyk**
alexdwor@gmail.com

## 1. INTRODUCTION

Colonizing Mars will require solutions to technical, economic, and social questions unique to the planet. Colonists will have to learn new techniques for living off the land on a planet without a large biosphere providing them the necessities of life and at the end of a supply line stretching millions of kilometers and months of travel time. They will need to find a way to finance their survival and growth independent of taxpayers' and politicians' fickle moods.

This report proposes a reference Mars colony design that attempts to answer the technical, economic, social, and aesthetic questions such an endeavor raises. The proposed colony design is called "New Jamestown," after the first successful English settlement in the Americas, where the successful cultivation of tobacco provided a luxury commodity that secured the colony's future. As the English colonies in America became centers of social, economic, political, and cultural experimentation, it is hoped that "New Jamestown," or something very much like it, can lay the foundation for a new and better human civilization on Mars.

## 2. RATIONALE AND BUSINESS CASE

### 2. 1. Exports and Other Income Streams

Where transportation costs are high, trade has historically been dominated by commodities with an extremely high price per unit mass or volume. The early-modern preoccupation with the spice, silk, and dye trades is an illustrative example. Such high-price commodities seldom have direct practical or utilitarian applications, but instead serve as status symbols. This "conspicuous consumption" as American sociologist Thorstein Veblen dubbed it, of luxury goods can be harnessed to finance the development of a Mars colony.

Luxury goods for conspicuous consumption on Earth can be manufactured and shipped back to the mother planet with substantially wider profit margins than even such high-priced commodities as gold. The small size of the early Mars colony limits the economies of scale that come with mass production. Goods produced on Mars will be made in small numbers using rapid-prototyping techniques and other methods well-suited to just-in-time production for a small local economy. These techniques can be easily transferred to the production of small quantities of high-price luxury goods.

An example of a luxury good that can be produced on Mars is a high-status wristwatch. The Rolex and Omega brands of watches have been marketed as capable of withstanding extreme conditions, from the depths of the ocean to the vacuum of space. Yet, the majority of their buyers do not use these capabilities on a regular basis (or at all; the Rolex Sea-Dweller is designed to withstand depths far greater than the record for the deepest human dive). These devices sell for thousands of dollars and mass well under 1 kilogram, producing a range of watch prices from $40,000 to $50,000 per kilogram, even for unadorned stainless steel watches. The products are therefore worth approximately their own weight in gold ($41,000 per kilogram, at the time of this writing).

The manufacturing cost of such watches is, of course, confidential, but it is possible to estimate the cost as under 50% of the retail price. This is comparable to the extraction cost of gold, which is closer to 65% of the commodity price [1]. A similar case can be made for other forms of jewelry, whose retail prices far exceed the cost of manufacture or materials. Martian landforms are in many cases analogous to terrestrial landforms, and many of the same volcanic, tectonic, hydrological, and chemical processes that shaped Earth's mineral bounty were active on the Red Planet [2]. It can be hypothesized, then, that gemstones and other decorative objects can be found on Mars. These can further augment the price of luxury goods exported to Earth.

The greatest challenge will not be the adaptation of rapid-prototyping and computer-aided manufacturing (CAM) techniques to luxury manufacture, but in creating a brand for which consumers are willing to spend a great deal. Rolex and its competitors achieved this by associating with famous explorers, divers, and celebrities. A Martian luxury brand might do the same by associating with adventurers on the Red Planet and elsewhere whose exploits might be viewed remotely by spectators on Earth. The first person to ski down a hill of dry ice in the Martian Antarctic or to summit Mount Olympus can do so carrying a Martian chronometer, optimized for dust- and vacuum-resistance. Conspicuous consumers on Earth, having vicariously shared the adventure, can also share in the aesthetic and technology. The rise of middle- and upper-income classes in China, India, and Africa can be a great opportunity for the next great luxury brand. What Rolex was to twentieth century American and European millionaires, New Jamestown watches can be to the status-conscious businessman of Mumbai, Lagos, or Shanghai. The increase of global wealth will increase the amount of market for such luxury goods, growing the pie for everyone. There is no reason that New Jamestown luxury watches cannot command in 2046 the same $4.7 billion in annual sales that Rolex did in 2016.

The exact amount of work done to manufacture the luxuries on Mars is negotiable, and depends on how much the consumer on Earth is willing to pay. Fundamentally, the product being manufactured here is not the luxury item itself but the status of its ownership. This premise has consequences for the distribution of work in the colony. At one extreme, if it is found that the place of the luxuries' manufacture has less to do with their value to the consumer than their association with a sense of adventure, or the exotic quality of their material, it may be more profitable to manufacture the watches on Earth from imported Martian materials.

In such a scenario, the workload at the colony would not be manufacturing *per se* but rather mining and providing logistical support for adventurers on the Red Planet.

If demand for finished luxury items from Mars cannot be created, it will be necessary to turn to the export of raw materials to Earth or to other markets in cislunar space. The high assumed cost of transport from Mars to Earth ($200 per kilogram) limits the range of products worth exporting to Earth to those whose market value is above $200 per kilogram. On the face of it, this ought to allow for the economical export of precious metals, like gold, platinum, and others whose market value can be tens of thousands of dollars per kilogram.

However, even on Earth, the cost of labor, equipment, exploration, and other expenses cuts into the profitability of mining. In 2012, the All-In Sustaining Cost (AISC) of gold mining was approximately $1,100 per ounce, or some 65% of the price of gold at the time [1]. The AISC of platinum mining was similar. On Mars, which presents unique heat rejection, maintenance, and irrigation problems, the AISC will be higher. The uncertain status of Martian indigenous life further complicates the question, as regulatory agencies on Earth will certainly demand assurance that mining operations will not harm a possible Martian ecosystem.

Still, the price of gold and platinum on Earth is high enough that, even with a higher AISC, these metals might form the basis for a successful interplanetary economy. At the time of this writing, the price of gold exceeds $41,000 per kilogram. Even if 80% of this price is lost to an elevated AISC, and a further $200 is lost to the shipment cost, each kilogram of Martian gold sold on Earth can return a profit of $8,000. To generate $1 billion, New Jamestown would need to export some 125 tonnes of gold. This is 4 times the annual production of the largest gold mines on Earth, each of which employs far more miners than New Jamestown's projected population. Economically-feasible gold mining on Mars will require significant advances in robotics, allowing more productivity per worker, or a particularly rich and easy mine.

Other precious metals command very high prices per kilogram at present, but are so rare in Earth's crust that their economic extraction is possible only as a by-product of the mining of bulk metals (like nickel) with which they occur. It is highly unlikely that 1,000 people will, in the near future, suffice to conduct a mining operation matching or exceeding in scale the largest bulk nickel mining concerns on Earth. Barring a revolution in autonomous or even self-replicating mining robots, extremely rare precious metals do not appear fit to support an interplanetary export economy.

The market is more open to cheaper commodities elsewhere in cislunar space. One can extrapolate the cost of launching payloads elsewhere from the $500 per kilogram assumed cost of delivering payloads to Mars using the rocket equation. Assuming a Low-Earth-Orbit-to-Mars delta-v of 3.9 km/s and methane-burning rocket stages with a specific impulse of 380 s, one finds that a rocket that delivers

100 metric tons to Mars can deliver 285 metric tons to LEO. The assumed $500-per-kilogram cost of delivery to Mars therefore divides by 2.85 into a $175-per-kilogram cost of delivery to Low Earth Orbit. From that $175 per-kilo-to-LEO cost, a similar approximation of costs to points in cislunar space can be calculated by calculating the mass fraction of a spacecraft bound elsewhere, as listed in Table 1. Many of these locations, including geostationary orbit (GEO), Low Lunar Orbit (LLO), and the surface of Earth's Moon come out to costs in excess of $300 per kilogram.

By a similar method to that described above, one can convert the $200-per-kilogram cost of cargo from Mars to Earth into a cost of cargo from Mars to points in cislunar space. The cost per kilogram to Low Earth Orbit is still roughly $200 (the propellant needed to circularize the orbit of a spacecraft after aerobraking approximated to equal that needed to land on Earth after aerodynamic descent), but that to other points is proportional to the fraction of payload lost to rocket propellant needed to adjust the final destination. Table 1 lists the extra delta-v needed to circularize at Geostationary Orbit, the Earth-Moon Lagrange Points, Low Lunar Orbit, and to land on the Moon after aerobraking through Earth's atmosphere.

The counterintuitive result is that Martian-produced raw materials and finished products can be competitive on a per-kilogram basis with terrestrially-produced ones in Earth's own cosmic backyard. This raises the possibility of colonists on Mars exporting geostationary-orbit communications satellites back to Earth. The approximate annual revenue of the commercial communications satellite manufacturing and launch industries was $2.4 billion from 2012 to 2016. Despite the long lead times inherent in interplanetary flight, the low cost of climbing out of Mars' gravity well can give New Jamestown a competitive edge in securing a substantial portion of this revenue stream. The colony is also superbly well-placed to provide logistical support to human and robotic expeditions and outposts on the lunar surface, so a program of lunar exploration can also open up a revenue stream for the colony.

| Destination | GEO | EML1/2 | EML4/5 | LLO | Moon |
|---|---|---|---|---|---|
| Additional ΔV from LEO/Mars (km/s) | 4.33/1.55 | 3.1/0.7 | 4.00/1.00 | 4.00/1.00 | 5.93/2.50 |
| Cost ($/kg) from Earth/Mars | 560/303 | 402/242 | 512/262 | 512/262 | 861/392 |

*Table 1: Additional Delta-v, and adjusted cost, to various orbits in of Cislunar Space.*

While they most likely cannot cover the entire expense, research-oriented income streams can supplement New Jamestown's operating revenue. The United States

Antarctic Research Program has an annual budget of some $350 million per year [3]. Since New Jamestown could provide logistical support for a similar program of geological, astronomical, meteorological, and potentially biological research on Mars, one can postulate a similar annual infusion of revenue to the colony in exchange for access to colony facilities and consumables for researchers from friendly nations or institutions. This extra funding will not be the dominant revenue stream, but in a lean year could make the difference between success and failure.

The four revenue streams outlined are not mutually exclusive. The same manufacturing equipment that the colony uses to manufacture luxury watches and jewelry from stainless steel, precious metals, or ceramics can be used to manufacture valves, tanks, structural members, and other components for satellites. The same equipment that smelt iron (for the colony's own use and for manufacturing for export) can be used to process gold ores. The life-support and logistics apparatus supporting the other efforts can also support a staff of geologists and other scientists bringing in research grant money. A notional sample distribution of income streams is presented in Table 2. Each citizen and their specialty can have a part to play.

| Revenue Source | Finished Luxury Goods to Earth | Precious Metals to Earth | Manufactured Products to Cislunar Space | Research Funding | Total |
|---|---|---|---|---|---|
| Annual Income (millions of dollars) | 4,750 | 250 | 2,000 | 400 | 7,400 |

*Table 2: Sample Revenue Distribution for New Jamestown*

## 2. 2. Start-Up Cost and Pay-Off Time

It is difficult to estimate the cost of establishing New Jamestown. While many of the constraints that regulate the hardware that will be sent to Mars (an emphasis on low weight and storage into small volumes) are shared with other aerospace projects, the nature of the hardware that will have to be sent to Mars (earth-moving equipment, megawatt-scale nuclear reactors, chemical processing equipment, etc.) is unlike anything launched into space before. Likewise, the colony's scale is orders of magnitude beyond the largest space projects in human history. It is also dissimilar to terrestrial products of similar scale. A large factory complex or mine or nuclear power plant does not have to manufacture provide its own food, and other resources that fall under the broad category of "externality" (like water, provided by municipal authorities at cents per ton, or oxygen, provided at no financial cost by Earth's ecosystem) must be painstakingly wrung from the Martian environment.

However, there are parallels between the hardware New Jamestown will need and that which industrial and scientific facilities on Earth utilize, and these can be used to give an order-of-magnitude estimate of the colony's establishment cost. The broad array of chemical processing equipment needed to sustain the New Jamestown colony is comparable to that required to run an oil refinery or similar chemical plant. Each requires a large number of pressure vessels, heating elements, compressors, turbines, cooling equipment, and filtration devices. Some processes (like the refinement of plastics from carbon monoxide and hydrogen through the Fischer Tropsch process) will be virtually identical between Earth and Mars but for their feedstock (coal and natural gas on Earth, carbon dioxide and water on Mars), strengthening the analogy. The cost of a coal-to-oil plant on Earth is on the order of $3 billion [4]. Other projects have similar possible analogies. Semiconductor factories are highly automated, and have a great deal of machinery optimized for extremely precise manufacturing. A semiconductor factory employing 4,000 people costs between $5 billion and $7 billion [5].

From an aggregate comparison of high-technology research and manufacturing facilities on Earth, an estimate of $10 billion for the hardware needed to build New Jamestown can be made.

The cost of transporting this equipment to Mars, of course, drives up the estimate. Unfortunately, immobile industrial installations on Earth are seldom measured in their mass. One of the few data points that found for the mass of an industrial facility broadly comparable in complexity and equipment to the proposed Mars colony is the Shell *Prelude* floating natural gas refinery. This facility masses 260,000 tonnes [6]. Of that 260,000 tonnes, most is likely the equipment needed to simply keep the vessel afloat, and not needed for the task of refining and processing natural gas. If half that mass can be disregarded as purely nautical, one finds that the mass of a chemical refinery is about 130,000 tonnes. To deliver this mass to Mars would cost some $65 billion. While this is only an estimate and can be shifted either way (industrial plant optimized for an interplanetary voyage might use more aluminum or composites than steel, to reduce mass), it does indicate that transportation costs will be several times the capital cost of the New Jamestown colony.

The aggregate result of this estimation process is that a 1,000-person colony on Mars with sufficient industrial plant to aspire to self-sufficiency and some export capacity would cost somewhere between $10 billion, with the cost of transportation about 80% of each. While a startlingly high capital cost, it is within the net worth of certain of the wealthiest individuals on Earth, and could be covered by a consortium of billionaires or corporations. Intriguingly, it is on the same order of magnitude as the postulated revenues that the colony can generate. With luck, if a particularly rich gold vein is found, or if a market for luxury Martian watches emerges, or if the Martians can corner the market for geostationary orbit satellites and induce a boom in that field, the colony might even swiftly (within a decade) cover its own startup costs.

There are, needless to say, many less optimistic scenarios where the colony fails to cover its costs, where disaster after disaster hampers the efforts to put down roots and to set up a thriving commercial relationship with terrestrial markets. However, New Jamestown's true pay-off in the optimistic scenarios is the success of a self-sustaining human outpost on another world, a proof-of-concept spurring further growth on the Red Planet and elsewhere, giving rise to a new society that hails the colony's founders as its physical and spiritual progenitors. It is indeed a risky venture. What birth isn't?

## 3. BUILDING NEW JAMESTOWN

### 3. 1. Colony Location

The optimal location for a colony on Mars provides ease of access, available raw materials, and readily-available energy. "Ease of access" here refers to the difficulty of landing spacecraft at the location. To maximize landing payload, it is desirable to maximize flight time in Mars's atmosphere. This maximizes the impact of atmospheric drag on the vehicle, reducing the need for breaking propellant. Locations below the Martian datum altitude are therefore preferable to high-altitude sites. This includes much of the northern hemisphere, the Valles Marinaris canyon system, and the Hellas and Argyre Planitia in the southern hemisphere.

While a detailed mineralogical survey of the planet for useful ores has yet to be conducted, the presence of water ice at certain locations on Mars can be proven or inferred from existing remote-sensing missions. Data from the Mars Reconnaissance Orbiter has indicated the presence of water ice at Utopia Planitia, in the Martian northern hemisphere [7]. It has also suggested the presence of water ice at Hellas Planitia in the southern hemisphere [8].

It is also possible to speculate on the availability of other mineral ores. On Earth, mineral ores result from volcanic and hydrothermal processes. It is possible that analogous processes took place on Mars, and that one would be well-advised to seek useful mineral ores in regions influenced by volcanic activity [2]. Utopia Planitia is close to the Elysium volcanic province, whose existence indicates that subsurface heating influenced the region's geology at some point. The possibility of subsurface heating at or near Utopia Planitia also supports the choice of that region for a colony. Hellas Planitia, too, is adjacent to regions of volcanic activity, at Hadriaca Patera and Tyrrhena Patera, both to the east of the basin [9].

Impact craters have been proposed as both ore-forming and ore-exposing features [2]. On Earth, impact structures have produced some of the richest mineral assemblages on the planet. They may produce similar wealth on Mars. While neither Utopia Planitia nor Hellas Planitia are particularly heavily cratered, they are each adjacent to more heavily-cratered highlands that may prove bountiful. A colony in either lowland site can be an effective base camp for surveying and providing logistical support to mining operations further up-country.

Further research is needed to characterize the mineral wealth of candidate sites at Utopia and Hellas Planitia, but their proximity to features analogous to sources of mineral wealth on Earth supports the presence of valuable industrial minerals at each location.

The last of the three main attributes a colony site must provide is access to energy. On Mars, a colony's energy needs must be met by solar power, nuclear power, or 'areothermal' power (subsurface heating, analogous to terrestrial geothermal power). While areothermal energy can, in principle, be harvested anywhere on the planet, given a suitably deep well shaft, it is much easier to tap it at locations with recent volcanic activity. Whether any location on Mars has significant areothermal energy potential remains an open question, but a positive answer would make such a site an ideal colony candidate.

This paper uses Utopia Planitia as the reference colony location, and does not assume the presence of easily-available subsurface thermal energy. The broad colony reference design can be applied as well to any site on Mars that combines access to subsurface ice with low altitude and temperate latitude.

### 3. 2. Securing the Necessities of Life: Oxygen, Water, and Food

Colonists and visitors to the colony will require a certain amount of oxygen, water, and food per day. These requirements, assumed to be identical to present NASA requirements for International Space Station astronauts, are 1 kg of oxygen, 11 kg of water, and 2.5 kg of food per day per person [10]. Simple arithmetic shows that supplying these consumable requirements through imports from Earth would grow extraordinarily expensive, averaging (at $500 per kilogram to Mars) $7.25 million per day. When the need for water and gas for industrial processes (from chemical synthesis to coolant for plasma-cutters and CNC machines), replacement parts for machinery, and replacements for broken everyday-use items is taken into account, the cost of maintaining a 1,000-person colony quickly spirals out of control.

Oxygen is the most urgent need of any human colony. It will be necessary to generate enough oxygen (and a suitable buffer gas, likely nitrogen and/or argon) to pressurize the New Jamestown living and working volume to a suitable atmospheric pressure, and to replenish this supply as it is consumed through human respiration and industrial processes that use oxygen (for example, concrete curing). The problem of supplying oxygen is closely related to the problem of removing carbon dioxide from the inhabitants' breathing mixture.

When the colony is first established, the breathing gas mixture will most likely be generated as a by-product of an In-Situ Resource Utilization infrastructure developed to generate propellant for interplanetary rockets. As described by

Zubrin, et. al. [11], oxygen can be produced from the Martian atmosphere through either direct carbon dioxide electrolysis (which yields carbon monoxide and oxygen) or through a combination of the Sabatier Reaction and water electrolysis. By the time New Jamestown is under construction, these techniques will be commonplace on Mars in support of propellant production for spacecraft returning to Earth. Either reaction would use, as its feedstock, the Martian atmosphere, so an additional step can be added to isolate the 4.3% of the Martian atmosphere composed of nitrogen and argon. The process of purifying Martian $CO_2$ through refrigeration and deposition [12] will yield a stream of nitrogen and argon anyway. They can be used to mitigate the fire hazard inherent in a high-oxygen atmosphere.

Once the colony is pressurized and populated, however, the problem of maintaining a gas mixture conducive to human homeostatis develops. Human respiration will convert oxygen into carbon dioxide. There are many ways to remove this $CO_2$, but the one chosen for New Jamestown is to use crop plants. Crop plants like wheat, potatoes, lettuce, and others appear to be an elegant solution, since they eliminate $CO_2$ while also providing nutrients for the inhabitants and supporting psychological well-being for the crew. However, as discussed by Do, et al. [13], they raise their own problems. Ironically, a life-support system that produces most of the colonists' caloric needs with plants can produce an excess of oxygen, which must be either diluted with nitrogen/argon or removed to prevent the risk of crew hyperoxia or fire hazards.

Fortunately, the problem of oxygen removal can be easily solved using technology similar to that used in hospital and industrial oxygen concentrators on Earth. These devices use pressure-swing adsorption (PSA), a method that involves trapping nitrogen in a zeolite bed, allowing oxygen to pass through [14]. On Earth, the oxygen is generally the desired product. On Mars, since the supply of breathing oxygen will be rapidly replenished by plant growth, the excess oxygen can either be stored for emergency or industrial use, or simply vented. A well-maintained and well-designed ventilation system is essential for moving air exhausted by humans over the colony's supply of plants, and for distributing emergency oxygen and scrubbing carbon dioxide through chemical and mechanical means in the event of an emergency.

After the problem of maintaining an atmosphere composition conducive to human life comes the problem of maintaining a supply of potable and industrial water. Water is necessary for human consumption, supporting agriculture, food preparation, and washing. These uses, from which water can be recycled with near-100% recovery through distillation, account for approximately 11 kg per person per day [10].

Water will also serve as a necessary source of hydrogen for the production of ammonia fertilizers, plastics, and other complex chemical products. Based on the mass fractions hydrogen in polyethylene and ammonia (1/7 to ⅙), and assuming a per-capita consumption of plastics and fertilizers equal to that in the US, one finds that the New Jamestown colony will consume 30,000 kg of hydrogen per year as chemical feedstock [15][16]. Much of this hydrogen will remain accessible in the system--plastics can be melted down or oxidized to recover their hydrogen content, and fertilizers will enter life-support loops), but at least some will be uneconomical to recover. At the very least, some quantity of hydrogen will be lost to rocket fuel. Producing enough rocket fuel for a single SpaceX "Starship" vehicle to return to Earth would consume 60 tonnes of hydrogen [17]. If two of those vehicles are assumed to return to Earth from New Jamestown per launch opportunity (an average of one per Earth year), the annual hydrogen consumption goes up to 90,000 kg (equivalent to 810,000 kg of ice).

As discussed in Section 3.1, New Jamestown will be located adjacent to a supply of subsurface water ice. If this is in the form of buried glaciers, water can be harvested through conventional mining techniques like excavation and simply cutting or smashing the ice into easily-transportable units. If the water is more evenly-distributed in the Martian regolith (which can be several percent water-ice by weight), supplementing the colony's water supply will require collecting the regolith and baking the moisture out. One advantage to the use of nuclear reactors for colony power is that their immense output of thermal energy and waste heat can be put directly to this purpose without conversion to electricity. If the regolith is assumed to contain 2% retrievable ice by weight (an estimate midway between the Viking 1 results and Zubrin's [11] more optimistic 3% estimate), extracting 810,000 kg of water ice will require processing approximately 40,500,000 kg. Per day, this is only 111 tonnes, or somewhere between 230 cubic meters and 300 cubic meters (depending on soil density). This is equivalent to about 200 loads of a consumer-grade Ford F-150 pickup truck, or just one load of a specialized mining haul truck. A handful of autonomous roving vehicles dedicated to the task of carrying dirt excavated by an autonomous backhoe (electrically-powered, by a cable connection to the colony's nuclear reactor) should suffice.

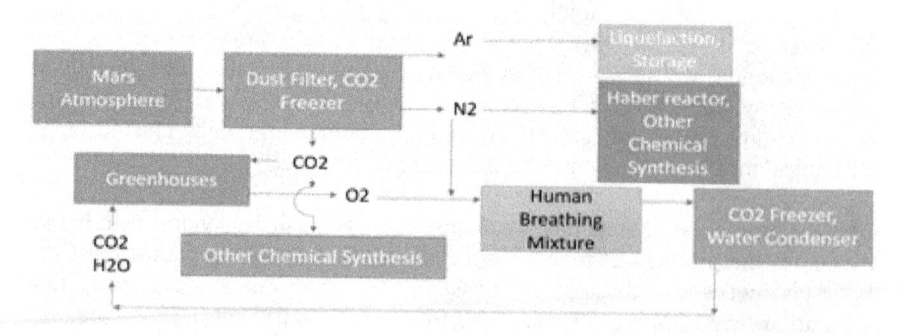


*Figure 1: Process Flow Diagram of Life-Support System Gases in New Jamestown*

Whether the water is extracted from Martian ice or the colony's own wastewater stream, it will need to be processed to ensure purity for human and industrial use. For the sake of simplicity, it may be most effective to process all colony water through a multi-flash distillation unit, similar to those used to desalinate ocean water on Earth [18]. The water stream can be heated modestly--at Mars's low ambient pressure, water will boil at temperatures below 10 degrees celsius. Indeed, solar heating alone, in pipes or tanks painted black, may suffice to boil all the water that flows out of New Jamestown. Since Mars lacks an ozone layer, direct exposure to ultraviolet light can also be used to purify biological contamination. Once boiled, the water vapor can be condensed back to a liquid or solid phase by simply exposing it to the ambient Martian temperature. The resulting frost can be melted down again, yielding pure water.

Of the "Big 3" consumables, food is the most complex to produce. Supplying the 2500 kilograms of food New Jamestown's inhabitants require per day will require specialized greenhouse areas. Lighting crop space sufficient to supply that much food would require almost 35 LED grow lights per person [13], or 35,000 for the entirety of New Jamestown. The current state-of-the art for LED grow lights guarantees their quality for slightly over 6 Earth years [19]. The intricacy of LED lightbulb manufacture suggests that this is not a process that would be economical to replicate on Mars in the near-term, so these lights would have to be imported from Earth. At a replacement rate, then, of approximately 5,000 lights (each with a mass of 8 kilograms) per year, New Jamestown would have to spend $20 million per year on lighting its crops (not including the cost of electricity, or indeed of buying the lamps).

Alternatively, New Jamestown can use crops grown in natural Martian sunlight. This introduces its own set of challenges, namely designing greenhouses that can sustain plant life in the face of a thin ambient pressure, dimmer sunlight than that available on Earth, and heat losses to the frigid environment. For this comparison, the AGPod design [20] is used as a reference design for a Martian greenhouse. This design uses a modular, 9.9-$m^2$-growing-area greenhouse with a low (25 kPa) internal pressure. The greenhouse is surrounded by concentrator mirrors that boost the intensity of sunlight and which, at night, double as an insulating thermal blanket protecting the crops within from heat loss. Each modular greenhouse is small enough to be brought inside a colony through an airlock for maintenance and for harvesting. To provide sustenance for the entire New Jamestown population, almost 5,100 AGPods would be required (to provide 50 $m^2$ of growing area per person [13]), with a total mass of 1,912 metric tons. While this seems like a much larger investment, it must be also considered that the greenhouses could be repaired on Mars to extend their useful lifetimes, or even manufactured on-site to give the colony a capacity for growth.

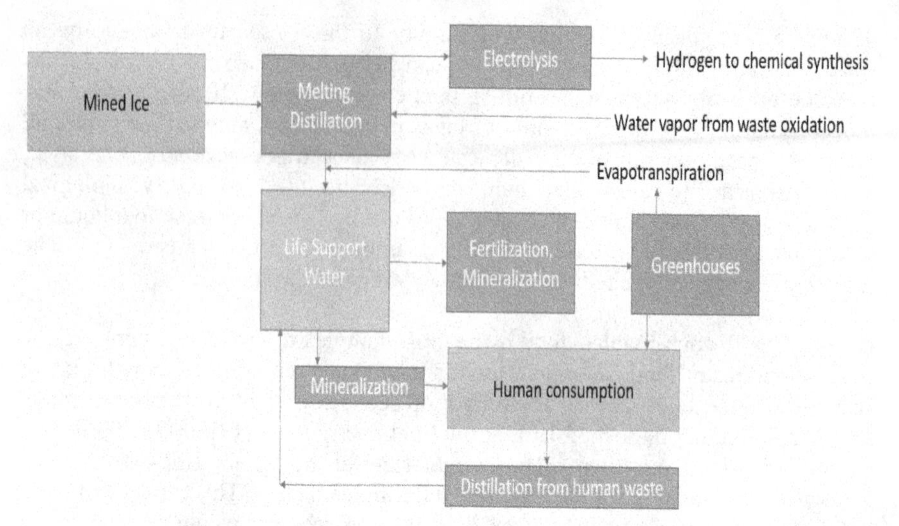


*Figure 2: Process Flow Diagram of Water in New Jamestown*

This capacity for growth leads the New Jamestown colony proposal to decide on the use of natural-light greenhouses. Manufactured by the hundreds by simple, automated procedures, the greenhouses free the inhabitants from a total reliance on a central LED supplier and push them to expand to new growing areas on the surface. As the colony grows, the AGPods will mark its expansion, a field of silvery petals closing at night over delicate green flowers, testifying to the spread of a symbiotic civilization-ecosystem relationship to a new world.

The AGPods will provide the majority of the inhabitants' calorie intake, but not the entirety of their diets. Some plants, like lettuces and fruit trees and herbs, can and should be cultivated in domestic and public spaces within the colony. These will supplement the inhabitants' diet and boost their morale through the positive psychological impact exposure to plants can have in a mostly-artificial environment. Similarly, cold-blooded animals (like tilapia, as discussed by Zubrin [11]) can be cultivated on the detritus from plant growth--stems, leaves, and other parts not suitable for human consumption. These can provide a useful protein supplement and some variety to the colonial diet.

Warm-blooded animals will be impractical unless AGPods can be manufactured at so low a cost as to make fodder cheap. The beasts of burden that accompanied the development of human civilization will find their meat and milk and hides outcompeted by plant-based alternatives. The inhabitants of New Jamestown will benefit from meat-substitution technologies invented to serve the growing vegetarian market. It is anticipated that, by the time the New Jamestown colony puts down roots, cost-conscious fast food manufacturers will have developed a method to imitate beef with black beans, like existing Impossible and Beyond Burgers.

### 3. 3. Manufacturing Metals, Plastics, and Ceramics

Mars has a wide variety of raw materials that can be used to fabricate the artifacts of an industrial civilization. However, not all of those artifacts are worth manufacturing at New Jamestown. Devices that are small, lightweight, and difficult to manufacture without specialized tools, like integrated circuits, would most likely be cheaper to import than to fabricate on Mars. Manufacturing on the Red Planet will focus on bulk commodities and objects that can be manufactured in large numbers by simple, automated assembly lines.

Building materials for the colony are the heaviest and "dumbest" products, and so they constitute the low-hanging fruit for the New Jamestown manufacturing base. Brick, mortar, and structural steel will constitute the vast majority of the material needed to build New Jamestown. Fortunately, the raw materials needed to make each are readily available on Mars. Boyd has demonstrated using Martian regolith simulant a "duricrete" that can be "over half as strong as terrestrial concrete." [11] Bricks, too, can be manufactured by simply wetting Martian soil and baking it in a mold. Since most of the New Jamestown habitable volume will be underground (for radiation protection and thermal insulation), these bricks can be used to build vaulted, arched, or domed ceilings enclosing large pressurized volumes. The process of brickmaking is mostly automated on Earth today. The techniques can be exported to Mars, and automated bricklaying robots can build walls with the products.

In addition to their structural utility, Martian bricks present an opportunity for decoration and beautification. As ancient and classical artists and artisans did, the inhabitants of New Jamestown can have their bricks colored and glazed. Colored bricks can be used to create mosaics or otherwise to introduce colors uncommon to Mars. An antiquarian-inclined architect might take cues from the ancient Babylonians and use glazed, cobalt-blue bricks in a Martian echo of the Ishtar Gate. Another might take inspiration from Medieval Brick Gothic churches, raising the Martians' eyes toward heaven with great pointed arches and elegant brick rib vaults. Different neighborhoods in New Jamestown might be built to different styles, and as time goes on new construction might follow new, even indigenous, architectural trends.

The process of making iron and steel is somewhat more involved. Iron constitutes roughly 16% of Martian soil by weight [11]. Iron can most easily be found on Mars in the compound hematite ($Fe_2O_3$), though magnetite ($Fe_3O_4$) and ilmenite ($FeTiO_3$) are also present. These compounds can be reduced with hydrogen or carbon monoxide to yield pure metal, and either water or carbon dioxide, which can be recycled into the process. Once reduce, the iron can be reacted with CO again to yield iron pentacarbonyl ($Fe(CO)_5$), a chemical that is liquid at room temperature (allowing for easy casting) but which, at just 120 degrees celsius [21], dissociates back into mostly-pure iron. This iron can then be forged and otherwise worked, or, if cast iron is really required (or there is some alloying element that must be introduced to molten iron), melted in an electric arc furnace. Carbonyl metallurgy lends itself well to automated rapid-prototyping techniques, since the pure-iron powder the process produces can be laser-sintered into a variety of complex shapes.

It may prove economical to begin construction of the New Jamestown colony as an ironworks. Colony construction will begin with the emplacement of the small modular nuclear reactor, an electric arc furnace, and equipment for reducing Martian hematite to iron. This ironworks can produce the pressure vessels, heat exchangers, and other elements needed to build the life-support systems for 1,000 people when the colony reaches its target population.

There are many different chemicals that all fall under the blanket term "plastic." Their unifying feature is that they are composed of long chains of repeating hydrocarbons. All of these hydrocarbons can be produced on Mars through well-understood chemical engineering techniques. The feedstock for all of them is carbon monoxide and hydrogen, produced on Mars from atmospheric $CO_2$ and mined water through the reverse water-gas-shift reaction (RWGS). Once these chemicals are produced, they are fed into a Fischer-Tropsch reactor, which can produce hydrocarbons of varying length, shape, and hydrogen-carbon ratio depending on the exact catalyst within. The catalyst is generally iron-based, but its doping agent influences the exact character of the exhaust product. For example, Pioneer Astronautics used a potassium-doped iron catalyst to force the reaction toward a higher alkene concentration [12].

These alkenes are the key to plastics production. The shortest of them, ethylene and propylene, can already be polymerized to make the common consumer plastics polyethylene and polypropylene. By way of benzene, they can also be converted into polystyrene, another common consumer plastic that, among other applications, is a good thermal insulator, and which itself is the feedstock for ABS plastic, commonly used in 3D printers. Through mastery of the Fischer-Tropsch process, the citizens of New Jamestown will be able to manufacture the overwhelming majority (by weight) of their consumer goods, clothing, and the equipment they will need to maintain their settlement.

One last material is worth noting, however: aluminum. This metal will most likely be needed to construct residential and industrial electrical wiring, and for a limited number of aerospace applications for which structural plastics or steel are either too complicated to manufacture or too heavy. It will not be so ubiquitous as it is on Earth, however. The energy cost of refining aluminum from aluminum oxide ($Al_2O_3$) is so great as to make it uneconomical compared to steel for most applications. Worse, the process of refining alumina ($Al_2O_3$) from minerals requires hazardous chemicals, either sodium hydroxide (for bauxite) or hydrochloric acid (for feldspar minerals) [22]. While a growing Mars colony will have to learn to deal with such chemicals sooner or later as economics favor more and more on-site manufacturing, the equipment and staff to handle them will drive up costs compared to competing iron (which uses only carbon monoxide and hydrogen, inevitable byproducts of the life-support system).

That aluminum which is produced at New Jamestown, however, will be produced by methods essentially identical to those used on Earth--the Hall-Heroult process, by which alumina is dissolved in a bath of synthetic cryolite (a compound itself produced from the reaction of alumina, sodium hydroxide, and hydrofluoric acid) and electrolyzed, with molten aluminum precipitating out of the vat. This process was discussed in great detail by Dyck in 2004 [22]. The complexity and power requirements of the aluminum manufacturing process will almost certainly drive the citizens of New Jamestown to seek alternative materials. If copper ore is found, that metal may be smelted for electrical conductors instead. Plastics and stainless steels might take its place in aerospace applications.

### 3. 4. Electrical and Thermal Power and Lighting

New Jamestown will require heat and electrical energy to power its life-support, manufacturing, communications, and other systems. As discussed in Section 3.1, an ideal location for the colony will provide access to subsurface thermal energy, which can be converted into electrical power as is done on Earth in geothermal power plants. However, the existence of sites suitable for thermal power plants on Mars has not yet been confirmed. In its absence, the remaining methods of powering New Jamestown are nuclear reactors and solar energy.

Solar power provides the advantage of being easily scalable (more panels can be deployed as-needed during construction of New Jamestown), an ever-declining cost, and relatively little in the way of regulatory hurdles. However, the recent loss of the Opportunity rover underscores a stumbling block for surface-based photovoltaic power. A severe planet-wide or local dust storm can severely reduce the available electric and thermal power to the colony. Workarounds to this problem exist, in the form of either space-based solar power (SBSP), whose microwave transmission of power from orbital arrays can be tuned to not be absorbed by Martian dust, and in the form of energy storage in batteries, capacitors, or chemical fuel. However, these introduce more cost and complexity to the colony's power grid, and the need to charge batteries or generate fuel for emergency generators acts as a drain on the colony's regular power production.

Because of the uncertainty of subsurface thermal power and the frailty and relatively anaemic productivity of solar photovoltaics, nuclear fission power in the form of Small Modular Reactors (SMRs) was chosen as the reference source of power for New Jamestown. SMRs are reactors designed for portability on ships and trains or ease-of-transport to remote locations, and whose size is generally much smaller than typical nuclear generating stations on Earth. For example, the SMR designed by Gen4 Energy has a design output of 70 MWt/25 MWe over 8-10 years and a mass of under 50 metric tons [23]. Even the reactor's waste heat can be used in a cogeneration scheme to make up for heat losses to the cold Martian environment, and its thermal power output can be used directly to heat chemical reactors. These qualities make SMRs ideally suited for transport to Mars to serve as power supplies for colonies. The Gen4 Energy SMR is chosen as a reference design for the New Jamestown Power Generating Station, but in principle any nuclear reactor that can fit on the interplanetary spacecraft delivering parts can be used.

How many such reactors will be required to power New Jamestown? The inhabitants' per-capita electricity usage will most likely exceed that of their counterparts on Earth, simply because so many processes powered on Earth by the combustion of hydrocarbons in the oxygen atmosphere will have to instead by powered by electricity or direct solar or nuclear heating. A useful terrestrial analogue is the nation of Qatar, which gets approximately half its water by purifying seawater [24] and which has the highest per-capita energy consumption on Earth [25]. The average per-capita power consumption in that country in 2013 was 25 kW per person. If the average New Jamestown citizen consumes exactly as much energy as the average Qatari, a single Gen4 Energy SMR would actually suffice to power the colony in electricity alone. If the reactor's waste heat can be harnessed for industrial applications, and if the power supply is supplemented by solar energy, then a safety factor emerges. The presence of a second reactor as a backup doubles the available power supply, allowing for a colony that consumes twice as much electricity per-capita as any terrestrial society today consumes in all forms.

Of course, Qatar is not New Jamestown, and the Persian Gulf is not Mars. The analogy will be imperfect. At the same time, with oxygen and food generated through natural solar illumination of crops, the situations are not so dissimilar as they might seem. Water must be purified through through osmosis or distillation in each case, and the Qatari fossil-fuel refining industry is not inherently so different in its equipment and the work it does from a Martian economy based on processing carbon dioxide and hydrogen and minerals into plastics, water, fuel, and metals. To provide redundancy and a margin for growth and unexpected power expenditures, two Gen4 Energy SMRs are chosen as the baseline for New Jamestown.

## 4. LIFE IN NEW JAMESTOWN

### 4. 1. Martian Demographics

A self-sustaining Martian colony will primarily be composed, in both its founding stock and its immigrants, of highly-educated, well-off people with a predominantly technical background. It has been demonstrated that increases in education correlate with lower birth rates and later ages of first childbearing in women [26]. This suggests that the birth rate in the Mars colony will be on the low end, and that the population will skew toward higher ages.

At the same time, the reality of a colony of only 1,000 people at the extreme end of a supply line across interplanetary space limits the availability of specialized medical care. Beyond a certain age, Martian citizens may find it desirable to relocate to Earth to retire or to seek assistance from specialized doctors. This factor to counterbalances the low expected birth rate, skewing the average and most common ages away toward productive adulthood.

These demographic factors have profound consequences for the needs of a Martian community. It can be expected that no more than 10% of the population will be between the ages of 0 and 18. A school organized with fairly strict division of classes by year of birth would be an inefficient use of pedagogical resources at the higher age range, when the 5-15 students per year would branch out into different specialties. Likewise, the absence of extreme old age and the difficulty of caring for invalids (who would most likely be returned to Earth on regular returning rocket flights) would leave the community with neither retirement communities nor invalid homes. These demographic considerations will inform planning for Martian society.

### 4. 2. Growing Up Martian

The small number of children in any age cohort at New Jamestown will have the most pronounced effect on the system of education. There will be so few that the test-heavy system of standardized classes used in the United States today would be a waste of time for all involved. Instead, beyond the most basic parts of primary education ("reading, writing, and arithmetic"), children will be directed into more hands-on methods of training, superficially similar to the pre-industrial system of masters and apprentices, where the worker gains specialized skills through direct, useful work. A student might apprentice herself, once she learns basic skills in programming and fabrication, to a roboticist and gain skill in this field by repair and iterative improvement on increasingly complex devices.

Like apprentices and journeymen of old, young Martians might demonstrate the completion of their education through some independently-managed project, a masterwork in the arts of robotics, or hydroponic agriculture, or construction, or some heroic work of surveying and exploration. The act of presenting this masterwork, of materially bettering the colony's condition or reputation, will supplant the traditional graduation ceremony as a coming-of-age ritual, and give the newly-minted citizen an emotional stake in New Jamestown's success.

Care must be taken that this system does not degenerate into a system of exploitative child labor, and that it does not produce citizens without any background in the philosophical and historical movements that culminated in their births on Mars. It is desirable to enforce some allocation of time toward academic pursuits in childhood and adolescence. There is no reason for this to take up an entire 8-hour day, however. Rather, the work of guaranteeing a firm grounding in more abstract math and science and in the humanities can be performed by a single specialist teacher on-site, or even on Earth through a remote connection. This teacher, or a small handful of them (for a student body numbering, at most, about 100), would assign readings, read submitted assignments, and interface with the students directly to answer questions, much as a college professor teaching an online class does on Earth today.

By integrating young Martians into adult economic and social life sooner, the New Jamestown colony will hopefully avoid creating an alienated and outright *bored* class of young adults. Such a class can be a nuisance to the citizenry on Earth, with their propensity toward idleness (at best) or petty crime (at worst). On Mars, where an act of teenage angst can lead much more easily to permanent injury or death not just for the culprit but for those around them, it is absolutely essential that young citizens recognize their societal role and societal dependence. At New Jamestown, the young will have their eyes turned outwards and upwards--to the untamed planet beyond, as they join in the heroic work of building a new branch of civilization.

### 4. 3. Building and Decorating for Human Health and Happiness

Like their terrestrial counterparts, most colonists on Mars will find that they spend the overwhelming majority of their time indoors. Whether they work in agriculture (on indoor, artificially-lit farms), manufacturing (in indoor shops), engineering and science (indoor laboratories, chemical refineries, and offices), or in the service sector (in indoor schools, infirmaries, kitchens, and offices), almost all of the work needed to maintain the colony will be in a pressurized, shirtsleeves environment. Even field geology or construction work will, on Mars, benefit from advances in teleoperation and artificial intelligence, so that even these specialists will not spend a great deal of time exposed to the Martian elements.

The amount of time the colonists will spend indoors underscores the need for homes, offices, laboratories, factories, and public spaces carefully designed to promote physical and psychological health. While the human mind and body have demonstrated a terrific resilience in the face of harsh living conditions throughout terrestrial history, this venture to spread human civilization outward ought to strive for, in the words of the scientist and science fiction writer Jerry Pournelle, "survival in style." [26]

Most of the colony's pressurized volume ought to be built underground to conserve heat, protect residents from background radiation, and to provide compressive force to counteract the large pressure differential between volumes habitable for humans and their Martian surroundings. Though the potential colony sites (Hellas Planitia and Utopia Planitia) have a more benign radiation environment than the rest of Mars, owing to their lower altitudes, the background radiation remains approximately 15 times greater than that to which the average American is exposed [27]. Solar proton events also present the danger of sudden increases in the radiation environment, and colonists will need to huddle in underground radiation shelters at such times. Because colonists will be spending most of their time indoors anyway, it is preferable to build the colony's habitable volume underground in the first place. Placing the colonists underground substantially reduces the deleterious effects of cosmic and solar radiation on their bodies and controls disruption to colonial activity during a solar proton event.

While the advantages of underground construction are substantial, it does present challenges which must be overcome. The absence of natural lighting can have adverse impacts on colonist psychology, similar to Seasonal Affective Disorder on Earth. This can be mitigated through the use of light tubes and other directional lighting devices, or through the use of copious amounts of artificial lighting [28]. To maximize the efficiency of underground lighting, it is advised to use light-colored and reflective materials for interior decoration. Bricks used indoors should be glazed or painted white to allow greater light reflection.

The large amount of time the colonists will spend indoors may have adverse impacts on their health, and particularly on the development of their children. Among other concerns, near-sightedness has been attributed to insufficient time spent outdoors [29], where the variety of distances across which the eyes must focus encourages healthier eye development. While recreational walks on the Martian surface ought to be encouraged for this and other reasons, it will not always be feasible to venture out. Public spaces should be designed to encourage healthy physical exertion and socialization among residents, and to give them a healthy exposure to vegetation, which has been correlated to better mental health [30].

The best way to channel light into underground living spaces may be to build the colony's residential quarters into the wall of a suitable crater or canyon. The living quarters themselves can be set back some distances, so that the overhanging mass of rock can serve as radiation shielding, blocking out more of the sky (though even a dwelling at the foot of a cliff already benefits from a 50% reduction in radiation flux). Light-tubes and other indirect light-transfer equipment can be used to provide illumination to back rooms.

Different living and working spaces at New Jamestown can be decorated using a variety of painting, carving, and brick arrangement styles. CNC machines can be used to sculpt rock into statues and relief carvings, either for historical and inspirational purposes (depictions of the heroic first Martians and other important figures in New Jamestown's collective memory) or for pure aesthetics. Colonists can take inspiration from every sculpting tradition on Earth here, or innovate, depending on their particular tastes. The availability of rapid-prototyping machinery will make available to every Martian the quality of statuary once reserved for mansions and cathedrals. Murals and frescoes will undoubtedly prove popular as ways of personalizing homes and adding color the the sepia-toned Martian backdrop. Glazed bricks can be used to line the outside walls of the colony, particularly if the bricks prove less susceptible than paint to erosion by Martian dust storms or color loss to ultraviolet light.

Public space should be designed to accommodate mass gatherings of the entire community and to serve small groups of citizens' needs. A communal park should be built to serve as a place of public recreation and assembly, purposes similar to those of urban green spaces. A larger excavation in the canyon or crater wall can be built for this purpose, shaped like the quadrant of an ellipsoid, with one flat side for the floor and the other a wall, lined with glass, arranged in cathedral-like pointed arches. The park should be liberally decorated with useful plants, like bamboo or fruit trees, and perhaps a pond where some of the colony's tilapia swim. The public buildings, too, should line the perimeter of this park--both government buildings and shops and businesses, to encourage mingling between the professions and a sense of communal solidarity. The ideal size of this great public atrium requires further evaluation, but a length of 800 meters and depth of 200 meters is arbitrarily selected to match the width of New York City's Central Park. A height of 20 meters is chosen, though the penetration of light to the back might be weak enough to require supplementary artificial lighting or light tubes there. The expense of its construction will depend on the degree to which automated equipment can be used in excavating and constructing it, and so such a project may remain aspirational for the time being.

### 4. 4. Exercise and Recreation

While regular walking will produce the impact stresses needed to stimulate bone and joint health, and the presence of a significant gravity field will address the problems of fluid accumulation observed in microgravity, it should be expected that the reduced-gravity Martian environment will cause some bone and muscle atrophy from disuse. Physiques trained in a 1 g environment will be up against much lower stresses. To address the physiological deterioration a low-gravity environment can cause, residents of New Jamestown will wear weighted clothing (lined with iron or heavier metals). This will force them to resist greater loads in their daily activities on Mars, maintaining bone and muscle density. The ability to lift large loads might come in handy during construction of the base or exploration of the planet, and is a boon of their earthly origins most citizens will be unwilling to yield. For this reason, weightlifting and gymnastics will also be popular, as will swimming in small pools with artificial currents (a space-saving alternative to traditional swimming lanes).

Further exercise can be provided through recreational ball sports. Large pressurized volumes of vaulted brick can be constructed underground to serve as multipurpose athletic facilities. Balls will have to be weighted compared to their terrestrial counterparts to keep them from going too far when thrown or kicked. The population is small enough that one reconfigurable court (with deployable rolls of artificial turf) should suffice for most major sports.

New Jamestown's citizens might also enjoy are reduced-gravity dancing (where new moves taking advantage of long jumps and easy lifting of partners might be invented) and human-powered flight in environments pressured to 1 atmosphere, as Zubrin [31] suggests.

Weather permitting (that is, when there is no immediate risk of a solar proton event or a dust storm), the surface of Mars can also become a site of recreational walks on the surface. Mountaineering and hiking in the exotic, unexplored terrain will be an activity of benefit both to the citizens and to the colonial corporation. At least some background in geology should be encouraged for all applicants to the colony, and prizes (including a stake in the resource) should be offered to any who discover resources of economic utility to New Jamestown. The prospect of personal financial gain might make surface exploration the most popular hobby on Mars.

### 4. 5. The Free City of New Jamestown

The citizens of New Jamestown will be, in general, much better-educated than the average electorate on Earth. Their mutual dependence on colleagues with different specialties to keep the colony functioning is another difference from their counterparts on Earth, who, in developed economies, can socially atomize themselves. The time delay in communications with Earth will exert a stronger pressure to socialize with fellow Martians. These three factors will drive a different political and social structure than is common on Earth today.

The typical citizen of New Jamestown will have an advanced education in a scientific or engineering field. All of them will be capable of understanding the importance of keeping critical systems operational and expanding the colony to provide ever more margin for unexpected breakdowns or unforeseen losses. Their collective expertise undermines one of the foundations of authoritarianism: the contempt with which the upper classes have historically viewed their peasant and proletarian compatriots. At New Jamestown, where every colonist comes from a higher educational background, there is no good foundation for isolating political power to a small elite, elected or otherwise.

Rather, decisions that impact the entire colonial population can be made by direct democratic consensus. The colony as a whole, even if it has investors on Earth, will by necessity be governed more like a workers' cooperative because it is not just quarterly profit at stake but the inhabitants' lives and health. Those closest to the struggle of extracting their livelihoods from Martian air and soil will know best what must be done to succeed in that struggle, and will have a greater personal stake in ensuring the colony's survival than most wage- or salary-laborers on Earth have in their companies. The citizens of New Jamestown themselves are the best-qualified to determine how to keep the crops growing, air flowing, and profits rolling in.

At the same time, there will remain specialists who know more about any given system than their neighbors will. These specialists will be organized into guilds or syndicates, according to their specialties. These guilds can offer a forum for those of the same trade to study problems common to their work (for example, a systemic defect in a design for an air reclamation system) and a united, louder voice to bring their concerns before the rest of the citizenry in an assembly meeting. They will also regulate their own members, ensuring that corners are not cut and that standards of professional excellence are upheld in every task necessary for New Jamestown's survival and growth. Standing together, the guilds will ensure that none of the necessary trades at New Jamestown will be unfairly marginalized, serving a purpose somewhere between a trade union and a political party.

The citizens will assemble in their guilds at public meetings, where concerns of interest to the guilds and the colony as a whole can be voiced, and a course of action can be debated and decided upon. The citizens' assembly also serves as a jury for trying crimes. At the low population level proposed for New Jamestown, it is not an undue burden to involve the entire community in hearing evidence and judging guilt or innocence, but in the future this aspect of the Martian constitution will need amendment. A simple majority voting system opens the possibility of resentment in the event of a very narrow vote, so a ⅔ majority voting system of all citizens is suggested to provide a better sense of consensus in decision-making.

In the system of professional guilds, there will be two special cases: the office of the Director, and the constabulary. The Director, like his namesakes appointed by the Dutch West India Company to represent their interests in New Netherland, serves as New Jamestown's investors' public face on Mars. The Director's role is somewhere between that of a corporate Chief Executive Officer and an ambassador. The Director represents the interests of the investors on Earth to keep the colony profitable and growing, organizing through persuasion the citizens in public works initiatives. The Director also handles relations with government and other organizations. For example, the Director's Office is the point of contact between New Jamestown and those bodies on Earth regulating launches and communications to the colony.

The Director is also the titular chief of New Jamestown's law-enforcement agency. Working together, the constabulary and the Director's Office ensure that regulations passed by the citizens' assembly are enforced. Punishment up to the level of confinement and exile (on the next rocket back to Earth) is within their purview. Incarcerating a substantial portion of the colony's population would drastically harm productivity, morale, and colony profitability, and a Director who allows the situation to deteriorate to that level would soon be replaced. It is in the Director's own interests to not degenerate into despotism, and to reform citizens rather than retribute where possible. The scarcity of manpower and the cost of maintaining nonproductive workers will check the growth of executive power, and keep power in the hands of free citizens.

## 5. CONCLUSION

Colonizing the planet Mars will require a great deal of work and innovation, but there is no problem involved that can't be solved with existing or near-term technology. The problems of extracting food, water, air, and tools from raw Martian materials rely on chemical synthesis methods demonstrated on small scales in laboratories or on large scales in commercial refineries. The reactors to power a colony are already in development, and will see widespread use on Earth before their interplanetary voyage. New Jamestown will be able to cover its own maintenance costs through export of luxuries to Earth and satellites to Earth orbit. While the development cost will still be substantial, on the order of tens of billions of dollars, that is a small price to pay for laying the cornerstone of a new civilization on the Red Planet.

## 6. BIBLIOGRAPHY


1. "Gold miners sustaining costs up 22% since the gold price bottomed." Mining.com, http://www.mining.com/gold-miners-sustaining-costs-22-since-gold-price-bottomed/, Retrieved 31 March 2019
2. West, M. D., and Clarke, J. D. A., "Potential Martian mineral resources: Mechanisms and terrestrial analogues." *Planetary and Space Science* v. 58 (2010), pp. 574-582.
3. "Office of Polar Programs," National Science Foundation, https://www.nsf.gov/about/budget/fy2019/pdf/30_fy2019.pdf, Retrieved 31 March 2019.
4. "Cleaner, Cheaper Liquid Fuel from Coal." MIT Technology Review, 6 January 2012. https://www.technologyreview.com/s/426551/cleaner-cheaper-liquid-fuel-from-coal/
5. Intel. "Intel supports American Innovation with $7 billion investment in next-generation semiconductor factory in Arizona." Retrieved 31 March 2019. https://newsroom.intel.com/news-releases/intel-supports-american-innovation-7-billion-investment-next-generation-semiconductor-factory-arizona/#gs.3uvedf,
6. "Construction of Prelude FLNG begins." Upstream, 18 October 2012. https://web.archive.org/web/20121124150508/http://www.upstreamonline.com/live/article1268028.ece
7. "Location of Large Subsurface Water-Ice Deposit in Utopia Planitia, Mars." NASA, 22 November 2016. https://mars.nasa.gov/resources/subsurface-water-ice-deposit-in-utopia-planitia-mars/
8. "Catalog Page for PIA11433," NASA, 20 November 2008. https://photojournal.jpl.nasa.gov/catalog/?IDNumber=pia11433
9. "Stop 4 at Hellas," Planetary Science Institute. https://www.psi.edu/epo/explorecraters/hellasstop4.htm
10. "Human Needs: Sustaining Life During Exploration," NASA, 16 April 2007. https://www.nasa.gov/vision/earth/everydaylife/jamestown-needs-fs.html
11. Zubrin, Robert. "The Case for Mars." Touchstone, 1996. 978-0-684-83550-1
12. Pioneer Astronautics, "Projects." http://www.pioneerastro.com/projects/, Retrieved 31 March 2019.
13. Do, Sydney, et al. "An Independent Assessment of the Technical Feasibility of the Mars One Mission Plan--Updated Analysis." Acta Astronautica, v. 120, pp. 192-228.
14. Friesen, R. M., Baber, M. B., and Reimer, D.H. "Oxygen Concentrators: A Primary Oxygen Supply Source." *Canadian Journal of Anaesthesia*, 1999. 46: 1185.
15. Cleetus, et al. "Synthesis of Petroleum-Based Fuel from Waste Plastics and Performance Analysis in a CI Engine." *Journal of Energy*, August 2013.
16. USGS, "Nitrogen (Fixed) -- Ammonia." Retrieved March 31, 2019: https://minerals.usgs.gov/minerals/pubs/commodity/nitrogen/mcs-2018-nitro.pdf,

17. SpaceX, "Mars." https://www.spacex.com/mars, Retrieved 31 March 2019.
18. Ghaffour, N., Missimer, T., and Amy, G. "Technical review and evaluation of the economics of water desalination: Current and future challenges for better water supply sustainability." *Desalination,* v. 309, pp. 197-207.
19. Heliospectra, "LED Grow Lights." www.heliospectra.com/led_grow_light_products, Retrieved 31 March 2019.
20. Frey, Aaron, et al. "Red Thumb's Mars Greenhouse." Mars Papers, http://www.marspapers.org/paper/frey_2002.pdf, Retrieved 31 March, 2019.
21. Moss, Shaun. "Steelmaking on Mars." Mars Papers, http://www.marspapers.org/paper/Moss_2006_2.pdf, Retrieved 31 March 2019.
22. Dyck, Robert. "Aluminum Extraction from Feldspar." Mars Papers, http://www.marspapers.org/paper/Dyck_2004.pdf, Retrieved 31 March 2019.
23. Irwin, Tony. "An Introduction to Small Modular Reactors (SMRs)." SMR Nuclear Technology Pty Ltd. http://www.smrnuclear.com.au/wp-content/uploads/2012/12/AN_INTRODUCTION_TO_SMALL_MODULAR_REACTORS_Feb_2013_FINAL.pdf, Retrieved 31 March 2019.
24. Khatri, S. "Research: Qatar's tap water could be harmful to your health." Doha News, December 23, 2016. https://dohanews.co/research-qatars-tap-water-harmful-health/. Retrieved 31 March 2019.
25. World Development Indicators, The World Bank. Retrieved 31 March 2019: https://datacatalog.worldbank.org/dataset/world-development-indicators,
26. Pournelle, Jerry. "A Step Farther Out." Ace Books, 1983. 0-441-78583-2.
27. Williams, Matt. "How bad is the radiation on Mars?" Universe Today, November 21, 2016. https://phys.org/news/2016-11-bad-mars.html, Retrieved 31 March 2019.
28. Howland, R. H. "Somatic Therapies for Seasonal Affective Disorder." *Journal of Psychosocial Nursing and Mental Health Services*, v. 47 (1): 17-20.
29. Pan, CW; Ramamurthy, D; Saw, SM (January 2012). "Worldwide prevalence and risk factors for myopia". *Ophthalmic & Physiological Optics*. 32 (1): 3–16.
30. Lee, Min-Sun et al. "Interaction with indoor plants may reduce psychological and physiological stress by suppressing autonomic nervous system activity in young adults: a randomized crossover study." *Journal of physiological anthropology* vol. 34,1 21. 28 Apr. 2015.
31. Zubrin, Robert. "How to Live on Mars." Three Rivers Press, 2008. 978-0-307-40718-4.

# 9: THE "CROWDSPACE" MARS COLONY

**Oleg Demidov**
Pattern Group, USA/Spain/Russia.
oleg@crowdspace.tech
**Ray Mercedes**
Pattern Group, USA
**Vitalii Pashkin**
Economist, Spain/Russia
**Nata Volkova**
MARCH, Russia
**Tatiana Schaga**
Utro, Russia
**Alexander Morozov**
Enbw, Germany/Russia
**Michael Denisov**
Engineer, Russia
**Annet Nosova**
Utro, Russia
**Kristina Karacharskova**
Utro, Russia

## INTRODUCTION

This document is developed to address the challenge of Mars Colony Prize contest by a group of entrepreneurs, economists, engineers, urban planners and designers based in multiple locations including such countries as the USA, Spain, Germany, Russia. Special thanks for the invaluable contributions to the group experts Cynthia Bouthot (Space Commerce Matters - USA), Daniel Satinsky (independent consultant - USA), Daniel Inocente (SOM Solutions - the USA).

Basic principles governing our team approach are the following:

- **Safety first** - the highest priority is the safety and comfort of colonists, maximal risk mitigation.
- **Self-supportive** - as low import from Earth as possible, maximal utilization of local resources.
- **Economically sustainable** - scalable business model, break-even reached with 1000 colonists.
- **Lean** - iterative construction of the colony by isolated modules with regular improvements.
- **Equal possibilities and rights** - every colonist has equal possibilities to develop their own project and earn money and reputation according to their skills and goals.
- **Social solidarity and openness** - the only way to achieve goals of colonizing Mars is to create social cohesion and a kind of brotherhood among the now, past and future colonists.

- **Social support** - every colonist should understand that in case of illness or any other problem he can find help either among other colonists or from the administration of the Mission.

Colony location selection:

Life on Mars for the first colonists will be highly restricted by the conditions given on the planet; that is why colony design should maximally oppose negative factors and leverage available resources. Colony location on Mars should consider a number of factors: Health and Safety (radiation level, volcanic activity, landscape), Energy availability (temperature pattern, latitude for greenhouses insolation and space launches), Resource availability (% of water in soil, liquid water availability, minerals and construction materials), Attractiveness (picturesque landscape, sightseeing availability). Based on available map data, including thermal, magnetic fields, radiation level, proton density and other relevant researches and news 4 promising areas were identified and compared according to selected criteria:

| Place | Lake Phoenix | Terra Noachis | Valles Marineris | Hellas Plantia |
|---|---|---|---|---|
| Temp.Day, min, K | 230 | 230 | 225 | 188 |
| Temp.Night, min, K | 170 | 170 | 203 | 153 |
| Rad Level, rem/year | 18-20 | 16-18 | 14-16 | 10 |
| Latitude (launches, sunlight) | -10...-15 | -20...-30 | -0...-20 | -40...-50 |
| Sightseeing availability | + | + | + | + |
| Predicted water availability | unknown | unknown | likely | likely |
| Water in the soil, % | 5-9 | 5-9 | 5-10 | 9-12 |
| Liquid seasonal water | No evidence | No evidence | Confirmed | No evidence |
| Geological zone features | Lava plateaus and lava hills | Lava plateaus and lava hills | Fluvial sediments Eolian sediments Exposed bedrock Breccia material | Eolian and fluvial deposits Lava plains |
| Minerals and Resources | Fe, Si | Fe, Si | Fe, Si, Alluvial minerals: Au, Pt | Fe, Si |

Table 1. Potential locations of the colony

Table 1 shows that the most promising area at the moment is Valles Marineris due to favorable temperature conditions (from 203 K in winter night to 313 K in summer day), lower radiation level (14-16 rem/year), low latitude (delta V gained compared to polar base is ~240 m/s and no need to change inclination plus more sunlight for greenhouses), evidence of liquid water and favorable geological features with potential for various natural resources including alluvial minerals like Au and Pt. Inside the Valles Marineris we are looking for large enough flat valley where space ship may easily land and be launched. This valley should be bordered by a slope of 30-40 meters high in order to accommodate the proposed base. A canyon in the slope of 40-50 meters wide could be beneficial for further base development. Pictures below show a potential location.

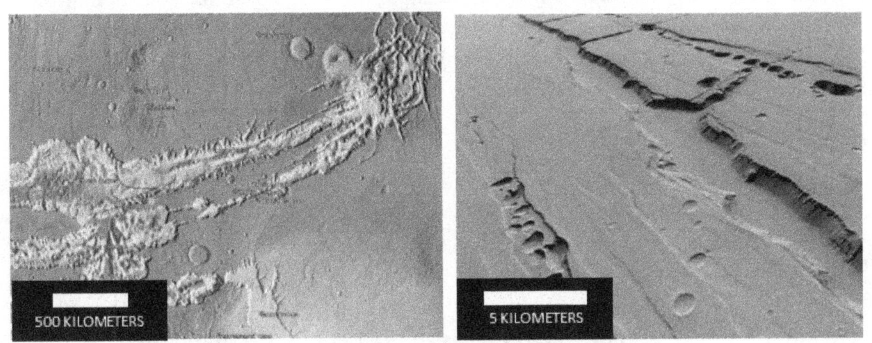


Figure 1. Potential locations of the colony

Colony development timeline:

The whole period of colony development from the first 10 colonists living in landing modules and building the colony from scratch to an economically sustainable colony of 1 000 colonists is going to take 10 Martian years. Every Martian year Space Transportation system will bring consistently increasing number of colonists. This 10-year period could be divided into 3 major stages of development that are characterized by various technical, economic, social, cultural, political, organizational and aesthetic conditions and capabilities: Setup stage 0-3 years, Early stage 4-7 years, Growth stage 8-10 years.

## TECHNICAL DESIGN

At the Setup stage, most colony elements are imported. Small local production is set up to provide major required resources and some elements to build the colony. The share of locally produced resources should reach 30-40%. Most station areas are buried right under the surface. Living and working areas are placed inside a slope to ensure proper radiation protection. 1-2 labs are established to bring colony-related research. Complex equipment is imported, robots are only assembled on Mars.

At the Early stage, a major part of the colony elements is produced locally with local construction materials. The share of locally produced resources should reach 60-70%. 2-level isolated colony modules with living, working and common areas are placed in a slope. The number of labs is increasing allowing external commercial research. Capabilities to produce some equipment out of local resources are improving. High automation of boring and construction processes is achieved based on the first two stages of experience.
At the growth stage, most colony elements are produced locally. A share of locally produced resources should reach up to 90%. The colony is becoming

multi-level to improve internal space which is flexibly adjusted to colonists needs. If technology allows (not in our basic scenario), the canyon is covered with a sealed roof and filled with an inflatable structure to allow creating Earth-like conditions inside the canyon space. Capabilities to produce quite complex equipment including robots are developed. These robots could build new colony parts almost autonomously. Mining of precious materials for export to Earth is started.

**Colony plan**

The total area of the colony is 290 000 sq.m. and has the following split by functions:

| **Purpose** | **Total area, sq.m.** | **Placement requirements** | **Construction materials** |
|---|---|---|---|
| Living | 32 000 | The lowest radiation, +25C, normal pressure, oxygen, home amenities | Tunnel bored in a slope, brick vaults |
| Education and healthcare | 4 900 | | |
| R&D labs | 13 600 | Low radiation, +25C, normal pressure/oxygen, office amenities | Tunnel bored in a slope, brick vaults |
| Canteens | 5 300 | | |
| Common, sport, entertainment | 35 000 | | |
| Manufacturing pressurized | 800 | Low radiation, +21C, normal pressure, office amenities | Tunnel bored in a slope, brick vaults |
| Storage pressurized | 20 000 | Low-medium radiation, +21C, normal pressure, no amenities | |
| Technical pressurized | 30 000 | | |
| Sub-surface power generation | 2 000 | Low-medium radiation, no amenities | |
| Greenhouses | 90 000 | Medium radiation, greenhouse parameters | Steel tubes with insulation |
| Surface manufacturing | 20 000 | High radiation | Steel frame and bricks |
| Surface storage | 40 000 | | |

Table 2. Colony areas

The proposed general plan of the colony incorporates features of the location selected with the functional split of modules.

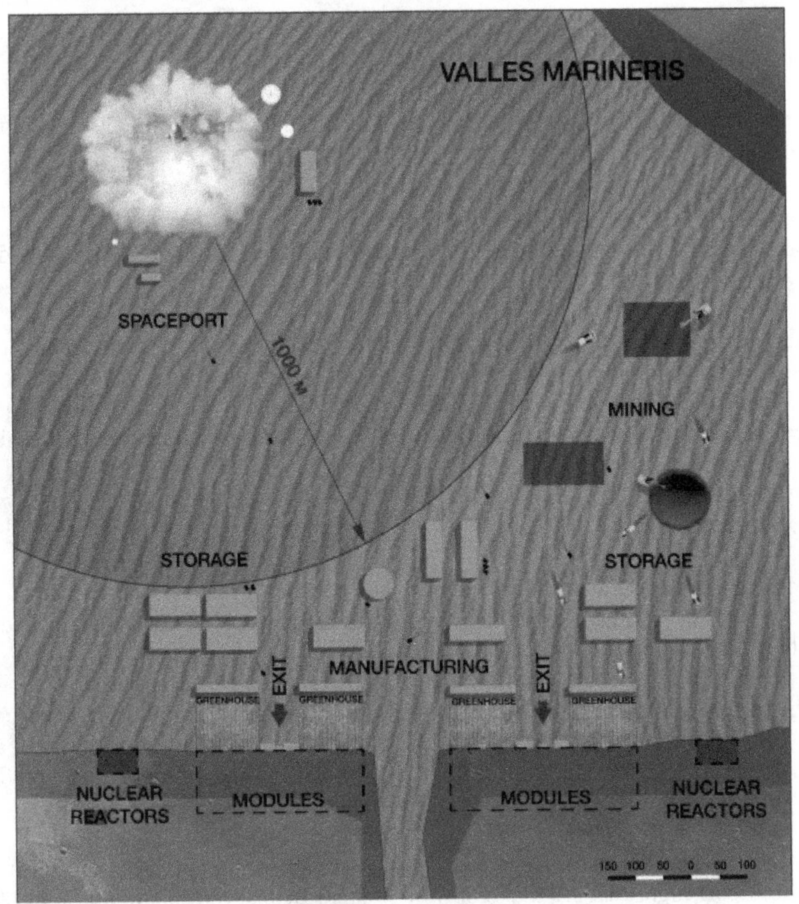


Figure 2. General plan

**Radiation protection**

The radiation environment in interplanetary space and on the surface of Mars is formed by galactic cosmic ray (GCR) fluxes and solar cosmic ray fluxes (SCR). According to the results of our calculations based on assumptions from the VNIIEM article [1]:

- Radiation level on Mars surface is at max 260 mSv/year
- Mariner valley (according to radiation map) average 150 mSv/year

Reduction of radiation by various levels of sheltering:

- 1 GeV- 540 g/cm^2, lowers the dose twice (2m of soil)
- 4 GeV- 4 000 g/cm^2, lowers the dose eight times (16m of soil)
- 10 GeV-40 000 g/cm^2, lowers the dose forty times (160m of soil)

The range of possible radiation levels depends on the visit duration. 20 mSv/year average is considered a safe dose of radiation internationally (no more than 50 mSv/year). That is why for all living areas we will focus on at least 16-meters soil protection.

A safe and reasonable pattern for human behavior on Mars (during 50-years stay) is:

- <4 hours/day at the surface
- 8 hours of sleep within the lowest radiation
- 12 hours other activities in living modules or somehow shielded environments

Radiation effect on plants varies depending on the plant stage: an elder plant is less affected than the young one. Effects are not well-studied but most papers state the effect of 260 rad/year won't get that much impact on a human beings [2][3]. **We will assume that plants could be grown in buried greenhouses right under the surface.**

**Thermal balance**

Since the average temperature of the surface of Mars (-70C lowest registered in Mariner Valley) is much lower than on Earth, the question of the thermal balance of the colony is very important.

1. According to our calculations, most energy for an open-air area surface is dispersed via thermal radiation in the infrared spectrum during night time [4]. Convective heat exchange with the atmosphere and thermal conductivity through an insulated wall are lower by the order of magnitude.

   For the structures of greenhouses with a relatively small open-air surface, losses at night of 70.9 kW (for 30 000 sq.m of plant segment, all is 90 000 sqm) throughout all pipes are distributed in the following way:
   - The share of infrared heat transfer is 73%
   - The share of convection heat exchange with the atmosphere is 15%
   - The proportion of heat transferred by thermal conductivity is 12%.

2. From both thermal and radiation standpoints it would be much more beneficial to build base modules that are completely underground or inside a slope.

3. According to our calculations, maximum heat losses throughout the station could reach 72 kilowatts, of which 68 are lost in the greenhouses (approximately 94%, but the sunlight energy collected during day time gives significant advantages in comparison to completely closed greenhouses).

**Food production**

For our food supply method, we chose hydroponics, because (1) they do not require processed and cleansed soil, (2) hydroponics is able to distribute fertilizer more economic, (3) hydroponics is more water-preserving, and (4) it is more productive in terms of yield per square meter [5].

To assess the need for food, we analyzed the nutritional value and yield of about 70 types of crops, as well as mushrooms, algae, insects, and fish. The diet of colonists will be based on rice, wheat, beans, soy, oyster mushrooms, champignon, seaweed spirulina, silkworm. Strawberry, tomatoes, leafy greens and other crops could be grown to make food more diversified. To produce soft fabrics, hemp will be grown. To increase productivity and radiation resistance, these organisms should be genetically modified. A balanced diet for 1 000 people requires 30 000 square meters of hydroponics in greenhouse plus 300 square meters of insect farms and recirculating aquaculture systems inside the station.

Greenhouse consists of two parts. The first one is a hollow insulated pipe (3 meters in diameter) with a small slot (not to scale in section, the real dimension is approx 20 cm in width)) in the upper part, covered with quartz glass. This tube is made of steel 3 mm thick, insulated with fiberglass, pressurized up to 0.3 atm and buried into the soil. For 1m of this tube, we need to produce 226 kg of steel spending 113 kWt*hour of energy. To build the same pipe by bricks, we need to expend 140 kWt*hour. Kevlar membrane for 1 m of the tube will weigh about 12 kg, and delivery costs will be $6 000 which is more expensive than to produce steel pipe.

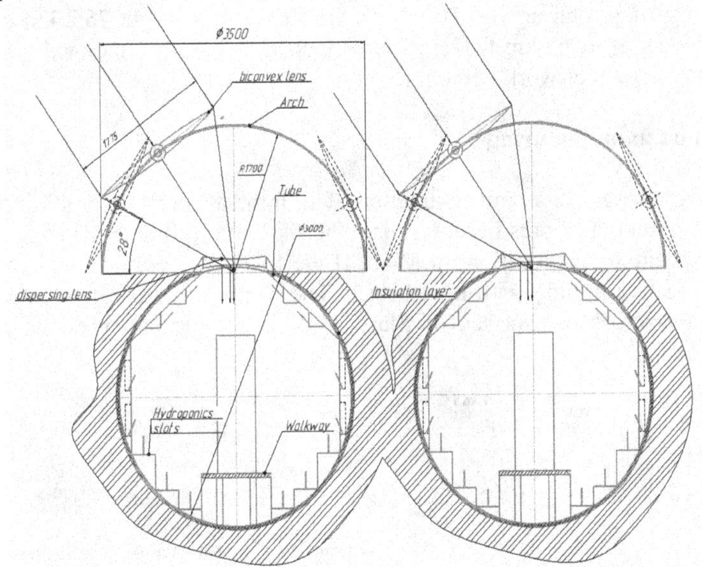


Figure 3. Greenhouse scheme

The second part of the greenhouse is semicircular arch (d=3.5 m) made from steel in the slots of which a profiled biconvex lens (convex lens radius can be found from the condition that the focal length is approximately equal to the radius of the arch) can move using a mechanism attached to it. The lens monitors the movement of the sun and collects the light into a single line, which is located at the focus of the second dispersing lens, and thus the light passes into the tube slot and is scattered inside it. The area of the lens is chosen in such a way that the radiation flux it collects is sufficient for normal plant growth. The advantage of this solution is that the light collection system is quite simple, and the mechanism does not require precise manufacturing and profiling (unlike parabolic mirrors, because we don't require making very precise and narrow light cluster. Moreover, we need to evenly disperse it over the pipe surface). The working mechanics involves the certain placing of the greenhouses (the section above left is west, and right is east). Thus lenses can efficiently track the Sun during day cycle and concentrate the light. Slight minus of this system is that minimal angle at which lens can operate is 28 deg. So, by using this system we effectively capture sunlight during day phases, but mediocre during morning and evening hours. So, from almost 12-hours a day cycle (due to equatorial location) we effectively can get approx 8-9 hours. The glass used in the manufacturing of lenses is in abundance in the regolith in the form of sands. 23% of tubes are used for growing mushrooms so it does not need this optic system and can be located near the slope in the shadowing part. Almost complete pipe immersion into the ground allows to minimize the heat flow through convection and radiation, and the arrangement of hydroponics in several levels reduces the required length of greenhouses to approx 8 735 meters of pipes with 3 meter diameter for 30 000 sqm of hydroponics, with 250 000 sq.m. of surface area including partitions. Applying a safety factor of 3 (and thus getting total area to 90 000 sq.m) we effectively get 26.2 km of pipes greenhouses plus 0.262 km for farms other than greenhouses – they will use either the same structure, or work without lensing.

**Mining and manufacturing**

In this chapter we will assume extraction of all minerals out of regolith instead of their extraction out of ores because it is rather difficult to find convincing data on ores composition and location on Mars. If we find some ores we can get much more productive mining with lower costs. The composition of the regolith itself varies little across the surface of Mars. Here is a table that presents the soil composition [6]:

| Component | % by mass | Usage |
|---|---|---|
| SiO2 | 43.60% | Glass, Ceramics |
| FeO, Fe2O3 | 16.70% | Iron and Steel production |
| MgO | 8.30% | Fireproof materials, cements |
| CaO | 5.80% | Cement |
| Al2O3 | 7.30% | Ceramics, Cement, Aluminium production |
| SO3 | 6.00% | Sulphur production, chemical synthesis component |

Table 3. Regolith composition

In order to excavate regolith on an industrial scale, a middle-size excavator with an operating weight of 10-30 tons and a bucket capacity of 0.6-1.6 cubic meters can be used. At the same time, the power required by the engine is about 50-130 hp. (37-96 kW). In order to crush regolith into pieces, a hydraulic hammer weighing from 1 to 3 tons can be used, it will require 200-300 kW of power. The crushing of pieces is carried out with a crushing plant with a capacity of about 220 kW. The output of the process could be filtered depending on further usage.

In the table below please find our estimates related to required production volumes and needs for the colony construction assuming that the current colony has a capacity to replicate itself by 30% in a matter of one Martian year:

| Resource | Measure-ment unit | Total current capacity p. a. | Production area, m^2 | Total power required, Wt | Comments |
|---|---|---|---|---|---|
| Power | Wt | --- | 300 | 4.8E+06- peak | Imported Nuclear reactors - see chapter Energy balance |
| Water | kg | 1.21E+07 | 500 | 6.00E+05 | Direct water extraction |
| Oxygen | kg | 4.25E+06 | | | |
| Hydrogen | kg | 4.86E+05 | 60 | 2.10E+06 | Electrolysis of water in industrial electrolyzer |
| Methane | kg | 1.55E+06 | 500 | 2.00E+04 | Photoactive catalytic CO2 dissolution in water |
| Nitrogen | kg | 2.27E+04 | 20-100 | 2.00E+04 | Adsorption on zeolite or using a membrane |
| Insulation | kg | 2.24E+05 | 2.94E+01 | 7.51E+02 | Fiberglass production in a reactor |
| Glass | kg | 3.29E+06 | 8.21E+02 | 5.01E+04 | SiO2 melting (quartz glass) |
| Iron and Steel | kg | 5.90E+06 | 9.83E+02 | 1.80E+05 | Using hydrogen reduction mechanism |
| Bricks | items | 3.52E+07 | 8.75E+03 | 2.36E+05 | Raw regolith pressing or 3D printing |
| Cement | kg | 1.04E+07 | 6.19E+02 | 1.58E+05 | From regolith components |
| Ceramics | kg | 1.70E+04 | 8 | 5.00E+04 | Heating in a separate furnace, partial use of energy emitted by a nuclear reactor |
| Plastics | kg | 4.00E+05 | 200 | 5.00E+04 | LPPE (Low pressure polyethylene) and PET production |
| Regolith | kg | 4.53E+07 | 7.55E+03 | 6.91E+05 | 3D printing |
| Food hydroponics | kg | 3.81E+06 | 9.00E+04 | 6.90E+05 | Safety factor 3 implied See: Food production |
| Carbon Fiber | kg | not specified | not specified | not specified | Production from atmospheric CO2 |

Table 4. Materials production

1. Water:

Major drivers for water requirements are colonists consumption of water – 3.27E+06 kg assuming 80% regeneration per annum (based on a value for ISS - 85% [7]), water electrolysis process – 7.45E+06 kg, other process – 1.35E+06 kg.

According to various articles about martian soils we can expect that soils of Mariner Valley contain 5-10% of water by weight [8]. The climate in the Mariner Valley is characterized by higher humidity compared to other points on Mars equator (water mist is often observed in the canyons), so it is possible to directly extract water from the atmosphere using, for example, Water Vapor Adsorption Reactor-WaVAR [9].

Among various options, we selected direct extraction as the most preferable for our colony because it is guaranteed and easy to control. In order to produce the required amount of water, we will need a station with a productivity of 18 t/day, that requires approximately 600 kWt and 500 sq.m. area.

Method efficiency is not so high in comparison to pumping that takes 0.1MJ/kg once established, direct extraction using sublimation will take approx 2.9 MJ/kg - estimated via phase transition coefficient [10]. If geological prospecting confirms the presence of a major water source like underground lakes in a nearby zone, it will make sense to use a pumping method that could be 30X more efficient.

2. Oxygen and hydrogen:

Capacity requirements for oxygen are based on the oxygen partial pressure requirements inside colony modules which are 14.7 kPa in pressurized quarters and 12.1 kPa in greenhouses. Level of oxygen loss is 8.12E+05 kg per annum [11]. Oxygen is also required in processes like Plastic production, chemical reactions, fuel production with overall needs of 3.58E+06 kg.

Capacity requirements for hydrogen are defined by processes like steel, iron production and methane production with overall needs 4.86E+05 kg.

Oxygen is partly produced by hydroponics in greenhouses with a total volume of 2.53E+05 kg. Water electrolysis is the best solution to oxygen and hydrogen simultaneously. In our calculations we used Proton On-Site M400 station [12] that allows producing a large amount of oxygen and hydrogen with an outlet pressure of 30 bar, provides reaction output close to one, and have energy costs equals 30 MJ/kg (for earlier years less powerful versions of that facility can be used). Therefore, 1 station that requires 30 sq.m. is enough to satisfy the colony needs. Energy requirements: 2.1 MWt.

3. Methane:

To make 1.55E+06 kg of Methane per annum we should use special Zn-Cu catalyst that can produce 290 liters / sol per 500 kg of catalyst (10 tons of catalyst will be 5.8 m^3 of methane at p=1 atm, T=293 K per sol under sun radiation requiring only pumping work, can be estimated as 2 kWt at most). Capacities of this production are based on opening the possibility of organic synthesis (5.93E+05 kg) and produce fuel for launch vehicles (9.60E+05 kg) (up to 4 BFRs). In the production of methane, a recently opened and unsurpassed method of photoactive production of methane using carbon dioxide dissolved in water can be used. CH4 is used to produce plastics with a rate of 0.657 of 1 kg of methane to 1 kg of PET, rate 0.857 to 1 kg of LPPE [13].

4. Nitrogen:

Will not be permanently consumed in any processes, but maintained in the colony on a permanent level. Nitrogen will be consumed mostly when new pressurized module is created and air is pumped in. For proper functioning we will need ~2.27E+04 kg of N2 per annum. Production can be performed by simple adsorption separation from the air by either of two methods: the first one is delivered by WaVAR-type device which is a chamber filled with a zeolite able to selectively adsorb nitrogen, the second one is based on passing atmospheric air through a polymer fiber membrane. In order to cover our needs zeolite flow reactors of up to 100 sq.m. area with up to 60 kWt power consumption should be built.

5. Insulation and glass:

In order to produce the required amount of insulation 6.72E+05 kg, we will need a station with a productivity of 1 000 kg/day, that requires approximately 0.7 kWt and almost 29 sq.m. area. Fiberglass is used as a thermal insulator for greenhouses and living quarters. Glass is a key component of making window panels and lenses for greenhouses.

The production of glass and fiberglass is essentially the same process - the fusion of silicon oxide, isolated from regolith (silicon oxide in the smelting of iron is an empty rock, and we do not need it). As for the glass we need to use small 20 kWt smelters to make up with 1.44E+06 per annum on year 10. That takes 360 sq.m. to accommodate.

a. In glass production, the molten mass will be formed either in the form of lenses of a certain structure (see the section Food production) or in the form of rolled products or other details that will be needed for the further design.

b. In the production of fiberglass, the melt will be sent to the reactor, where thin threads will be drawn out of it - interweaving such threads, even without a binder polymer solution or resin, can provide a good insulating material that can greatly reduce heat loss through the wall in the colony and, accordingly, save us most of the important energy.

c. Part of sand can be used to print out ceiling parts and then reinforce them with steel bars [14].

6. Iron and steel:

Amount of steel and iron produced is 5.90E+06 kg per annum at year 10 with energy expenses of 1.80E+02 kWt and area of 9.83E+02 sq.m. Steel and iron both are used in the construction process for buildings and frame elements of frames of equipment and devices.

A mechanism will be used to reduce iron oxides with hydrogen at a temperature of about 900-1 000 degrees Celsius, followed by casting steel or making iron structural elements and/or other parts [15]. Arc furnaces can be used for doping processes.

a. The volume of one furnace is 4-5 tons.

b. The energy consumption of the furnaces was assumed to be equal to the arc, which is about 160 kW for furnaces of small volume.

7. Bricks:

To make 3.52E+07 bricks per annum, we need manufacture that consists of couple punching stamps and molds. It is used primarily for building purposes. Production is relatively easy, but it makes the main wall material for our colony. We compress regolith into bricks [16]. Also, the Boring Company technologies can be implemented [17]. With proper algorithm and specialized punching device assessed power requirement of 2.36E+02 kWt over an 8.75E+03 sq.m. area.

8. Cement:

To make 1.04E+07 kg of cement per annum we need 1.58E+02 kWt of Energy (mean parameter) and area of 6.19E+02 sq.m. Main usage: building purposes. Can be reinforced with steel to make concrete. It is obtained by fine grinding of clinker and gypsum. Clinker is a product of uniform calcination prior to sintering of a homogeneous raw material mixture consisting of limestone and clay of a certain composition, ensuring the predominance of calcium silicates. Necessary numbers are given in the table higher.

9. Plastics:

To make plastics (approx. 4.00E+05 kg per annum) we need manufacture that consists of several flow reactors with the assessed power requirement of 50 kWt over a 200 to 300 sq.m. area. They are used for furniture (partially), making 3D printable items with PET and so on. In the production of plastics, you can give priority to thermoplastic ones: the simplest for us in production will be low-pressure polyethylene (LPPE), and lavsan (PET, polyethylene terephthalate), which can then be converted into mylar if it is applied to it with a metallic coating.

10. Regolith:

Our goal is to make 4.53E+07 kg of construction materials out of regolith per annum by year 10. Those materials could be either regular bricks, cement, iron, ceramics or a filament for a 3D printer to form robust and rigid structures [18].

For processing we should dig regolith chunks, grind them and then they can be processed into construction materials or 3D structures. That will require 6.91E+02 kWt of peak energy to construct. Area covered by production is 7.55E+03 sqm.

Energy costs calculation for regolith mining is the following. Assuming that the extraction is made by an excavator with a hydraulic hammer, several dozen jackhammers, and a crushing plant is used for grinding and subsequent processing, the cost are ~ 2 (doubling machinery) * (130 + 220 + 250 + 4 * 20) = 1 360 kW of energy.

11. Carbon fiber:

It is important to add that there is a possibility of producing carbon fiber and, consequently, of cheap composite materials directly at the colony's home site using the methods proposed by chemists at George Washington University [19].

**Energy balance**

Almost all the processes described above require a certain amount of energy to maintain. That is why it is important to assess what level of peak load we must ensure in order to maintain the simultaneous functioning of the colony with 1 000 people for all the industries indicated above. Here are some assumptions and implications:

1. The use of solar panels can be cost-effective only in cases of some private devices or installations. Up to 40% panels efficiency, coupled with the integrated solar flux at least 2 times lower than the Earth's one and effects of dust storms makes solar panels less preferable choice than nuclear reactors. Wind power is not effective due to low atmospheric pressure. Geothermal power could be a very useful source of energy in case of availability of warm underground lakes that is not the case in the moment.

2. Total production and regolith mining costs sum up to 5.13 MW of energy which is ~ 5 nuclear reactors with a capacity of 1 MW each. The mass of a single reactor can be estimated at about 13 tons, i.e. total mass is 65 tons.

3. It makes sense to split this power distribution in order to increase reliability while slightly increase the amount of imported mass. Also, we should have a reserved capacity equal to the capacity of the most powerful reactor (assuming that we can quickly fix any of the reactors).

Our proposition is the following: 3 reactors of 1 MW each, 5 of 0.5 MW each, 5 of 0.1 MW each.

**Communication and navigation**

Communication and navigation is a significant infrastructure for safety, science and development of the colony. For communication the colony will use a sat#1 which will be at MSO (martian-stationary orbit) above the station. Sat#2 which is located at MSO at an angle of 120 degrees to the sat#1 will be used for retranslating signal to Earth when sat#1 will be eclipsed by Mars. For retranslating the signal at the conjunction of Mars and Earth we are going to use spaceship coursing between these planets. We estimate the annual amount of information exchanged by the 3rd stage at 3-5 PB.

At the 1st and 2nd stage of the colony, a system similar to Japanese QZSS [20] will be used for navigation. It consists of sat#1 (the same used for communication) and 3 sats at Tundra orbit (sat#3, sat#4, sat#5). At every moment 4 navigation sats can be observed from the station neighborhood. At the 3rd stage, global martian navigation system which consists 24 sats will be created - like GPS or GLONASS on Earth. These 5 sats will be enough for permanent connection with Earth and navigation for research expeditions in the station neighborhood.

**Aesthetic**

According to the Lean principle the colony will be built by modules (200 inhabitants each) with the following timeline:

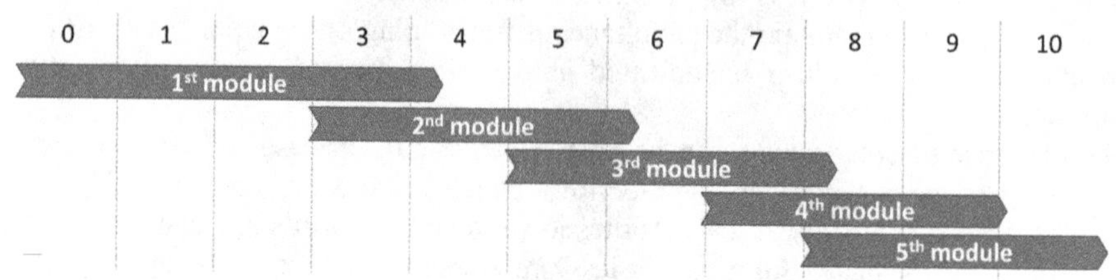


Figure 4. Modules construction timeline

At the setup stage when the colony is built in a very harsh environment all facilities are maximally functional and interior is not focused on beautiful design. The main target is to solve pragmatic issues and utilize available space in the most economically beneficial, safe and efficient manner. This concept is implemented in the phase 1 module that mostly consists of tunnels on underground and one ground levels.

At the early stage minimalistic design with highly compact architectural and interior solutions will dominate. To make colony more beautiful and friendly it is possible to use creative solutions like fashionable furniture made out of regolith [21] or lighting of internal areas by natural sunlight [22]. Common areas get multifunctional use and are slightly adjustable to people needs. Implementation of this concept starts in phase 2 module and develops in phase 3 module.

At the growth stage it is possible to achieve a broader scope of facilities functionality due to more advanced equipment and integrated solutions. Transformable mix of living, working and common areas will be organized. Areas where people spend their time will become much nicer and vary by comfort level addressing evolution in social organization of the colony. Overall the phase 4 module will be as comfortable as buildings on Earth for upper middle class in developed countries. At the same time, hotel zones will be equal in comfort level to a 5 star hotel on Earth in order to accommodate VIP tourist.

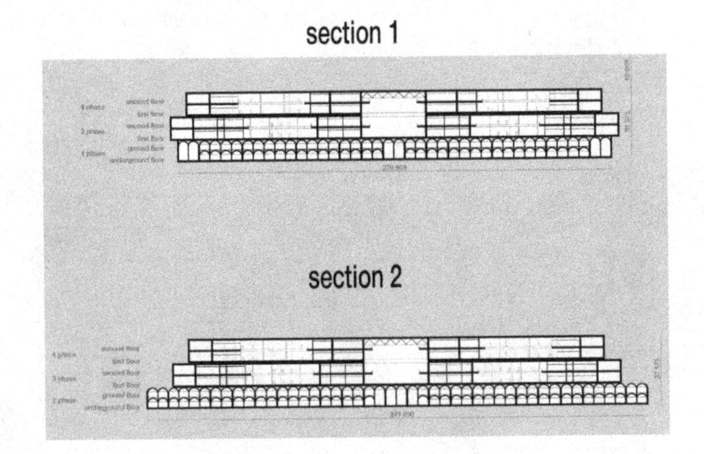


Figure 5. Modules compositions

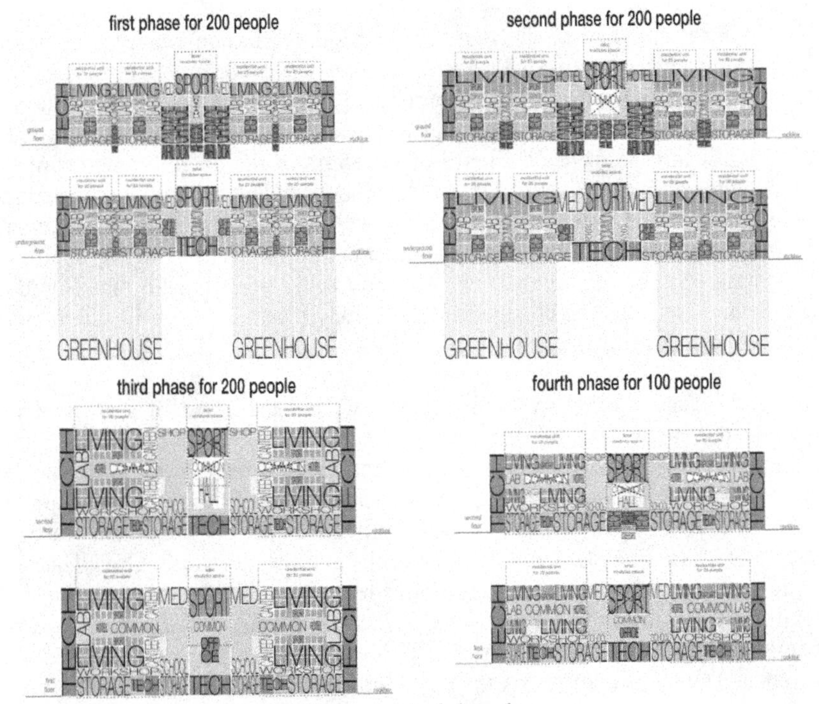


Figure 6. Modules plan

Finally, it could become technologically possible to cover larger spaces on Martian surface and transform them into human-friendly environments. One of the best suitable places are canyons that could be covered by a roof and isolated from outer space using inflatable constructions placed directly inside the canyon. The roof could be filled with water to ensure an access of sunlight and radiation protection. This water could contain oxygen generation weeds that glow during night time creating a marvelous picture for canyon habitants. Some future colony development may look something like the following image.

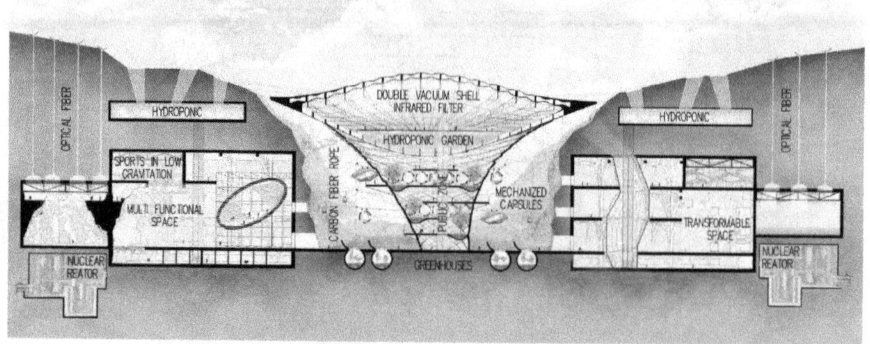


Figure 7. Futuristic vision on the colony design

**POLITICAL/ORGANIZATIONAL**

At the setup stage, the mission is managed by a single Mission Leader with clearly defined safety instructions. Every team member has a contract with the colony owner (or consortium of owners). The contract fixes all conditions of living and working, emergency cases includes insurance and costs of body transportation back to Earth in case of death. Board of Directors of the operating company on Earth includes owners, investors, and selected independent experts. It is supported by the Board of Trustees that includes entrepreneurs, engineers, scientists, architects. Organizational structure is horizontal: colony staff is highly qualified with a broad profile of competencies. Mission Leader is assigned by the Board of Directors and could not be changed by the colony team. 2-3 team members need to have advanced medical skills.

At the early stage, Colony Administration team is established. It sets rules and regulates the activities of researchers (from ethics standpoint as well), entrepreneurs and tourists. Colony Administration consults on rules with selected Colony Council. Colony Council resolves disputes inside the colony and supports the interaction of local members with organizations on Earth. In case of emergencies, Mission Leader possesses all executive rights. Organizational scheme is matrix: 20 colonists (Team Leaders) from Setup stage establish 20 specialized teams by types of activities, lead these teams, develop roadmaps for operations and scaling. The basis for hierarchical organization system is developed. The number of teams increases gradually based on topics complexity. The separate group consists of medical specialists including surgeons.

At the Growth stage, the colony may either keep its political structure or potentially gets a political status of the autonomous territory. In the second case, Colony Administration is transformed into Colony Government, and Mission Leader position is exchanged by Prime Minister position. Prime minister represents the colony in negotiations with the Board of Directors, UN and all nations on Earth. Board of Directors creates Operating Company that owns all assets created by the colony team except for administrative buildings. By that commercial operations are separated from the colony administration. The government collects taxes and import/export fees, grants Martian citizenship, ensure social protection and internal security. Elected Colony Council plays a legislative role – develops constitution and laws independent from national ones. Supreme court has a jury trial model with randomly selected members. Organizational structure is hierarchical but stays very meritocratic in order to ensure less bureaucracy and higher efficiency. All government processes and state services are highly automated based on blockchain technology and highly transparent.

The model above is the result of a review of two the most relevant examples on Earth: Hong Kong and Makao. In addition, other relevant and economically successful cases are Singapore and Taiwan.

Based on the review there are three possible scenarios of relations between the Earth governments and Mars colony:

1. The colony is under the full control of the Board of Trustees and the UN.

It can have its local laws, but the main framework is the part of the Earth juridical system. Colony Administration performs complete execution power. Colony Council plays an advisory role only and supports business and social initiatives. Internal security is ensured by the Colony Administration. The local currency is the currency used on Earth for international trade operations.

2. The increase of the independence of the colony from Earth government - first through economic independence.

The main framework of laws is the part of the Earth juridical system, but local laws play an important role. Colony Administration performs complete execution power. Colony Council may play either advisory or a legislative role in developing local laws. Internal security is ensured by the Colony Administration. The local currency is the currency used on Earth for international trade operations.

3. The colony establishes its political independence and develop its own Constitution.

Colony constitution is based on Earth juridical system but with a large part of local specifics. Government performs complete execution power and re-elected every Martian year in order to reduce corruption level. Colony council is also re-elected every Martian year. All elections are universal, and electronically provide the highest transparency of election processes. Colony Council plays a legislative role, develops a constitution and local laws. Amendments to the Constitution should be supported by two-thirds of the Martian population. Internal security is ensured by the Government. Local currency could be independent and based on blockchain technologies to ensure seamless economic transactions.

## SOCIAL&CULTURAL LIFE

The first generation of colonists will be people from all over the Earth. All of them have different cultural backgrounds, speak different languages. Social and cultural life will develop step by step. At the 1st stage, almost all the population is colony staff. Cultural life is limited - mostly media and translations from/to Earth. Development of requirements to colony staff and researchers competences to create a program of selection and adaptation of newcomers. At the 2nd stage, colony staff is still a majority but the number of independent residents and temporary visitors is increasing. All social institutes are created and developing. Cultural exchange with Earth evolves due to broadcasts and temporary visitors.

By the 3rd stage of the colony development, a unique martian mix of culture, education, and healthcare will start forming the frontier society with high motivation and abilities to achieve and advance. Authentic sport and art, the advanced education system would form a next-step society for new generations of Martians. Using media tools for permanent contact with Earth and developing tourism (including art and education tourism) the colony will play a key role in promoting an idea that human beings are multiplanetary species.

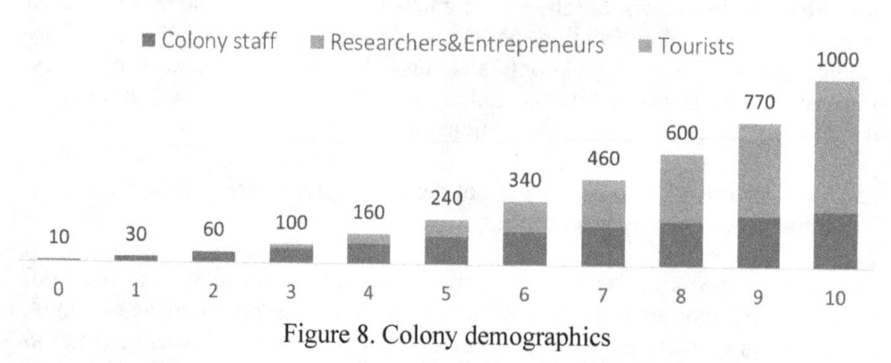


Figure 8. Colony demographics

There will be 3 social groups on the station - staff, permanent residents and temporary visitors:

1. Staff - management, engineers and scientists who will service the station - approximately 300 people. They will be recruited on Earth by analogy with the selection of commercial companies. Accommodation and food for free. They will be given a loan to buy tickets to Mars and back (assuming that every traveler needs 2 ton of cargo for 9 month flight including food and water ~ about $1M to Mars, about $0.2M to Earth). The minimum annual salary (per Martian year) for them will be $0.4M, average - $0.5M. It allows to pay back the loan in 3 years. This policy will reduce staff turnover and exclude candidates who are not motivated to stay on Mars for the long-term.

Accumulated savings due to high salary will allow colony staff to become researchers or entrepreneurs in the future.

2. Permanent residents - independent researchers, entrepreneurs, space explorers, workers, trainees, artists – approximately 600 people. This colonists pay for their stay at the station – $2M per year for apartments, food and other services. These people are sponsored by states, private firms or by themselves. This social group conduct research work and create a business environment on Mars that is critical for the success of sustainable colony development.

3. Temporary visitors who are mostly tourists and some artists - approximately 100 people. These people can stay in the colony for a few years at the station hotel. Cost of 1-year tour is $10M. It includes transfer to Mars, accommodation, food, and entertainment. Visitors can participate in the research work at the station labs. At the 3rd stage of the colony development, tourists will be able to choose a VIP tour with luxury accommodation and exclusive excursions to remarkable sites on Mars surface which costs $20M per annum.

The colony architecture allows the presence of public multifunctional zones which could be used for sport, entertainment, and education. Sport will be an important part of the social life of colonists - in limited space of the station and in low gravity it's important to have physical exercises. Some new kind of sports could be invented in martian conditions. Also, low gravity will affect spectacularity of such sports such as basketball, volleyball or wall climbing, so that MBA (Martian Basketball Association) broadcast could become very popular on Earth.

Healthcare system has the potential to reach high sophistication level due to available research on how various external conditions affect human bodies. Such areas as automated surgery, genetic modification, oncology, and organ transplantation may see major breakthroughs on Mars.

At the growth stage, advanced medicine would allow colonists to have children. Majority of people migrated from Earth will be of fertile age. The average birth rate in developed countries is about 10 births per 1 000 people. In the case, if pregnancy is safe on Mars, we can expect the rate of natural population growth about 20 - 30 births per Martian year on the 10th year. Annual payment for a child up to 16 years old is $0.5M including healthcare and education. Such payment will have to stimulate the natural birth rate in the colony and help to create a new generation of truly Martian humans.

Social connectivity will also be on a high level because Martians need to cooperate a lot to solve multiple joint challenges. Online communication can have a strong development due to high automation and involvement of robots in everyday life. Human-to-robot interfaces, AI, AR/VR, are also among the most promising areas to develop.

The vast majority of the colony population will be scientists and high-level professionals in multiple areas of expertise. This combination will allow a synergy effect to develop state-of-the-art education system. Every student will have access to cutting-edge labs and the opportunity to join advanced research works. Working together with universities on Earth through a broad range of exchange programs will positively affect education systems on both planets. The education center will include a school for children who were born at the station or came from Earth with their parents. The unique scientific environment could provide the colony with a status of educational leader and top research institute in the Solar System.

Martian nature and environmental conditions will also affect even the social and cultural life of colonists. These people will think, unlike earthlings. For example, this applies to the language. We cannot use terms containing "geo-" like "geology" or "geography" in relation to Mars. Time counting on Mars differs to Earth's as well. Duration of day and year is not the same as on our home planet, so Mars calendar (like Darian) cannot overall synchronize with Earth one - it will affect financial accounting, duration of sports events and other activities. In a nutshell, first colonists should think out of the box and constantly adjust to the new environment starting a new branch of human civilization.

**ECONOMICS**

At the 1st stage of the colony, major investments should be made. We need to develop, produce and deliver equipment - temporary living modules, boring and construction equipment, 3D printers, manufacturing plants, nuclear reactors, vehicles, navigation and communicating systems and other infrastructure. The overall weight of the equipment is 3 500 ton. Payroll for colonists and mission-related staff and infrastructure on Earth is another cost driver. Overall investment amount is ~$11.1B. Revenue sources include advertising, usage of the brand for commercial projects, merchandising revenues, media projects, pre-sale tickets for first tourists and researchers. The additional revenue stream is the export of various exotic artifacts. Overall revenue amount is ~$1.1B.

2nd stage of the colony - a period of active development. Most required manufacturing lines are established. The share of imported goods reduced dramatically, but complex equipment and chemical catalysts are still imported Earth. Investments are significantly lower than on the previous stage – $5.7B. New revenue streams are from tourists, rent and other services provided to independent researchers, entrepreneurs, and workers. First Martian entrepreneurs start their business to provide services for colonists or work with Earth. Overall revenue is ~4.4B. By the end of this period, the colony may reach break-even.
By the 3rd stage, the colony will start to make a profit. It becomes self-supportive with an import of only high-quality equipment and electronics. Marsology expeditions allow the colony to rise marginal revenue from export to Earth to $400 per kg due to sending rarer geological formations. Complete economic

chains appear with the introduction of taxation but still with preferable conditions. The colony is building the business park in addition to labs. Private investments are attracted both from Mars and Earth. The colony sells real estate and patents on land to build commercial real estate adjacent to the colony. At this stage, the colony is economically sustainable (overall profit is ~$4B) providing a business model for further scaling. Here is the colony dynamics on revenues and profit/loss.

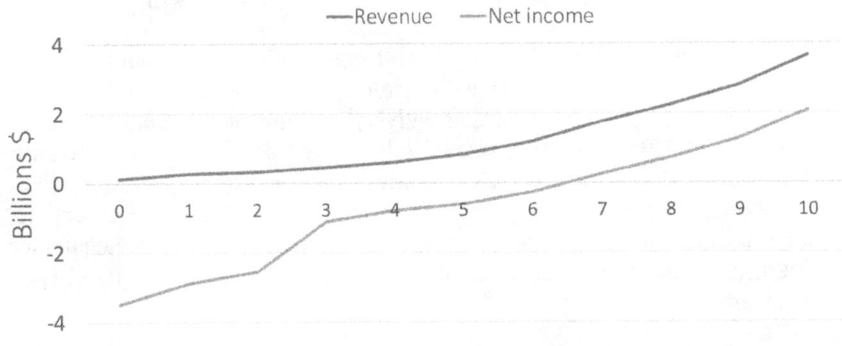


Figure 9. Colony Financials

The colony may be sponsored by private and government investors. Also, national space agencies can invest their technologies in the project - like existing equipment using on the ISS and researches made for their lunar and martian programs.

As for the financial model itself, we estimated revenues and costs using data from public sources. Our assumption regarding revenues per Martian year is the following:

1. Advertising based on ad revenues of Olympic games and NBA: $70M at the colony started with 25% growth rate.
2. Merch based on merch revenues of such brands as Ferrari, Star Wars and Real Madrid: $50M at the start with a 25% growth rate
3. Research and technology transfer based on financial statements of several world-leading scientific centers such as MIT, Stanford and NASA laboratories – 20% of stuff do research work, annual revenue per person is $1M
4. Tourism based on cases of Space Adventure, SpaceX, Blue Origin and Virgin Galactic – $10M for standard year tour, $20M for VIP year tour
5. Taxes were evaluated using financial data of the US budget and considering the size of a population and average salary assuming preferable conditions – $40 000 per person in the growth stage
6. Price for exported martian regolith is assumed as $300 per kg in the setup and early stages and $400 per kg in the growth stage. All vehicles launch as soon as possible.

The current business plan does not take into account the sale of fuel for the carrier company because it depends on many factors and may affect the price of transportation. However, it is also a potential source of income and the colony can produce and supply fuel if agreed with a carrier.

Our assumptions regarding costs per Martian year are the following:

1. Payroll - the average salary for staff on the colony is $0.5M, for staff on Earth – $0.2M
2. Import of supplies – food and water in the first years, some unique items in the future – from 6 000 kg per person at the 1st year to 100 kg by the 10th year
3. Earth support services - command and communication network - $500M annually based on NASA financial statement
4. Development, manufacturing, and delivery of equipment - from $3B in the 1st year to $0.7B by the 10th year. Total mass and cost of equipment required were calculated based on ISS metrics, papers from SpaceX, Bigelow Aerospace, Boring Company, national space agencies, and many others. The cost has been adjusted in accordance with the trends of the space equipment market to use COTS components and price dumping by private space companies.

All financial calculations were made in real prices; the base year is 2019.

The final important part is the internal economy of the station that will be determined by market rules. Internal economy is a dynamic structure and will be affected by changes in technologies, science, and social life. Colony administration can stimulate the environment for sustainable growth – creating a special economic zone and business hubs.

## CONCLUSION

The major outcome of the paper is a fact-based description of how the colony of 1 000 people could be developed and become economically sustainable on Mars with most of critical technologies available on a horizon of 3-5 years. The major implications of this result are the following forward-looking strategies which have to be developed in more detail to increase chances of overall mission success: pre-setup stage, colony scaling, Mars terraforming, technology readiness level improvements. Here are major ideas.

### Pre-setup stage

The following actions should be performed in advance to setting up the colony:

- Exploration for water sources and underground lakes
- Exploration of available mineral resources
- Deep on-site analysis of locations preselected for the colony establishing
- Missions to test models of various colony parts, materials to be used, mining and manufacturing processes, interactions human-to-human and human-to-robot in Martian environment

Colony scaling

Important implications for scaling are:

- Initial location should allow easy spatial scaling - along the slope or the canyon
- The first colony P&L should show the most promising revenue streams to focus on
- Proven business model should be scaled via new investments and/or franchising network.
- Mars colonies should allow to build a new economics for asteroid mining with Mars as a hub

Dimensions of scaling:

- Scale current base horizontally or vertically
- Establish new bases close to the current one in order to improve regional economy
- Establish new bases in other regions to take advantage of resources diversity and build planetary-wide economic links

Sources of funding:

- Current investors of the colony to develop and grow the project
- Venture investors interested in new Martian and interplanetary businesses
- Corporations that want to put a footprint in a new economy
- Governments to increase areas of influence and address some challenges on Earth

## TRL Improvements

Finally, we have looked at key technology areas where improvements may lead to significant cost reduction and acceleration of colonization process. Here are the major topics:

- Transportation: travel time reduction, lower costs per kg
- Biotech: influence of different gravity, radiation, insolation conditions on human bodies
- New materials: temperature/pressure resistant, low mass/high robustness
- 3D printing: buildings, interior, spare parts, robots, filament options
- Radiation shielding: multiple layers, artificial magnetic fields
- Food production: effects of radiation, low gravity, CO2 level, insolation, gene modification
- Closed-cycle systems: efficiency, robustness, safety, mobility, replicability
- Power and heat generation: nuclear, solar panels, geothermal, fusion
- Full autonomous mining, manufacturing, construction
- Terraforming hacks

## REFERENCES


1. http://jurnal.vniiem.ru/text/138/53-57.pdf
2. http://large.stanford.edu/courses/2015/ph241/miller1/
3. https://pdfs.semanticscholar.org/8855/2ce617b3a5999238601abd88780caefc369d.pdf
4. http://www.marspapers.org/paper/Teasdale_2018_pres.pdf
5. https://www.nasa.gov/missions/science/biofarming.html
6. https://www.lpi.usra.edu/meetings/LPSC98/pdf/1690.pdf
7. https://www.nap.edu/read/9892/chapter/4
8. http://www.geokhi.ru/Lists/List1/Attachments/6542/2015_Demidov_Bas_Kuzmin_Mars_soil_AV.pdf
9. https://apod.nasa.gov/apod/ap971017.html
10. http://www.marspapers.org/paper/Crossman_2010_contrib.pdf
11. http://www.marspapers.org/paper/Kotliar_2001_1.pdf
12. https://www.protononsite.com/products/hydrogen-generators
13. https://www.nature.com/articles/s41467-017-01165-4
14. https://kayserworks.com/#/798817030644/
15. http://camcofurnace.com/camco-hydrogen-furnaces.html
16. https://www.nature.com/articles/s41598-017-01157-w
17. https://www.nature.com/articles/srep44931#ref20
18. https://www.teslarati.com/elon-musk-boring-company-bricks-price/
19. https://www.kurzweilai.net/diamonds-from-the-sky-approach-to-turn-co2-into-valuable-carbon-nanofibers
20. https://ru.wikipedia.org/wiki/QZSS
21. https://www.dezeen.com/2018/09/27/video-driade-moon-mission-installation-furniture-movie/
22. https://raadstudio.com/project/the-lowline/

# 10: ENDEAVOUR: 1000 MARTIANS AND THEIR STORY

**Silviu-Vlad Pirvu**
contact@silviu.uk
**Mateusz Portka**
mateusz.portka@gmail.com
**Eduard-Ernest Pastor**
edupas001@gmail.com
**Sławomir Tyczyński**
slawomir.tyczynski@hotmail.com
**Ibok Kegbokokim**
ibokkegbo@gmail.com
**Roxana Lupu**

## ABSTRACT

The paper describes key elements of the first human colony on Mars, called Endeavour. Being a new home for one thousand people, its initial proposed location is St Maria crater. We provide key spatial planning and engineering assumptions as well as an overview of the systems and resources management, economical plan and socio-cultural form. We distinguish and characterize four phases of Endeavour's growth until it reaches the targeted size.

The structural engineering concepts for cylindrical habitats protected by deployable origami shields are provided, with suggestions on the designs of novel multifunctional composites. Additionally, we specify the risk assessment of the colony's systems, flow of the resources as well as key assumptions on various engineering such as transportation, robotics or manufacturing. This is accompanied by an outline from the economical perspective highlighting various revenue streams as well as a chapter on social and cultural consistency of Endeavour's community. The paper is a proposal how a sustainable and self-sufficient Mars colony could look like in the nearest future.

## ENDEAVOUR DEVELOPMENT PLAN

### The First Brave 5

A team of five people will descend from the Mars Mothership to test the first habitation modular concepts. They will collaborate with robotic arms and rovers controlled remotely from the orbit by their colleagues via VR/AR systems (Fig. 1). They will experience firsthand the conditions on Mars and will be monitored for a better understanding of what could be the best designs and human-machine interactions with flows of energy and resources.

This first mission will end with the confirmation of a viable cluster of craters, lava pits or/and Martian valleys, whose terrain profile, environmental and geotechnical characteristics will be appropriate to host around 100 to 200 people in each of those.

Fig. 1 The initial modular habitats will be deployed, monitored and tested remotely

We have identified a particular crater as an ideal host for the first phase of our colonisation process. The Santa Maria impact crater is located at 2.172°S, 5.445°W close to a larger crater called Endeavour, which the colony took its name from. Santa Maria has a diameter of 90m with the depth of around 15-16m.

**The First 100**

The first 100 colonists will reside in cylindrical structures inside the crater. The cylinders will be five-storey inflatable structures, which will be divided into two groups called "bullets" and "bubbles".

Bullets will be habitation modules, where people will eat, sleep, work and spend their leisure time. Bubbles will be designed to host hydroponic systems to provide food for the colony. Each repeatable module of these two elements will be protected from outside by anti-impact and anti-erosion origami structures with self-healing capabilities. They will form an autonomous, smart shielding system for the habitats.

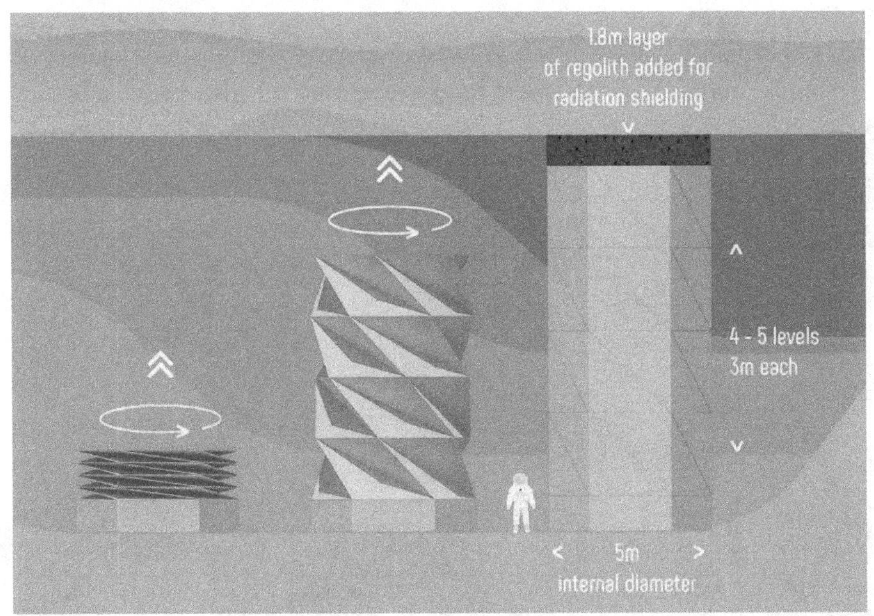


Fig. 2 The deployment mechanism of the tubular origami anti-erosion self-healing shields

To complete our triad, we assume a third type of building called “bagel” (Fig. 3). Although also constructed step-by-step in a modular manner, once the crater is filled with bullets and bubbles (the next phase) it will finally be a continuous type of structure. Located on the outside edge of the crater, it will form a characteristic industrial ring, where resources gathered from outside will be stored and equipment/robots will be maintained. Additionally, the bagel will be used as a shield for the entire colony against the Martian environment and a barrier between them.

The bagel will also serve as an airtight entrance to the colony, with decontamination chambers located in certain intervals. At this level, we assume the start of erection of the ring, so once the crater is filled, the bagel’s buildings located outside will enclose the crater.

**500 People: Halfway There**

The crater should support the habitation and food production for a minimum of 100 people, but no more than 200 in order to avoid total mission failure in case of catastrophic contingencies in one of the craters. At this phase, bullets and bubbles will start filling in the crater effectively, as the whole system is based on modular design and allows for flexible arrangement.

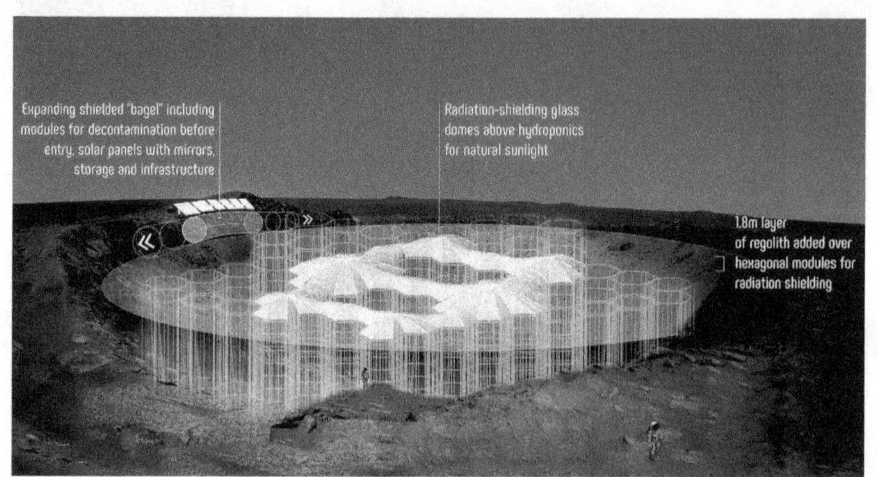


Fig. 3 'X-ray' view of the inhabited Santa Maria crater (photo credit: NASA)

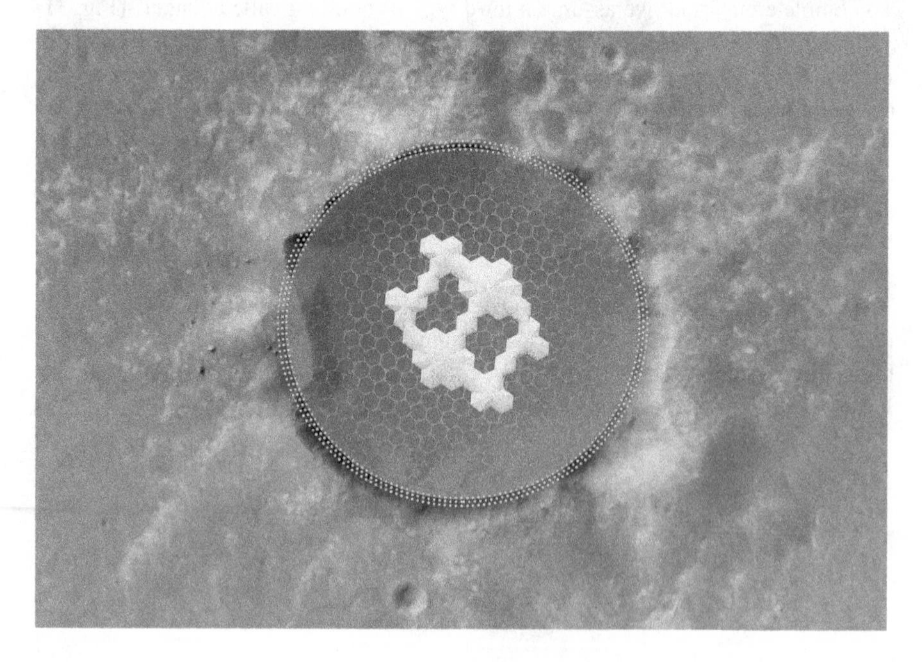

Fig. 4 Aerial view of the inhabited Santa Maria crater

The crater should support the habitation and food production for a minimum of 100 people, but no more than 200 in order to avoid total mission failure in case of catastrophic contingencies in one of the craters (Fig. 5).

Fig. 5 The concept of linked craters to separate potential contigencies (e.g. system failures, epidemics, social unrest)

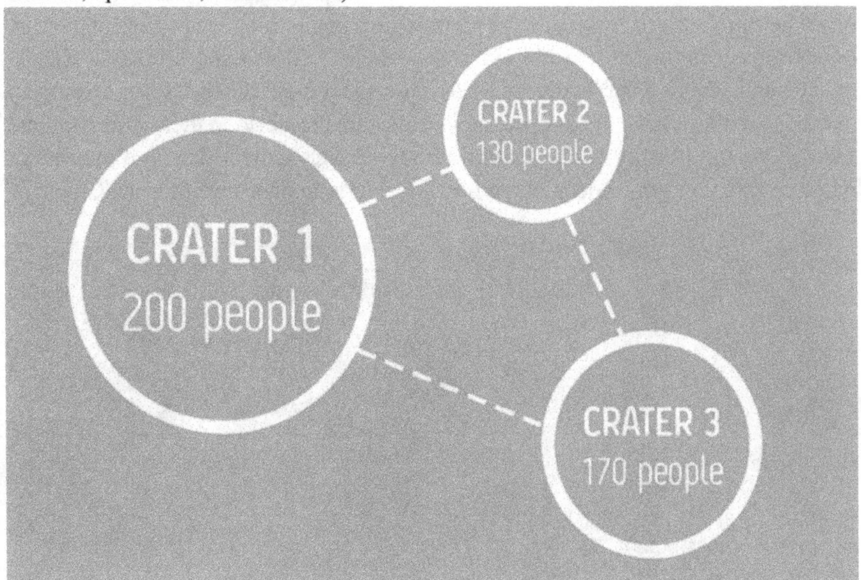


During this phase, bullets and bubbles will start filling the crater effectively, as the whole system is based on modular design and allows for flexible arrangement. A star-shaped arrangement (described later in the text) can provide the flexibility in allocating different uses in irregular shapes of craters, solving the differences in elevation at the edges and make sure the crater is air-tight sealed faster.

Once the optimum food and resource production is achieved in one crater, the further habitation modules should be deployed in the next one using similar techniques or developed 3D printing. A building of the second storage and decontamination ring will start around the second crater - the formation of the second bagel will commence.

To avoid exposure to radiation and regolith contamination, surface mobility will be used at first but underground mobility between craters is planned in later stages once it becomes more viable. A system of underground tunnels between craters will be gradually developed by extending one habitation module towards the other crater and converting it to a "dry activities" area.

**1000 and growing**

At this point, the colony should be self-sufficient in terms of food and energy requirements. Doubling the population might be convenient in a dense settlement as economies of scale apply (i.e. doubling the population usually require just +15% more infrastructure [1]). However, developing inside one single cluster might be unreliable in case of dust storms, micrometeorites and social unrest. A second cluster is vital in case of contingencies as well as a more reliable way to manage and secure a stable growth. Having a second cluster that's not dependent on the initial one will empower the new inhabitants to develop their own path and socio-economic identity, while learning from the first settlers. The two clusters of linked craters will grow and at some point converge in one large Martian City.

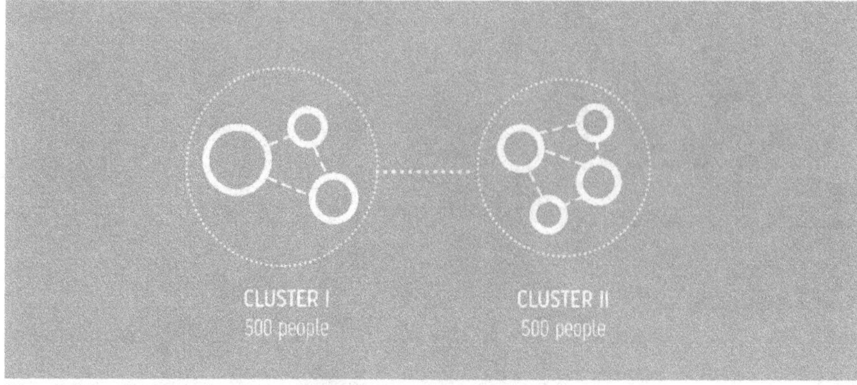


Fig. 6 Two linked clusters of craters

## THE "BBB" ARCHITECTURE: BULLETS + BUBBLES + BAGELS

**Safety first**

Bullets and bubbles are inflatable structures in cylindrical shape, with 6m external diameter (Fig. 7). We assume that their height will be around 15m. With such dimensions, the inflation should be carried out in secure environment, so that the cylinders will not be affected by external environment. Event such as dust storms or micro-asteroid impacts may interrupt or even ruin the process. Therefore, we assume that the inflation process will be executed in special protective origami coats, which will automatically deploy with a growing cylinder. After the cylinder is fully inflated, the coat could be reused. However, we suggest leaving it as a permanent shielding system of habitats.

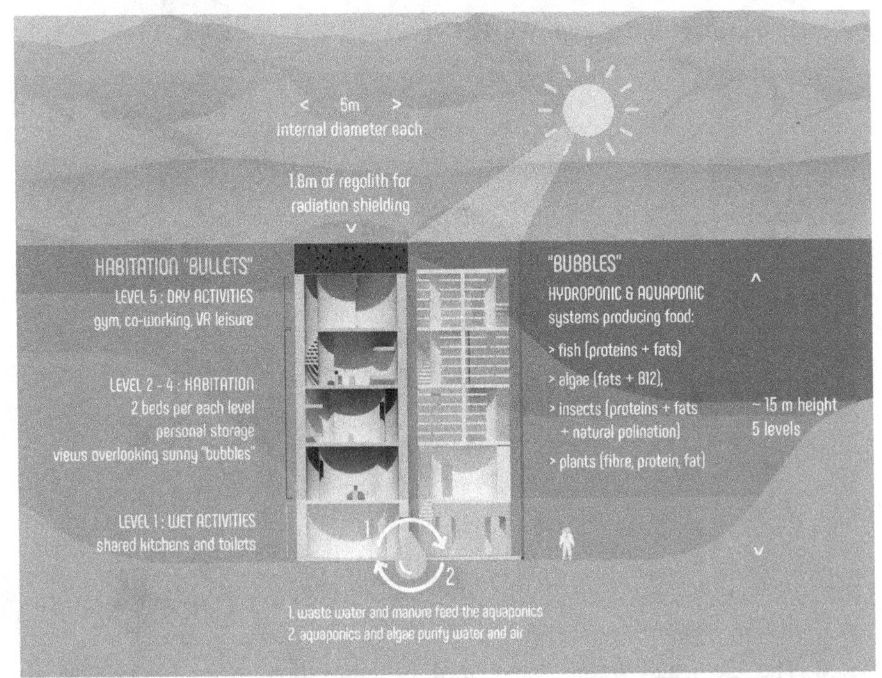


Fig. 7 The "bullets and bubbles" design

Suggested origami coats are inspired by a tubular origami pattern which allows the structure to deploy in longitudinal direction. The coats will be made of specially designed composites. To obtain a composite resisting random impacts and particle erosion because of always present winds and sand, we assume the use of plain woven aramid (kevlar) fibres. Additionally, to mimic the nature's ability of recovering after damage, the composite polymer matrix will have special self-healing intrusions. These will be microcapsules filled with chemical compounds of formaldehydes or alternatively the matrix will have multiwall carbon nanotubes inside. For more details about the composite, see section Composites Design.

**Bullets in a crater: Eat. Sleep. Work. Repeat**

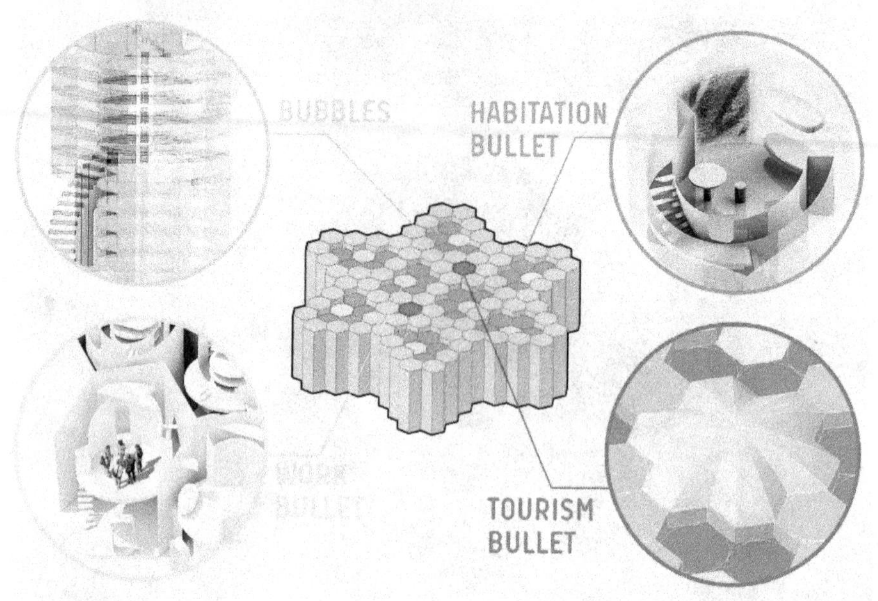


Fig. 8 The elements of the star-shaped agglomeration, types of cylindrical buildings

The first habitation bullets will be deployed on the inner edge of the crater. We assume that a typical residential bullet will have 15m of height. This means five floor levels with various uses as described in Fig.7. Access between different modules (illustrated in Fig. 8) can be ensured either by stairs or zero energy elevators (with a system of ropes). The bullet module will be covered by a 1.8m layer of regolith, which will protect the habitat against radiation. A typical bullet will have an external diameter of 6m with the internal one measuring 5m. The 0.5m gap will be a space inflated with the same atmospheric mixture that will be used inside cylinders. This is to provide a safety reservoir in case the habitats unseal and the internal chambers are depressurize. In such an event, the pressure from the safety gap can give some time to colonists to move safely to other modules.

Alternatively, for edge bullets, the gaps could be filled with crushed ice. In contact with the temperature inside habitats, it will at least partially melt from the internal side allowing tight feeling of gaps. This ice layer will provide an effective anti-radiation shield. If newer bullets are added, the room temperature on both sides of the walls will cause full melting of the ice. It could be then removed in the form of water and the gaps could be filled with the mixture of the atmospheric gases to obtain the safety layer.

**Bubbles in a crater: Eat. Work. Relax. Repeat**

In terms of production, hydroponics will be installed at the ground level at first, until specially designed for that bubbles are deployed. They will have five levels as well, destined for food production, with the lowest one serving as a storage space. In cross section, they will have the same dimensions, what will allow the use of the same type of origami Kevlar coats. However, they will be higher by 1.8m. The height of 16.8m will mean that the top of the last floor of a bubble will be at the same level as the top of regolith covers of bullets. This is because no regolith cover should cast shadow on bubbles to let as much light as it is possible inside.

To do this and simultaneously to protect hydroponic environment against radiation we assume the use of the characteristic element that gave name to bubbles. They will be covered by transparent domes, which will be made from another special composite that will be produced by the colony. The mixture of glass fibres and a matrix based on silsesquioxane will result in a transparent composite with fine mechanical properties allowing the dome to withstand even micro-asteroid impacts. Additionally, if the hollow glass fibres are filled with self-healing agent, the bubbles' cover will recover from damage. More on these materials in section titles 'Composites Design'.

An upgraded water system will be put in place to connect the aquaponics to bullets' "wet" modules. Some minerals from waste water will be consumed by fish and plants and the remains will be turned to soil and food. The resulted filtered water will re-enter the modules for drinking and cleaning. Once the habitats form locally a group of four residential cylinders, a special bullet should be erected. This one will be used as a co-working space, where people living in the neighbourhood will carry out their professional activities - medics working in hospital, education facilities, it-officers in server rooms, scientists living close to their labs etc. We will call it "work bullet". Its dimensions will be similar to those of the normal bullets and it will be covered with a 1.8m regolith layer as well.

In order to obtain coherent and organised living and working environment, the three types of buildings can form locally a repeatable cluster. It should consist of four residential bullets aligned in a crescent shape, with a work bullet inside it. The five bullets will be surrounded with bubbles filled with hydroponics providing food for the cluster. At a macro scale, the clusters can be repeated and arranged so that they form a star-shaped agglomeration. Additionally, the inside of the formed star can be filled with two more clusters. To sum up, such an agglomeration will consist of eight 5-bullet clusters. They will be surrounded by bubbles and will share food produced inside.

If we assume that the lowest level of bubbles that contains magazines will be also used for communication and transport, bubbles will serve as the "corridors" of the colony. This will mean that Martians willing to walk around Endeavour will not have to enter and disrupt bullets where people work and live needlessly.

If the star shape arrangement, we are left with two free hexagonal spaces between the eight local clusters (Fig. 8, red hexagons). They will be used for the third special type of cylindrical buildings called "tourist bullets". Similar in arrangement to residential bullets, they will serve as hotels for space tourists. Located in the centre of the crater and surrounded by hydroponics it will provide aesthetic experiences to newcomers while separating their day-to-day activities from working and living environment of colonists.

This bullet will be the only one to mach the height of bubbles. This is in order to cover it with a transparent dome as well, so that not only space tourists but also permanent colonists could have access to the Martian surface to admire its views without affecting the operation of the hydroponics. As both bubbles and tourist bullets will be covered with transparent composites on the surface, bearing in mind their organised arrangement, some special variations of the dome covers can be designed. If we analyse the proposed star shape of the agglomeration, we can notice that the biggest aggregates of domes will be around tourist bullets. This will give the most impressive design and the biggest area for experiencing Mars, what will only increase the experiences of both the tourists and colonists. Additionally, between clusters there are three bubbles adjacent to each other, so some 3-unit domes can be designed. As the buildings covered with bigger domes will be connected underground by single bubbles, the colony will look like a system of bigger glass-like composite domes connected by transparent unitary corridors from the outside.

**Bagels around the crater: Work. Store. Clean. Repeat**

During the erection of bullets and bubbles inside the crater, an external ring at its edge will be gradually constructed. Cylindrical in cross section, it will be air-tight and covered with thick layer of regolith. This will not only stop radiation from affecting people inside bagel but also will provide additional cover for the whole colony against both the radiation as well as dust storms.

The top of bagel will be covered with retractable system of solar-mirror panels (Fig. 9). When deployed to the vertical position, one side of the cover with solar panels will be facing outside of the crater to gather additional portions of the energy. Meanwhile, the second side, facing the centre of the crater will be coated with a mirror-like material to refocus sun rays towards the colony's bubbles to increase the efficiency of hydroponic system.

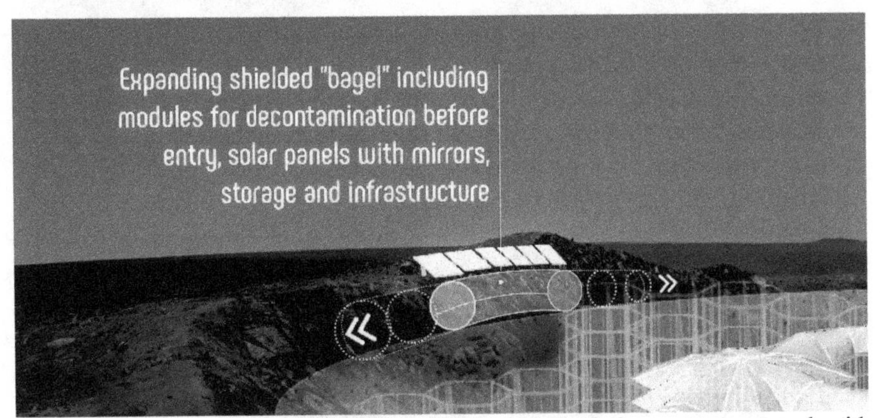


Fig. 9 Expanding shielded "bagel" developed around the crater, covered with regolith and directional solar panels

It will have many uses for the colony. First of all, this will be a place where resources gathered from outside will be stored and processed for further use. Secondly, mechanists and engineers there will supervise machinery that will clean and provide maintenance to robots working outside. As some of the robots will not be autonomous (the discovery of unknown) and will be remotely steered via VR/AR, the stations for engineers controlling them will be located inside bagel to enhance coordination between teams. Most importantly, the ring with its gates and in-built decontamination chambers what will make a characteristic entrance to the colony.

## COMPOSITES DESIGN

### Erosion resistant composites to decrease material reduction due to solid particles impact

The Mars atmosphere contains suspended solid particles (dust, fine regolith particles), which with always present winds, storms and hurricanes are going to directly affect the habitats. The tiny particles, once put in motion and transported at full speed by gusts, will be hitting the structures causing abrasive effects and the risk of erosion by removal of the material due to impact. Therefore, the composites for the habitats and other objects exposed to the Martian winds should be ideally designed in the anti-erosion manner.

As the directions of the winds are difficult to foresee, it is recommended to use balanced composites with woven fibre structure (e.g. plain weave) [2]. It has been reported that generally aramid fibres outperform carbon fibres when used e.g. with epoxy resin [3]. Glass fibres, being the most brittle, are not the best idea. Therefore, one should opt for e.g. plain woven Kevlar composites. As the required chemical elements such as H, Fe, Cl, S, Al as well as compounds such as ammonia are possible to obtain on Mars, Kevlar production can be considered.

One of the key ingredients will be Benzene, and/or its derivatives – Toluene and p-Xylene. These hydrocarbons can be produced out of the atmospheric CO2, e.g. by using the Sabatier's reaction [4]. Alternatively, both Benzene and p-Xylene can be obtained out of the Toluene, which can be distilled from the Balsam of Peru – a substance harvested from a tree native to South America called Myroxylon balsamum [5]. If the tree could be grown in the colony's greenhouses, the balsam could be used also for medical purposes, food and drink flavouring or even for fragrance.

Kevlar fibres should be embedded in a polymer matrix to provide protection from environment, reduce permeability and increase airtightness. The choice of the composite's matrix should be determined based on the environmental testing of the possible materials. The fine particles in the Martian atmosphere suggest the tendency to ductile form of erosion. However, due to lengthy abrasive processes their round shape may suggest the brittle form. Therefore, if erosion is the main concern for the composite design, the matrix choice should be based on research on how the composite is affected by the Martian environment. While thermoset matrices (BMI or polyester) are susceptible for the brittle type of erosion, the thermoplastic matrices (PEEK, PI) show ductile type of material loss [2].

**Self-healing composites to automatically repair habitat damage**

To simulate the nature's ability to transport liquid healing agents to damaged areas, like during the skin regeneration, we could use transport assisted self-healing autonomous systems [2]. A set of microcapsules filled with Urea-formaldehyde (UF), Melamine urea-formaldehyde (MUF), Melamine formaldehyde (MF) or Polyurethane/Urea formaldehyde (PU/UF) could be introduced into the polymer matrix. These capsules would release the self-healing agents once damage happens. The above components can be obtained on Mars. Urea could be product of the reaction of ammonia and CO2. Ammonia could be also produced out of hydrocarbons, which can be created using atmospheric CO2. Formaldehyde also could be obtained from hydrocarbons, out of their oxidation (e.g. oxidation of methane or methanol). Melamine production in industry is often integrated additionally into urea production, as it uses ammonia [6].

It has been shown that similar systems can reach up to 67% of healing efficiency [7]. If some "smart" self-healing systems are used, the efficiency can be boosted. The introduction of piezoelectric layers and electrically conducting materials would allow creating an electronic skin that could detect the damaged area and self-heal with the efficiency up to 90% within 10 minutes after damage [8]. If (Multiwall) Carbon Nanotubes (CNT) are used as a healing-reinforcing agent, due to identifying the damage to composites and secondary interactions with the matrix, they could surpass healing efficiency of 100% by even toughening the damaged area [9].

Before the systems are used on Mars, we need to test them on Earth and in the target environment. While woven fibre composites have been tested with self-healing injections, this has been done for glass and carbon fibres [2] - Kevlar-based systems still need confirmation. This should not be a problem as the self-healing composites have seen tremendous advance over the past 14 years. Also, once some systems are chosen, it is crucial to confirm healing efficiency and the number of healing cycles possible on Martian environment.

**Lightweight transparent anti-impact composites for domes**

The assumed hydroponic design includes transparent domes to let sunlight inside while protecting the internal environment. Such domes, while as light and transparent as possible, should also have excellent mechanical and ani-impact properties to resist e.g. micro-asteroid hits. Recent advances in material science suggest it is possible to obtain lightweight and low-cost composites that are superior in terms of mechanical properties and simultaneously exhibit up to 84% of transparency [10]. Such properties were achieved with the design including cross-woven (0/90) S-glass fibre mats impregnated with specially designed transparent matrices based e.g. on polyfunctional silsesquioxane.

On Mars, glass fibres could be produced relatively easily as the required chemical components i.e. aluminium (Al) and silicon (Si) can be found in the environment (Martian soil). Silsesquioxane is usually synthesized by hydrolysis of organotrichlorosilanes [11]. For their production, again silicon (Si) as well as chlorine (Cl) is needed [12].

Ongoing research allows expecting more advanced composites with superior properties in the near future. If glass ribbons are used, researchers estimate to obtain transparency up to 90% while delivering even better mechanical performance [13]. Additionally, if hollow glass fibres are filled with self-healing agents, Mars colonists could obtain transparent, lightweight and resistant composites capable of recovery from damage [2].

## ENGINEERING AND KEY INFRASTRUCTURE

### Critical Systems Rating

To determine the level of severity of an accident, each technical system is assigned a critical level value. Systems with Level 1 have higher priority compared to systems with higher numbers.

Tab.1 Critical System ratings

| LEVEL 1 | LEVEL 2 | LEVEL 3 |
| --- | --- | --- |
| These system failure will have immediate catastrophic effect on the whole colony or cluster. The survival of a colony is reliant on these system. | These system failure will have near term effect on the whole colony or cluster. The near term survival of a colony is reliant on these system. If a failure occurs there is about 2-3 weeks safety window to repair fault. | These systems failure will have a long term effect on the whole colony or cluster. The long term survival of a colony is reliant on these system. If a failure occurs there is about 3 months -1 year safety window to repair fault. |

**Fault Rating**
Fault rating value to determine the level of severity for any subsystem.

Tab.2 Fault system Rating

| GREEN | YELLOW | RED |
| --- | --- | --- |
| Subsystem operates as expected | Subsystem will require repair or servicing | Subsystem will need to be changed |

**Water**
Water is critical to the survival of humans on Mars. Martian surface has water in the form of water ice buried under the surface of Mars. Water is found at the polar regions, subglacial liquid water and ground ice. Depending on the location of the cluster, water will have to be extracted, purified, stored and distributed across the colony. A ground penetrating laser will be used to determine the amount of water present at a site.

**Water Extraction**
Water extraction will be done using a rodriguez well. A hole will drilled using a mechanical drill into the ground till ice table is reached. An electro thermal drill is used to drill through the ice table. Boring through the ice causes melt water to

form a pond. The pumping can begin when sufficient pond water has formed within a cavity. The size and shape of the ponding cavity depends on the relative rates of melting water removal by pump and upon the rate of heat application to the pump.

The cavity can grow laterally rapidly with a large heat supply and small pumping rate. If the pool is over pumped, the cavity will develop downward rapidly due to high temperature of the reservoir. If the rate of water extraction exceeds the rate of heat input needed to maintain the liquid pool, the well will collapse.

Water is transported from the well to storage tanks for distribution and purification.

**Water Distribution and Storage**

Water will be heated within the needed storage tank. The water flows through heated pipes into living areas. This water is used for drinking and preparing meals. Boron nitride nanofibers filters are used to filter the water during distribution.

Water Recycling: wastewater is collected and recycled for hydroponics,aquaponics, cleaning and other human operation. Waste water is converted into fuel and recycled for cleaning.

**Oxygen generation and Storage**

Oxygen is generated from the process of electrolysis of wastewater and processing of $CO_2$ and water to produce Methane. The Oxygen is stored in fuel tanks at the launch complex and also passed into the living areas to provide breathable air to the colonist.

**Food production**

This covers all hardware and software subsystems involved in the production of food. This includes sensors to monitor the health of plants and actuators pumps to control the flow of gases into the hydroponics, aquaponics, food and seed storage, silos and all other supporting systems.

**Power distribution and Generation (L1)**

Kilo power reactor will be used to generate electrical energy. It consists of a radioactive isotope undergoing fission decay, the heat generated is transferred via pipes to drive a stirling engine that converts the heat into electrical power. The system is self regulating. If the reactor overheats, the system can draw more energy that cools it back down, if it cools the core contracts increasing the rate of fission again.

Solar power uses photons from the sun to knock electrons from atoms. The electrons are harvested and provide electricity to the system. With the current improvements in the solar cells technology, Mars colony will have solar panels with an efficiency of over 60% . Solar power satellite will be placed in orbit to transmit microwave signals to a rectenna ground station. Station keeping for the satellite will be used to maintain the areostationary orbit of the satellite to correct for the effect of orbital resonance due to phobos and deimos on in Mars Clark belt. Solid state batteries and Ultracapacitors will be used to provide energies for rovers, cargo rovers, Mars exploration rovers and 3d mapping drones.

The Ultra capacitors enable fast recharge of battery powered units. Solid State batteries will be able to operate in the -20 degrees temperature of Mars. Its high energy density of 1000kW/kg will give rovers a range of over 1000km on a single charge. An auxiliary power unit called the DroneGod will provide emergency power for mobile units and space suits. The drone flies over the mobile unit to provide wireless power to the space suits or other mobile units. Electronics used inside the habitats will all be powered wirelessly.

**Communication System (L2)**

The Mars colony will use a constellation of cubesats designed and developed on Mars. Similar to the One web constellation architecture, global broadband will be available all around the planet. The communication satellite will be able to provide uplink and downlink to Mars, uplink and downlink to earth, communication with other satellites and attitude adjustment.

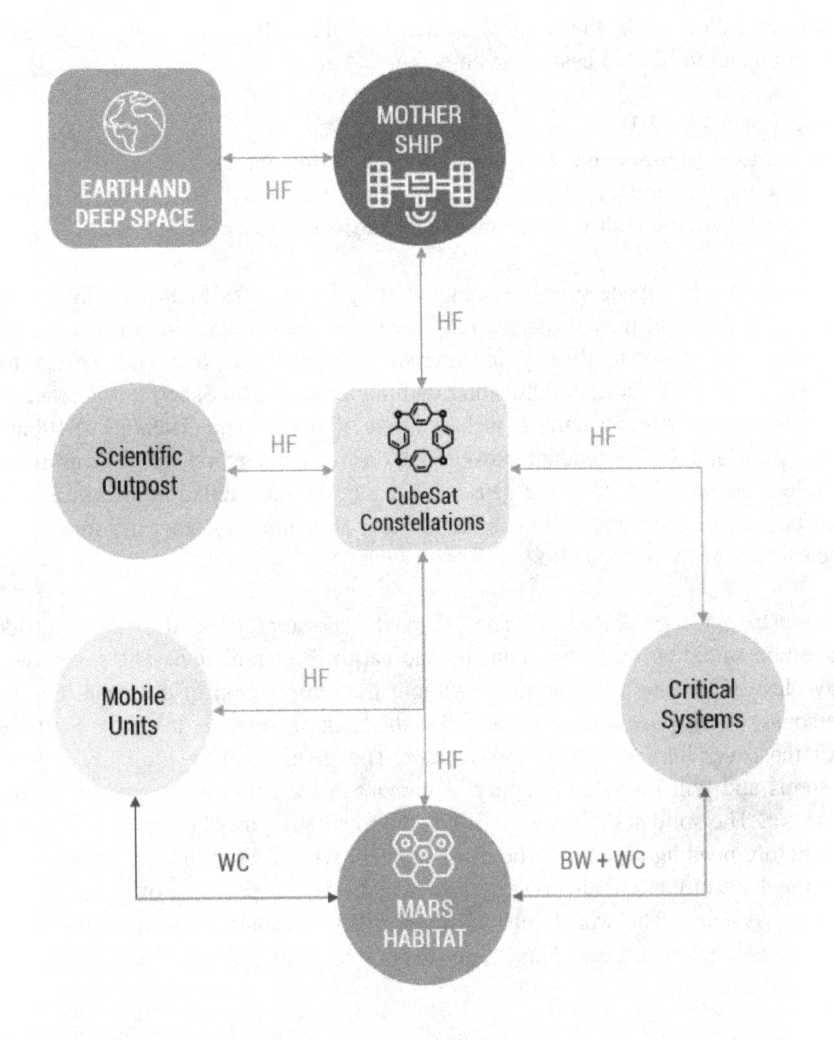


Fig. 10 Communication Lines

With advancements in communications technology and light atmosphere that prevents signal degradation data transfer rates will be in terms of thousands Gigabits/sec. Wired communication will be used as backup to enable constant communication with the critical systems. This will allow open and direct communication lines in case of an emergency.

**Transportation (L3)**

Mars colony transport network constitute rovers for human and cargo transport, space suits, jets and explorers. The transport network will have achieve level 5 autonomy with the ability for a human override as needed.

*Space Suits* - The space suits are designed using boron nitride nanotube to prevent from radiation and heated using . the space suits comes with an AI as personal assistant, AR visor to display information from the surroundings, sensors for tracking and VoIP for communicating with the habitat. The Space suit is powered with solid state batteries and has a charge time of 10mins and discharge of 100hrs. The space suit has redundant power unit that can power up all systems using wireless power. With tracking sensors and Satellite constellation all space suits can be tracked from the command center and all critical systems information of the suit can be monitored also.

*Human and Cargo Rovers* - The personal and cargo rover use boron nitride nanotube shielding to protect humans and cargo from radiation. The rovers will have level 5 autonomy to be drive without the intervention of a human. On the personal rover, space suits are attached to the back of the rover to enable martians exit the rover into the Mars environment. The rovers will have all life support systems and will be able to support 2 humans for 2 days before oxygen supply runs out. The solid state battery will be able to supply power for a range of 1000 km before needing recharge. The power system will be used to provide power to run the drive motor and an auxiliary power unit that will provide power for the life support systems. The wheels on the rover will be adaptable to enable the rover trod across rocky surfaces. The Cargo rovers will be able to carry load about 500 kg.

*Autonomous Explorers* - The Autonomous Explorers will be used to continue exploration of the martian surface. 3D scanners onboard the explorer will help surface Mapping of the entire planet. The explorers will be operated via teleoperated and fully autonomous agents. Power will be provided using a radioisotope thermoelectric generator to provide power for a long time without needing recharge. It will be an improved version of the Mars 2020 rover.

*Mars Jets* - Mars jets will provide high speed air transport across the surface of Mars. This allows access scientific outposts at various sites on the surface of Mars. The jet engine uses Carbon dioxide as oxidizer and magnesium as fuel will be able to carry up 1000 kg[16].

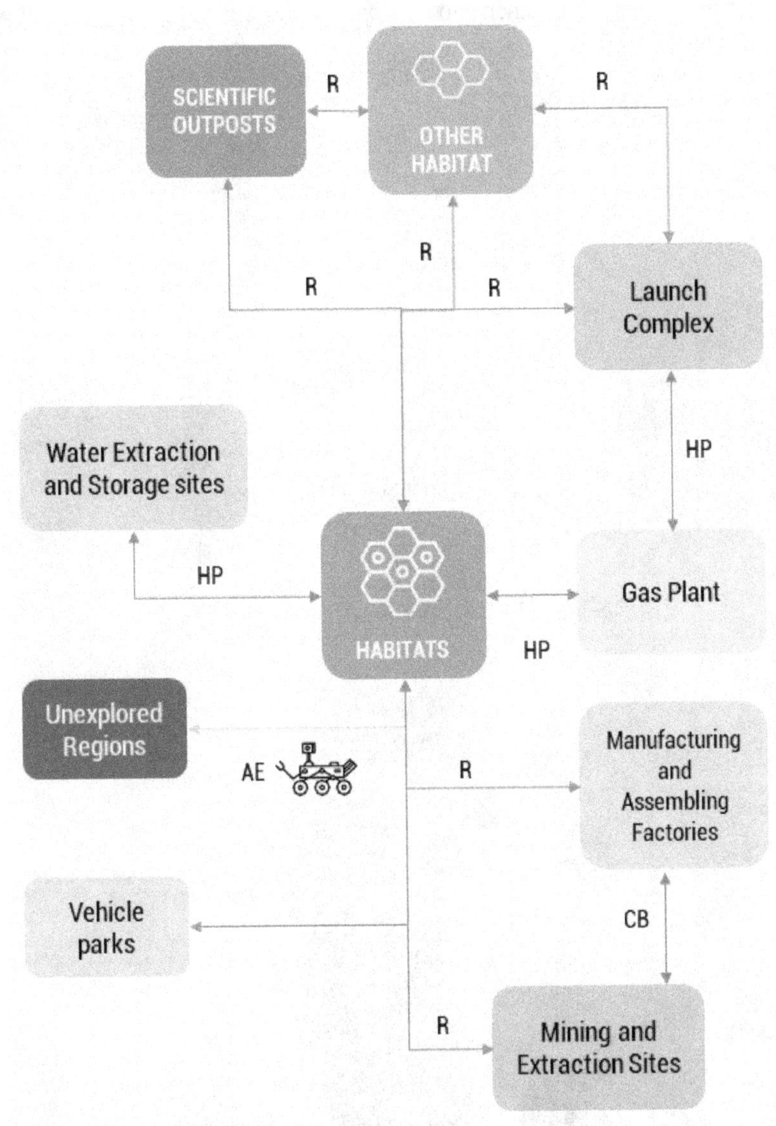


Fig. 11 Transport Network

*Industrial processing (L3):* Manufacturing and Processing - 3D printing will play a major role in the manufacturing process as the technology will advance to speed up printing time and increase the quality of the print. The manufacturing process will be undertaken by a symbiotic intelligence architecture composed of autonomous systems with human supervision. Monitoring and quality control will be done using AI models.

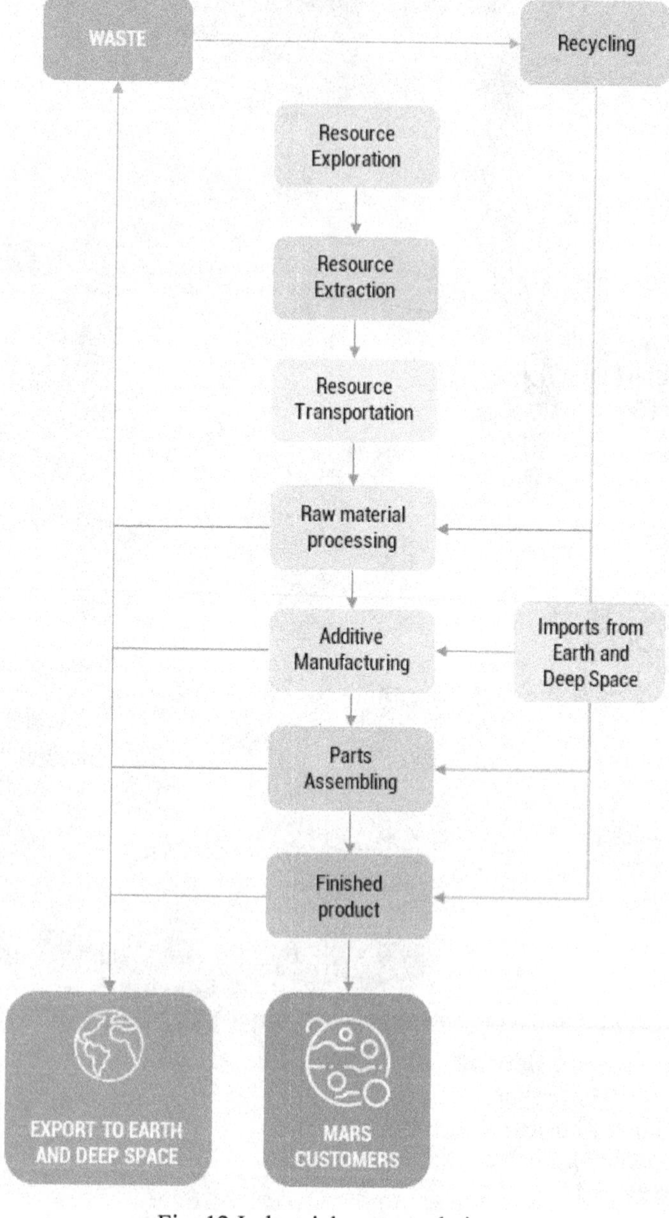


Fig. 12 Industrial process chain

There will be 4 classes of robots used in the Industrial process:
*Class 1 Robots* - part of the mining and extraction process. These are automated heavy machines used in mining operations. They are supervised remotely and operate autonomously. Power is provided by a radioisotope thermoelectric generator.

*Class 2 Robots* - part of the manufacturing and processing line. These will include robots used in factory lines similar to the Kuka robotic arm. They operate autonomously with little human interaction.

*Class 3 Robots* - bipedal robots used for heavy lifting. They can carry between 1000 - 200 kg depending on use case and were used for setting up the manufacturing, processing and extraction robots. They can be operated with a human driver onboard or teleoperated. Power will be provided using a radioisotope thermoelectric generator.

*Class 4 Robots* - teleoperated robots used to carry out repair task outside in the Mars atmosphere. They are humanoid in shape and size to allow maneuverability and easy.

**Waste Management**
Recycling: for efficient resource utilization, recycling will be an important part of the industrial process. Below are some example of waste and their recycled by products

| WASTE | BY-PRODUCTS |
|---|---|
| **Human and food waste** | Compost, Manure |
| **Water** | Rocket Fuel, Hydroponics, Cleaning |
| **Electronics** | Raw material Extraction e.g gold, silicone |
| **Machine Parts** | Melting and reworking e.g Aluminum |

**Control Computing (L1)**
IoT - A network of sensors and actuators are placed everywhere across every system across the colony and used to gather data for monitoring and control.

AI - This is has an advanced avatar interface to enable easy interaction with it. It monitors, control and optimize the systems in the colony, including the planning and design process. A symbiotic intelligence that connects machine processing capabilities with human intuition and responsibility will support on monitoring, simulating and providing advice to the Martian population and will communicate with Earth's other AI's to reveal and unlock new economic, cultural and social opportunities.

Electronics - These are radiation hardened electronics. Compared to current electronics on earth, these electronics systems are 3 years behind.

**Military (L1): Classified**

**Launch (L1)**
The launch system is designed to control the launch of spacecrafts to the Mothership. This system can be computer controlled or human control. Using AI to monitoring critical systems for and categorize the safety of the launch process. The launch system is not regarded as not part of the transport network because of its high critical system rating. In the event of an emergency evacuation, the launch system must be operational.

Tab. 3 Critical system rating for all systems

| SYSTEM | CRITICAL SYSTEM RATING |
|---|---|
| Water | L1 |
| Power | L1 |
| Oxygen | L1 |
| Communication | L2 |
| Control Computing | L2 |
| Transport | L3 |
| Industrial Processing | L3 |
| Launch | L1 |

**KEEPING BATTERIES FULL AND STOMACHS HAPPY**

In terms of food production for the assumed star shaped agglomerate, there are 75 hydroponic bubbles per 32 residential habitats and 2 tourist bullets. Each bullet is assumed to be a home for 2 colonizers per storey. As there are three residential floor per bullet This equals to 32x2x3 = 192 colonizer per agglomerate. Assuming the tourist bullets to provide bedrooms for e.g. four travellers per bullet (1 normal shared floor and two double-sized floors) we end up with 200 people to be fed with what is produced locally.

With bubbles having the internal diameter of 5m, the assumption of using 4 floors for food production (the ground floor for magazines and communication) provides around $70m^2$ of food production area per bubble (after taking into account some corridors at each level). We assume safely this can be increased to 120% of that value due to the use of artificial lights. Stacking the plants on shelves with led lights over each of the levels will allow to increase the area at least on the lower floors, where sunlight will be weak. Therefore, the increased food production area per bubble will be $70m^2 \times 1.2 = 84m^2$.

The suggested start-shaped arrangement assumes 75 bubbles what equals to the total food production area per agglomeration of $84m^2 \times 75 = 6300\ m^2$. Assuming the needs per person of 30 $m^2$, the assumed system should allow to feed 210 people.

Therefore we obtain safety margin even with relatively cautious assumption of increasing the growth area of each level by using shelving system only by just 20%. It should be noted that the star shape of the agglomeration is still a suggestion. The structural concept of bullets and bubbles is based on modularity and each of the elements can be erect depending on estimated needs or in face of random events

Tab.4 The estimation and overview of the colony's resources.

| | | OXYGEN | ELECTRICITY | HEATING |
|---|---|---|---|---|
| **Requirements per day** | One person | 0.55m³ | 2kWh | 3.5kW |
| | 1000 Population | 550m³ | 2MWh | 3.5MW |
| | Other activities | Aquaponics oxygenation | Oxygen production 1.4MWh | Ice Melting 0.2MW |
| | Total | 550Nm³ | 3.4MWh | 3.2MW |
| **Production process** | | Electrolysis | Sterling engine | Nuclear reaction |
| **Alternative source** | | Photosynthesis | Photovoltaics | Hydrogen burning |
| **Infrastructure for production** | Device | Electrolyser | Nuclear reactor | |
| | Output | 385m³/day | 0.1MWe | 0.5MW |
| | Required quantity | 2 | 35 | |
| | Total mass | 1.5Mg | 140Mg | |
| **Infrastructure for transport** | Device | Ventilation plant | Transformers | Water pumps |
| | Required quantity | 4 | 6 | 4 |
| | Total mass | 2Mg | 6Mg | 0.6Mg |
| | Transport system | Ventilation system | Electric grid | Central heating system |
| **Infrastructure for storage** | Device | Liquid oxygen tanks | Li-ion Batteries | - |
| | Capacity | 550m³ - one day reserve | 1500kVA - safety backup | - |

| | | CLEAN WATER | WASTE WATER |
|---|---|---|---|
| **Requirements per day** | One person | 0.1m³ | 0.08m³ |
| | 1000 Population | 100m³ | 80m³ |
| | Other activities | Plant watering 5m³ | Compost production 20m³ |
| | Total | 105m³ | 100m³ |
| **Production process** | | Ice melting | Sewage filtration |

| Alternative source | | Hydrogen burning | Sewage sedimentation |
|---|---|---|---|
| Infrastructure for production | Device | Water purificator | Sewage treatment plant |
| | Output | 8.5m³/day | 50m³/day |
| | Required quantity | 13 | 2 |
| | Total mass | 2Mg | 20Mg |
| Infrastructure for transport | Device | Water pumps | Sewage pumps |
| | Required quantity | 4 | 2 |
| | Total mass | 0.6Mg | 0.5Mg |
| | Transport system | Water distribution system | Sewage transport system |
| Infrastructure for storage | Device | Water tanks | - |
| | Capacity | 100m³ - one day reserve | - |

## ECONOMY: MAKING SURE THAT 'TIME IS MONEY'

### Achieving economic sustainability

It is important to quantify the reasons for colonising Mars and preparing the future pioneers undergoing this challenge with a better overview of having a sustainable environment in which the idea of colonising Mars goes beyond imagination and enters a phase of planning and execution.

As human ventures of achieving space explorations were once driven by the government owned agencies, today these are being replaced and driven forward by commercial enterprises filled with entrepreneurial spirit and innovative capabilities ready for space exploration. The challenges around colonising Mars revolve around providing a viable commercial enterprise that is capable of functioning independently in the end and provide a return on investment back to its originals donors and organisations that are willing to undergo the project of funding and colonising the red planet.

We hope to clear some of the uncertainties around the economic prospects the red planet has to offer and open a discussion to understand better the real economics and present a business case for Mars that will open easier access to funding.

### Viable economy and the foundation for thriving and growth

The new economic resources found in the space exploration programmes will potentially contribute vastly to the human development, as it will push new concepts of sustainable living, better technologies and also access to resources once were deemed impossible to reach [13]. Currently, we have developed technologies from achievements in the field of AI and Quantum Computing to new breakthroughs in 3d printing technologies and bio-tech, all creating exponential growth and future developments.

Majority of the proposals studied to date were indications of governmental agencies and financial organisations trying to make sense and understand how the economics of colonising Mars would work. On the basis of having to use real-time scenario within a realistic time-frame, we decided to focus on a circular economy concept where Mars would supply goods and services to Earth and in return alongside investments, Mars would receive labour, high-technology goods and resources, see Fig. 13.

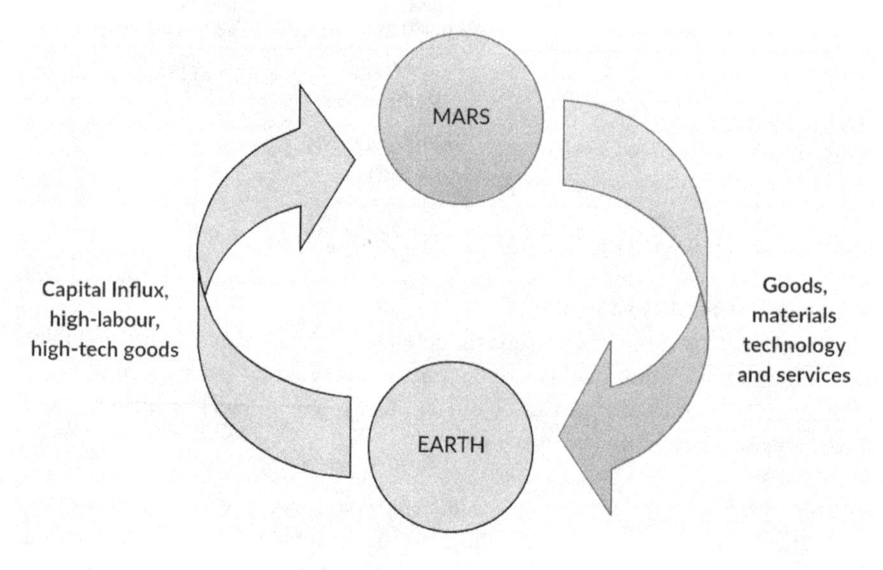


Fig.13 Circular economic flow between Mars and Earth

Whilst the 1st phase of Mars colonisation, with 1000 people we are expecting to have a settlement status and the economic model to rely on exports of goods, licensing of feature technologies to new companies and providing them help to conduct operations on the surface of Mars. Examples of such exported goods and services are modified crops, renting rovers and rv's, selling rare gems and martian rocks for various industries on earth and patent licencing.

**Revenue stream from advertising**

Using solar powered electric rovers and RV's for space exploration, could follow a similar business model as Uber did since renting the vehicles to carry out specific missions on the surface of Mars, or even commercial uses for companies are interested in exploring new advertising businesses. Japanese startup ISpace has raised $90 mill to put a satellite on the Moon and project small billboards, which gives shape to a new space economy. On the other hand in 2012 James Cameron took his film mission to the deepest part of the Ocean spending over 2 hours in deep sea levels and spending a few million dollars in cost. Furthermore, Red Bull had Felix Baumgartner to jump from an altitude of 123,491 feet in an attempt to break the sound barrier and be used as an example of how private companies think about breaking innovation barriers and push the exploration of new frontiers. The total global advertising industry has surpassed in £563 billion in 2019 and this could allow Mars to be able to capitalise on some of the total revenue. In simpler terms if we would have the colony established at the time we wrote this paper, if we managed to capitalise on 1% of the total industry we would have an income stream of £5.6 bilions yearly only from the advertising industry [15].

On the other hand, new business models could span out , such as a B2B renting vehicles which would allow easier access to Mars without the capital intensive requirements in order to have missions deployed on the red planet. The two brave NASA Rovers cost to date around $800 mill to deploy and maintain on the Mars surface. With access to resources and being able to build rovers in-situ, gives way to a whole different level of access to the Mars surface for earth-owned companies, as they can provide cheaper access to Mars resources, data gathering, site observation or even search for molecules or other bespoke deployments and industrial applications.

**Martian Patent Licensing**

The billions that will be spent on technologies to get humans and establish a colony on Mars will provide significant opportunities not just for Mars but Earth applications as well. One field that requires mentioning is biotech and how different studies made on Mars could be sent back to Earth for further research. Under the assumption that sending payloads to earth would only cost $200/kg, this type of research could bring early return on investments and further the interest of different companies to join the ente, which will stimulate demand and further growth. To give an example we can look at the current medicine environment where antibiotic resistance is becoming a health crisis and this could endanger the lives of millions of people. A hybrid biology created in a Mars environment could tackle this and other areas that could span out to be a trillion-dollar health technology market.

**Other Goods for export**

Crops for Export, IP patents in self-replicating robots, automation, biotech, farming, sustainable materials and other advanced robotics, remote power systems, space tracking and guidance systems, remote telecommunications and information technology systems.

Licensing on Earth inventions created under harsh conditions on Mars could bring vast amounts of revenue that will support the R&D on the Red Planet, and these would continue to raise the standards of living on Earth and push technological advancements.

**Mining Operations and Belt Economy**

Over 1000 mill asteroids are available around the Earth and the Asteroid belt near Mars. These vast amounts of resources are rich in mineral deposits, and the close proximity to Mars has drawn the attention of an extremely lucrative business. As of today, there are many prospecting missions initiated with many more to come with the specific focus of mining these resources. The first colony will rely heavily on mining carbonaceous chondrites which is a category of rocks which will enable the extraction of oxygen, water and hydrogen as the first commodities that will support the off-Earth settlement.

This will enable not only faster development of resources but potentially a prime motivator in getting private and state-owned companies to buy and use these resources to further expand their exploration of space and reveal an economic engine. Most achievable would be to target water-rich minerals that are easier to extract and could be done with existing technologies allowing companies establish on Mars trade-routes with Moon bases and Earth private and public companies in exchange for materials and other technologies that could span out these companies.

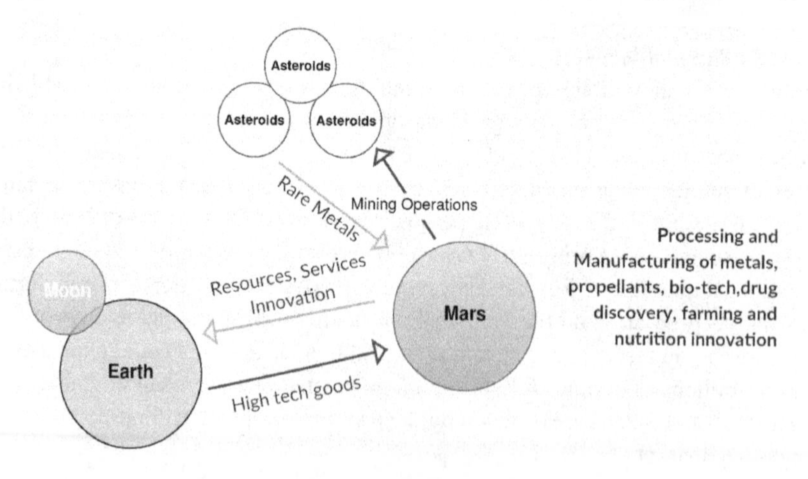


Fig. 14 Conceptual trade route between Mars, Earth and the Asteroid belt

When we talk about space commodities, an important caveat that must be mentioned here, is the possibility of the mined resources to stay up in space as the economic incentive of returning such assets to earth will be expensive and inefficient. Moreover, reusing the mined resources on Mars will fuel growth, help build colonies and sustain the new economic viability in space.

**Mining Resources and trade options**

This model takes into account the principal currency of trade on Mars is most likely to be water and hydrogen as it would be used not only as a propellant but also as a mean to decrease the cost of space exploration. With having an early colony to be able to produce that fuel at the source, it will not only help boost local infrastructure but also opening better access to longer missions and cheaper transportation costs. Considering the colony have already established these miniature refineries and the cost of having unmanned missions to overcome the challenges have been achieved, we forecast the market to be substantial and account for one of the main drivers of the economy where being able to offer the service to refuel on Mars using available resources is comparable to having control over the energy supply for an entire planet.

Table 5. Total Number of Launches Historically (source: spacelaunchreport.com)

| Year | Unmanned Launched (Failed) | Manned Launched (Failed) | Total Launched (Failed) |
|---|---|---|---|
| 2010 | 67(4) | 7(0) | 74(4) |
| 2011 | 77(6) | 7(0) | 84(6) |
| 2012 | 73(6) | 5(0) | 78(6) |
| 2013 | 76(3) | 5(0) | 81(3) |
| 2014 | 88(4) | 4(0) | 92(4) |
| 2015 | 82(5) | 4(0) | 86(5) |
| 2016 | 80(3) | 5(0) | 85(3) |
| 2017 | 86(6) | 4(0) | 90(6) |
| 2018 | 110(2) | 4(1) | 114(3) |

If the present value of a rocket to be fuelled is considered to be less than 1% averaging about $200,000 of the total launch mission cost, having an in-situ facility could decrease this by a factor of x2 but the on control the market price due to high barriers of entry and expensive process to replicate the concept could keep the price to stagnate. As presented in Table 5 the number of mission launches have grown on a year-to-year basis, and this provides the basis for the mission forecast presented in Fig 15.

The total addressable market for Mars fuel station will depend on the level of competition in the current space race with the companies and nations that will take part and set goals and objectives similar to mining the asteroid belt and establishing processing and manufacturing facilities on Mars. If a company uses $100k worth of fuel to deploy mining operations and does 10 trips per month, this will result in $1mil / month / company. If we are the only station on Mars and we multiply that by a factor of 10x, revenues would increases exponentially, driving long term plans and immediate benefits to the Mars economy.

**Coins, Luxury Items and other Martian Material Exchange**
Historically, fragments from a Martian meteorite was sold in 2012 for $22,500 an ounce. Assuming this is due to the scarcity of the resource and our colony will have access to a more stable trade route where we can ship various resources more regularly and with the cost of just $200/kg per cargo sent, this could turn into a multi-billion opportunity.

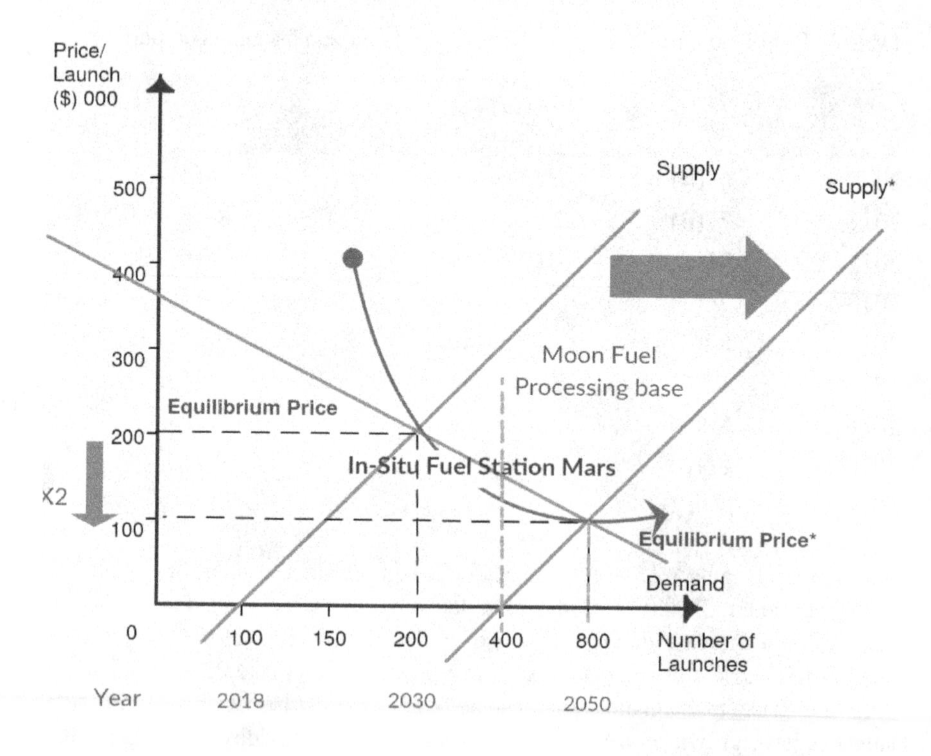


Fig 15 Supply and Demand Diagram

The market for Diamond Jewellery in US alone has surpassed $82 billion in 2017 [16]. Fig. 16 shows the potential route from the extracting facilities on Mars to manufacturing companies on Earth and how this could be potentially distributed to the front-end selling point. We could potentially explore an pre-sale auction based model on a $50.000/kg ( a theoretical framework in which labour,extraction, and distribution are covered). If that would be the case, if we bring back 1 kg of various Martian rocks and gems to the processing facilities, we could end up with a profit of $700,000 / kg, if demand is high and supply is controlled and well regulated.

Under these assumptions Mars rocks could turn into exotic must-have items, where gems and other precious metals could recoup some of the early investments costs if we build an auction type of business around the samples returned on earth. The diamond market is a trillion-dollar economy on Earth and Martian gems and metal could be worth far more than their Earth counterparts, not necessarily due to the scarcity but point of origin could help boost the value.

Sending them as raw materials to manufacturing bases on Earth where they could be processed and auctioned could behave as a secondary market for capital intensive companies who will boost in the short term the Martian economy. We believe that applications are extremely diverse for these goods and prices might vary but will still remain an extremely profitable business for decades to come.

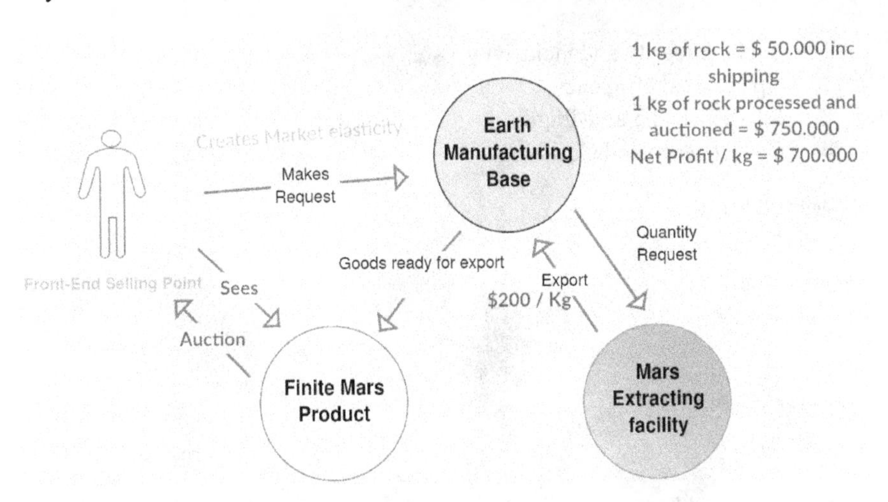


Fig.16 Example of the pricing model based on historically similar auction model of Mars fragments

We have examined the prospect of colonising Mars and although its economic viability remains very much an unknown factor with many variables in place, we decided that the best way of approaching this was through business models that are already working on Earth and that the parameters of establishing and growing such businesses are very well known. However, the biggest wealth that Mars can give to Earth is innovation that is beyond the realm of being able to quantify it and monetise it. We assume Mars will help push the boundaries of exploration giving us a door into the universe which no economic model would be able to comprehend and quantify that.

## SOCIETY: LIVING THE MARTIAN DREAM. TOGETHER

Mars will become the new frontier, as the New World was considered when the first settlers reached the shores of America, so will Mars be for the new colonists who will reach its shores. Any discussion around designing a social and political framework for the first colonists will have to be carried out from Earth as it will take decades before Mars would potentially gain its independence and draw its own system of governance and principles.

### Designing a human-centric society on Mars: The initial phase

We assume that the first colony of 1000 people will be part of the initial phase and it will be split into a number of settlements governed by the companies who operate on Mars and be populated with their employees / colonists. The main principles that will support the design of a social-political framework would have to rely on the following principles:

1. Social integrity and inclusivity,
2. Operational efficiency,
3. Resilience and anti-fragility,
4. Co-operation and Co-living.

Fig.17 Conceptual Framework for decision Making on the Mars Colony

The habitats will foster an even number high specialised and trained individuals where health, intelligence, emotional stability and skills will be a decisive factor in choosing the candidates for the colony.

### Designing a human-centric society on Mars: Implementation

Learning from the data obtained on Earth on how to build social, political, cultural systems based on training facilities and periods of testing the candidates in Mars-

like environment could be the first implementation on the most promising frameworks and selecting one for in-orbit (and then colony) deployment.
The social framework will likely revolve similar to a village with appointed community leaders, in our case each settlement will appoint a representative that will try to improve the experience, deal with issues and collaborate with other settlements. They will hold periodic meetings and same as a council meeting, the issues and concerns around the colony will be addressed. As an addition to this, a dedicated small team of specialised people will deal with planning meaningul activities, run therapy sessions and promote scientific-based training and education. In terms of political framework, the initial phase will very much rely on their Earth counterparts as they will be the ones funding and providing supplies and technical help to the colony.

## ONE THOUSAND TO ONE MILLION

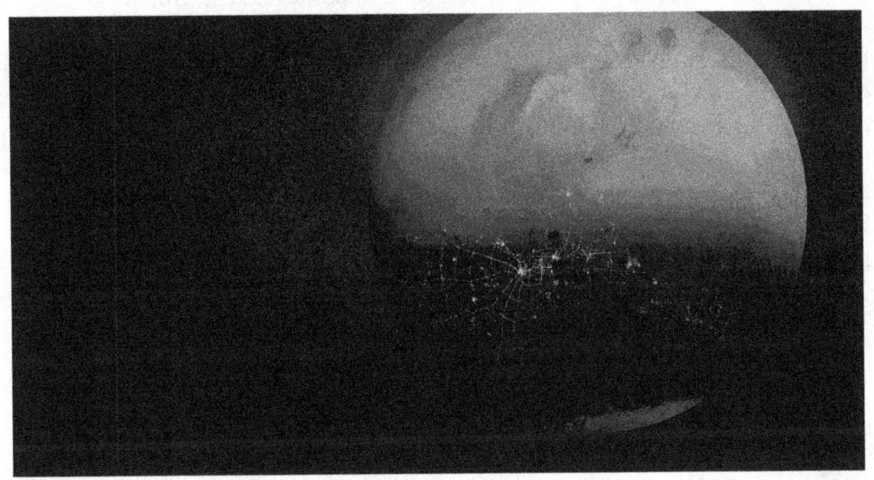

Fig.18 Greater Endeavour – 100 years after first settlements

Endeavour is a small but ferocious leap towards the stars. Many global cities started with one thousand people and then grew into millions. With responsible commitments stretching beyond our human lifespans, humanity has the chance to settle and reborn as a star faring civilization.

## ACKNOWLEDGEMENTS

We would like to give our thanks to the British Interplanetary Society, Andy Gray, Stanley Starr, Ellen Harrison, Gracinda Ferreira and Daniel Costache for making the Endeavour vision plausible and meaningful.

**REFERENCES**

[1] “Scale: The Universal Laws of Growth, Innovation, Sustainability, and the Pace of Life in Organisms, Cities, Economies, and Companies” (2017), Geoffrey West.
[2] Barbero, E. J., ed. (2015). Multifunctional Composites, CreateSpace, North Charleston, NC, US.
[3] Tsiang, T-H. (1989). “Sand erosion of fiber composites: testing and evaluation.” Test methods and design allowables for fibrous composites., 2:55-74.
[4] Heidmann, R. (2015). " A Mars Colony – A Tentative Technical Analysis." Planete Mars, http://planete-mars.com
[5] "Balsam of Peru", Wikipedia (20 Mar. 2019)
[6] "Melamine", Wikipedia (20 Mar. 2019)
[7] Kessler, M. R. and White, S. R. (2001). “Self-activated healing of delamination damage in woven composites.” Composites Part A: Applied Science and Manufacturing, 32(5):683-699.
[8]Tee, B. C-K., Wang, C., Allen, R. and Bao, Z. (2012). “An electrically and mechanically self-healing composite with pressure- and flexion-sensitive properties for electronic skin applications.” Nat Nano, 7(12):825-832.
[9] Qian, D., Dickey, E. C., Andrews, R., and Rantell, T. (2000). “Load transfer and deformation mechanisms in carbon nanotube-polystyrene composites.” Appl Phys Lett, 76(20):2868-2870.
[10] Krug III, D. J., Asuncion, M. Z., Popova, V. and Laine, R. M. (2013). “Transparent fiber glass reinforced composites.” Composites Science and Technology, 77, 95-100.
[11] "Silsesquioxane", Wikipedia https://en.wikipedia.org/wiki/Silsesquioxane (21 Mar. 2019)
[12] "Trichlorosilane", Wikipedia https://en.wikipedia.org/wiki/Trichlorosilane (21 Mar. 2019)
[13] Velez, M., Braisted, W. R., Frank, G. J., Phillips, P. L., Day, D. E. and McLaughlin, M. D. (2012). “Impact strength of optically transparent glass ribbon composites.” J Compos Mater, 46:1677–95.
[14] Space Launch Report: Orbital Launch Summary by Year (2019) [online] Available at https://www.spacelaunchreport.com/logyear.html (Accessed 31.03.2019)
[15] Growth of advertising spending worldwide from 2010 to 2021 (2019) [online] Available at https://www.statista.com/statistics/272443/growth-of-advertising-spending-worldwide (Accessed 31.03.2019)
[16] Diamond Market Overview (2019) (Accessed 31.03.2019).
[17] NASA concept for a Mars Airplane. Available at http://www.wickmanspacecraft.com/Marsjet.html/ (Accessed 31.03.2019)

# 11: THE NEWMARS COLONY CONCEPT

**Steve Theno P.E.**
Retired Engineer
stevetheno@pdceng.com

## INTRODUCTION

This report presents the initial concept for NewMars, the first permanent colony to be established on Mars, initially the home for 1,000 colonists. Many will participate in the Earth-side planning and the construction on Mars. Others will join as the colony develops bringing ever-expanding diversity, expertise, innovation and opportunity.

The NewMars colony concept is founded on several key underlying philosophical principals:
NewMars is not attempting to recreate Earth on Mars, but rather embraces the uniqueness, beauty and richness of a new world.
NewMars embraces the need to be a spacefaring society—to explore, to discover, to learn and to expand our understanding of the universe we are a part of.
NewMars will not be a remote work center supporting a "two on – two off" concept. Colonization is a one-way trip! NewMars will be a permanent home community.
NewMars motto: Go Red – Live Red – Thrive Red. NewMars will embrace and sustainably leverage the locally available natural resources; engage and foster Martian ingenuity; and unleash the spirit, adventure and inspiration of Mars!

The NewMars concept recognizes the harsh environment within which it will exist. However, it is developed with the knowledge that humankind has survived and thrived in extreme environments on Earth and continues to do so. And not by employing "rocket science" so to speak, but with proven technologies, practical ingenuity, creative solutions and a willingness to adapt and change with lessons learned. A great example is the US Amundsen Scott South Pole Station located at the South Pole, high on the Antarctic Polar Plateau, a science support community varying from 50 to 150 souls depending on the season. The Station has evolved over its life, adapting with lessons learned and leveraging innovation and developing technologies. It has been doing so successfully and uninterrupted since 1956. Consider a comparison of the challenges:

| NewMars South Pole vs Mars | *Mars* | *South Pole* |
|---|---|---|
| Average Temp Range | 60F to -150F | 0F to -100F |
| Length of Day | 24 hrs 39 mins | 8076 hrs |
| Resupply Timeline | Every two years | Once per year |
| Water Source | Ice | Ice |
| Natural Vegetation | None | None |
| Endemic Life | ? | None |
| Personal Gear Required Outside | Heated, pressurized, suit w/breathing supply | 20# cold weather gear |

The NewMars concept also recognizes that it is not a science mission. It will not be staffed with the Earth's leading experts (although there might indeed be a couple amongst the colonists), nor supported by a legion of specialists and technicians standing by on Earth to plan, research, study and model every situation and dispatch commands. NewMars will be a living, thriving community operated and maintained by craftsman, tradesman, technologists and practitioners living at the colony and managing practical technologies leveraged by Mars grown ingenuity to Live Red!

## TECHNICAL DESIGN

The NewMars technical design solution is based on proven technologies applied to successfully address the extreme Martian environment. The systems that will be used make use of leading edge, as opposed to bleeding edge technologies if you will, selected based on their ability to be readily operated, maintained, repaired and upgraded by the colonist population and the resources they will have available.

The final technical design for NewMars will be evolving in parallel with both ongoing Mars exploration and technological advancement on Earth. As such, a guiding principal for NewMars is that all systems will be planned for flexibility to both adapt to the latest information and integrate a range of technologies. This will enable the optimum solution available during pre-launch fabrication and assembly to be selected. It enables the systems to integrate new technologies as they emerge, to be flexible in accommodating growth and change, and to more readily adapt to lessons learned once in place on Mars.

All NewMars systems are developed based on the following key principals:

- Baseline systems are founded on what is known and do not rely on potentially what might be available.
- All primary central systems are provided in a minimum 331/3%+331/3%+331/3% configuration. This does not necessarily provide full redundancy (N+1), or double redundancy (N+2), but it does economically meet primary system requirements while avoiding the potential for catastrophic failure.
- All primary distribution systems (water, electrical, thermal, etc.) are duplexed (N+1).
- Individual components of central systems and components of local systems may be duplexed (N+1) based on detailed operability, maintainability, reliability and failure risk analysis during detailed design. In some instances local systems may be readily backed up by direct connection to the central system in lieu of the need for redundancy. In all cases, the guiding philosophy will be fail safe, that no single point of failure can compromise any infrastructure, system or facility.

## Energy

Both electrical energy and thermal energy will be required for the colony. A number of energy sources are available and are considered.

### Solar

The average solar flux at the Mars surface is approximately 500 w/m2, about half that of Earth. Generation in the range of 10-100kwe using photovoltaics is viable and equipment could be transported from Earth. It would be suitable for small remote sites and portable applications. This technology would be applicable to larger scale utility class generation however installed capacity sufficient to support the colony in a significant way would need to be in the 500-1000kwe range, uneconomical to transport from Earth. Solar could provide a component of the colonies energy needs, installed initially at 100kwe capacity and expanded over time as manufacturing capabilities mature on Mars.

### Wind

While the average near surface wind velocities on Mars are compatible with conventional wind generation equipment, the density of the Martian atmosphere is quite low. As such, the potential wind energy is marginal. Meaningful wind application would likely require placing the wind generation equipment at altitude, in the region of higher wind velocities. This could be accomplished with fixed wind generation equipment placed at higher elevations, or wind generation equipment could be floated on tethered balloon platforms at optimum altitudes. Wind generation in the range of 10-100kwe could be transported from Earth and would be viable for small remote sites and portable applications. However larger scale utility class generation would need to be in the 500-1000kwe range to support the colony in any significant way, uneconomical to transport from Earth. Some form of wind energy generation could provide a component of the colonies energy needs in the future, dependent on the nature of the colony site and the development of on planet manufacturing capabilities.

### Geothermal

It is anticipated that geothermal energy is available on Mars, although it has not yet been proven, nor the extent of the resource defined. If available its value for the colony would be highly dependent on the proximity. Geothermal would be a source for both electric and thermal energy. Development would be appropriate only in the utility class range (1000kwe) given the infrastructure necessary to utilize it. If a developable utility class geothermal resource was available in close proximity to the colony, it would be an optimal energy source. Given the source development, transmission network and infrastructure development required to implement such a source, it would not likely be a first energy source, but would be developed as the colony matured.

**Nuclear Fission**

Nuclear fission reaction based utility class generation systems are a mature technology on Earth, however, only in scales orders of magnitude above the demand of the colony. Transportation from Earth would be unrealistic.

Micro-modular Reactor (uMR) units have been developed at the research level for space vehicles and space stations. The size range has been focused on 10-100kwe. Generally they involve a reactor with a heat pipe structure to transfer heat to a secondary working fluid that then generates electric energy via a Stirling engine or an open or closed loop Brayton cycle. Examples include the SAFE400 and more recently the Kilopower Reactor. Such units would be ideal for initial colony construction electric and thermal energy. They would be transportable, packaged, shipped from Earth.

Very Small Modular Reactor (vSMR) units are now in research and development. This research is targeting the 2-10MWe size range. US DoD is very interested in this technology and would require many of the same attributes as needed for NewMars. Such units would be in the packaged, transportable range from Earth. Current research models are focused on self contained, plug and play units; Halos and MegaPower are two current examples. The units typically utilize heat pipe heat transfer from a nuclear core to a working fluid that generates electricity via an open or closed loop Brayton cycle. This technology would be very applicable as a primary energy generation system for the mature colony.

**Fuel Cells**

Fuel cells are a well developed energy technology for both terrestrial and space applications. Typical commercial sizes range from 1kwe to 2mwe and would be transportable from Earth. The application of fuel cells on Mars is viable using a hydrogen and oxygen feedstock. However, they cannot be considered a baseline energy source. Energy is required on Mars to generate the hydrogen and oxygen, and as such, an energy loop involving fuel cells would function at a net loss. Fuel cells could be a supporting component of the colonies energy infrastructure applicable for transient remote camps, surface transportation and short term back-up energy.

**Selected NewMars Energy System**

The baseline energy system for NewMars will be a mix of PV solar, uMRs and vSMRs grouped into Energy Centers. Three separate Energy Centers will be provided, separated, but linked by a distribution grid, creating a distributed energy network. Each Energy Center will be capable of supplying a minimum of 331/3% of the peak colony demand.

The electrical energy distribution grid will be conventional AC electric. A duplex arrangement is provided for redundancy. The thermal energy distribution grid will be a closed loop, pumped hot fluid system, also duplexed for redundancy.

Load centers will be provided at points of use; tapping the central energy distribution systems for local distribution. Secondary systems will be decoupled and isolated from primary systems for failure risk mitigation.

The application of PV solar, uMRs and vSMRs establishes a baseline energy system that is universally applicable at any selected colony site. Furthermore, it allows the energy system to grow efficiently with colony development. uMRs can be readily transported and setup to support initial construction. As the colony matures and vMSRs are brought on line, the uMRs can provide load matching, can serve as backup, or can be repurposed for remote site applications. Similarly, small scale PV solar installations can provide a ready form of back-up generation during construction. Then, they can serve for supplemental input, backup generation, be repurposed for remote site applications, or serve as the basis for an expanding solar component.

The distributed energy network maximizes reliability and can readily adapt to growth and change. Should geothermal or wind resources prove to be viable at the colony site, they can be readily integrated into the system at any stage.

Invertor systems with battery backup, coupled with hydrogen oxygen Fuel Cells will be employed to provide grid stability, maximize the energy generation system efficiency and provide short term backup. During periods of excess power generation (especially from PV solar and wind sources), excess energy will be used in the electrolysis of water to generate and store hydrogen and oxygen. During short term energy demand peaks or longer term energy generation outages, the batteries coupled with the Fuel Cells provide both backup electric and thermal energy. Invertor/batteries/Fuel Cells would be integrated into the Energy Centers and also be located at key load centers providing local backup.

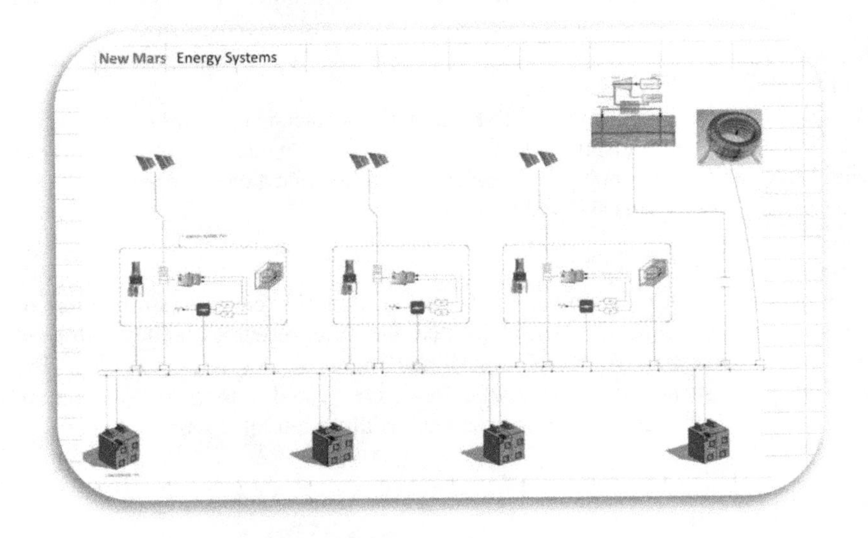

| NewMars | Power Demand | | |
|---|---|---|---|
| Core Conditioned Space | 2,000 | kw | 2.0kw/person; McMurdo Station 1.8kw/person; South Pole Station 4kw/person US average 1.3kw/person |
| Industrial and Manufacturing | TBD | kw | Baseline in above; construction seperate; future Manufacturing and Industry TBD |
| Food production | | | Food Production is controlling load. Actual load depends on final greenhouse-chamber mix |
| Greenhouses | 0-3,850 | kw | Range varies from 0% to 100% greenhouse based production |
| Growth Chambers | 0-32,000 | kw | Range varies from 0% to 100% growth chamber based production |

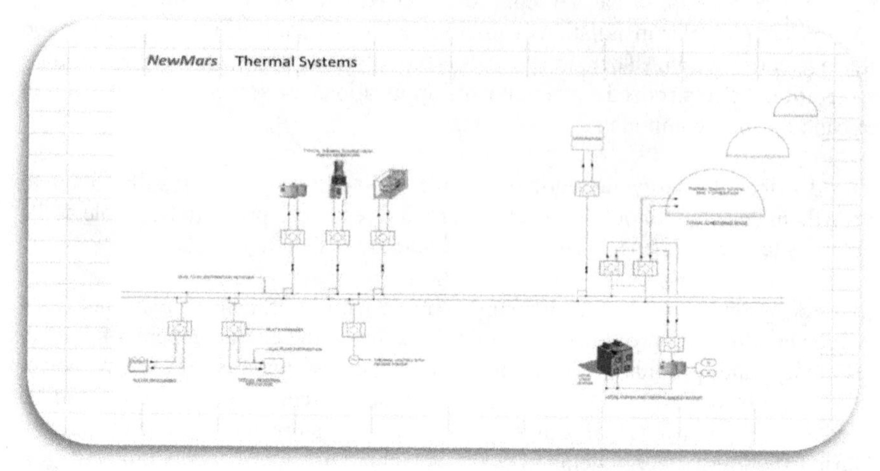


## Water

A reliable water supply is a critical resource, necessary for fuel production, oxygen production, human consumption and agriculture. A number of in-situ water resources have been identified for NewMars and are considered.

### Polar Ice Caps

Water in the form of ice is known to exist on the surface in polar ice caps. This ice could be harvested, transported to the colony site and melted to generate a water supply. However, it is anticipated NewMars will be quite a distance from the polar ice caps challenging this as a viable source.

### Subsurface Aquifers

It is postulated that there are subsurface water deposits, either liquid aquifers or ice lenses. If such deposits could be located, the water resource could be extracted and either processed on site or transported to the colony and processed. The proximity of such sources to the colony site would impact the extraction, processing and delivery mechanism, as well as the extent of resources and energy required.

### Ice/Regolith Mix

It is expected that ice is held in the near surface Mars regolith. The near surface regolith could be mined, be processed by batch heating in an enclosed volume to sublimate the ice and then the condensate collected for water supply. The concentration of the ice in the regolith will control the quantity of material that must be processed and the resources and energy required.

**Water Recycle and Reuse**
Processed water used in the colony cycles can be recycled and reused; reducing the net in-situ water resource required. Such cycles generally involve physical-chemical treatment processes and can be both maintenance and energy intensive. The extent of recycle and reuse will be determined by the available in-situ water resources located for the colony and the resources and energy required for that extraction and processing.

**Selected NewMars Water Supply System**

The baseline water supply system for the colony will be harvesting in-situ water from the near surface ice/regolith mix. This is the highest probability water resource that will be available within a reasonable proximity. The system will be flexible to accommodate alternative in-situ water sources should they be identified. The baseline system will also utilize recycle and reuse to reduce the reliance on ice/regolith mix harvesting.

Areas on the outer perimeter of the colony site will be identified for ice/regolith harvesting. The near surface regolith will be extracted, stockpiled and heated within enclosed volumes to sublimate the ice. The enclosed volumes could be portable, pressurized domes. Solar energy would provide some of the energy for heating and sublimation. The process would be supplemented with thermal energy from the colony systems. The water would be collected as condensate within the domes, stored in water supply tanks and pumped to the colony. Multiple harvesting/water collection/water processing sites would be employed for maximum reliability. A duplex water transmission/distribution system would transmit the water from processing sites to the colony and distribute to points of use.

**Oxygen**

Oxygen is required for human consumption, for fuel cell operation and as an oxidizer in energy production cycles using fuels. The colony source for oxygen will be the colony water supply. Oxygen will be generated where required using electrolysis.

**Wastewater**

Wastewater will be generated in the colony predominantly from human processes. Wastewater can be treated using biological processes or physical-chemical processes. Generally physical-chemical processes are reserved for smaller systems and are generally maintenance and energy intensive. Treated wastewater can be discarded or it can be further treated and reused. To a great extent, the more challenging the NewMars in-situ water source is, the more rigorous the wastewater treatment and reuse approach must be. The NewMars wastewater systems are selected to maximize reuse (conservation) but minimize complexity, energy use and maintenance requirements. A combination of biological and physical-chemical treatment systems will be used dependent on the waste stream.

## Selected NewMars Wastewater System

Grey water waste from hand washing, showers, laundry and similar is relatively easily treated and recycled for reuse using physical-chemical processes. There is extensive experience with this application on the International Space Station as well as conventional applications on Earth. Grey water waste will be recycled and reused locally using physical-chemical treatment processes.

Urine is high in nutrients. With relatively minor treatment, it can be used to fertilize crops. Urine waste will be extracted from urinals and water closets, will be treated locally, and then transmitted to the Food Production Centers for crop fertilization. Black waste is high in nutrients, but must undergo treatment prior to application for crop fertilization. Black waste discharge from water closets will be treated locally using an aerobic biological treatment processes. Treated liquid product will be extracted, mixed with treated urine and used for crop fertilization. Treated sludge will be composted, and then utilized for crop fertilization on a batch basis.

Back-up Systems will rely on direct waste discharge to in-situ storage chambers.

## Food Production

NewMars will be a predominantly agrarian colony. A significant surface area under cultivation will be required to achieve self-sufficiency. The cultivation process needs to deliver high density crop yields to minimize the square footage of constructed, conditioned space required and the cultivation process needs to conserve critical resources; water and energy.

Solar energy levels on Mars are sufficient to support photosynthesis for crop production. The Mars diurnal cycle is similar in duration to Earth. Mars is tilted on its axis, closely matching Earth and experiences seasonal cycles. However, the Mars year is nearly twice that of Earth and its orbit around the sun is more elliptical than is Earth, both conditions contributing to unique factors to contend with in crop production.

The key minerals needed to create a viable soil material (nitrogen, phosphorus, potassium, calcium, magnesium and sulfur) can be extracted from deposits available in the Mars regolith, with perhaps the exception of potassium. The regolith itself is suitable to be processed for a base material for soil, but processing and introducing key minerals will be required to generate a viable soil mix.

Unsophisticated open air soil based crop production on Earth now creates crop yields that would support 1-2 people/acre/year; commercial farming techniques produce yields of 3-4 persons/acre/year. Intensive cultivation processes achieve yields approaching 6 persons/acre/year and research now underway hints at near future production technologies that could boost yields by 40%; approaching 8-9 persons/acre/year. A baseline assumption for a soils based NewMars concept might be 6 persons/acre/Earth equivalent growing season.

While the vast majority of crop production on Earth is open air soils based, alternative technologies offer the opportunity for increased yields. Greenhouses extend growing seasons, increasing yields and provide the vehicle for recycling and reuse. Other technologies including hydroponics, aquaponics, and high density indoor vertical stacked farming offer further increases in yields on a per square foot basis; but recognizing they also require a much higher degree of control and energy use. Regardless of the technology selected, agriculture on Mars will require the use of a conditioned enclosed environment. And, given the two-year seasonal cycle of Mars, it is assumed both year-round food production and bulk food storage will be necessary.

**Selected NewMars Food Production System**

Food production will be accomplished using a mix of large scale sealed, conditioned greenhouses and enclosed, conditioned high density indoor stacked growth chambers. A baseline yield of 36 person-Earth years/acre/Mars year is used; resulting in approximately 2.5 million square feet of enclosed growing facilities.

Using a large mix of greenhouses and growth chambers allows the crop production, harvesting and food storage to be optimized, simplifies construction logistics, provides flexibility in crop rotation, minimizes the risk of catastrophic failure and allows the food production infrastructure to be phased with colony development.

Soil will be generated from locally processed regolith, enhanced with minerals extracted from in-situ mineral deposits. Treated urine waste will be used for fertilizer, injected into the soil. Treated black waste sludge will be used for fertilizer, mechanically mixed with the soil after composting. Food wastes will also be used for fertilizer following composting. Collected condensate from evapo-transpiration within the enclosed environments will be locally recirculated for crop irrigation. Make-up water will be provided from the central water supply.

Greenhouses and growth chambers will be environmentally conditioned spaces and artificially lighted. The indoor environment will be maintained at nominal 5 psi to support occupants. It will be a high $CO_2$ atmosphere to enhance plant growth. Excess oxygen will be recovered and diverted to the Residential Clusters to supplement oxygen supplies.

Food processing centers and processed food storage centers will be co-located with groups of greenhouses and growth chambers to economize processing and storage functions.

**Environmental Systems**

NewMars colony occupied spaces will require environmental conditioning. Key elements include pressurization, oxygen, carbon dioxide removal, temperature control, ventilation and filtration.

Humans have responded well to occupied spaces pressurized to 5 psi from past experience with space vehicles, Skylab and the International Space Station. This pressurization level is also compatible with lightweight dome structures and other enclosed construction that can be easily sealed to support this pressurization condition.

The minimum required oxygen level is 1.8 psi for human respiration. 100% oxygen atmospheres are highly flammable. As such, an atmosphere of 60/40 oxygen/nitrogen, pressurized to 5 psi has proven to be an optimal condition. The nitrogen serves as an inert filler gas, and reduces the flammability. The oxygen level is well above the 1.8 psi minimum.

**Selected NewMars Environmental Systems**

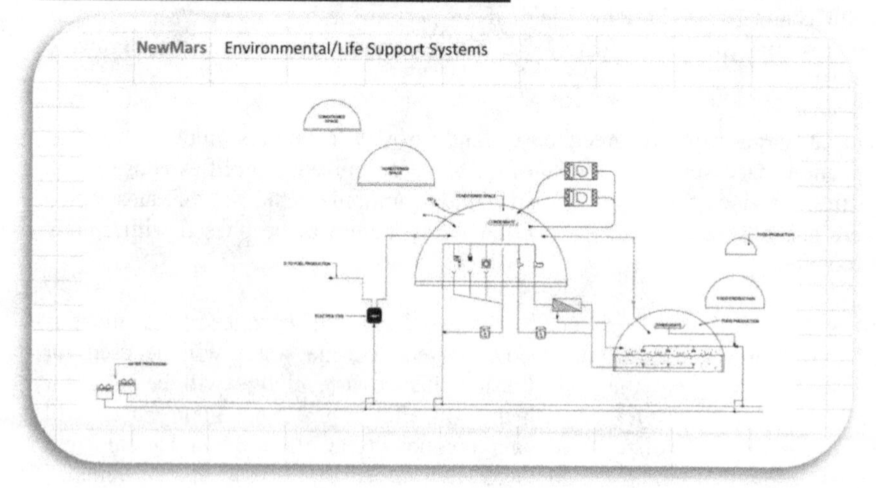


Enclosed occupied spaces will be pressurized to 5 psi with a 60/40 oxygen/nitrogen mix. Oxygen will be generated locally via electrolysis of water with water provided by the central water supply system. Multiple electrolysis units are provided at load centers for redundancy.

Nitrogen will be extracted from the Mars atmosphere and make-up oxygen and nitrogen will be metered into the occupied space. Excess carbon dioxide will be extracted locally from the occupied space using scrubbers and vented to atmosphere.

Thermal energy for space heating will be provided by the central thermal energy distribution system in the form of a circulating thermal fluid. Local backup is provided by fuel cells. Thermal energy distribution within the occupied space will be by relatively conventional HVAC equipment and systems adapted to Mars requirements.

Ventilation of the occupied space will be provided for contaminant removal, local cooling, ventilation, filtration and odor control. HVAC equipment and systems will be utilized. Systems will generally be fully recirculating, with enhanced treatment for contaminant and odor removal.

Excess moisture in the conditioned space will be extracted through the HVAC system, treated and recycled as potable water through the grey water waste recycle and reuse system.

## Fuel Production

Fuel will be needed for fuel cells, for combustion engines and for rocket engines. Fuel cells are an integral component of the colony energy system. Combustion engines may be used for surface vehicles and portable power generation. Fuel for rocket engines will be essential for the colony's role as an inter-planetary transportation center.

Hydrogen and oxygen are the most readily generated fuel for fuel cells. Both can be generated via electrolysis of water, with water feedstock from the central system water supply.

Methane and oxygen have been identified as an effective and relatively easily generated rocket engine fuel supply from in-situ Mars resources that are abundant; hydrogen, oxygen and carbon. Hydrogen and oxygen feedstocks can be generated in the same process as for fuel cells. The Mars atmosphere provides the carbon source. Methane can be manufactured from carbon and hydrogen using the Sabatier process; $CO_2 + H_2 = CH_4 + H_2O$. The water product can be recycled for reuse, supplementing the hydrogen and oxygen feedstock.

## Selected NewMars Fuel Production System

Hydrogen and oxygen for fuel cells will be generated via electrolysis of water, with water feedstock from the central system water supply. The water discharged from fuel cell operation is recycled for fuel production.

Methane will be generated for combustion engine and rocket engine fuel using the Sabatier process. Atmospheric carbon dioxide will be pretreated and used in a Sabatier reactor. Hydrogen will be generated via electrolysis of water, with water feedstock from the central system water supply. The oxygen, also generated in the electrolysis process will be stored to serve as the oxidizer in the combustion cycles. The water by-product resulting from the Sabatier reaction is recycled for continued fuel and oxidizer production.

## Colony Layout and Construction

### Layout

NewMars is arranged in a circular configuration with common functional elements grouped in sectors. The circular arrangement optimizes logistical movements and utility distribution efficiency while maximizing the ability to accommodate growth and change. The primary functional elements are; Central Services, Residential/Commercial, Operations, Industry and Manufacturing, Food Production, Energy and Outdoor Living. A key remote functional element is the Space Port. Additional remote functional elements may include; Water Supply, Geothermal Energy, Wind Energy, and In-situ Resource Extraction.

The focal point of the circular arrangement is Central Services. This multi-functional facility houses administration, education, human resources, government, medical and similar types of programmatic elements. Co-located with Central Services are transient quarters. The generous open area surrounding Central Services supports vehicular surface movement and allows for growth and change.

The Residential sector is subdivided into multi-tenant Residential "neighborhoods" called Residential Clusters. Four clusters are planned, providing a level of diversity and safety from catastrophic loss. The Residential Clusters take advantage of terrain variations the site offers to provide views and vertical development. Co-located with the Residential Clusters are Commercial Centers, housing dining, recreational, cultural and entrepreneurial spaces.

The Operations Sector houses Public Works. It includes the various shops, spare parts storage, garage/shops, vehicle and equipment storage and warehousing needed for colony operation. The generous sector spacing accommodates vehicular movement and growth and change.

Beyond the Operations Sector is Industry and Manufacturing. It readily accommodates growth while its adjacency to Operations facilitates related activities. Fuel production and bulk storage would be located here, as well as stockpiles of raw materials needed for processing and manufacturing feedstocks. The constructed facilities and configuration of the industry and manufacturing area will be a dependent on the entrepreneurial activities that develop. The baseline infrastructure is envisioned to include flexible, multi-use conditioned structures housing fuel processing, chemical processing, materials processing and manufacturing equipment supporting the colony. It is shared with, flexible and expandable to meet entrepreneurial opportunities.

The Food Production Sector encompasses a significant surface area. Its adjacency to the Residential Sector facilitates the waste recycling process. Constructed space totals 2.5 million square feet in a mix of greenhouse type structures and enclosed conditioned growth chambers. Additional facility space is devoted to food processing and processed food storage. Greenhouse structures will be lightweight pressurized transparent domes. Initial structures will be transported from Earth, ready for installation, and will be nominal 300' in diameter; 70,000 square feet each. As manufacturing capabilities develop on planet, greenhouse structures will be manufactured locally. Larger sizes will be used and shapes will be optimized for site conditions. Future food production growth is readily accomplished with outward expansion of the Food Production Sector.

The three Energy Centers are distributed equidistant around the circumference of the colony. This arrangement optimizes the distribution distances between energy supply and load centers as well as maximizing system reliability. As the colony grows outwardly, the siting of the Energy Centers continues to optimize and equalize energy distribution distances.

The Outdoor Living Area is terrain set aside for predominantly outdoor recreation. It will include some scattered enclosed conditioned facilities supporting such activities but it is largely natural and undeveloped. It is for the welfare, enjoyment and creativity of the colony community, intended to spark exploration, adventure and synergy with the Mars environment. It is located adjacent to and beyond the Residential Sector to promote ready access.

**Construction**

Initial facilities supporting colony construction will be self-contained, modularized, transported essentially complete from Earth. The approach is adopted from the Mars Direct concept. As the colony develops and matures, the initial facilities will be repurposed for transient quarters and as emergency backup. Utility components (fuel processing units, power generation equipment, etc.) will be relocated and integrated into the permanent colony infrastructure.

The basic building block for the colony will be a "concrete" mixture, using in-situ regolith and additives extracted from in-situ mineral deposits. The mixture will be applied using 3-D construction printers. Constructed forms and shapes will be basic for simplicity, repeatability and ease of setup. The concrete structures will be used to create structural shells or chambers. Interiors will be lined to create a pressure seal, allowing for pressurized interior environments. Within the chambers, conditioned spaces will be finished using a mix of panelized and modularized assemblies transported from Earth.

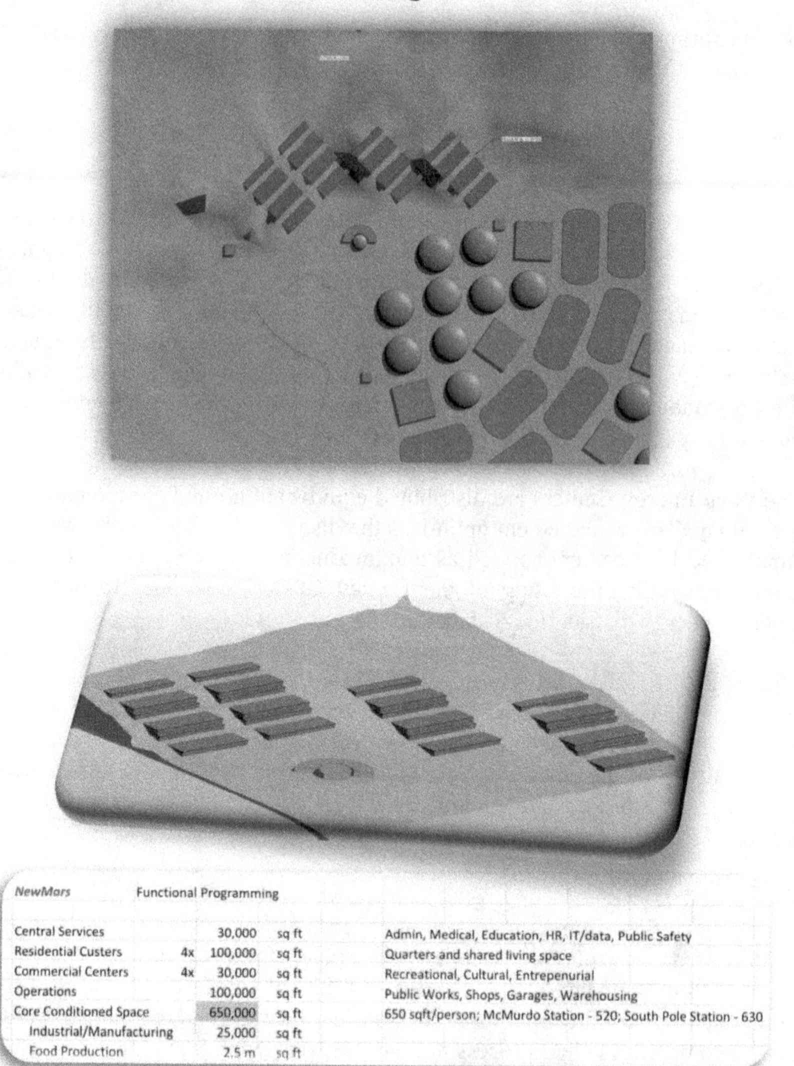

| NewMars | Functional Programming | | | |
|---|---|---|---|---|
| Central Services | | 30,000 | sq ft | Admin, Medical, Education, HR, IT/data, Public Safety |
| Residential Custers | 4x | 100,000 | sq ft | Quarters and shared living space |
| Commercial Centers | 4x | 30,000 | sq ft | Recreational, Cultural, Entrepenurial |
| Operations | | 100,000 | sq ft | Public Works, Shops, Garages, Warehousing |
| Core Conditioned Space | | 650,000 | sq ft | 650 sqft/person; McMurdo Station - 520; South Pole Station - 630 |
| Industrial/Manufacturing | | 25,000 | sq ft | |
| Food Production | | 2.5 m | sq ft | |

Construction in the Residential Sector would take advantage of the vertical terrain relief using a "cut and cover" approach. Structural chambers would be embedded in the hillside with the top exterior recovered with additional regolith material or formed regolith blocks. The arrangement takes advantage of the available regolith mass for radiation protection as well as thermal performance.

As the manufacturing capabilities of the colony developed (during the construction time period), the extent of panels and modules transported from Earth would diminish. More and more materials and components would be manufactured at the colony using in-situ extracted resources, on planet processing and AI driven fabrication and manufacturing machines. Sub-assemblies of specialized, sophisticated components would continue to come from Earth, integrated into the construction.

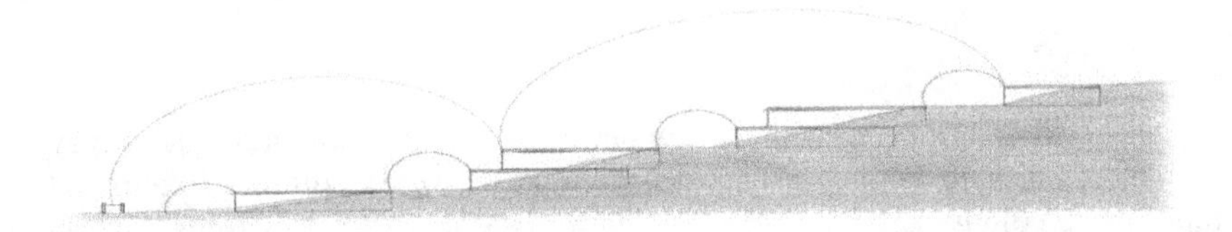

The Residential facilities would necessarily be multi-tenant, high density and communal in nature, using shared spaces. However there would be a variety of diverse choices to suit the colonists. A representative example includes a quad arrangement with say four "efficiency" type apartments with an adjoined shared "great room". No-host residential kitchen/dining centers would be serve groups of quads. "Dining out" facilities, recreational and hobby space and community cultural spaces would be established in the Commercial Centers, which would also house entrepreneurial space (local services, markets, entertainment, etc.).

The Central Services facility would also be constructed using printed in place structural chambers, finished and outfitted in the same fashion as the residential construction. The Central Services would be bermed using regolith for enhanced thermal performance. The structural roof would be layered with regolith blocks for radiation protection.

Both the Residential Clusters and the Central Services facility would be provided with "shirt sleeve outdoor environments" to enhance the colony space. Such spaces would be constructed with lightweight pressurized transparent fabrics, transported from Earth. As the manufacturing capabilities developed at the colony, the initial domes would be replaced with much larger structures.

Operations, Industry and Manufacturing and Food Production facilities would all be constructed using techniques similar to the 3-D printed structural chambers and lightweight pressurized domes discussed. Shapes, degree of sealing and interior conditioning would be a function of the space use. Pre-manufactured modules performing specialized functions (i.e. fuel processing units, materials processing units, manufacturing equipment, etc.), would be transported from Earth and integrated into the site-constructed structures.

Regolith working equipment (dozers, tractors, etc.) and construction equipment (self-propelled cranes, etc.) will be transported from Earth. Given the Mars gravity, it is anticipated construction personnel using exoskeletons will be applied with significant benefit.

## ECONOMICS

NewMars will require a viable economic foundation in order to repay development costs, to fund on-going (near term at least) resupply from Earth and to fund future colony activities. The good news is Mars offers NewMars a world of opportunities!

The NewMars approach will be focused on establishing sustainable long term economic engines. It is anticipated that the development of NewMars will be at the forefront of a progressive expansion of space activities. Government interest, driven by exploration and research; and Private Enterprise, driven by market opportunity will underlie that growth. NewMars will be focused on servicing the needs of that growing space activity. An analogy might be the California gold rush of the 19th century. NewMars isn't digging for the "gold". Others are coming for that. NewMars is supplying the "blue jeans".

Asteroid Mining and Moon Activities
There is significant discussion in the reference materials about the viability of mining asteroids for resource extraction and transporting those resources to Earth. There is also significant discussion regarding the strategic position of Mars in that extraction operation; that, from an energy standpoint, materials to support asteroid mining can much more efficiently be delivered from Mars then Earth. It is similarly more energy efficient to service even the Moon from Mars. NewMars then would function as a key service and supply center for asteroid mining providing the following services:

- Produce bulk food supplies, manufactured routine consumables, and fuel.
- Provide warehousing for consumables, spare parts and critical backup equipment strategically staged on Mars for rapid response when needed.
- Manufacture and supply on demand spare parts and components using 3D printing and value added manufacturing technology and raw materials available on Mars.
- Support initial mining infrastructure construction by receiving sub-modules and components from Earth, integrating them into finished modules on Mars, including the use of local raw materials, 3D printing and value added manufacturing and then packaging complete modules for delivery to the mining station.
- Provide servicing and refueling for transiting space vehicles between Earth, Mars and the asteroids.
- Provide an R&R center for crews transiting between Earth, Mars and the asteroids.

Similar services could be provided for commercial activities on the Moon. A hypothetical trade route has been suggested in the reference material with space vehicles making round trip circuits from Earth to Mars to asteroids to the Moon to Earth. The NewMars asteroid and Moon service sector would fill a key role in this trade circuit.

Mars Exploration, Science, Industry and Manufacturing Activities
There will be growing interest in expanded exploration of Mars and conducting science on Mars by Governments, the Private and Public Education and Research Institutions, and others. Those activities will require infrastructure on Mars to support the activities, infrastructure that would require massive investment if undertaken on an individual basis. NewMars will be in a position to leverage colony infrastructure to more cost effectively support individual users.

Similar to the way the US Antarctic Program provides the underlying infrastructure for a host of science activities undertaken in Antarctica, via McMurdo Station and South Pole Station; NewMars would do the same on Mars. NewMars will provide flexible and adaptable multi-use science space, energy, communications and other infrastructure to support a diverse range of science missions undertaken by other parties. In effect, NewMars would provide the turnkey "lease space" hosting visiting science missions; a flexible, commercial science platform for hire.

The range of supporting services is endless. NewMars functions as the Space Port of entry for the missions and provides on station assembly and value added manufacturing to finalize installation of mission equipment and setup. NewMars provides the transient quarters for mission leaders and technicians required to travel with the activity and provides on going technical support on station, allowing mission specialists to return home and monitor activities remotely from Earth.

Similarly, there will be a great deal of interest in testing new manufacturing and industrial processes in the Mars environment. NewMars would provide a flexible, commercial industrial and manufacturing center for hire. Similar to providing turnkey support for science activities, NewMars would provide facilities, energy, communications, technical support and other infrastructure and resources to host small scale industrial and manufacturing test programs.

**Mars Gateway Center**
As science, exploration, industry and manufacturing activities concentrated on Mars grow, those activities will expand out across the Martian surface. Additional science and exploration stations will be established. Commercial scale manufacturing and industrial processing facilities will be established. NewMars will be perfectly positioned to serve as the gateway service and supply center for such activities. NewMars will provide the Space Port; function as the surface transportation center; provide transient quarters; function as a staging and assembly center, a resupply center and a technical assistance center.

There will also be other colonists looking for a new beginning on Mars. NewMars will be the ideal Gateway Center for them as well. And, as the science, commercial and colonization activities grow, NewMars positions itself as the Planetary Center of Commerce.

**Local High Value Resource Extraction**
NewMars has the opportunity to undertake resource extraction, processing and commercial sales to Earth directly, the most likely high value resource according to various references being Deuterium. However, the commercial viability of this economic activity must wait for the demand to develop on Earth. When it does, it is anticipated the NewMars opportunity might still best focus on supporting this industry as opposed to competing with other commercial interests pursing the resource.

**Longer Term**
NewMars will be perfectly positioned for longer term growth markets in space activity. As humanity expands as a spacefaring community, more extensive interplanetary travel and colonization will occur. NewMars will serve as a major Space Port supporting interplanetary commerce and travel. An analogy would be the old Roadhouses along the ALCAN (Alaska-Canada) highway. This frontier highway traversed extremely remote regions of North America. The Roadhouses along the way served as the rest stops and resupply outposts for the adventurous travelers. NewMars will do the same on an interplanetary scale.

**Short Term**
NewMars' role as a support and service center will take time to develop. The economic activity will grow in response to the demand. There are a couple of opportunities available to NewMars almost immediately however.

The first is on demand exploration and sampling. Earth entities can take advantage of NewMars' immediate access to areas of interest for rapid response surveys and in-situ sampling and analysis. All work would be performed on Mars and data transmitted to Earth.

The other is documenting life at NewMars. In todays social media craze on Earth, there is a huge market for near real time coverage of life. The extreme example is Reality TV. That would not fit with the culture of NewMars. However, creating documentaries in near real time of the development of and life at NewMars produced at NewMars and distributed digitally to Earth would be viable. Documentaries could be tailored to the interest of specific market sectors; education, research, design and construction, social, adventure, etc. Creative pseudo-interactive modes could perhaps be explored for additional value.

**Internal Economic Model**
NewMars will function as a collective community. The colonists will be recruited for NewMars and will arrive with defined responsibilities and primary job

assignments. All colonists will receive equal compensation for the job they perform. That compensation will be set at value that is sufficient to cover all baseline living expenses plus some discretionary spending. Health care and educational services are provided as an inherent NewMars service. NewMars colonists will have choices in housing options, dining options, entertainment, cultural activities, etc.; and will apply their compensation in exchange for the choices they make.

Revenue that is earned by NewMars belongs equally to the collective community. The community will set the direction and priorities for how the revenue is to be invested. A portion of the revenue will be set aside as an Enterprise Fund. Individual colonists can apply for Enterprise Funds to pursue entrepreneurial opportunities in addition to their job responsibilities. These entrepreneurial activities would be internal to the community. Opportunities targeting outside markets would be undertaken as a collective colony endeavor.

## SOCIAL/CULTURAL

The society and culture of NewMars will ultimately be shaped by the interaction between the colonists and between the colonists and the Martian environment. It will be grounded in the underlying philosophical principals that drive the colony concept.

It will be an open and diverse society, embracing new and creative ideas, continuously innovating, full of spirit and adventure and inspired by its unique surroundings. The colonists, given the demands of a new world and the recruiting process will be well educated, well versed in technology, practical ingenuity, creative problem solving and a willingness to adapt and change with lessons learned. That will lead to an exciting exchange of ideas, critical thinking, and engaging debate.

NewMars will be operating on the leading edge of technology, adapting, innovating and pushing boundaries out of necessity. It will be on the leading edge of discovery, where there might be something profoundly new just over the horizon. It will be an exciting environment!

NewMars will also expose the colonists to a harsh, extreme, foreign and unforgiving environment. It will require teamwork, cooperation and the collective expertise of the colonists to survive and thrive. But, it will shape a close-knit, caring and sharing culture. A family community—One for all and all for one!

NewMars is a permanent home, the first spacefaring community for humanity. While the population will be diverse it will also most certainly include families. And as the colony matures, we should expect the population age distribution to expand with it. The colony will have pre-K through 12 educational programs. Innovative education vehicles will evolve to support higher education. And, given the range of expertise, and the new experiences driven by the new environment and the diverse background cultures of the colonists, there will be an engaging community education program.

The NewMars design establishes space within the Commercial Centers in the Residential/Commercial Sector for recreational and cultural activities. It is envisioned that this will fill with the creative ideas of the colonists: theater and arts, unique culinary options, perhaps the first "Red World" craft brewery. All inspired by the unique social, cultural and environmental setting.

Likewise, the colony design establishes an Outdoor Living Sector, relatively undeveloped and set aside for recreational opportunities. One can imagine the liberating benefits the 0.38g Martian gravity will offer for outdoor activities. Perhaps ultra-trampoline, boulder leaping, and long range precision darts will be big hits. Certainly traditional Earth activities like fat tire biking, canyoneering and rock climbing will be popular. And, if dunes are adjacent to the site, perhaps dune skiing will become the craze. Given the adventurous nature of the colonists, we would anticipate recreational exploring with drones as well as "backcountry exploro-camping" using self-contained mid-range Rovers.

How can life on Mars be better than life on Earth? How could it not be?! It's a whole new unexplored world!

## POLITICAL/ORGANIZATIONAL

NewMars will be the size of a very small US town. But it will be extremely remote, extremely isolated and must be extremely self-sufficient. It will also be unique; autonomous, self-governing, the first human colony on another planet.

NewMars will govern itself as a secular democracy. It will draw on the unique and admirable ideals that formed America on Earth. It's governing structure and guiding principals are intended to be fair, functional, succinct, resilient, and lasting. It will abide by the Golden Rule and reflect the philosophical principals under which the colony is founded.

Statement of Ideals and Guiding Principals – NewMars Constitution - The ideals and guiding principals from which NewMars is founded will be documented in a Constitution. The template shall be the US Constitution and Bill of Rights. The NewMars Constitution will be formalized collectively by the key leaders of the NewMars colonization effort in concert with the colonists prior to first launch.

### Governance

- The colony shall be governed by an elected Council and a Director (analogous to a city council and mayor). Each shall have defined authority and responsibility. Generally, the council shall enact laws, rules and regulations, establish priorities and set policy. The Director shall manage the day to day execution.

- The Council members shall be elected by the colony citizens. The Council members shall select the Director. The Council members serve at the pleasure of the colony citizens and the Director serves at the pleasure of the Council.
- A Citizen at Large shall also sit as a Council member. The Citizen at Large is a rotating seat. It shall rotate amongst the general population on a random selection process, similar to Jury Duty in the US legal system.
- An Advisory Board shall be established. The Advisory Board shall have no governing authority, but exists to offer guidance, share institutional knowledge and function as a resource, not unlike the Council of Elders in traditional indigenous societies. The Advisory Council shall be made up of key planners, organizers and leaders involved in the original establishment of NewMars.
- All actions of governance shall be in accordance with the NewMars Constitution.

**Justice System**

- The colony shall abide by the overarching principal of the Golden Rule: do unto others as you would have others do unto you, and the statement of ideals and guiding principals of the Constitution. The Council may from time to time enact laws, rules and regulations to clarify the Golden Rule.
- A Council of Peers shall mediate disputes, infractions and legal actions.
- A designated Magistrate shall oversee the proceedings.
- The Parties in the dispute or legal action shall plead their cases.
- The Council of Peers shall hear the case, may ask questions of the Parties, then adjudicate the settlement.
- The Council of Peers shall be filled by members of the general population on a random selection process, similar to Jury Duty in the US legal system.
- A designated Legal Counsel, trained in the US legal system, is available to provide historical background and Earth based legal context to the Council.
- Any judgment may be appealed. Appeals shall be heard by the colony Council.

**Public Safety**

- The Council appoints a Safety Officer.
- The Safety Officer and his staff act under the direction of the Director to assist in colony public safety.

## AESTHETICS

The aesthetics of NewMars will center on showcasing the unique natural beauty and wonder Mars offers and that is being experienced for the first time. It will be different from Earth, but uniquely different and equally inspiring. Not unlike experiencing the stark beauty of the vast Arctic Plain or the amazing grandeur of the Grand Canyon on Earth.

The Site Selection process for NewMars considers the importance of what the local environment offers in terms of an array of natural beauty and stimulating "surfacescapes." Location adjacent to rising terrain offers the opportunity for vertical relief, creative site layout options and visual interest. These are all factors that will be instrumental in integrating the environment into the beautification of the colony.

The colony site design integrates the Residential Sector into available vertical relief. The goal is to create opportunities for views of the Martian scape and horizon. The colony design envisions "shirt sleeve" conditioned environments outside of the Residential Clusters using pressurized transparent domes. It is envisioned this neighborhood space would be landscaped, integrating an interesting mix of Earth origin vegetation with Martian features and backdrop. Furthermore, strategic views from the Residential Sector will focus on the quite large Food Production Sector. This includes extensive "fields" of cropland under transparent pressurized domes. Again, integrating a mix of Earth origin vegetation with a Martian surface scape.

The Colony Design locates in the Outdoor Living Sector an elevated conditioned space that will offer sweeping views of the Martian surface scape to enjoy. This should help bind the natural beauty of the Martian environment with the colony. Similarly, the Martian atmosphere would frequently afford unique night sky viewing opportunities, again binding the richness of the Martian environment with the projected beauty of the NewMars place.

The colonists will undoubtedly possess a range of artistic talent, and the cultural and social environment of the colony will enable that talent to flourish, incorporating the rich offerings of the new environment into art creations that will become an integral part of the community.

As the colony matures there would be opportunities to host an Artist in Residence program. Imagine the rich artistic work that would arise from artists of all types coming from Earth for a year long in residence engagement in NewMars. This opportunity has been an integral part of the US National Science Foundation Antarctic Program for many years, with stunning results.

Similarly, NewMars could host a virtual art exchange program; creatively using data and communication links between Mars and Earth to bring Mars as a rich subject to artists confined to Earth and at the same time NewMars sharing in the creative interpretations of the works those artists might be inspired to create.

## ACKNOWLEDGMENTS

I would like to acknowledge the creative work accomplished by a host of others that ultimately illuminated strategies, theories, ideas and potential technologies that could be applied to the creation of a Mars colony. I found such information in an abundance of published books, reports and articles. I have included sources that informed my creative design below. As is the strategy of a practicing engineer, I have applied that which is available in the public domain to build my vision of an integrated solution. I would also like to acknowledge and thank PDC Engineers, Inc. of Alaska, and Michael Bartell, BIM Manager for the excellent graphics design support.

## REFERENCES

Agrilyst, A report giving an overview and analysis of the indoor farming industry. A follow on to a 2016 report.

Buchholz M, "Closed Greenhouses, A Tool for Productive Water and Land Management in Arid Areas", Berlin University of Technology

Cassidy R, "Almost Everything You Wanted To Know About Industrial Construction", Building Tech, March 13, 2019

Crossman F, "4Frontiers Earth Mars Moon Asteroids, Building a Permanent Mars Settlement", ASM-SAMPLE Talk 14 April 2010, Published by the Mars Society

Crossman F, Mackenzie B, "Agriculture For An Early Mars Settlement", 2018, Mars Foundation, Published by The Mars Society

Department of Defense Science Board, "Task Force on Energy Systems for Forward/Remote Operating Basis", Final Report, August 1, 2016, Office of the Under Secretary of Defense for Acquisition, Technology and Logistics, Washington DC

Devine M and Baring-Gold EI, "The Alaska Village Electric Load Calculator", NREL/TP-500-36824, October 2004

Greenmatters, "The First 3D-Printed Neighborhood Will Be Sustainable and Affordable, It's Breaking Ground This Year"

Office of the President, "Science and Technology Highlights in the First Year of the Trump Administration"

Shutterstock, "3D Concrete Printing Could Free the World From Boring Buildings", The Conversation, November 13, 2018

Skocik C, "NASA, Department of Energy Testing 'Kilopower' Space Nuclear Reactor, SpaceFlight Insider, November 26, 2017

Wikipedia, "ISS ECLSS"

Wilson A, "Urine Collection Beats Composting Toilets for Nutrient Recycling", Building Green Blog Post, April 2, 2014

World Nuclear Association website, "Nuclear Reactors and Radioisotopes for Space", June 2018

Zubrin R with Wagner R, "The Case for Mars; The Plan to Settle the Red Planet and Why We Must", Free Press, New York, NY, June 2011

Zubrin R, "The Economic Viability of Mars Colonization", Lockheed Martin Astronuatics, Denver CO

# 12: A SELF-SUSTAINABLE SMART CITY DESIGN ON THE RED PLANET

**Hiroyuki Miyajima**
International University of Health and Welfare
h.miyajima@iuhw.ac.jp
**Reiji Moroshima and Tomofumi Hirosaki**
Space Systems Development Corporation
**Shunsuke Miyazaki**
University of Houston
**Mayumi Arai**
The National Museum of Emerging Science and Innovation
**Takuma Ishibashi**
Faculty of Medicine, The University of Tokyo

## 1. INTRODUCTION

The year is 2400. 431 years have passed since 1969, when the first human beings landed on the lunar surface. 120,000 people now live on the Moon, Mars, and asteroids, with a majority of 117,000 on Mars. Mars Colony has become a space oasis for people working in the Mars and Asteroid Development Public Corporation (MADC), providing food, water, medical services, fuel, machine maintenance, and entertainment.

Back in 2020, press releases introducing MADC and announcing it would be established in 2028 were released by 30 countries and 50 private companies for the purpose of sustainable development of Earth. Capital funding for MADC was made up of 40% from the 30 countries' investment, 40% from the proceeds of government-run lotteries around the world, and 20% from private funds. The total development cost was US$100 billion dollars. The goal of MADC was to sustain a global civilization on Earth and protect the environment by utilizing space resources. In addition, the project aimed to foster a creative human civilization in space spanning a variety of fields including art and culture, and science and technology. In 2050, the Mars Education Policy and the R4 (Recycle, Reuse, Reduce, and Refuse) Policy was enforced by the Mars Colony government to promote these activities.

In 2034, construction of the first Martian base was started by robots at Endeavour crater. The first 12-person crew began life on Mars in 2038. A mission of colonists was sent to Mars every two subsequent years. Infrastructure on Mars was set up to support the first gateway city. When the first 72-person non-professional colonist group arrived in 2050, the population rose to 243 people. New immigrants continued to arrive every two years, and the population exceeded 1,000 in 2070. The present report aims to make such a story feasible.

## 2. DESIGN PHILOSOPHY

Mars Colony is intended to support 1,000 people living and working on Mars, with emphasis on economic self-sustainability, cost recovery and operational simplicity, beginning in 2034 and growing to full capacity over the course of 36 years. The outpost was designed with human health and safety in mind, to meet current NASA standards. Technologies for in-situ resource utilization (ISRU) proposed for Mars Colony are evaluated based on technology readiness levels (TRL), reliability, maintainability, operational simplicity, and cost. The Mars Colony design approach was derived from mission requirements to realize a self-sustainable society: conceptual operation and development plans were formed, followed by detailed bottom-up strategies for design of each subsystem.

Financial feasibility is key to achieving a self-sustainable Martian outpost. It is important to develop an economic model to understand the costs of construction, operation, and management, and to develop a business model to enable profitable commerce between Earth and Mars. Mars and asteroid resource development and tourism are incorporated into the Earth-Mars-Asteroid (EMA) economic model, developed by Miyajima. Economic factors in the EMA model are based on Meadows' WORLD II model [1].

## 3. BUSINESS MODEL

The goal of Mars Colony is to promote and support commercial space business between Earth, the Moon, and Mars, that will be open to all nations. The project will: create investment opportunities; share the benefits of space development and distribute operations costs between governments and the private sector; and expand international partnerships, widening participation of emerging space-faring countries. Mars Colony will be managed and operated by the MADC, which is supported by multinational governments, the commercial sector, and the public.

Figure 3-1 shows the business model of MADC. Primary commercial revenue will be generated by: scientific research and technology development, Mars' natural resources and asteroid mining, manufacture of valuable industrial materials, entertainment, and a space tourism lottery. The core customers of commercial activities on Mars (and beyond, in deep space) will be governments, space agencies, public institutes, and the private sector. The commercial space sector will provide services both for construction of Mars Colony and its continued operation, such as transportation, crew training, and hardware development. The technology used to develop Mars Colony will be commodified to maximize its value to society.

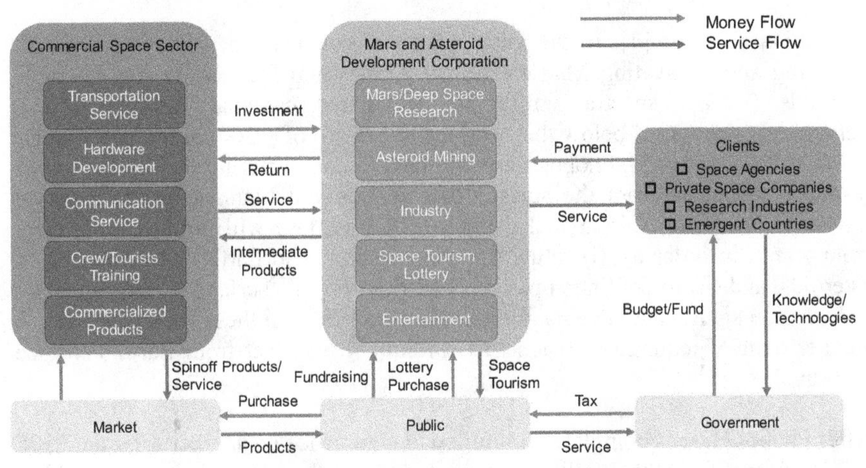


Figure 3-1. Business Model of MADC.

### 3.1 Science Research and Technology Development

MADC will provide a research facility on Mars to conduct various types of research related to physics, biology, chemistry, material science, physiology, pharmacology, and technology development. Field research will extend across the surface of Mars, Martian moons Phobos and Deimos, near-Mars asteroids (NMAs), and additional outer planets or asteroids. In-situ manufacturing and recyclable satellite systems will enable cheaper planetary exploration.

### 3.2 Space Resources Development and Manufacturing

MADC will make use of resources on Mars and asteroids to generate stable high profits. Many asteroids contain enough valuable raw materials (nickel, iron, cobalt, and platinum) to meet current human demand for millions of years. It may also be possible to produce water from asteroid volatiles in order to supply propellant by electrolysis. In the near future, space resource utilization will become more cost-effective than Earth resource utilization. Mining Earth's mineral supplies, such as copper, gold, or even cobalt, is inefficient and unprofitable when comparing total investment with the actual value of resulting discoveries [2]. By contrast, the value of space resource utilization is clear from NASA's plan to spend $1 billion on sending a probe to asteroid Psyche 16 (between Mars and Jupiter), estimated to be worth 700 quintillion dollars [3]. MADC aims to explore and extract useful resources effectively with reusable probe satellites and mining systems. Near-Earth asteroids (NEAs) have long synodic periods, and time intervals between closest approach to the Earth can exceed three years. Thus, MADC will target multiple asteroids that cross or approach Earth's and Mars' orbits to augment accessibility opportunities, allowing access from Earth and Mars to create constant profit.

In the preparation phase for Mars colonization, MADC will aim to explore asteroid Ryugu, one of the most cost-effective asteroids near Earth, to develop excavation methods [4]. In the initial phase of colonization, MADC will begin exploring and excavating Martian moons Phobos and Deimos, which resemble asteroids (surface spectra similar to dehydrated carbonaceous chondrites). Permafrost is expected below their surface at a depth of < 20m at the poles and > 200m at the equator [5]. Phobos Base will have a manufacturing facility to process resources collected from the surface of asteroids and Mars, to produce useful industrial products. Also, a fuel depot on Phobos Base will produce propellant from water, in order to: (i) refuel Starships returning to Earth or transferring to asteroids, and (ii) refuel Starships launched from Earth's surface directly in LEO. Indeed, one significant advantage of conducting ISRU on these moons is a much smaller delta-V required to reach LEO (1.8km/s) than that from Earth's surface (9.3km/s).

After Phobos Base is completed, mining will commence on another asteroid, 1998 UT 18. A profit of $99.62 billion is expected [4]. In the maturation phase, MADC will commence excavation of a large asteroid, Anteros, yielding an estimated profit of $1.25 trillion and creating a self-sustaining economy [4]. MADC will also send probes to asteroids in the main asteroid belt. Figures 3-2 to 3-4 show the orbit of each targeted asteroid. Estimated profits are summarized in Table 3-1.

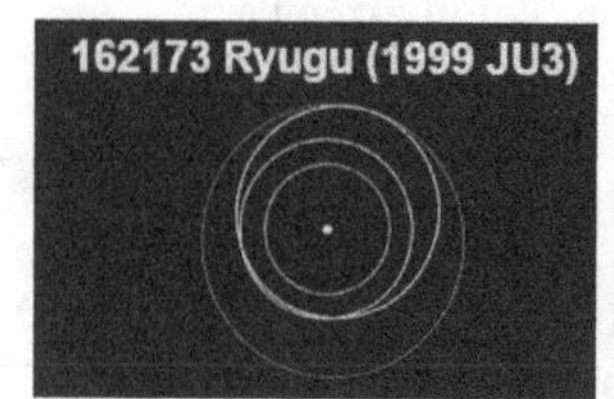


Figure 3-2. Orbit of Ryugu [4].

Figure 3-3. Orbit of 1998 UT 18 [4].

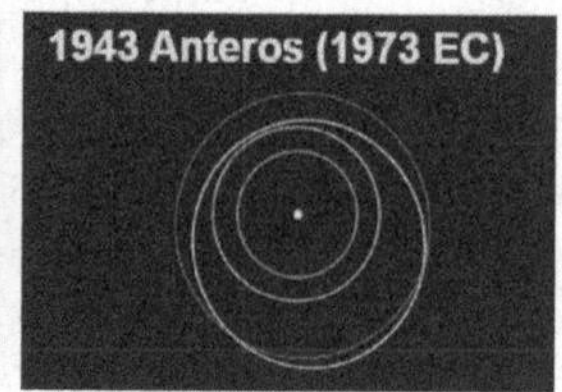


Figure 3-4. Orbit of Anteros [4].

Table 3-1. Targeted Asteroid Value and Estimated Profit.

| Asteroid Name | Type | Value ($) | Est. Profit ($) | Delta V from LEO (km/s) |
|---|---|---|---|---|
| Ryugu | Cg | 82.76 billion | 30.08 billion | 4.663 |
| 1998 UT 18 | C | 644 billion | 99.62 billion | 6.221 |
| Anteros | L | 5.57 trillion | 1.25 trillion | 5.44 |

### 3.3 Space Tourism Lottery

Space tourism is an attractive commercial tool to capture and maintain public and market enthusiasm for space development. However, most people may not enjoy long interplanetary trips, nor desire to live on Mars for long periods. Also, profit from space tourism on the surface of Mars is limited because the pool of acceptable customers is restricted. In the interest of safety, Mars Colony initially needs trained professionals to develop and mature Mars' society. In light of these points, lunar tourism is therefore a more suitable alternative.

MADC will offer a space tourism lottery with equal opportunities for space tourism between Earth and the Moon for all citizens of United Nations member states. Prizes will include: Moon surface stays, lunar orbital trips, low Earth orbit

(LEO) space hotel stays, and sub-orbital space travel. The space lottery will also offer an option that allows winners to choose between space tourism or a monetary prize. The future reduction of launch costs to 1/10$^{th}$ of current levels will dramatically increase the population in cislunar space and allow the development of a space economy.

People in the United States spend about $80 billion per year on the lottery [6]. This statistic shows the feasibility of meeting part of MADC's operational and management costs through a space lottery. Income from annual lottery sales will be distributed: 40% to winners, 42% to the MADC fund, 15% to education or public welfare, and 3% to lottery operational costs. MADC expects marginal profits of $3.5 billion to $5 billion every year with ticket prices of $20-$30.

## 4. TRANSPORTATION SYSTEMS

The SpaceX reusable Mars transportation system, consisting of a Starship payload and a Super Heavy first stage, is capable of transporting crew members on long-duration interplanetary missions. The Cargo Starship can deliver up to 100mT (metric tons) of cargo with orbital refueling by a propellant tanker [7]. Our proposed mission scenario was designed based on SpaceX's Mars transportation architecture, shown in Figure 4-1. Multiple craft carrying crew members and cargo will be launched every two years. Crew Starship can deliver crew members and small cargo to Mars directly with one or two refueling stops in Earth orbit. Cargo Starship will need to be refilled by four refueling tankers in Earth orbit to transfer 100mT of cargo to the surface of Mars. Starships for returning to Earth will begin operating during Phase 1b using fuel produced by ISRU on Mars. Cargo Starship will return to Earth with valuable industrial resources.

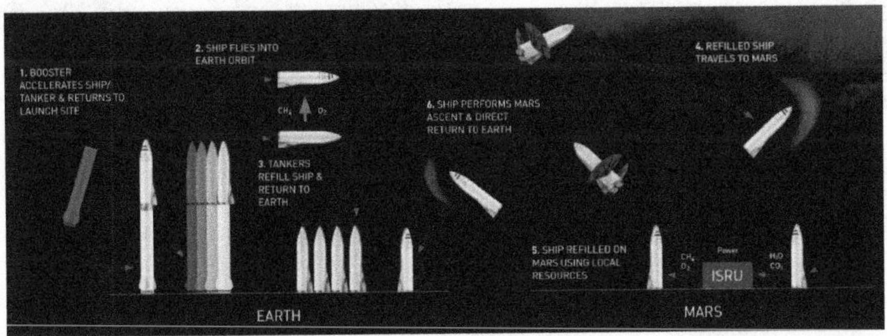


**Super Heavy (1st stage)**: Gross mass 3,065 mT.
**Crew Starship (2nd stage)**: Empty mass 85 mT, Payload to Mars surface 100 mT, Cabin volume 825 $m^3$, Propellant mass 1100 mT ($CH_4$: 240 mT and $O_2$: 860 mT).
**Cargo Starship (2nd stage)**: Empty mass 85 mT, Payload to Mars surface 100 mT, 1000 $m^3$ pressurized volume and 88 $m^3$ unpressurized volume, Propellant mass 1100 mT.
**Tanker (2nd stage)**: Empty mass 85 mT, Max ascent payload to LEO 150 mT.

Figure 4-1. Starship Mission Architecture for Mars Mission [7].

A Starship requires large ISRU systems on Mars to produce enough propellant for the return trip to Earth, increasing the burden of power consumption and cost. The schedule for reutilization must be managed to minimize operational expenses. Mars Colony will require around 50 launches to carry a total of 5,000mT of crew and cargo from Earth to Mars over 36 years, but it is not cost-effective to return all Starships to Earth every time. Thus, the most cost-effective and sustainable method is to return fewer than three Starships to Earth from Mars at every launch opportunity, and to repurpose remaining Starships as Mars/asteroid exploration habitats, as cargo ships to transfer resources from NEA, or as storage tanks for propellant and life support systems. Chapter 10 gives a detailed cost analysis.

## 5. LANDING SITE

Mars Colony will develop the Endeavor crater area for exploration and as a permanent habitation zone. The location of Endeavor crater is at longitude 4°31'2.33"W and latitude 3°10'26.25"S, with a low elevation of -1.5km below datum [8]. The advantage of this near-equatorial and low-altitude exploration zone is high density of the atmosphere, which provides 50g/cm$^3$ of $CO_2$ as radiation protection [9]. In addition, the high-density atmosphere can be used for aerobraking during spacecraft descent and landing. Launching spacecraft near the equator also reduces propellant consumption by assistance of Mars' rotation for added acceleration. Low latitudes will allow operations to benefit from additional hours of sunlight, more solar energy, and higher temperatures compared to high-latitude zones, facilitating machine operation, solar power generation, and food production.

Some areas near Endeavor crater contain polyhydrated magnesium sulfates (PHMS). These constitute 10-35% of bedrock sediments, and may yield 5-20% of their mass as water on heating [8]. Endeavor crater may contain abundant metal and ice resources, as well as regolith material. In sum, choosing Endeavor crater is promising to make the entire colonization process more sustainable.

Figure 5-1 shows the Southern Meridiani Planum Region from various presentations at the 2015 First Landing Site/Exploration Zone Workshop for Human Missions to the Surface of Mars. A single Exploration Zone (EZ) contains a landing site, a habitation site, and several nearby Regions of Interests (ROIs), such as resource regions and science regions, extending over approximately 100km Endeavor crater is located in region 8.

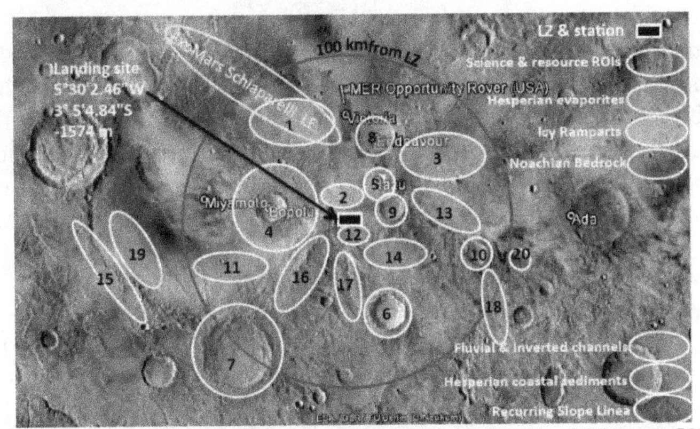


Figure 5-1. Exploration Zone of Southern Meridiani Planum [8].

## 6. CONCEPT OF OPERATION

Mission architecture is composed of three major phases. Each phase can be separated into a former and latter step. The following mission scenario describes the process of Mars colonization. The settlements growth plan is described in Table 10-1 in Chapter 10.

### 6.1 Phase 0: Preparation for Initial Exploration (2034~2038)

The first mission in the initial phase is to send 15 small CubeSats for communication and GPS into Mars orbit, allowing autonomous robots on the surface to carry out stand-alone tasks. Cargo Starships will send power modules, an ISRU module, and autonomous machines. In this phase, robots will commence excavating and processing Martian regolith to produce water and oxygen. The fundamental infrastructure will be built before the first manned mission.

### 6.2 Phase 1: Initial Crewed Mars Exploration (2038~2048)

The first crewed mission will consist of a team of 12, who will commence science exploration within the exploration zone (100km radius around the habitation site). After the first crewed mission, the population will gradually grow, reaching 120 crew members by the end of this period. Starships will be repurposed as habitation for the crew, and 5 inflatable greenhouses will be deployed to supply food. In the later part of this period (starting 2046), construction of the first ice dome will begin, followed by an underground city. After completion of the first dome, crew members will move into the dome and begin work on other domes in 2048. After completion of dome habitat construction, the first phase of Starship habitation on Mars will be utilized to explore other Martian regions, or the Martian moons Phobos and Deimos.

### 6.3 Phase 2a: Intermediate Mars Colonization (2048~2060)

During this phase, the population will reach 557, with human reproduction on Mars fostering a growth rate of 2 percent. Three new domes and an expanded underground city will be constructed in step with population growth. The domes will provide sufficient food, water, and living space to support life on Mars. From 2048, Phobos Base construction will begin. It will function as an industrial facility to manufacture valuable products in microgravity using resources extracted from asteroids, and Phobos and Deimos. Manned scientific exploration will expand to various regions of Mars to search for extant life and map out raw material deposits.

### 6.4 Phase 2b: Maturation Mars Colonization (2060~2070)

Bolstered by reproduction, the population will reach more than 1,000 by the end of this phase. Three new domes and the expanded underground city will be able to support 1,000 people, fostering new value and a new identity. MADC will begin to tap into $1 trillion worth of asteroid resources to realize a self-sustaining economy and return a profit to Earth, in the form of both technology and money. In this phase, the MADC will send reusable exploration probes to asteroids that lie between Mars and Jupiter.

## 7. URBAN DESIGN

Assumptions for our Mars Colony and conditions for urban design are shown in Table 7-1. Required areas for habitation and biomass production in Phase 2 are both taken as 100 $m^2$/person. Required power for living and biomass production are assumed at 10 kW/person and 25 kW/person respectively. Assumptions for the life support system and infrastructure are also shown. Assumptions for architecture and other subsystems are described in Chapters 8 and 9.

Table 7-1. Mars Colony Assumptions and Conditions.

| Items | Phase 1a | Phase 1b | Phase 2 |
|---|---|---|---|
| **Area, power, and architectural structure** | | | |
| Area for habitation [10, 11], $m^2$/person | 25 | 25 | 100 |
| Area for biomass production [10, 11], $m^2$/person | - | 50 | 100 |
| Power for living [10], kW/person | 10 | 10 | 10 |
| Power for food production [12,13], kW/person | - | 25 | 25 |
| Mass of shielded module [11], kg/$m^3$ | - | 133.1 | - |
| Mass of unshielded module [11], kg/$m^3$ | - | - | 9.16 |
| Mass of dome (CNF and CFSR) [14, 15], kg/$m^2$ | - | - | 1.064 |
| Mass of LED [13], kg/$m^2$ | - | 7.5 | 3.75 |
| Mass of biomass production system [13], kg/$m^2$ | - | 12.5 | 1.25 |
| **Life support** | | | |
| Water consumption for living, kg/person | 10 | 30 | 100 |
| Water production by crops, kg/person | - | 100 | 100 |
| Oxygen [11], kg/person | 0.84 | 0.84 | 0.84 |
| Food [11], kg/person | 2.51 | 2.51 | 2.51 |
| Recycling ratio | 0.9 | 0.99 | 0.99 |

The organizational structure of Mars Colony – including the municipal government, social and public services, culture, sports, and entertainment – are described in the following section. 70% of the total population is assumed to be in employment. Major occupational percentages at the end of Phases 1 and 2, and in the U.S. in 2016 (for reference), are shown in Table 7-2 [16].

Table 7-2. Occupation Percentage.

| Occupational group | U.S. | Phase 1 | Phase 2 |
|---|---|---|---|
| Office and administrative support | 15.7 | 10 | 10 |
| Construction and extraction | 4 | 20 | 10 |
| Food preparation and serving related | 9.2 | 5 | 8 |
| Education, training, and library | 6.2 | | 8 |
| Production | 6.5 | 10 | 7 |
| Transportation and material moving | 6.9 | 10 | 6 |
| Arts, design, entertainment, sports, and media | 1.4 | | 6 |
| Sales and related | 10.4 | | 4 |
| Healthcare practitioners and technical | 5.9 | 5 | 4 |
| Management | 5.1 | | 4 |
| Business and financial operations | 5.2 | | 4 |
| Farming, fishing, and forestry | 0.3 | 10 | 4 |
| Installation, maintenance, and repair | 3.9 | 10 | 4 |
| Computer and mathematical | 3 | | 3 |
| Healthcare support | 2.9 | | 3 |
| Building and grounds cleaning and maintenance | 3.2 | 5 | 3 |
| Personal care and service | 3.2 | | 3 |
| Architecture and engineering | 1.8 | 10 | 3 |
| Community and social service | 1.4 | 5 | 2 |
| Protective service | 2.4 | | 2 |
| Life, physical, and social science | 0.8 | | 1 |
| Legal | 0.8 | | 1 |

### 7.1 Community, Politics, and Social Service

2% of the employed population works in the community, politics, and social service fields. The Mars government is independent from all countries on Earth and the MADC. The MADC pays the Mars government a tax to support its employees when on Mars, which makes up approximately 20% of all Mars government revenue. The Martian government is obligated to protect the people's welfare and health and maintain stable Martian economic growth.

### 7.2 Education

8% of the employed population works in education in primary and secondary schools. Higher education is mainly conducted through distance learning from Earth. The Mars Education Policy and the 4R (Recycle, Reuse, Reduce, and Refuse) Policy are enforced by the Mars Colony government.

### 7.3 Arts, Leisure, Entertainment, Sports, and Media

6% of the employed population works in arts, design, entertainment, sports, and media. These are important occupations for a creative human civilization of space origin and contribute to a variety of fields including culture & society and science & technology.

The unique Martian environment with 38% of Earth's gravity inflates people's imaginations and creativity. Every sport on Mars looks like an extreme sport or action movie. For example, martial arts such as karate or kendo are more exciting. In addition, the Martian environment enhances expressive performances of dance, music, and entertainment shows like Cirque du Soleil. Original films on Mars are delivered to Earth via Netflix. Cultivating plants is a precious leisure activity where people can feel in harmony with lives other than that human. The third floor of a greenhouse dome (see Chapter 8) allows people to enter and spend time in nature.

### 7.4 Medicine and Healthcare

7% of those employed work as healthcare practitioners and technicians, such as doctors, nurses, and medical service personnel. All colonists learn core medical skills, such as basic life support (BLS). A preventive medicine program is the principal method for keeping the colonists healthy. If it is impossible to find suitable treatment in a Mars hospital, a patient can return to Earth. In addition, Mars hospitals provide medical care to workers based on asteroids or in deep space.

### 7.5 Protective Service

2% of the employed population are police officers, fire fighters, and security staff. Wireless security cameras are installed in public places to record and monitor troubles in real-time. A fire detection system kicks in at an early stage, warns those affected, and an autonomous sprinkler system performs first-aid firefighting. The Protective Service Department is obligated to inform the Martian public of daily radiation levels and any social troubles, and to conduct evacuation drills.

### 7.6 Transportation

6% of the employed population works in transportation. Almost all of these workers are involved in interplant transportation and material moving. In the colony, all people and materials are moved around by self-driving electric cars in a smart mobility system.

### 7.7 Sustainability and Supportability

Sustainability and supportability for the Mars Colony consists of: a logistic system, a maintenance system, and a repair system. The logistic system contains a production system, a recycle system, and a storage system. The maintenance system consists of a defect detector system (enhanced by AI) and a spare parts production system. The repair system involves production of repaired parts. Almost all materials are recycled and reused. ISRU compensates for shortages, and repaired parts and spare parts are produced on Mars.

## 8. ARCHITECTURE

This section describes the details of Starship habitation in Phase 1 and the inflatable ice dome concept in Phase 2.

### 8.1 Phase 1: Starship Habitation, Greenhouse, and Plants

Habitat in the first stage of colonization is based on repurposing a Starship itself, with nearby inflatable greenhouse structures for subsistence farming. A Starship's cabin provides 825m$^3$ of habitable volume, and five starships (4,125m$^3$) are needed to give 34m$^3$ of habitable volume per crew for a long-duration stay on Mars [17]. Each spacecraft can operate on a stand-alone basis, has a research facility, accommodation, and life support system. A greenhouse has 1,200m$^2$ of cultivation area.

Each Starship's landing point will be very close (e.g. 1 km) to the northern edge of Endeavour crater, which will host construction of the future Mars Colony City. This will provide feasible access to resource regions of interest. Starship fuel tanks will be repurposed to store propellant.

## 8.2 Phase 2: Large Inflatable Ice Dome and Underground City

After a decade on Mars, the growing Martian community will likely require a permanent living base instead of the Starship habitat. A colossal inflatable ice dome with underground 3D printed regolith concrete chamber modules is one solution for an emerging Martian city (see site plan in Figure 8-1). A habitable space without spacesuits will provide freedom for human activity in otherwise restricted life on Mars. More domes and regolith modules will be added as the population grows, each bringing specific functions and characteristics of an urban city.

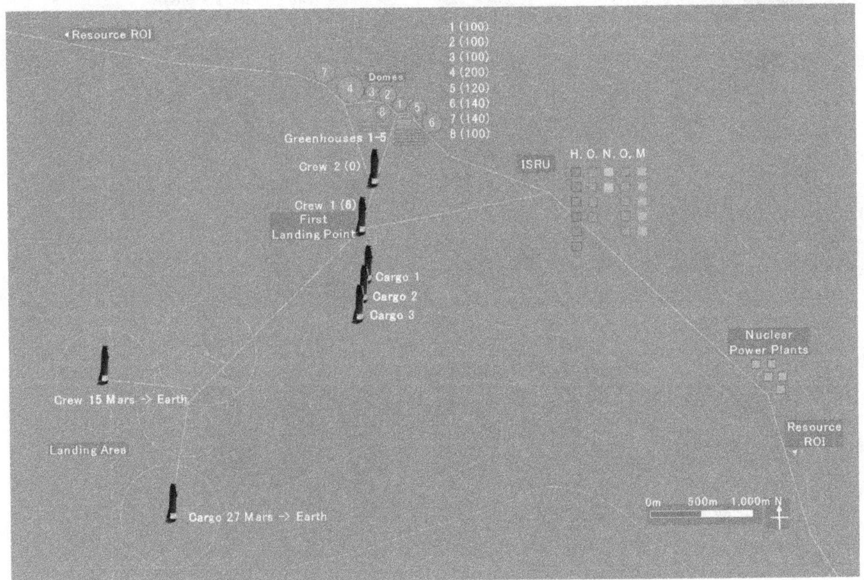


Figure 8-1. Mars Colony Site Planning.

The domes are interconnected through underground tunnels, so that people can walk to every habitable corner of the city without wearing spacesuits. Two types of domes are used, with different roles. One is for agriculture: food crops are produced on the topmost of three layers of regolith concrete modules, housed within the habitable space of a dome. Another is for health and recreation: an area such as a park with trees, flowers and various plants is located on the third-floor level in this case. The first and second floor chamber modules in both domes are utilized for residential, commercial, business, industrial, medical, educational, governmental, cultural, and amusement functions (Figures 8-2 and 8-3). The layers of different sized honeycomb chamber modules and openings create organic connectivity between spaces, providing a nature-like atmosphere in the living base.

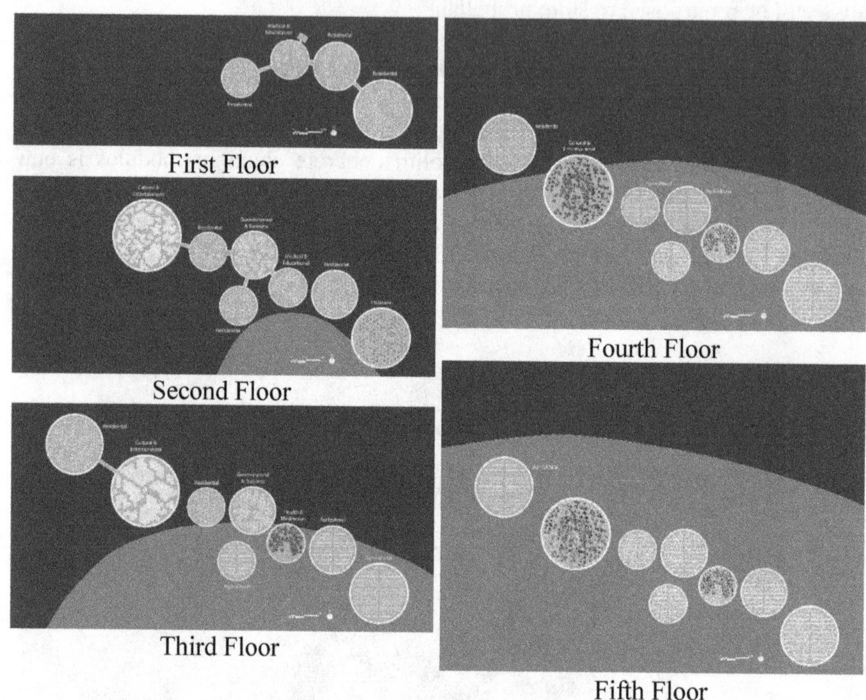


Figure 8-2. Floor Plans After 32 Years.

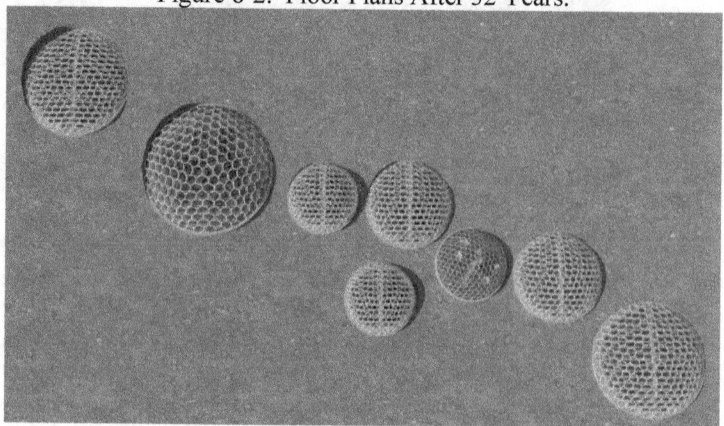

Figure 8-3. Top View of Mars Colony.

8.2.1 Inflatable Dome Concept and Structure

The domes are located along the steep, sloped northern edge of Endeavour crater, which provides natural protection from a portion of micrometeoroids and space radiation (Figure 8-4). The air-supported structure of the dome consists of an inflatable membrane of cellulose nanofiber (CNF) with hexagonal surface cells, supported by ultra-light and strong carbon fiber strand rods (CFSR) to restrain expansion by internal pressure [14, 15]. In order to give sufficient resistance to space radiation, each hexagon has a 50 cm-thickness cell of CNF to store water, mostly in solid form (given the outside atmospheric temperature on the Mars surface), with outer and inner cells of CNF filled with $CO_2$ as insulation layers (Figure 8-5). The membrane of the agricultural domes is integrated with a flexible LED lighting system, utility cables, and flexible ventilation passages to provide appropriate climate control for food crops.

The advantage of the ice dome concept is its durability, thinness, and light weight, unlike existing inflatable modules such as BEAM, BA330 [18], or Trans-Hub which consist of heavy and thick laminate walls to protect from the harsh external environment. Also, an inflatable dome equipped with ISRU-produced water ice for radiation shielding and as structural components does not require deep burial below Martian regolith, which is a common challenge when constructing a large dome.

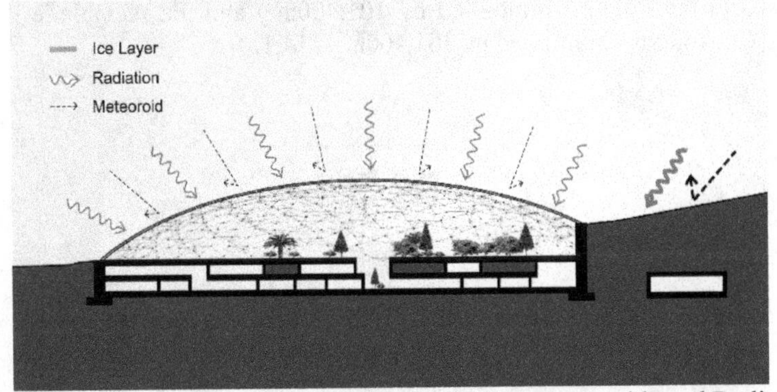


Figure 8-4. Section View for Protection from Micrometeoroids and Radiation.

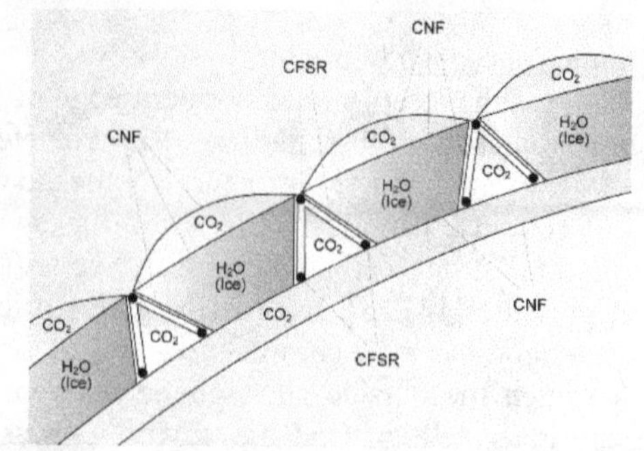


Figure 8-5. Detailed Section of Inflatable Dome.

The transparent cellulose nanofiber membrane allows energy from the Sun to enter the dome, and also provides structural strength to resist the combined load of self-weight (structural elements and $H_2O$) and pressure differential (higher inside pressure than outside) acting on the dome, through its characteristic of aramid-level strength (Figure 8-6). The strength of CNF increases probability of withstanding micrometeoroid impacts, and in addition each membrane cell is easily replaceable in case of damage and deterioration, resulting in a sustainable environment for life under the dome. As an example, for a greenhouse dome with total area 101,400m$^2$ constructed using CFSR, CNF and water, the roof structure mass is 8,112kg (80g/m$^2$ multiplied by 101,400m$^2$) and the membrane mass is 1,622kg (0.016kg/m$^2$ multiplied by 101,400m$^2$) [14,15].

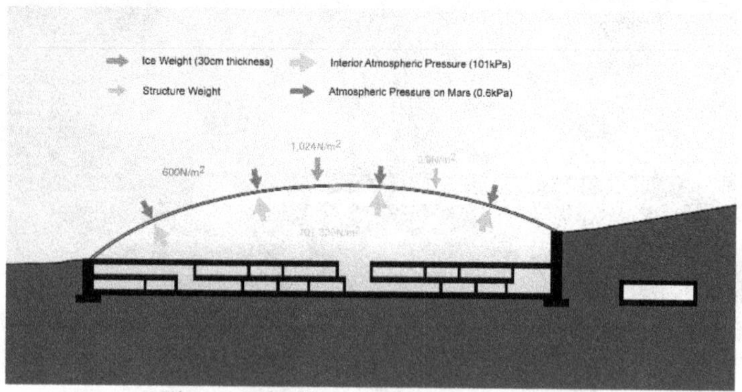


Figure 8-6. Structural Stress Concept.

8.2.2 Underground City Concept and Structure

The underground regolith concrete chamber modules are fundamental spaces for Martian urban city life. The honeycomb structure used for the complex of hexagonal module units draws upon the same structural geometry as the dome surface, minimizing usage of materials. The surrounding wall beneath a dome's circular footprint, and the first and second floor structures, constitute a primary part of the foundation of the dome. On the first floor, hexagonal units with 60cm thickness and 6m side-length walls carry all the weight from the second and third floors. A hexagonal unit provides around 80m$^2$ of space, which can be kept as one middle-sized space, or divided into a couple of smaller room-sized spaces (e.g. in a residential area). On the second floor, hexagonal units with 8m side-length provide a larger space, more suited for office, commercial, industrial or social usage. The third floor of a health & recreational dome holds large spaces of greenery, with openings to lower floors. The white walls decorated with plaster of Paris mined around the site make a powerful contrast with the green leaves on the red soil. The difference between hexagonal units' size, and open spaces with sectional connectivity, create a terraced elevation between floors which offers nature-like views of landscapes in the underground urban city (Figure 8-7).

The inflatable ice/water dome that mimics the Earth environment is a protective cradle for burgeoning life on Mars. Primitive life emerged in liquid water, a universal solvent which can dissolve every substance and provides an environment for them to react and bond together to enable more complex life to form. The ice/water wall design expresses harmonious integration of Mars and Earth, or lives on Mars and lives which originated on Earth, in the hope of triggering new bonding reactions, to create a brand-new civilization and forge personal identity as a Martian.

Figure 8-7. View from Second Floor of Dome Habitat.

## 9. SUBSYSTEMS DESIGN

### 9.1 Biomass Production System (BPS)

Biomass production is supported by LED lighting in Phase 1b, and by a combination of sunlight and LED in Phase 2. Assumptions for the hydroponic biomass production system (BPS) are shown in Table 9-1 and Table 9-2 [13].

Table 9-1 shows an example of eight crop candidates for Mars Colony. Biomass requirements for each crop are based on nutrient requirements for a single person. Required cultivation area per crop is obtained from the biomass requirement divided by biomass production per day. Total crop cultivation area required per person is 80.2 $m^2$.

To estimate power needed for the LEDs, the following formula is used:

2,600 kcal x 4.2 J/kcal /43,200 s (12 h) /0.01 = 25 kW.

An assumed energy intake of 2,600 kcal/person-day from biomass origin is first converted to kJ/person-day (4.2kJ/kcal). This energy is divided by 43,200 seconds (12 hours of light per day), and then by photosynthetic efficiency (ratio of chemical energy to light energy, taken as 1%). The result shows that 25 kW/person of LED power is needed for biomass production.

Table 9-1. Crop Candidates for Mars Farming.

| Crop | Energy | Biomass requirement | Growth cycle | Harvest Index *1 | Biomass production per day | Cultivation area |
|---|---|---|---|---|---|---|
| | kcal/CM-day | dry-g/CM-day | days | - | dry-g/$m^2$/day | $m^2$/CM |
| Rice | 1421.1 | 335.2 | 90 | 0.50 | 8.5 | 40.0 |
| Potato | 58.6 | 15.6 | 100 | 0.82 | 4.6 | 3.4 |
| Sweet potato | 206.2 | 52.0 | 120 | 0.65 | 7.0 | 7.5 |
| Soybean | 646.9 | 131.9 | 100 | 0.52 | 5.6 | 25.0 |
| Lettuce | 26.3 | 7.4 | 30 | 0.91 | 4.2 | 1.8 |
| Tomato | 48.8 | 12.4 | 100 | 0.70 | 13.9 | 0.9 |
| Cucumber | 18.3 | 4.8 | 80 | 0.70 | 9.7 | 0.5 |
| Strawberry | 12.2 | 3.1 | 60 | 0.70 | 2.8 | 1.1 |
| **Subtotal** | **2438.5** | **562.4** | | | | **80.2** |
| Animal protein | 211 | 50 | | | | |
| **Total** | **2649.5** | **612.4** | | | | |

*1 Harvest index = edible portion mass / total biomass

Table 9-2. Mass and Power Assumptions for Biomass Production Subsystem.

| Component | Initial mass | Volume | Power | Thermal cooling | Logistics mass | Crew time |
|---|---|---|---|---|---|---|
| | kg/$m^2$ | $m^3$/$m^2$ | kW/$m^2$ | kW/$m^2$ | kg/$m^2$-year | CM-h/$m^2$-year |
| Crops | 5.76 | 2.6 | 0.14 | 0.14 | 0 | 1.3 |
| Lamps | 7.54 | 0.4 | 0.691 | 0.138 | 0.19 | 0.0027 |
| Ballasts | 2.77 | 0 | 0.02 | 0.02 | 1.07 | 0.0032 |
| Mechanization System | 4.1 | TBD | TBD | TBD | TBD | TBD |
| Secondary Structure | 5.7 | - | - | - | - | - |
| Total | 25.9 | 3.00 | 0.86 | 0.31 | 1.25 | 1.31 |

## 9.2 Environmental Control and Life Support System (ECLSS)

An International Space Station (ISS) type environmental control and life support system (ECLSS) that recycles air and water is used in Phase 1a. A fully-closed

ECLSS that produces biomass is used after Phase 1b. The ECLSS mass distribution are shown in Table 9-3 [19]. The total mass and power of the 1,000-person colony are linearly scaled up from the baseline mass and power by 1,000 times.

Table 9-3 shows for each dome: its radius; population; and mass budget of BPS, LED, and ECLSS, which consists of air revitalization, thermal control, water processing, waste management, and water storage. Water for protection against radiation is used on the roofs of domes 1 and 6. Dome 1 is protected by 660m$^3$ of water and Dome 6 by 346 m$^3$. Other domes are not protected by water. The biomass production system can regenerate air, water, and food through photosynthesis by crops.

The thermal control system (TCS) was designed to collect, transport, and reject excess heat from the dome habitat and other infrastructures. The air circulation system maintains comfortable temperature and moisture levels for living or food production in the internal environment of the dome. Indoor heat generated by daily life and production is transferred along with water between habitats, greenhouses, and facilities.

Table 9-3. ECLSS Mass Distribution [19].

| Dome Number | 7 | 4 | 3 | 2 | 8 | 1 | 5 | 6 |
|---|---|---|---|---|---|---|---|---|
| Radius, m | 100 | 120 | 65 | 80 | 65 | 65 | 80 | 100 |
| Population, person | 140 | 200 | 100 | 100 | 100 | 100 | 120 | 140 |
| Subsystem | | | | | | | | |
| Biomass Production, mT | 7.5 | 10.7 | 5.4 | 5.4 | 5.4 | 5.4 | 6.4 | 7.5 |
| LED, mT | 62.7 | 89.6 | 44.8 | 44.8 | 44.8 | 44.8 | 53.7 | 62.7 |
| Air Revitalization, mT | 15.2 | 21.8 | 10.9 | 10.9 | 10.9 | 10.9 | 13.1 | 15.2 |
| Thermal Control, mT | 9.1 | 13.0 | 6.5 | 6.5 | 6.5 | 6.5 | 7.8 | 9.1 |
| Water Processing, mT | 3.3 | 4.7 | 2.4 | 2.4 | 2.4 | 2.4 | 2.8 | 3.3 |
| Waste Management, mT | - | - | - | - | - | - | - | 57.9 |
| Water storage, m$^3$ | 14.0 | 20.0 | 10.0 | 10.0 | 10.0 | 10.0 | 12.0 | 14.0 |
| Water for roof, m$^3$ | - | - | - | - | - | 660.0 | - | 346.0 |

## 9.3 In-Situ Resource Utilization

ISRU consists of a gas collection system to collect carbon dioxide from the atmosphere of Mars, and a water mining system to produce water from the planet's soil. ISRU could reduce the need to import propellant or consumables by producing oxygen ($O_2$) and methane ($CH_4$) from the Martian atmosphere and soil, as shown in Table 9-4. The chemical conversion equation has two processes: a Sabatier reaction with water electrolysis (WE) and electrochemical reduction (ER).

Table 9-4. ISRU Process.

| Production | Process |
|---|---|
| $O_2/CH_4$ production with Mars $H_2O$ | Sabatier: $CO_2 + 4H_2 \rightarrow CH_4 + 2H_2O$<br>2nd Water Electrolysis: $2H_2O \rightarrow 2H_2 + O_2$<br>or Electrochemical Reduction: $CO_2 + 2H_2O \rightarrow CH_4 + 2O_2$ |

The ISRU, Life Support System (LSS), WE, and propulsion systems should be optimized in conjunction with one another when considering system operation, because those functions are interconnected and complementary as shown in Figure 9-1. $O_2$ is produced by ISRU from $CO_2$ in the Mars atmosphere. $H_2O$ is produced by ISRU from the surface soil. $CO_2$ is produced by ISRU from trash and waste. $CH_4$ is produced from $CO_2$ and $H_2O$, or $H_2$. A portion of $H_2O$ is decomposed into $O_2$ and $H_2$ by an electric current. $O_2$ is consumed and $CO_2$ is produced by humans, and then $CO_2$ is regenerated into $CH_4$ by the LSS. Waste $H_2O$ is regenerated into clean $H_2O$ by the LSS. Trash is transferred to ISRU from the LSS. $O_2$ and $CH_4$ are used for propulsion.

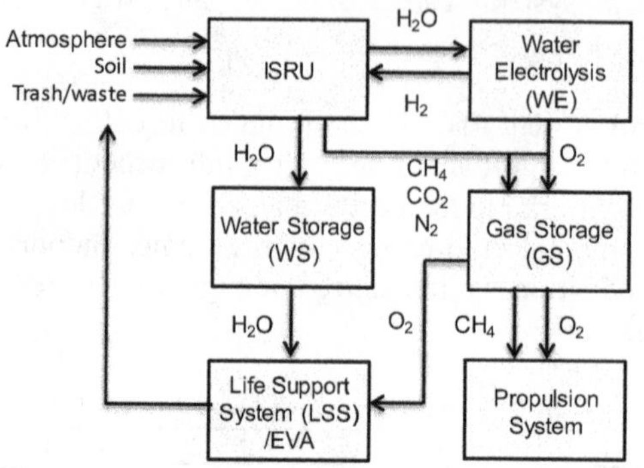


Figure 9-1. Integrated Fluids and Interfaces.

Table 9-5 shows required mass and power for ISRU hardware using Sabatier and WE based on a design by Donald Rapp [20]. If oxygen and methane in a quantity of 1,100 mT are to be produced in 720 days for an interplanetary return flight to Earth, the required production rate is 61.8 kg/hr (49.5 kg $O_2$ and 12.3 kg $CH_4$) from 34.0 kg/hr $CO_2$ and 27.8 kg/hr $H_2O$. Then, required ISRU mass is estimated to be 4,170 kg for $O_2$ production and 20,535 kg for $H_2O$ production. Required power for ISRU is estimated to be 467 kW for $O_2$ production and 260 kW for $H_2O$ production.

Table 9-5. ISRU Mass and Power Baseline [20].

| Processes | Unit feedstock rate | Mass, kg | Power, kW |
|---|---|---|---|
| $CO_2$ acquisition | 1 kg/hr of $CO_2$ | 50 | 1.2 |
| Sabatier conversion | 1 kg/hr of $CO_2$ | 12.5 | 0.16 |
| Water electrolysis | 1 kg/hr of water | 11.2 | 2.4 |
| Liquefying $O_2$ | 1 kg/hr of $O_2$ | 35 | 1.1 |
| Liquefying $CH_4$ | 1 kg/hr of $CH_4$ | 85 | 3 |
| Regolith excavation and water extraction | 1 kg/hr of water recovered | 700 | 8 |

Table 9-6 shows the ISRU mass and power required for producing carbon dioxide, water, nitrogen, oxygen, and methane when assuming 24 hours of nuclear power plant operation at Endeavour Crater [19].

Table 9-6. ISRU Mass and Power Requirement.

| Production | Mass, kg | Power, kW |
|---|---|---|
| ISRU $CO_2$ | 958 | 50 |
| ISRU $H_2O$ | 84,976 | 971 |
| ISRU $N_2$ | 958 | 23 |
| ISRU $O_2$ | 2,088 | 158 |
| ISRU $CH_4$ | 1,200 | 37 |

## 9.4 Electricity and Power System

Power generation must meet the consumption needs of habitations, life support systems, biomass production systems, ISRU, and construction machines. Biomass production makes use of direct sunlight, in addition to LED lighting. Solar energy at Mars orbit is $590W/m^2$, and net solar energy available on Mars is assumed to fluctuate between 0 to $400W/m^2$. The base load in the power grid is provided by nuclear power stations installed nearby habitation sites. During the night and dust storms, it is assumed that all power is provided by nuclear stations. The baseline power requirement is estimated by using the space nuclear design 4,000kg/100kW (SP-100 as shown in Table 9-7) [21]. For example, it is assumed that the power plant mass is 4,000 x $\sqrt{(10,000/100)}$ = 40,000kg, a mass of 200,000kg is required for 50MW by using the scaling rule of a nuclear reactor [22].

Table 9-7. Power Supply System Design.

| Nuclear Reactor SP-100 | | √SCALING |
|---|---|---|
| Capacity, kW | Mass, kg | Mass, kg |
| 100 | 4,000 | 4,000 |
| 500 | 12,000 | 8,944 |
| 2,000 | 25,000 | 17,889 |
| 10,000 | - | 40,000 |

## 9.5 Guidance, Navigation, and Communication

The Guidance, Navigation, and Communication (GNC) system enables surveying and navigation, relays communication signals, and supports prospecting decision-making with a constellation of 15 small remote sensing satellites or COMPASS, providing uniform coverage of the entire Martian surface [23]. GNC architecture with small satellites can lower costs because they are compact, lightweight, and easy to recycle. The COMPASS constellation supports more accurate and reliable work by autonomous ISRU excavation, construction and research-oriented exploration.

CubeSats transmit information in the X-band or Ka-band to two ground stations in the equatorial region, located 180 degrees apart. The ground stations can communicate with Earth via the Deep Space Network (DSN) and directly with the entire navigation network. Surface infrastructures use wireless (RF or Laser) communications between habitats, mobile surface systems, and other facilities.

Big data from surface observation collected by spectrometers on CubeSats will be sent to analytical platforms in ground stations, training their machine learning algorithms for site mapping and hazard avoidance pathing. AI can help find the most promising mineral deposits faster by sifting through the massive amounts of geological data obtained by CubeSats and sample collectors.

## 9.6 Robotics and Artificial Intelligence

Robotic service is crucial to enable efficient operation of ISRU, construction, and food production. Robots will need to perform most tasks autonomously, communicating with each other to work towards a common objective. Tele-operation by human crew must also be possible in the event of complex procedures or malfunctions.

Common platforms, such as the Mars Electrical Vehicle (MEV) and All-Terrain Hex-Limbed Extra-Terrestrial Explorer (ATHLETE) [24], can be configured for multiple specifications, with modular attachments to reduce costs and enhance maintainability. For ISRU, mining operations will be conducted by an excavator with a dozer blade, as shown in Figure 9-2 [24]. Cargo carriers or haulers will provide transportation services to deliver cargo or excavated Martian regolith to an ISRU factory. Small and large manned MEVs can transport 12 and 20 crew members, respectively. A large MEV will be utilized from Phase 2. 3D printers equipped with material tanks will perform human habitat construction. If a materials tank is empty, another tank will connect to the 3D printer module unit.

Robotic AI can improve efficiency of a given task using machine learning technology. Robots can iterate given tasks to optimize performance and learn from cloud information accumulated during routine operation.

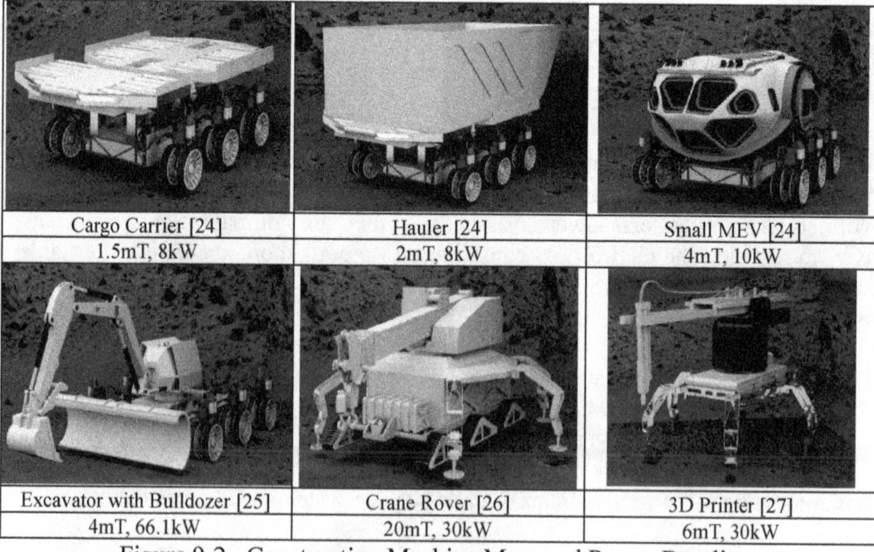


Figure 9-2. Construction Machine Mass and Power Baseline.

## 10. OPERATION AND COST ANALYSIS

Our settlement scenario was made using the Mars Colony Development (MCD) model, and is shown in Table 10-1 [19]. The population experiences 2% natural growth, and exceeds 1,000 at Year 36. In year 36, the habitation areas, greenhouse areas, total mass, mass transported from Earth, and required power are respectively: 100,600 m$^2$, 107,400 m$^2$, 11,570 mT, 2,390 mT, 43,937 kW when sunlight cannot be used, and 22,393 kW when 100% of sunlight can be used.

The cargo manifest and launch manifest satisfying our settlement scenario, summarized in Table 10-1, are shown in Table 10-2. There are five unused ships on Mars between Year 0 and Year 6: two Crew Starships and three Cargo Starships, for habitat use in Phase 1.

Table 10-1. Cumulative Numbers in Settlement Scenario.

| Phase | Year | FY | Population | Habitation, m$^2$ | Greenhouse, m$^2$ | Total mass, mT | Mass from Earth, mT | Total power (0% Sun used), kW | Total power (100% Sun used), kW |
|---|---|---|---|---|---|---|---|---|---|
| 0 | 0 | 2034 | 0 | 0 | 0 | 172 | 172 | 1,525 | 1,525 |
| | 2 | 2036 | 0 | 0 | 0 | 218 | 218 | 1,804 | 1,804 |
| 1a | 4 | 2038 | 12 | 300 | 0 | 329 | 317 | 3,142 | 3,142 |
| 1b | 6 | 2040 | 24 | 600 | 1,200 | 567 | 496 | 5,237 | 4,996 |
| | 8 | 2042 | 48 | 1,200 | 2,400 | 1,060 | 919 | 9,807 | 9,325 |
| | 10 | 2044 | 72 | 1,800 | 3,600 | 1,264 | 1,052 | 10,683 | 9,961 |
| | 12 | 2046 | 120 | 3,000 | 6,000 | 2,309 | 1,295 | 12,389 | 11,185 |
| 2a | 14 | 2048 | 168 | 16,800 | 18,675 | 3,473 | 1,419 | 14,095 | 10,348 |
| | 16 | 2050 | 243 | 24,300 | 31,350 | 4,370 | 1,523 | 16,760 | 10,471 |
| | 18 | 2052 | 319 | 31,900 | 31,350 | 4,758 | 1,612 | 19,481 | 13,192 |
| | 20 | 2054 | 397 | 39,700 | 44,025 | 5,654 | 1,704 | 22,253 | 13,422 |
| | 22 | 2056 | 476 | 47,600 | 56,700 | 6,554 | 1,796 | 25,061 | 13,687 |
| | 24 | 2058 | 557 | 55,700 | 56,700 | 6,977 | 1,900 | 27,960 | 16,586 |
| 2b | 26 | 2060 | 640 | 64,000 | 69,375 | 8,254 | 2,007 | 30,910 | 16,993 |
| | 28 | 2062 | 724 | 72,400 | 82,050 | 9,175 | 2,100 | 33,895 | 17,436 |
| | 30 | 2064 | 810 | 81,000 | 82,050 | 9,592 | 2,179 | 36,971 | 20,512 |
| | 32 | 2066 | 898 | 89,800 | 94,725 | 10,545 | 2,288 | 40,099 | 21,097 |
| | 34 | 2068 | 987 | 98,700 | 107,400 | 11,488 | 2,383 | 43,262 | 21,718 |
| | 36 | 2070 | 1,006 | 100,600 | 107,400 | 11,570 | 2,390 | 43,937 | 22,393 |

Table 10-2. Launch Manifest and Starship Operation.

| Year | Crew ships | | Cargo ships | | Stationed ships on Mars | Stacked ships on Mars | |
|---|---|---|---|---|---|---|---|
| | Earth to Mars | Mars to Earth | Earth to Mars | Mars to Earth | | Reusable | Single use |
| | flights | flights | number | number | number | number | number |
| 0 | - | - | 2 | 0 | 2 | 0 | 0 |
| 2 | - | - | 1 | 0 | 1 | 0 | 0 |
| 4 | 1 | 0 | 1 | 0 | 1 | 1 | 1 |
| 6 | 1 | 0 | 2 | 1 | 1 | 2 | 3 |
| 8 | 1 | 1 | 5 | 0 | - | 7 | 4 |
| 10 | 1 | 1 | 2 | 2 | - | 7 | 6 |
| 12 | 1 | 1 | 3 | 2 | - | 8 | 9 |
| 14 | 1 | 1 | 2 | 2 | - | 8 | 11 |
| 16 | 1 | 1 | 2 | 2 | - | 8 | 13 |
| 18 | 1 | 1 | 1 | 2 | - | 7 | 14 |
| 20 | 1 | 1 | 1 | 2 | - | 6 | 15 |
| 22 | 1 | 1 | 1 | 2 | - | 5 | 17 |
| 24 | 1 | 1 | 2 | 2 | - | 5 | 18 |
| 26 | 1 | 1 | 2 | 2 | - | 5 | 20 |
| 28 | 1 | 1 | 1 | 2 | - | 4 | 22 |
| 30 | 1 | 1 | 1 | 2 | - | 3 | 23 |
| 32 | 1 | 1 | 2 | 2 | - | 3 | 25 |
| 34 | 1 | 1 | 1 | 2 | - | 2 | 26 |
| 36 | 1 | 1 | 1 | 1 | - | 2 | 27 |
| Total | 17 | 15 | 33 | 28 | 5 | - | - |

Operating plans for both reusable and single-use mode Starships are shown in Figures 10-1 and 10-2 respectively. Given the large amount of $O_2$ and $CH_4$ required as fuel for a Cargo Starship returning to Earth, the operation plan must assume a constant level of production volume. A reusable Starship to return crew to Earth is operated every two years after Phase 1b. As many as eight Starships will be accumulated on Mars for 2 years in Year 12 and Year 16. To reduce this number, a larger scale of ISRU and number of power plants would be required in order to produce the required fuel. However the number of Starships carrying cargo would also increase for shipping ISRU and power plant hardware. Starship operation, and in particular launches, have the biggest impact on mass transported from Earth.

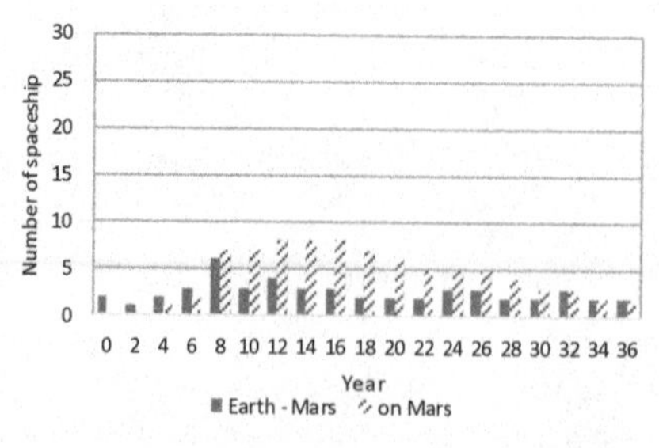


Figure 10-1. Number of Starships in Reusable Cargo Ship Operation.

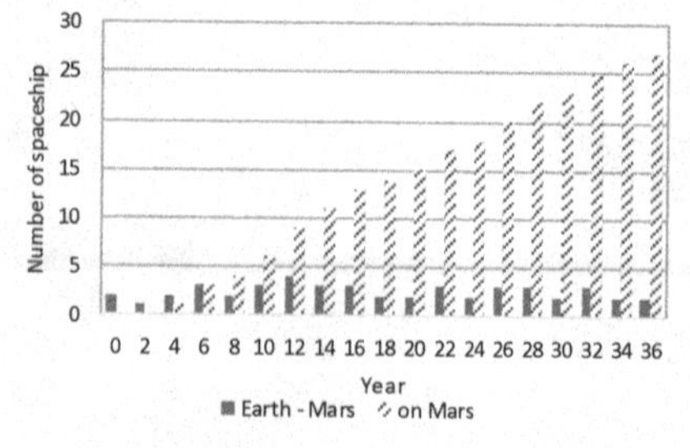


Figure 10-2. Number of Starships in Single-use Cargo Ship Operation.

A cost comparison between reusable and single-use spaceships to transport the cargo of the manifest is shown in Table 10-3. The costs consist of Starship development, Starship production, tanker production, crew launch to Mars from Earth, cargo launch to Mars from Earth, and launch to Earth from Mars. The number of spaceships and the number of launches were chosen to satisfy the necessary cargo volume for a reusable Starship and single-use Starship based on the cargo manifest.

Table 10-3. Cost Comparison of Reusable and Single-Use Starships.

| | Unit price (a) | | Cargo ship reusable Number (b) | Cargo ship reusable (a) x (b) | Cargo ship single-use Number (c) | Cargo ship single-use (a) x (c) |
|---|---|---|---|---|---|---|
| Starship development | 5.0.E+09 | $ | 1 | 5.00E+09 | 1 | 5.00E+09 |
| Starship production | 3.4.E+08 | $/ship | 14 | 4.69E+09 | 32 | 1.07E+10 |
| Tanker production | 3.4.E+08 | $/ship | 22 | 7.37E+09 | 12 | 4.02E+09 |
| Crew launch cost Earth>Mars *1 | 7.0.E+06 | $/launch | 51 | 3.57E+08 | 51 | 3.57E+08 |
| Cargo launch cost Earth>Mars *2 | 7.0.E+06 | $/launch | 165 | 1.16E+09 | 150 | 1.05E+09 |
| Crew and cargo launch cost Mars>Earth | 7.0.E+06 | $/launch | 43 | 3.01E+08 | 15 | 1.05E+08 |
| Total cost, $ | - | | - | 1.89E+10 | - | 2.13E+10 |

*1 A crew starship is refueled by 2 tankers in LEO, 17 crew ships land on Mars. 17 x (1+2) =51.
*2 A cargo starship is refueled by 4 tankers in LEO. Reusable cargo ship: 33 x (1+4) = 165, Single-use cargo ship: 30 x (1+4) = 150.

The change in population is shown in Figure 10-3. Though 100 people arrived at Mars Colony after Year 16 in Phase 2, only 72 people permanently stay in the colony. 2% natural growth of the population was assumed after Year 16.

The total mass required for Mars Colony is shown in Figure 10-4. The solid circle (●) shows the Mars colony total mass by means of reusable Starships. The solid square (■) shows mass transported from Earth by means of reusable Starships. The empty circle (○) shows the Mars colony total mass by means of single-use Starships. The empty square (□) shows mass transported from Earth by means of single-use Starships. For reusable Starships, both the Mars Colony total mass and mass transported from Earth are larger than for single-use Starships, due to the increase in ISRU mass to produce fuel for return to Earth. However, the total cost for reusable Starships is less than that for single-use (ratio of single-use to reusable = 1.13 (=2.13/1.89)), as shown in Table 10-3.

Mass production of $H_2O$, $CO_2$, and $N_2$ by ISRU is shown in Figure 10-5. Production rates of $H_2O$ for oxygen generation and $CH_4$ (methane), which are used as fuel for return of Starships to Earth, drastically increase after Year 10 in Phase 2.

The evolution of power consumption is shown in Figure 10-6. The solid square (■ ) shows required power on a sunny day. The solid triangle (▲) shows required power during a dust storm or at night (i.e. no sunlight). The solid circle (●) shows available power. It is sufficient to supply power required for photosynthesis in the case of no sunlight.

The expected budget profile for Mars Colony is shown in Figure 10-7. The solid line, the dotted line, and the dashed line respectively show the revenue, expenditure, and capital of MADC. US$5 billion in revenue is accumulated annually (40% government funding, 40% lottery funding, and 20% private funding) for the first 20 years of the Mars Colony development phase. The accumulation of capital decreases after 2037. Expenditure decreases after a peak in 2047. Lottery revenue and private funding continue during the start of the self-sufficiency phase which begins in 2047. Asteroid mining will commence around 2047 and revenue generated by selling precious materials on Earth will gradually replace the lottery and private funding sources. The MADC will achieve a budget surplus in 2057. Accumulated capital will exceed US$50 billion in 2080. The total development cost for the entire project will be US$100 billion.

The Mars Colony site plan was previously shown in Figure 8-1, and contains: 8 greenhouse domes, 5 inflatable greenhouse modules, 2 Crew Starships, 3 Cargo Starships, ISRU, power plants, and roads. 5 Starships for habitation and 5 greenhouse modules are installed in Phase 1b, and 8 greenhouse domes (lower level contains habitats; upper level contains greenhouses) are constructed in Phase 2. The final mass budget is shown in Table 10-4. 84% of total mass at the final stage of Phase 2 is structural mass. 5% of structural mass, which cannot be produced on Mars, is transported from Earth. All mass for power plants, biomass production systems, LED, life support systems, and construction machines is transported from Earth. Overall, 20% of total mass is transported from Earth.

Table 10-4. Mass and Budget in Year 36.

| Subsystem | Total mass in Phase 2 (a), mT | a/b | Mass from Earth (c), mT | c/a |
|---|---|---|---|---|
| Structure | 9,650 | 0.84 | 471 | 0.05 |
| Power plant | 200 | 0.02 | 200 | 1.00 |
| Greenhouse | 209 | 0.02 | 209 | 1.00 |
| LED | 448 | 0.04 | 448 | 1.00 |
| ISRU | 574 | 0.05 | 574 | 1.00 |
| LSS | 363 | 0.03 | 363 | 1.00 |
| Construction machines | 126 | 0.01 | 126 | 1.00 |
| Total | 11,570 (b) | 1.00 | 2,390 | 0.20 |

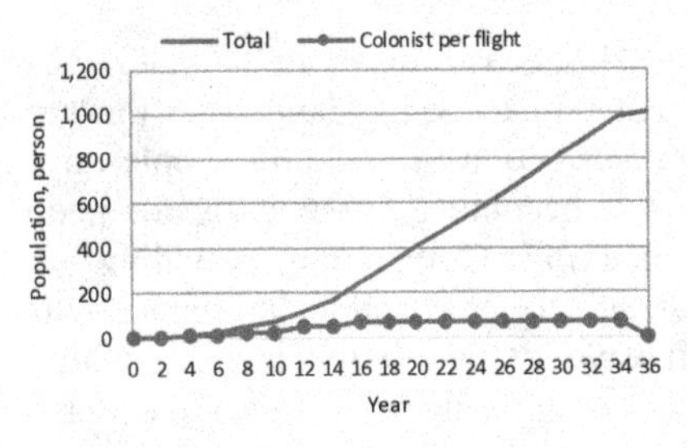


Figure 10-3. Change in Population.

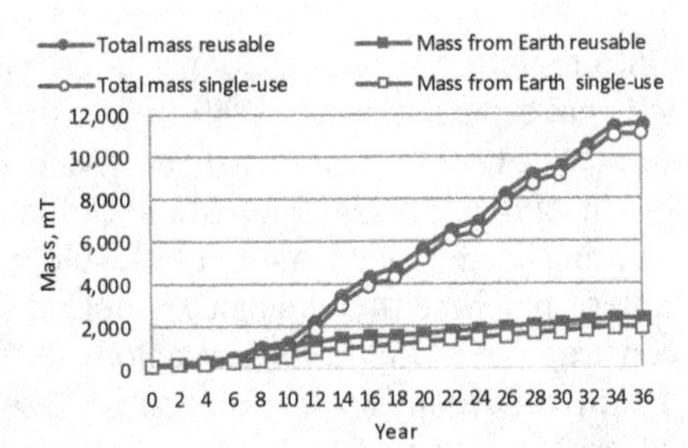


Figure 10-4. Cumulative Mass of Mars Colony.

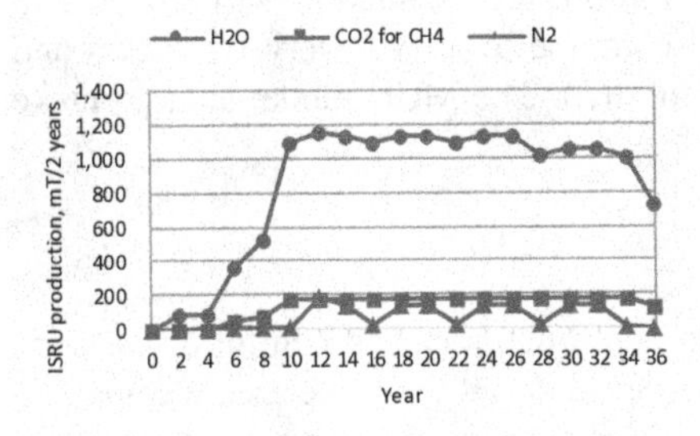


Figure 10-5. Production Mass by ISRU.

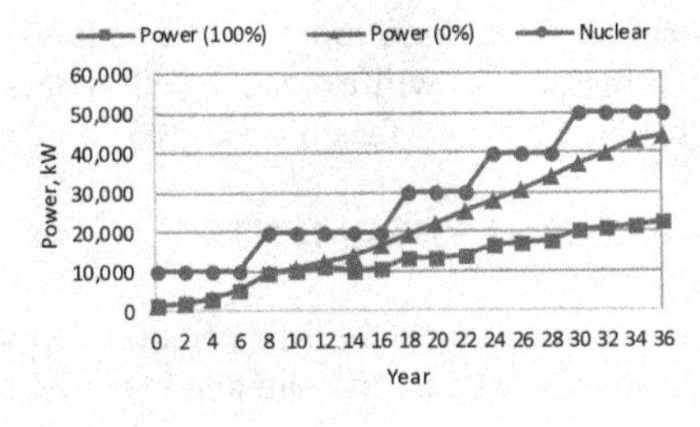


Figure 10-6. Change in Power Consumption.

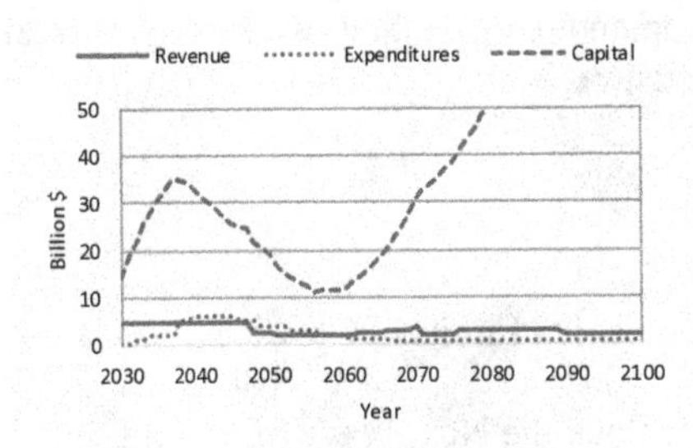


Figure 10-7. Expected Budget Profile for Mars Colony.

## 11. SUMMARY

The MCD model was developed to propose a Mars Colony development schedule, taking into consideration mass transported from Earth, resources on Mars, total energy, and total cost (transportation cost). The population is expected to exceed 1,000 people at Year 36. The habitation area, greenhouse area, total mass, mass transported from Earth, required power when sunlight cannot be used and required power when 100% of sunlight can be used at Year 36 are respectively: 100,600 $m^2$, 107,400 $m^2$, 11,570 mT, 2,390 mT, 43,937 kW, and 22,393 kW. The available power supply is sufficient to provide power on a dust storm day or at night. It can supply the power required for photosynthesis when there is no sunlight.

Our prediction for Mars Colony in the year 2400 is as follows. In 2400, 431 years will have passed since 1969, when the first human beings landed on the lunar surface. 117,000 people will be living on Mars as well as 3,000 people on the Moon and asteroids. The Mars Colony will become a space oasis for people working in the Mars-Asteroid Development Public Corporation, providing food, water, machine maintenance, fuel, medical services, and entertainment. The construction of the first Mars colony at Endeavour Crater will commence with the use of robots in 2034. The first 12-person colony will begin living on Mars in 2038. A mission of colonists will be sent to Mars every two years. Infrastructure on Mars will be set up to support the first gateway city. When the first 72-person non-professional colonist group arrives in 2050, the population will rise to 243 people. After that, more immigrants will continue to arrive every two years and the population will exceed 1,000 people in 2070. The MCD model aims to make this kind of story feasible.

## ACKNOWLEDGEMENTS

We wish to thank the Mars Society Japan board members for helpful discussions on selecting a landing site and using resources on Mars for space architecture. We wish also to express our gratitude to Yoji Ishikawa, senior engineer at Obayashi Corporation, for his very helpful advice. A special thanks goes to Maximilien Berthet, an intern at Space Systems Development Corporation, who gave us great suggestions and corrections for the final revision.

## REFERENCES

[1] Meadows, D. H., Randers, J., and Meadows, D., Limits to Growth, Chelsea Green Pub Co, 2004.

[2] LePan, N., How AI and Big Data Will Unlock the Next Wave of Mineral Discoveries, https://www.visualcapitalist.com/ai-big-data-mineral-discoveries/, accessed Feb. 26, 2019.

[3] Psyche Mission – A Mission to a Metal World, https://psyche.asu.edu/, accessed Feb. 26, 2019.

[4] Asteroid Database and Mining Rankings, https://www.asterank.com/, accessed Feb. 26, 2019.

[5] Muscatello, A. C., Mueller, R., Sanders, G. B., & Larson, W. E., Phobos and Deimos Sample Collection & Prospecting Missions for Science and ISRU, https://www.lpi.usra.edu/meetings/marsconcepts2012/pdf/4296.pdf, accessed Feb. 26, 2019.

[6] State lotteries: Sales U.S. 2016 Statistic. (n.d.), https://www.statista.com/statistics/215265/sales-of-us-state-and-provincial-lotteries/, accessed Feb. 26, 2019.

[7] SpaceX, https://www.spacex.com/mars, Starship and super heavy, accessed Feb. 27, 2019.

[8] Clarke, J., Willson, D., Smith, H., Hobbs, S., & Jones, E., Southern Meridiani Planum - A candidate landing site for the first crewed mission to Mars, *Acta Astronautica,133*, 195-220, doi:10.1016/j.actaastro.2016.12.039, 2017.
[9] Abston, L., Amundsen, R. M., Bodkin, R. J., Benshabad, A., Boyer, C. M., Clowdsley, M. S., and Senter, I., Ice Home Mars Habitat Concept of Operations (ConOps), http://bigidea.nianet.org/wp-content/uploads/2018/07/IceDome-ConOps-2017-12-21v-reduced.pdf, accessed Feb. 26, 2019.
[10] O'Neill, G. K., et al., Space Resources and space Settlements, NASA SP-428, 1979.
[11] Hanford, A. J., Advanced Life Support Research and Technology Development Metric – Fiscal Year 2005, NASA/CR-2006-213694, 2006.
[12] Kozai, T., Sunlight Plant Factory - Sustainable Design of Advanced Plant Factory, ohmsha, 2009.
[13] Miyajima, H., Life Support Systems Trade Study for Lunar Habitation with Greenhouse Food Production, 48th International Conference on Environmental Systems, ICES-2018-173, 2018.
[14] CFSR: CABKOMA Strand Rod, https://www.komatsumatere.co.jp/cabkoma/en/, accessed Feb. 27, 2019.
[15] Nippon Paper Group, Cellulose nanofibers (CNF): cellenpia, https://www.nipponpapergroup.com/products/cnf/, accessed Feb. 27, 2019.
[16] U.S. Bureau of Labor Statistics, Major occupational groups as a percentage of total U.S. employment, May 2016, https://www.bls.gov/oes/2016/may/featured_data.htm#largest3, accessed Mar. 16, 2019.
[17] Kitmanyen, V. A., Disher, T. J., Kobrick, R. L., and Kring, J. P., *Human Factors for Small Net Habitable Volume: The Case for a Close-Quarter Space Habitat Analog*, 47th International Conference on Environmental Systems, ICES-2017-203, 2017.
[18] Bigelow Aerospace, https://bigelowaerospace.com/pages/b330/, accessed Feb. 27, 2019.
[19] Miyajima, H., Self-Sustainable Smart City Design on the Red Planet, 49th International Conference on Environmental Systems, ICES-2019-158, 2019.
[20] Rapp, D., Human Missions to Mars: Enabling Technologies for Exploring the Red Planet, Praxis Publishing Ltd., 2008.
[21] Truscello, V.C., SP-100, the US Space Nuclear Reactor Power Program, Technical information report, JPL D-1085, 1983.
[22] Mole, A., Baseline Design for a Mars Colony, http://marscolony.marssociety.org/MARSColonyi2.pdf, accessed Feb. 27, 2019.
[23] Kelly, P. W., & Bevilacqua, R., Constellation Design for Mars Navigation using Small Satellites, *2018 AIAA Aerospace Sciences Meeting*. doi:10.2514/6.2018-1538, 2018.
[24] NASA *Space Exploration Vehicle Concept* (Tech.), https://www.nasa.gov/pdf/464826main_SEV_FactSheet_508.pdf., accessed Feb. 27, 2019.

[25] YANMAR,Co., Ltd, "Innovative concept Backhoe, http://advertisementfeature.cnn.com/2015/yanmar/new-content-model-products.html, accessed Feb. 27, 2019.
[26] Tadano Ltd, https://www.tadano.com/, accessed Feb. 27, 2019.
[27] Machines-3D 3D Constructor review - house construction 3D printer. https://www.aniwaa.com/product/3d-printers/machines-3d-3d-constructor/, accessed Feb. 27, 2019.

# 13: BUILDING A SUCCESSFUL MARTIAN COLONY

**Vincenzo Donofrio**
Michigan State University
donofr19@msu.edu
**Meghan Kirk**
Michigan State University
kirkmegh@msu.edu

## ABSTRACT

It is obvious that the implementation of a feasible colony on Mars has been discussed and analyzed for some time now. The main topics of debate have been in regard to the practicality of a fully self-sustaining colony as well as the economic viability of the process. At this point, it is noted that Mars offers all necessary in-situ resources to allow for self-sustainability, with the most efficient techniques and systems to produce water, food, metals, plastics and other materials to be discussed. However, the solution to the economic viability of colonizing Mars appears to still be in question, mainly because many leading nations still display great ignorance to this matter and are blind to the prosperity that Mars offers financially. Perhaps an initial mission of upwards of hundreds of billions dollars may be too expensive and intimidating for these groups, which is generally the average to high end price settled on in most discussions, so it is necessary to generate an economic plan that requires a small fraction of this cost, as well as detailing exactly how quickly the colony can return on this investment and begin to profit. Throughout this paper, the focus of most topics will be split upon three general points in the population: one thousand, hundreds of thousands, and millions. For this scenario, one thousand small, fit, and intelligent professionals will comprise the crew initially inhabiting Mars. This will be an important point in the population for obvious reasons, these people will research the territory, set-up crucial support systems, as well as ignite the colonies' economic prosperity. At hundreds of thousands of people, the colony may perhaps multiply into many, with societal factors really taking off at this point. When the population reaches the millions, this is when the prospect of terraforming arises, as well as when complete self-sustainability and absolute economic success comes to fruition. Important decisions involving government and society will be discussed in this paper, as well as the need for an aesthetically pleasing colony.

Additionally, our given scenario assumes that prior missions lacking human presence on the planet will take place, which will allow for the set-up of nuclear generators by way of robot and the continued robotic research that has taken place for decades now. An in-depth discussion of the distribution of this source as well as any other form of power will still be necessary. We also can assume that irrigation water and fertilizers have been found on Mars, with the water assumption being highly reasonable at this point considering the planet is almost certain to contain reservoirs of ice and perhaps liquid water and the fertilizer narrative leaves us with a nice bonus. Furthermore, we cannot forget that the expected costs of shipping goods from Earth to Mars and then from Mars to Earth will be $500/kg and $200/kg, respectively. As mentioned previously, the remaining necessities will be thoroughly discussed, as well. However, we shall offer some reasoning for the case of Mars, as well as some important pre-arrival factors beforehand.

## WHY MARS?

Before we dive into the building of a successful colony, we should outline why Mars is the best available option in the entirety of the solar system. More specifically, the discussion should be why Mars is factorially better than the second-best option humans realistically have: The Moon. The advantages the Moon possesses over Mars are almost non-existent barring a few. Earth's Moon quite clearly has the distance from Earth as well as the required velocity to escape it in its favor, with maybe a couple other advantages that serve no real purpose. While the distance notion really has no counter to it, the issue with escape velocity does. While it is easier to escape the moon, that comes at a huge cost. For a massive body to have this advantage, it means that its gravity well is generally more lenient, which is the case for the moon. With the moon being protected by literally no atmosphere whatsoever and the gravity on the surface coming in at an excruciating 1/6 of Earth's compared to Mars' manageable atmosphere and 1/3 of Earth's gravity, the advantage of an easier time leaving isn't worth the negatives. In fact, just a glimpse at the moon using the naked-eye will leave its holder fascinated by the massive craters that likens earth's moon to an "asteroidal dartboard", so-to-speak. The plethora of advantages Mars has over the Moon only rests this case. The combination of a workable atmosphere comprised heavily of carbon dioxide as well as other useful gases and manageable gravity makes Mars the most comparable candidate to Earth in the entire solar system. The necessary resources a colony will need to thrive are all available as well, with the moon being comprised of very little in comparison. The biggest advantage will be its water; while the Moon has recently been discovered to contain water, its amount hails in no comparison to the amount on Mars. Perhaps this discussion should be geared towards a comparison between an exoplanet like Kepler 438b and Mars, but its category alone should indicate why there isn't one to be had as of today.

## PRIOR TO HUMAN ARRIVAL

Another conversation to be had before a human steps foot on the planet is in regard to what will be brought. We will discuss the necessities and their respective quantities, with the approximate costs of the totals exemplified in the more relevant economic section.

If we want to really lower the cost, then it is important to choose the correct items and how much of each will be brought. It is crucial to think about what will essentially be available to make right away on Mars and then what will not. Also, the crew will need to factor in the probable eight-month journey they will spend traveling through space, which will be discussed first. The colonists will need a considerable amount of food and water to survive the flight. While what types or kinds of foods and beverages aren't of upmost importance, it can be assumed that a resounding amount will in fact just be water and various healthy greens for the small, fit crew. It is best to simply base how much should be brought around the recommended amount these sized people should consume for eight months. This means that for an eight-month period, around 800,000 kilograms of food and water should be brought to meet those requirements. An additional 250,000 kilograms of various plants and foods should be brought to give the crew enough food for a few months until the first imports arrive.

The crew will also have to bring resources that will serve great purpose on arrival and some time after, too. While they should and will not bring any metals, plastics, glass, or fabrics, these can be made simply using in-situ resources on Mars, the crew should consider bringing shelter, spacesuits, and advanced electronics. This will also contain the nuclear generators as well as any robots that will provide great assistance to the humans, but as previously stated these will be brought prior to this mission with the amount (approximately 100,000 kilograms) still being factored in to the total cost.

Although the colony will initially be placed underground with the colonists being able to walk freely in their tubes, they will still want spacesuits. Considering that there will be no available shops that will allow the crew to buy their suits, the only option for the initials will be to bring them. There are two types of spacesuits to choose from as well: one that emphasizes comfort and one that specifies safety. In technical terms, these two options are referred to as elastic and pneumatic suits, respectively. While the design and obvious comfort of the elastic suit will appeal to most, the wise decision is to go pneumatic. The use for a spacesuit on Mars is to protect a colonist from the cold, allow for breathable air, while also offering sufficient protection when traversing the surface prior to terraforming the planet. And with current suit technology, pneumatic is the option that more effectively meets these needs. A pneumatic spacesuit, specifically a Mars Suit[1], allows for an overall warmer experience, especially for the lower half of the body, as well as superior protection. Elastic suits will tear easier than pneumatic which practically will have no problems in that regard. In terms of pricing, elastic does offer the much lower price (a factor of 10 less), and as previously mentioned is more comfortable, but these advantages do not outweigh those that the pneumatic suit offers. If the crew wants comfort, they will make their own synthetic clothing upon arrival, in addition to the sweats they will undoubtedly bring. The total price of the suits will be only $32.5 million, anyway, with the advantages they give valued at priceless.

The crew will also have to bring their own shelter. The other option that is available is to bring enough shelter for maybe half the colonists, or simply not bring any at all and rely on instantaneously manufacturing tubes upon arrival, but to lower risk and boost safety, there will be enough shelter for everyone brought from Earth. Future structures will be made using in-situ resources, but for now the various greenhouse, living, and research tubes, as well as any helpful manufacturing items, will come in at a mass of 445,000 kilograms. Any advanced electronics that will be used will weigh in the range of the tens to hundreds of thousands, as well.

It is important to consider leaving some room for error, as well as understanding the idea that full self-sustainability will not be possible until the population reaches the hundreds of thousands, at least 25 to 50 years later. This means that the colony will still need to rely on and map out imports from Earth for at least a few decades, with the loads that comprise them decreasing at a likely exponential rate. Water can be exempt from this list, as it is fairly straightforward to produce on Mars. The materials required for shelter and structures will also not need to be imported, say, a few years after initial settlement as the advancement of 3D printing technology will offer great aid to the colony. Food (mainly plant) imports will most likely be a mainstay for a considerable amount of time because the experimental procedures and testing on Mars that has to be done using terrestrial plants will probably offer the biggest challenge. Synthetic biological techniques are currently in the works[2], but until this can be perfected for Mars, food imports will be required. Factoring in this hopeful success, we will deem these imports to gradually decrease, but still be a large part of the load. Suits can easily be made on Mars for future colonists, and the generators should be made using in-situ resources, as well. Advanced electronics may be a mainstay, too, but not as long as food will be.

If the following parameters are followed, then there will need to be around $500 billion to $1 trillion invested if we are trying to be as efficient while offering some risk as possible, with numbers easily reaching into the double-digit trillions if a safer approach with more funds is available. While these projections are very high, keep in mind they are the totals until self-sustainability. As the colony reaches this point, the cost will continually decrease until virtually nothing will to be spent on imports, anymore. In a later section, we will discuss how to deal with these costs, as well as how much the initial mission will be valued at. The initial mission and the future cargo shipments should be separated financially, as they both coincide with their respective periods of economic success that will see profits starting right away in the initial phase as well as beyond these later shipments. As stated, specifics will be discussed.

**LOCATION**

The ability to pick the desired location of a colony is something that is unique to Mars. All the major human explorations of the past didn't allow for the ability to really scope out the targeted terrain. This can even be said for the moon. Although there were no real intentions to start a colony on the Moon, the technology that was used to study it before arrival doesn't even come close to the technology that has studied Mars, specifically over the last couple decades. This means that picking the perfect location for the colony is crucial and shouldn't be taken lightly.

There are many options that would allow for a successful colony to be built on Mars, but quite frankly only a few should seriously be considered. Factors that should be of highest value when discussing location are the availability of necessary resources, the condition of the terrain (i.e. not on top of volcano), and the opportunity for economic success that may come from tourism. Considering this, the two specified and prominent locations to be compared are the areas near Gale Crater and Valles Marineris (Mariner Valley).

Gale Crater would be a wise place to choose in terms of tourism. Imagine the attraction that would be centered around this massive 154-kilometer-wide crevice. Additionally, a colony placed near the edge will overlook the crater's enormous mountain, Aeolis Mons, which is located near its center and rises an incredible 5.5 kilometers, a sight incomparable on Earth. The depth of the crater may also allow for easier access to the possible huge water reservoirs beneath its surface when necessary. Gale Crater also offers a level of relatability per se to humans because it is thought to once be a lake many years ago[3], with the idea that it will be the first lake the humans engineer upon terraforming Mars. However, while this spot is a very reasonable landing place, Valles Marineris simply beats out Gale Crater in every way.

Figure Valles Marineris (Mariner Valley)

Valles Marineris is quite comparable to Gale Crater in that it also has great depth but reaches down to an even more extraordinary 8 kilometers in many areas. This will in turn allow for easier access to extradite water beneath the surface, especially considering that the valley is hypothesized to possess spring-like deposits[3], as opposed to the standard ice beneath it, bingo! Located near the equator, the valley will experience average highs of zero degrees Celsius. While not so much should be focused on temperature, location near the equator will maximize the sunlight available that will be used as solar power. As for another measurable, the valley is an astonishing 4,000 kilometers long which is a staggering five times longer than earth's Grand Canyon, its closest comparison, and as long as the United States itself! The Grand Canyon is one of the most attractive tourist destinations on earth, which only means that the Valles Marineris could be one of the most attractive sites in the solar system. Exact location should be pinpointed on either of its shorter sides, in the need to go around it, with the exact side being towards the east near the region Eos Chasma, which is thought to be pretty resourceful.

## UPON ARRIVAL–VARIOUS TECHNICAL SYSTEMS

After the crew endures the long trip and eventually arrives on Mars, it will be important to start off on the right foot. This means being able to construct the various systems and conditions in place to raise the chances of survival and efficiency. Immediately, the crew will need to set up shelter, begin creating their systems to store energy, undergo the retrieval of water and production of food, and start to research and explore with the aim to quickly produce all remaining materials.

### Shelter and 3D Printing

The crew can use the provided robots to dig out holes to place and assemble the inflatable tubes and greenhouses along the valley, a relatively quick and easy process. Some things to consider, however, are where and how far apart the shelter should be placed to avoid any shadowing problems with the eventually implemented solar panels. The crew can simply spread out the tubes, solving this problem. Using the reference provided by MarsSociety[4], it is safe to map out an area of three million square feet; this will allow for the colony to be a square with 1,700 feet on a side with six city blocks of 300 feet which will allow for the colonists to easily walk between labs and shelters when needed. Again referring to the reference, the tubes should be 16 feet in diameter to limit costs of shipping. They will be pressurized at eight PSI to provide a safety factor of four against the pressure of two PSI that is evident at about eight feet underground on Mars. There will be a transparent strip along the top, with aluminized mylar reflectors being used to reflect sunlight into the tube.

3D printing has opened a range of possibilities that will improve a colony on Mars. The colonists could easily print out any object or tool quickly, instead of relying on imports. More importantly, this concept will thoroughly be used in the construction of above ground structures.

As the colony expands, massive domes, perhaps made of ice[5], will be made. But, the cheapest and most efficient option will be to make use of this ever so interesting concept of 3D printing. A competition[6] held by NASA that looked to find the best concept for an above ground home ended with many exciting designs chosen as finalists. The winning group, Team Zopherus, designed a moving printer that would deploy rovers to collect local resources and produce the intended habitat. The use of technology like this will pay dividends for the colony.

**Energy/Power**

Picking the correct sources and then efficiently distributing them to power the colonists' various systems will be key. On Mars, the two main sources of energy will be through solar and nuclear power.

Solar power will be extremely useful but shouldn't be relied on. Considering the devastating dust storms that could engulf the colony for months at a time, only a minimal amount of power that the colony uses should be solar. Additionally, dust storm or not, the amount of dust that will settle on the panels during a normal day will be a hassle to clean off and will lower the amount of sunlight actually received. We will also have to factor that solar irradiance on Mars is 595 watts per square meter, nearly a factor of three less than Earth's, which comes as a result of Mars being further away from the Sun. So, solar power should exclusively be used in processes like water electrolysis as well as heating the living spaces and greenhouses.

Nuclear power will be relied on heavily by the crew, which will require the instillation of a plant above ground. Both nuclear fusion and fission reactions will be used, with the isotope deuterium, commonly found on Mars, being applied to both (deuterium will be discussed more later). Nuclear fission is and will be the preferred source, as it the most compact and least complicated to deploy. The crew can specifically use NASA's Kilopower reactor[7] for fission and one bigger generator for the rest of the needs. For a colony of 1,000 people, 2MWe is needed[4]. This means that 100 Kilopower reactors (10 kWe) can combine output with the one MWe generator, which will be a cheaper option to bring when considering import taxes. The total mass for the previously stated scenario will be around 40,000 kilograms, half that of a scenario that involves twenty 100 kWe generators each weighing 4,000 kilograms: a total of 80,000 kilograms.

Nuclear power will definitely be able to keep the colony warm and running during dust storms, as the power plant will produce a thermal output of 27 million btu per hour (British thermal unit), compared to the nearly 20 million btu per hour the greenhouses will lose[4].

**Water**

For our scenario, it is a given that irrigation water has been found. As mentioned before, this is a wise assumption, but an overview of the process to retrieve more water is necessary. The most efficient way would be to assign the robots to extract the ice water underground by way of mining and melting of ice.

The availability of water also allows for some crucial gases to be obtained, too. Water provides access to hydrogen, though the electrolysis of water, which is an important component of fuel (liquid hydrogen). Oxygen is also a product of this process, the remaining component in the liquid hydrogen fuel. Additionally, hydrogen can be used to make certain plastics, with oxygen of course allowing the colonists to breathe.

**Food**

The production of food may be the most difficult process to complete for the colonists. Realistically, food (plants) will have to be imported for a while until the colony understands how Earthling plants grow in Martian soil. For food to grow using Martian soil in the greenhouses, the regolith that contaminates it can be cleaned using the water the colonists produce. On Mars, the practice of plant synthetic biology can be further researched[2], with the goal of engineering plants to adapt specifically to the Martian environment and end the dependence on shipping food from Earth. Of course, the plants the crew grow will provide them with the oxygen they need in the tubes.

**Glass**

The production of glass, and specifically fiberglass early on, will be useful for the crew. About 40% of the weight of the soil on Mars is made up of silicon dioxide: the basic component in glassmaking. Unfortunately, Martian soil is also heavily made up of iron oxide as previously mentioned, which will leave the colonists will a lower quality of glass. However, this isn't really a problem for the colonists who will want to opt for the most efficient processes, and the need to remove the iron oxide through a tough process isn't necessary. Therefore, the tinted silicon dioxide glass can be made using the same sand-melting techniques used on Earth.

As the colony grows, a good shout would be to build an above ground city and dome out of this glass, preferably with the iron oxide removed. This would be wise as something like glass will be more available and most likely easier to make than cement, but that is something that will be interesting to discover.

**Vehicles and Transportation**

As the colony grows and walking becomes inefficient and tiring for the crew, there will be a need for an upgrade in transportation. The first need will be in the form of a simple bicycle. Mars offers two materials that can make up the frame of the bike: steel and aluminum. Additionally, the seat and handlebars and what not can be made of the plastic that the colonists can produce, as well. The wheels should be of required quality to traverse through the thick sand using technology already present on earth.

Eventually, as the colony expands into millions of people and inhabits a variety of regions across the globe, more sophisticated technology will be preferred. This is also where things will get a little interesting. But first, the obvious choice of transportation, as is on earth, will be by car. If we stick to the idea of using clean energy, then specifically we want electric cars. And on the rocky and bumpy surface Mars will provide inhabitants, a sort of jeep modification will be the type of car selected. While roads will eventually be placed, they are not of top priority for travel on a planet that will not have any vegetation or bodies of water to get in the way for a while, but simply just a whole bunch of red sand.

The interesting part is when the discussion of air travel on Mars is brought up. While travel by the earth's standard commercial airplane is quite bad for the environment, a less threatening option is by way of rocket hopper. More specifically, the most viable option that is available currently is Spacex's Starship[8]. The ability to travel around the planet in an hour will be vital for the expansion of the civilization on mars, but the big concern that lies with this method still resides in its ability to use a clean, cheap source of fuel. This is because on earth, the production of the hydrogen and oxygen sources used to make the required liquid oxygen fuel mainly are generated using non-renewable forms of energy. However, renewable sources like wind and solar power could be used in the process, but at much higher costs. If on Mars comes a breakthrough in this technology that lowers costs and in turn allows these renewable sources to be efficient and useful, then a rising form of travel that will completely alter the aerospace industry may just be perfected and explode.

A form of transportation that will be wise to look out for on the future of mars will be by way of passenger-carrying-drones. Mars will certainly be a place where the risks that are just too much for Earth will be tried instead. While not necessarily a viable option until the dust storms are settled, which means probably not until the planet is terraformed, the use of this technology will surely be explored. Although these could easily just turn out as glorified helicopters, a slick Martian or two may just unlock the keys to a new, unique way to travel.

**Communication**

Communication during the human settlement of mars will be key. The routes that need to be covered are communication between Mars and Mars, and Mars and Earth and locations like the asteroid belt.

The most important form of communication, especially early on, will obviously be between Martians. Ideally for the few who are still left with exploration jobs, but really useful for everyone, will be an accessible way of communication that is at the push of button and attached to one's spacesuit. To allow the radio waves to be transmitted between Martians, satellites will need to be implemented, something not entirely too difficult, as long as there is the means of transportation to get the required position in Mars' low orbit. This is something that can be implemented during prior, unmanned missions, with the cost being virtually nothing when added to the total amount of the entirety of these first missions to mars. Communication between Martians and Earthlings and those at other locations can be done through these satellites as well, specifically using laser transmission techniques to limit the transmission distance handicap. Another option for this long-distance communication would be to use smartphones or the like to simply message between planets. This will of course require an internet connection to be founded on mars, something that may actually impose greater difficulties than setting up satellites, but once invented will be used as another form of communication unfortunately again with a delay. Instantaneous communication between planets will perhaps be unearthed, but for now a slight delay will have to do.

**POLITICAL**

For the first 1,000 colonists, it may be quite difficult to establish a complete government solely on Mars, but basic laws and rules can be put in place. In a perfect scenario, many of the leading powers on earth will be united by this great cause, and with thorough negotiation, the first 1,000 colonists can be governed by this newfound united front, with possibly one key leader with a few other higher-ups to lead the other colonists.

Unfortunately, this will unlikely be the case, and either a private company (like SpaceX) with help from a nation's government and its space program (like the US and NASA) will lead the crew, or the crew will simply be guided entirely by said government. Using the former, highly likely scenario, there will be a local government on mars comprised of a democratically selected leader and board of officials as well as a strong presence from, in this scenario, the US government back on earth. This is seen in many other instances in colonialism throughout recent history, most notably, colonial America. Like our plan with the colonization of mars, colonial America was comprised of a few forms of local government, with also a great deal of interference from the British government. Of course, however, it is important to not impose the strict and unfair taxes and laws that eventually led to the breakaway of colonial America, something that absolutely cannot happen on mars in terms of unification. While seemingly this will be simply avoided because of the positive track record the democratic government of the US possesses in terms of the capability its citizens possess to express basic freedoms, there is still a need to specifically address these necessities. If Mars wants to ultimately end up as the massive economic success envisioned my many in the short-term future, mainly because of the unique industries and opportunities for great inventions and achievements, it will need to be allowed to freely express these avenues. This means that, unlike colonial America, the Martians will need to be given the ability to apply for and at least retain the rights to their creations.

The goal is to allow the chosen Earth government to oversee the colony, protect the rights and interests of its colonists, collect on the fairly priced export and import taxes, all with the confidence that Mars will quickly become economically, technically, and socially successful which will then lead to eventual self-governing metropolises. There will initially need to be a high-level of interest from Earth towards Mars, with a relatively steep decline that coincides with the expansion of the colony.

As the colony expands and soon is comprised of multiple cities with its population in the millions, this will coincide with changes in governmental structure. Like with many nations on Earth, the cities on Mars should consist of multiple types of initially healthy governments, based on the various necessities and needs for that society. The strongest and most successful ones will survive, and the ones with poor ideas will fail. While the objective isn't to deem cities as utter failures completely right away, as seen on earth, this will eventually happen as the planet becomes heavily populated. However, with the goal of keeping peace and unity between these cities that will arise from a culture that is built on togetherness and survivability, hopefully the planet will not experience much catastrophic failure and unhealthy competition that is prominent on earth. But, hopefully a great level of support for those failing cities will be expressed, because at the end of the Martian day, it will be incredibly tough to succeed.

In order to ensure success and sustainability on Mars, these specific policies should be instilled or granted. First, the need for property rights will be crucial. As the colony expands, and people begin to look to structure their own housing, the last thing they will want is for some Martian pirates stealing it away. So, the Martian government will need to establish a basis for these property titles to avoid this problem. Additionally, the need for this process may even be called for sooner as undoubtedly businesses on Earth will want to stake an early claim in territory before the eventual skyrocketing of the price for the land.

As stated before, the rights of any invention created by a Martian should stay with the Martian. If the colony wants to succeed, they will have to retain the entire value of their products to hold up in the trading business. The Martians should also be able to apply for patents from Earthling marketplaces if they desire to grow their business independently, and not just be limited to only being able to bargain their inventions.

Efforts from nations that will try to disrupt the colony will surely be acted out, so it is important to have a politically diverse planet, with key regulations.

## SOCIETAL/CULTURAL

Initially, the society will be made up of many values from earth. With only 1,000 on the planet, the colony will tend to be hierarchical, with one leader who has a board comprised of multiple officials. As the colony expands, there will inevitably be those that want to venture out and create new colonies, which will eventually lead to massive cities and metropolises. As the planet reaches into the millions and eventually gives birth to its third and fourth generations, perhaps these Martians will finally consider Mars their true home and prefer never to go back to Earth. This will raise the question of if and how the planet can hold values that are better than those of earth's society, with the need to avoid war and congressional conflicts. This can be done through thorough education, in both technical and cultural aspects in how to be and what it takes to be a Martian. Also, the forming of a society that can make better use of its time when it comes to their social, personal, and professional life, as well as the important need of increasing physical, mental, and emotional health is necessary.

Martian society will always be compared to the only other society in existence: Earthling society. While Mars will never be a utopia many claim it could be, it will most certainly be a place that will build off those values that Earthling society presents and build as close to a perfect society as possible, while in turn giving Earth something to compare itself to in order to better its own society.

As the colony quickly progresses into the hundreds of thousands and multiple branched-off colonies begin to arise, the need to keep essential values that will instill the peace and creativity the planet was founded on will be crucial. While no need of formal education will be necessary yet, the implementation of certain groups and activities will be. On earth, this would be compared to one's religion, but of course the inhabitants of mars are allowed to keep their faiths. This expression will represent religious qualities in the sense that they will give many a sense of hope and unity, which will be much needed on a planet that is quickly expanding and distancing itself proximity wise. Perhaps a new form of social media will arise, maybe in the form of virtual reality, that will be able to link those who have grown apart and allow to keep those bonds that were formed out of pure love to lead a life of creativity and survivability, or simply to start a new, better one. Maybe on Mars the need for in-person interactions, something ever so disappearing on earth, will arise, with many civilians longing to actually take a visit to their friends or even the other citizens of different cities, and spread the hope and creativity they've found on the planet. The expression of this strong, loving culture will most definitely be present in each respective city, with life almost being tribal like as each member knows how risk taking this life was and is, which will allow great support to everyone alike.

As the colony grows into the millions, and these colonies turn to major cities, the need for formal education will arise, as at this point there will be Martians who have been born on the planet. Martian school should consist of the technical and cultural necessities that allows for the survivability on mars. While this schooling for young Martians should of course still retain the basic mathematics, language (perhaps a new one will arise), and history of Mars and Earth alike, the education should be integrated in the students' lives. There should be great emphasis on what brought them to this wonderful planet, why being a Martian is so important. The technical studies should be in touch with the environment around them, teaching these kids how and why what they learn will be useful for them to perhaps one day colonize a planet in another solar system, and not just learn about what a concept is. As on earth, the need to educate youth properly to allow them to uphold values as well as knowledge that will let them use all of their potential to positively impact society is a must, so hopefully Mars will display that more so than earth, which does a better job that most gives it credit for, anyway.

The day society on Mars breaks away from its core values will be a sad one, perhaps it will eventually occur, but if these steps are followed and kept in the back of the mind of future Martians, this won't happen.

An important mindset to keep will be the need to be as environmentally friendly as possible on Mars. This kind of goes without being said, but a fine job in this area will result in possibly the greatest difference the Martians will possess over the Earthlings. The focus can simply be split up into phases: pre and postterraform. In the early stages of the colony, the crew is essentially dependent on clean sources of energy: solar and nuclear. And if after a little while it is absolutely certain that Mars were dominant of life in the past and contains oil reserves beneath its surface, and the colony is tempted to use it as another source to power their stations, they must refuse at all cost. Or if they find reserves of a completely unknown but plentiful resource, and it is deemed to emit "dirty" gases, then another use that doesn't involve the burning of the resource will have to be found if the colonists want to implement it into their lives. The precedent of relying solely on clean sources of energy over a substantial period of time that will be set on Mars will prove to be of great importance and set a much-needed example for Earth.

In terms of post-terraforming, this same notion will have to be set. As the population disperses and people start to break off and inhabit all regions of the planet, they will be tempted to use their own energy as well. And if there are indeed oil reserves as mentioned before, rules will have to be placed that prevent the emission of the dirty gases that are released once these sources are burned even if they offer convenience for that group. With a fresh, completely clean earth-like atmosphere, there can and will not be any way that humans will make the same mistakes as they did with earth's atmosphere.

One of the most attractive narratives of Martian society that will sell many on Earth to make the trek to Mars will be its unique but promising job market. This also can be applied to the initial colonists. It is true that a labor shortage will quickly commence for the first inhabitants, as any remaining maintenance and exploring can more effectively be done by robots, with a handful number of humans needing to be hired to perform the odd repair on said robot. However, these colonists will be in great luck as there will be plenty of other jobs that will arise after this very early phase. For one, research on Martian plants, or the adaptation of Earthling plants, will need to be continued. Any heavy-hitting physical tasks can be performed autonomously, but for now humans should still be the primary conductors of the actual research. And not just plant-based research will be conducted, but plenty of other avenues all the way from terraforming to examining possibilities of life on Mars. Additionally, many colonists will be tasked with engineering and design of future above surface habitats and systems, or at least how to get the most out of them.

As for future colonists, as stated there will be a variety of options that will available. The most obvious should be the need for teachers. Unfortunately, on earth, teaching is not a very financially attractive profession, but serves a great purpose. However, on Mars, the job of a teacher will for some time be rewarding as well as important. As mentioned previously, the need to educate future Martian children will be of highest order in order to instill the values as well as the necessary skills to survive on the planet. But it won't be easy. A teacher on Mars will be someone who has lived generally at least 10 or so years on the planet and therefore has the necessary experience to convey crucial knowledge. Like with a good amount of jobs available on Mars, a degree from a university, perhaps on

Mars, will be necessary, and combined with the required experience will liken a Martian teacher more towards a college professor on Earth. So, the starting pay of the Martian teacher should be in the low six figures a year, especially considering that the school year will most likely be sanctioned at one full Martian year (687 Earth Days).

A plethora of other jobs prevalent on Mars can be discussed, too. On a planet where cryptocurrency will likely dominate, and reasonably should, the need for relevant data scientists and financial analysts will be prominent. An already popular industry on Earth, the "Airbnb industry" per se will be massive on Mars. Certainly, by this time, the production of unique Martian food will be perfected, which will only lead to a stream of restaurants on the planet. Of course, it can be inferred that many will be comprised of menus that focus on Earthling foods, too. The timekeeping industry will surely need to be reinvented, with a new line of expensive smartwatches to be released. Oh, and doctors, plenty of doctors, will need to be required, as well.

The colony will need to pay the salary of every job one way or another. Wherever the money comes from, whether if the job is with a private company or the Martian government is paying largely out of their own pocket with maybe some help from Earth, money will need to be generated through key industries.

**ECONOMIC**

Perhaps the most important discussion to be had when it comes to sending humans to Mars is the cost, and more specifically how to make a profit.

Realistically, the crew will need to be supported financially as, although these 1,000 people will be some of the most eager and willing to go, they are of course intelligent beings and will not want to go for absolutely free. Much of the work done on Mars can be compared to tasks done on Earth, with the obvious difference being the environment. Considering this rigorous environment the crew on Mars will have to work in, their pay should be increased by a factor of 2 compared to the average rates on Earth. However, we will also consider that after a few months, labor will significantly decrease along with the pay. This would still leave the average colonist with a cool pay of around $20,000 per the first six earth months, which this rate dropping to $10,000 per earth month after that. This would leave any space agencies and partners to shell out another $200 million.

Given these initial investments that will cover one earth year worth of time, a goal to meet, say, one earth year after settling will be to break even or possibly even begin to profit on this whole ordeal, with heavy focus towards the latter. The good news is that Mars offers us quite a few minerals and ores that will be valuable trading options with earth. The bad news is that export costs will leave us with very little profit for most of these trading options, but not all.

While there is simply no real use to mine resources like gold or platinum on Mars, there is an incentive to mine deuterium, the heavy isotope of hydrogen. Deuterium is a very valuable source of fuel for nuclear reactors, while also having the capability to be used in the production of heavy water which in turn can be used as fuel for fission reactors. It is deemed valuable because of the incredible advantage in regard to abundance it has over the earth. On Mars, deuterium[9] occurs 833 times out of every million hydrogen atoms, while it only comprises 166 out of every million hydrogen atoms on earth giving mars a near five-to-one advantage. Additionally, considering that we will have majority of our power used in water electrolysis to run our various life support needs, we will be able to produce deuterium at virtually no extra cost. Deuterium is a by-product of electrolysis which would produce around one kilogram of deuterium for every six tons of water. Now, this is where we can make a case as to why the mining of these resources isn't necessarily worth it. A colony of 1,000 people would produce about 6,000 tons of water in the initial year, which means that only 1,000 kg of deuterium will be produced. Considering that pure deuterium is valued at approximately $10,000 per kilogram, this leaves us with a profit of $9.8 million after export costs. While it is a nice chunk of change for practically no additional work, the colony will only see an extreme value in deuterium when its numbers reach the hundreds of thousands. Until then, this unique isotope shouldn't be a trading priority for the crew, and even should seriously be considered to solely be used as an input to the nuclear reactors on the planet.

Sticking with the same sort of theme, though, a large portion of the debt can be made by using something simpler than a rare isotope: Martian sand. It is inevitable that collectors will want to get their hands on the very first samples of Martian sand and rocks brought back to Earth. The moon landing provides a precedent event of this scenario. Moon rocks[10] brought back from the various missions between 1969 and 1972 were valued at around $50,000 per gram with a grand total of around 400 kilograms of the stuff up for auction. One could only imagine that sand from the first planet colonized by humans in their attempt to make the species multiplanetary could fetch five or ten times that amount, but we'll stick with the precedent that was set by the moon samples, especially if our goal is to bring back say 1,000 kilograms which may decrease the value slightly. However, a high enough price will still be set, so a sensible amount to export will indeed be around 1,000 kilograms which would total to just under $50 billion in value after export costs. Let's say that the collectors on earth will not exhibit as much interest in our sand and only shell out $5 billion; well, that alone should cover the costs for this initial mission over the course of one year.

So, after one year we can continue to rely on sand and greedy collectors to make our easy money and supply future missions, right? Of course, the answer is no. It will be obvious to see that people can only handle so much sand and it can be easily assumed that after one auction of our 1,000 kilograms of it, the price will decrease exponentially and will soon cause the colony to lose money. Besides, there must be more sophisticated ways to make money on Mars than using its own sand, right?

The answer to this question will be... of course, yes! A huge industry will reside in the ingenuity of our own colonists. Using the example of nineteenth-century America, we can compare this frontier to our own. As seen with nineteenth-century America, our colony of 1,000 will quickly experience a labor shortage after all maintenance as well as trivial tasks will be in the hand of robots, with the only job the human will have will be to recalibrate said robot. In America, their shortage lead to a flood of inventions in a society that didn't have nearly the same technological culture as us humans today and certainly the colony. So, will this explosion of ingenuity apply to the colony? Of course, it will! Unlike this period in America, Mars offers the crew a unique environment that will lead to some of the most extraordinary, cutting-edge technology in existence. Whatever field these inventions reside in, the licensing and patents for them will soon be worth perhaps trillions of dollars. Specifically, robotics and energy production will be amazing areas to profit from.

So, as it stands, we seem to have our financial needs covered, especially over the years as the colony, while continuing to exponentially expand, will also exponentially increase their ability to survive only using the in-situ resources Mars offers thus decreasing costs.

Well, as the colony reaches into the hundreds of thousands of people, say some 20 years after initial settlement, the ingenuity will level off. And as the explosion of ingenuity levels off, there will need to be more reliable industries and markets that will allow the colony to meet the costs of the remaining imports, needs, and salaries of its employees.

This calls for the need of some industries that could possibly work initially but are definitely much better suited quite a few years down the road, with the most prominent being the mining of asteroids.

The mining of asteroids for high-grade metals will undoubtedly be a trillion-dollar industry, and it already has started on Earth, but the clear advantage would to focus any company interested in this endeavor on Mars. Obviously, many of the mined asteroids will lay in the asteroid belt, a region that is significantly closer to Mars than Earth and contains around 99 percent of the 800,000 or so known asteroids today. Additionally, the Near-Earth-Objects that make up the remaining one percent exclusively consist of asteroids that actually orbit closer to Mars, instead. The issue, as stated before, isn't if this industry will be successful, but rather how. Previously, critics would mention the risk of sending people to retrieve samples from these massive rocks, but fortunately humans will not have to go at all. The Japanese Hayabusa2[11] has already snagged a small sample from the asteroid Ryugu by firing a tantalum "bullet" as it neared. While the technology to mass mine multiple asteroids at large quantities may not be present today, surely there will be huge incentive to develop this technology on Mars. Nonetheless, this inevitable industry will undoubtedly lie in the hands of Martians, with Earthlings willing to pay a steep price for the riches that it will bring.

Another industry that will be very successful on Mars will be tourism. With the colony being positioned near one of the most fascinating tourist sites the solar system has to offer, a trip to Valles Marineris will trump all others on earth. While the tourism industry most likely will not explode until the permanent population reaches the tens or hundreds of thousands with the implementation of massive domes or structures specifically designed with hotels and resorts for tourists, it will definitely be up there with mining if asteroids in terms of financial success for the colony and planet, generating billions if not trillions of dollars, as well. Of course, the engineering of technology that significantly reduces the time it takes to get to Mars will help the tourism industry succeed even more, but this will certainly be a huge market regardless.

So, for a financial round up, the colonists will require an initial input of $1.25 billion dollars. There will need to be around $400 million spent on transporting enough food and water for the colonists to survive the trip to Mars with another $125 million of additional food, a total of $32.5 million in suits, and another $400 million for the basic shelters, generators, robots, and advanced electronics.
Also, a good $100 million can be added to maintenance and miscellaneous costs. We cannot forget that about $200 million needs to be set aside to pay the salaries of the brave 1,000. Slightly inflating the total budget to around $1.5 billion, this number is definitely on the lower side, but still respectable and covers all of the necessary aspects the colony and colonists need. As stated previously, this first year total with be met and exceeded with the sale of the in-demand deuterium ($9.8 million) and intriguing Martian-sand-collectibles ($5 billion). Additionally, as the amount of labor decreases, the rise of profitable inventions will take place which will allow those ingenuous colonists to make a fortune (billions and eventually trillions) and really allow us to see some serious return on investment.

As the colony expands, we need to pay the average citizen on Mars for their average job as well as a good possibility of $500 billion to $1 trillion in additional imports. When the population reaches a million, a good estimate is anywhere between half to three quarters will need to look for a job. While some jobs will pay better than others, the average salary is around $100,000 factoring the rigorous environment in. While say half works with a company that obviously pays their salary, another half will work in jobs funded by the government (the very trivial, busy work kind of jobs that will be needed and useful for newly arrived Martians). In a given year with this number of Martians receiving their salary through the government, another $37.5 billion will needed to be generated. This will call for industries to take off in order to contribute to the pay for these jobs. Namely, the asteroid mining and tourism industries will explode coincidentally, generating billions and eventually trillions of dollars, as well. While not as explosive as before, the invention and creativity niche will still pull in massive numbers. A representation will be provided below to visualize all the financial factors and how each debt will be overcome over a respective length of time.

**AESTHETIC**

The need for the colony to look beautiful will be very important for the crew. During those times of struggle, the crew will need some psychological relief. Well, Mars presents all the psychological attributes one will need. Imagining looking out towards Valles Marineris and taking in the layered bands of rock which reveal millions of years of geological history. Additionally, the warm colors of the ground and sky give a lucid life. The colonists should also bring mementos from home and decorate their respective homes with them, which will provide a deeper level of beauty to the individual colonist, something of great importance, as well.

As the planet becomes fully terraformed, the beauty of the planet will really come to fruition. Imagine the nearby crater, previously dry for millions of years, soon to be completely filled with water and becoming a lake. Or maybe ponder how beautiful it will be to see the once empty lot of red land soon covered with rich forests and wildlife. The colony can even begin to bring back the once massive oceans the planet was once covered by. The dust storms can be traded in for a nice cool breeze that can relax anybody.

The real beauty can be found when the planet resembles the colonists' home of once upon a time. This accomplishment will offer great emotional appeal and finally allow the inhabitants to feel at home, a feeling that will be unmatched. The beauty that resides in the process to get to Mars will one day reveal itself on it.

**REFERENCES**

1. Macdonald, Fiona. https://www.sciencealert.com/nasa-s-released-a-prototype-of-the-spacesuit-astronauts-will-wear-on-mars. Accessed February 21, 2019.
2. Llorente, Briardo. https://theconversation.com/how-to-grow-crops-on-mars-if-we-are-to-live-on-the-red-planet-99943. Accessed February 10, 2019.
3. Lavars, Nick. https://newatlas.com/great-places-to-live-mars/45654/. Accessed March 1, 2019.
4. Mole, Alan. http://marscolony.marssociety.org/MARSColonyi2.pdf. Accessed January 29, 2019.
5. Gillard, Eric. https://www.nasa.gov/feature/langley/a-new-home-on-mars-nasa-langley-s-icy-concept-for-living-on-the-red-planet. Accessed March 25, 2019.
6.Harbaugh, Jennifer. https://www.nasa.gov/directorates/spacetech/centennial_challenges/3DPHab/five-teams-win-a-share-of-100000-in-virtual-modeling-stage. Accessed March 20, 2019.
7. Anderson, Gina. https://www.nasa.gov/directorates/spacetech/kilopower. Accessed March 1, 2019.
8. Cavendish, Lee. https://www.space.com/spacex-starship-hopper-elon-musk-explained.html. Accessed March 29, 2019.
9. Zubrin, Dr. Robert. *How to Live on Mars.* Accessed December 18, 2018.
10. Pearlman, Robert Z. https://www.space.com/11804-nasa-moon-rock-sting-apollo17.html. Accessed March 9, 2019.
11. Chang, Kenneth. https://www.nytimes.com/2019/02/21/science/ryugu-asteroid-hayabusa2.html. Accessed February 27, 2019.

Aquaponics Sector and Prototyping Hall Illustrated Concepts from Chapter 22

# 14: THE COLONIZATION OF MARS

**Christopher Wolfe**
Space Development Network
Christopher.Wolfe.Public@GMail.com

This competition seeks designs for the **first** human colony on Mars. That distinction is important since the first colony is faced with a host of unknowns. We know little about the surface and even less below it. We cannot be certain humans can thrive long-term in hypogravity. The technologies required are either theoretical or experimental, at least at scale and on another planet. Still, there is reason for optimism.

Described in a word the colony is diverse. We will provide the means to build almost anything in small amounts. The first colonists will traverse the solution space in person, devising local answers to local problems. They will have the power to try ideas and invest in what works rather than planning everything in detail from Earth.

The project will require a vast array of capabilities and expertise, which means a very large financial investment. In return a body of knowledge will be developed that allows future outposts to specialize and to be built at significantly less risk and cost than the trailblazers. These advancements may be useful on Earth in the fight against climate change and scarcity as well as providing inspiration for future generations of STEM students. In the process a significant industrial base will be established on Mars, with the eventual goal of becoming independent from Earth resupply.

For detailed analysis of this plan I have assumed that the colony is operated as a corporation and supported by a consortium of private investors with national space agencies. Colonists and investors share ownership of the colony itself and its operations. The colony's charter emphasizes permanence over short-term gains and cooperation over competition; the goal here is to establish a new branch of our civilization and push the boundaries of our knowledge, not just to make a buck.

Since the defining feature of this proposal is its adaptability, I must make guesses about which approach will prove most effective at various stages of development. As a result - although I will present a comprehensive picture of the colony surpassing one thousand residents - the actual implementation of this plan will deviate as local conditions demand. Fortune favors the bold and chance favors the prepared mind. We will need both.

**Assumptions**

The request for submissions specifies a 'shipping cost' of $500 per kg to Mars. That number derives from conversations with SpaceX personnel on the target price point of the Starship vehicle (previously BFR, ITS or MCT) currently under construction. For a bit of realism I've included limited ship production and used

Elon Musk's estimates as of 2019-08 for numbers like propellant load (1100 tonnes), maximum payload (100 tonnes), dry mass (85 tonnes) and Isp (355 s). Any financing is at 10% APR and all dollar amounts are before inflation.

One key component of this effort is that ships should return to Earth in the same launch window whenever possible. Depreciation will be priced into the cost of flights that stay on Mars between windows. Under the average condition on the ideal day the return trip could be only a little over six kilometers per second. Real conditions vary and the actual day will be considerably later than ideal, which puts us closer to 7.5 km/s. Allow 300 m/s for Earth landing and 200 m/s for course correction and that's an even 8 km/s.

The baseline ship in a poor-case window such as 2024 can return 38 tonnes with a full tank (1100 tonnes propellant) or can return with no payload at all for 760 tonnes of propellant. That 2024 case is to depart Earth on day 60600 MJD, 6km/s TMI, 115 days transit, 7 days surface, depart Mars on day 60722, 3.6 km/s to orbit + 3.9 km/s TEI. A good-case window like 2033 only saves about 300 m/s on the return, which pushes max payload to 51 tonnes and zero-payload fuel to 690 tonnes. I've chosen to assume 1,000 tonnes of propellant returning a minimum of 25 tonnes, which conveniently allows a single ship's payload to refuel two ships in one synodic period.

Entire flights are bought outright at the given cost of $50 million each, so there are no partial discounts for flying less than the maximum payload. Every passenger flight is half colonists, half guests; any additional payload mass available is used for in-transit supplies and personal effects rather than cargo. I also assume that the build cost of each ship is $100 million, which is based on a scale-down of the ITS cost estimates with a further discount for the switch to stainless steel from carbon fiber. Elon has publicly suggested a number less than half that high, so it should serve as a starting point. The design life is assumed to be ten flights to Mars, so a one-window depreciation fee is $10 million.

Life support in transit and temporary housing uses ALS estimates with storable food, a 'salad machine', laundry wash, and wastewater processing via SCWO. Permanent housing uses a bioregenerative life support system at scale with molecular sieves, heat pumps and SCWO systems as backups. ISRU operations generate a staggering excess of oxygen, so the hydroponics are only required to produce food and can be optimized for that goal. Habitable volume per person is 50 m³ temporary and 300 m³ permanent, which includes life support and hydroponics.

Lastly, the 'rule of ten'. A complete ISRU system (power, cooling, excavation, purification, electrolysis, Sabatier, liquefaction) can generally produce ten times its mass in propellant each year and requires a tenth its mass in spares. In Martian terms each tonne of hardware produces 20 tonnes of propellant in one synodic period (24 months operational, 2 months offline) and consumes 200 kilograms of spares. There are a number of studies with prototype hardware that produced results in this range, so this is a reasonable assumption. 'Spares' is an abstraction but it includes consumables like lubricants, seals and filters plus field-replacement units like pumps, motors, cables, bucket teeth, microprocessors, wheels/treads, hoses and actuators.

**Establishing the Colony**

A settlement of a thousand people does not appear out of nowhere. It requires time and effort to build. I have projected a history of our colony to explain how we get to our 'snapshot' point; let's explore in detail what the colony looks like and how it functions at scale. The numbers underpinning this section are covered in the final section and references.

**Location**

One difficult question is where to land. There are many good options, but since we intend to use solar power the ideal latitude is within 15° of 10° N. Water is the next most important resource, but it is most abundant where solar energy is scarce. Fortunately there are hydrated clays available at low latitudes, and there may be deeply buried ice deposits even near the equator.

We aim for Tikhonravov Crater, roughly 13° N, 36° E. Soils in the area were measured by Mars Odyssey[1] to contain hydrogen equal to 8-10% water (by mass), which means a bake-out oven and simple dragline excavators can produce large quantities of water. The crater itself is early Noachian, perhaps four billion years old, heavily modified by water and later impacts. Banded structures suggest the current soil is formed of ice and dust accumulated over time and then eroded. The presence of pedestal craters (HiRISE[2]) strongly suggests icy subsurface material. It is thought this crater was once a groundwater-fed lake, which means there should be evaporite deposits at the low points of the original surface.

The combination of age and water means the area is scientifically interesting for geology and potential evidence of life. We should find industrially meaningful mineral concentrations thanks to the action of water and meteorite impacts.

**Transport**

The first colonial flight is actually the very first passenger flight to Mars, supported by a total of three cargo flights. The first two ships are flight zero (NET 2022) carrying ISRU equipment plus food, spares, scientific instruments and industrial tools like 3d printers. They are not expected to return to Earth but might be used on Mars for suborbital flights or Phobos/Deimos missions.

The third cargo ship and first crew ship are flight one (NET 2024). A crew of 12 includes six astronauts from partner space programs and six colonists. Colonists are responsible for setting up the initial ISRU plant while astronauts commence exploration. This sets the pattern for the bootstrap phase: the burgeoning colony hosts personnel from various funding agencies while investing effort in expansion.

Each window after the first brings several cargo flights and one or more additional passenger flights. There is a bit of delay as the first wave of ships cannot return same-synod (not enough fuel), but by synod 5 every ship returns in time for the next transfer window. Passenger count grows from 12 to 24 to 40 to 80 per ship as the local facilities improve. The Mars Starship fleet grows at about two ships per year initially, with the rate of production improving over time.

Colonists account for half of all passengers so the colony passes the 1,000 resident mark during synod 6 (1,204 colonists about eleven years after first crew launch). An equal number of visiting astronauts, scientists and tourists will pass through by then as well. By flight ten / year 20 we approach ten thousand residents. From synod 11 onward ships would start aging out, so the Martian immigration rate will depend on the rate of Starship construction back on Earth. A detailed history is provided in the final section.

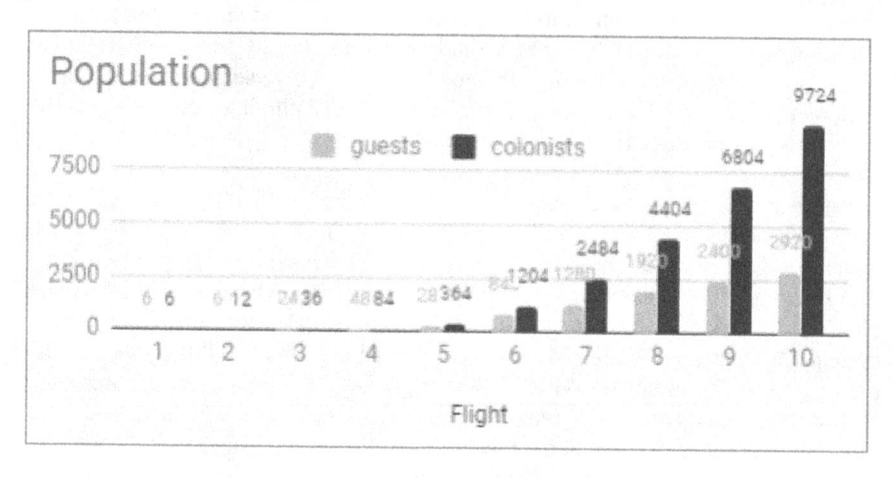


## Infrastructure

The first few crews will face several immediate challenges. The most urgent priority is to deploy ISRU equipment (including huge PV fields) and start producing propellant. For technical details see the section on using Martian resources. In practical terms it means crew moving steel tubes, running fluid and power lines, rolling out PV blankets and monitoring autonomous excavators.

Power storage is a concern. For short-term needs an array of lithium ion battery packs will be used. Long-term storage will be via gravity: an electric crane will stack (to charge) and unstack (to discharge) heavy regolith blocks. This is similar to a hydroelectric reservoir but without the water or the complex turbines.

Once the chemical industry has established a foothold and power is abundant, the next order of business is to build temporary housing. This will be expandable habitat modules with rigid airlocks (almost like concrete tents), rolled out into trenches and covered with regolith block arches. These arches allow inspection and repair of the hull without excavation, while a thin layer of foam in the trench provides insulation and reduces abrasion. There is enough space on the ships themselves to accomodate people for the first few synods if there are difficulties.

Next up is expanding on the basic industrial capabilities of the propellant plant. We add product lines for other chemicals and for metals. We build all manner of tools: extruders, wire draws, motor winders, mills, lathes, 3d printers. Each tool builds on the ones before to add functionality and reduce reliance on Earth.

With this foundation we can start building heavy construction equipment for the next phase, which is permanent housing. Several viable options exist; I prefer deep tunnelling for living space with surface greenhouses doubling as heat rejection. We build and maintain everything from ISRU gear, excavators and mining equipment to crewed rovers, hab modules and life support systems all with minimal mass from Earth.

**Food**

People require a varied but balanced diet. It is often assumed that food in space will be vegan by necessity, but certain species of animals are effective at converting waste biomass to high-quality protein. Even a small colony can sustain fish, while larger ones can manage chickens. If you absolutely cannot live without dairy, cattle become viable somewhere around 5,000 people. Insects out-produce all of the above if you're willing to work past cultural mores.

We are not very good at food efficiency. At least 30% of food grown on Earth goes bad before it can be eaten. To help offset that, most food would be prepared centrally. That might look like cafeterias or it might look like a string of restaurants, but they will be run by nutritionists and chefs who can coordinate with the hydroponics team to plan menus. There could be at least one meal prepared at home so people have the option of personalizing their diets, but kitchens are risky and expensive; anything fancy would have to be done at a communal kitchen space. That also means a market or exchange for food would still exist; necessities would be free but exotics, imports and goods with limited availability would be up for sale.

A controlled-environment intensive aquaculture system (expensive sushi machine) provides the primary food source as well as primary life-support $CO_2$ scrubbing. ($O_2$ production is incidental thanks to the ISRU excess.) Approximately 40,000 $m^2$ of tray space is devoted to peanuts, sweet potatoes, beans, broccoli, strawberries and a variety of other fruits, vegetables, greens and berries. That's ~16,000 $m^2$ floor space with 2.5-meter ceilings, a total of 40,000 $m^3$ in volume of which 7,000 $m^3$ (17.5%) is access. 21 MW of power is required for lighting (17 MW), heat pumps (3.9 MW), nutrient pumping and air handling (0.1 MW together). (Note that all numbers in this subsection are per thousand people.). Steel racks and plastic trays are used for structural support; trays can slide on racks for ease of handling. Add LED strip lighting and PE tubing for nutrient solution. Heat exchangers control temperature and humidity, concentrating waste heat for transport to the surface.

Secondary production comes from surface greenhouses with thin plastic walls and reduced pressures; these are used to grow approximately 80,000 m² of wheat, barley and shell beans under ambient lighting conditions. Supplemental LEDs are provided to allow crops to survive long dust storms, but are not normally used. Temperatures are maintained by storing daytime heat from the colony in thermal mass, which means these greenhouse modules double as radiators. Internal atmosphere is balanced with molecular sieves; CO2 is taken from outside and O2 is vented, while a bit of N2 is extracted to make up for any leaks. The modules are initially hydroponic but could transition to soil over time. These grow areas provide excess storable food for reserves, bioindustry or export. They are independent and can be used for experiments or to prevent crossbreeding of different plant lines. Power requirements for pumping nutrient solution, heating fluid and air are minimal, a few tens of kW. Productivity is low due to poor lighting conditions, but the return on mass invested is still good.

Only about half of harvested biomass is edible; the rest is waste and must be recycled. Biomass wastes from all growing areas and food prep are converted to animal feed via multiple streams. Some inedible biomass is suitable for animals and is used directly. Other biomass can be fed to insects which in turn yield high-fat/high-protein meal. The remainder can be degraded and metabolized by algae which provide further high-protein input that is rich in vitamins and minerals. These three streams are recombined in appropriate ratios to feed the aquaculture species and chickens. The result is a means of efficiently recovering low-grade biowaste into high-protein foods.

**Living Space**

While the ships themselves and some expandable surface units can handle the demand at first, a colony eventually needs durable permanent housing. There are a wide variety of ideas floating around, but I prefer underground construction.

If the geology is favorable we will use cut and cover. Heavy equipment will excavate large trenches to bedrock, a minimum of 22 meters depth. Removed soils will be reinforced and compressed into blocks, preferably with cement if available but sintered if necessary. Those blocks will be used to build arched vaults and halls. These walls are sealed inside and out with fiber-reinforced aluminized plastic and then backfilled with excavated material. This allows very large structures to be built deep underground without needing a TBM. The backfill at 18 meters depth exerts just over 1 bar of pressure, which means the structure is lightly loaded in compression even with Earth-normal atmosphere inside. That's important because it is much easier to handle compressive loads with local materials than it would be to build a bunch of 1-bar pressure vessels loaded in tension. Larger structures require going deeper for pressure balance; a ten-story structure would need a 47-meter trench so the ceiling is still >18 meters deep.

If the geology is not favorable (unstable sediments too deep to remove, icy soils, near-surface bedrock) then the alternative is to build a large-bore TBM and drive habitable tunnels directly into the rock. It's possible these tunnels would need only minimal reinforcement but we would be prepared to fully line them anyway.

Either way, access to the surface would be by vertical shaft and cable elevator with double airlocks on both ends. Stairs or ladders work too, but we need the ability to move heavy equipment back and forth. The surface airlocks would open to an arched structure made of regolith blocks which provides radiation protection and a safe space for ambient-pressure maintenance or experiments. There may be a pressurized garage at the surface for shirtsleeves maintenance of rover/excavators if that capability is helpful.

Housing would be built inside the excavated volume using local concrete, steel and plastic. As the biological systems become fully established, foam-core paper panels would be used to partition spaces and absorb sound. A hall or tunnel would be designated as the central assembly, large enough for at least a thousand people to gather at one time. This is public greenspace with fruit and nut trees along the center. The volume is reasonably productive while doubling as recreational space and a psychological connection to Earth's biosphere.

Individual homes would be built as apartments inside the primary volumes or cut out of the surrounding rock as expansions. Each one would have an ECLSS unit used primarily for regulating temperature and humidity, but in the event of a colony-wide environmental problem they would provide CO2 scrubbing as long as there is power. Doors should be vacuum rated, but that might not be feasible. Heat removal would be via circulating coolant, probably water for safety reasons, with a central site boosting heat to the surface as supercritical CO2 (same working fluid as the heat pumps).

Businesses, work areas and growing areas would also have independent systems as backups. The primary life support system is hydroponics, which means moving quite a bit of air to maintain circulation. I suspect hallways would have relatively high ceilings to provide enough volume for moving air. The overall structure of the colony would be a bit like city streets: long tunnels or halls of a kilometer or two with cross connections, housing along the sides and a layout that encourages walking.

**Society**

A colony off-Earth is science fiction today. The field of literature by that name is a rich source of thought regarding how people behave in groups, what manner of conflicts they are likely to encounter and how they can overcome. Our history here on Earth informs certain political decisions, suggesting that a Western-style democratic government is preferable. The harsh realities of the Martian environment will require a society that is largely communitarian, meaning behaviors that maintain or benefit the colony are incentivized even if the rules are not always 'fair' by our standards.

First and foremost, space will kill you. Space is the penultimate adversary (second to death itself), never tiring, always around every corner and behind every panel.

A mistake, an oversight, a concealment or fraud could kill not just you but everyone on-site. This reality applies unusual pressure to a society with two likely outcomes: authoritarian surveillance or enlightened self-interest. Let's agree that a Martian dictatorship is undesirable; that path includes options such as 'military-style outpost' that might seem reasonable at first but do not scale or endure.

How do we arrange for every person's interests to align? Equity, and not just in the financial sense. The first step is ownership: every permanent resident owns a share of the colony and every resident is guaranteed a basic existence. This necessarily includes equality under the law, a common ideal of Western societies. There will be no resident air tax or meal credit. Exotics, imports or private products are fair game for market adventures, but the basic existence of a colony is incompatible with unrestrained capitalism. That also means native-born children automatically gain a share as part of their citizenship. People start on the same footing, while estate transfers would be sharply limited so that wealth and power cannot accumulate to excess.

The success of the colony as a whole improves the position of all residents; individual profits can be made from private investment and use of excess resources, but ultimately if the colony fails everyone is done. We have at our disposal the tools of several economic systems; the challenge is to take advantage of the useful parts while minimizing the destructive ones.

Next is to understand that survival is job one. If a person makes a mistake or intentionally breaks a rule that threatens safety, the justice system must prioritize their **information** over their punishment. If someone has been lying about maintaining an airlock, the punishment for that can't be so severe that they choose never to tell.

It is very important that people be willing to accept that someone has paid their dues after their punishment is complete, otherwise the social stigma of personal failings can override one's commitment to the community and increase risk all around. In other words, colonists should intentionally focus on a person's recent improvement or decline rather than past highs or lows, which also helps reduce the influence of charismatic or heroic people based on their past deeds. Sober, careful thought is a better guide than passion or instinct. Demonizing people for not appearing to 'pull their weight' will lead to destructive acts.

It would be best if colonists generally hold a commitment to the group's success, but we must expect outliers. That means intentionally designing rituals and norms which promote group coherence and contribute to the success of the colony by improving the average colonist's 'buy-in'. If that sounds weird, think of how summer camps build camaraderie through oddball traditions that generate shared experience; this is intentional community design.

Since we are building a human colony rather than a corporate outpost, it is just as important to allow for personal expression and private interests. Ideally a colonist would spend no more than half their working hours in a week on 'civic shifts', the work that is required to keep the colony running. Those would be tasks assigned based on certification, aptitude and preference; typically these would rotate so every resident knows every critical system well enough to be helpful in an emergency. I suggest that no able person should be exempt from civic shifts, partly for equality and partly for safety. This policy in concert with comprehensive education should mitigate the 'three generations' rule as well.

The other half of a person's working hours would be spent on something useful to the colony and of interest to the individual; examples might be filling research contracts for Earth agencies, breeding crop varieties, practicing medicine or cooking. Sufficient time would be allowed for hobbies (or small businesses) that are fully by personal choice so long as they are not a safety risk. Some system of allocating resources will be necessary for these activities if they use public goods like power or water. A productive hobby could develop into one's 'day job' if successful.

**Economy**

Mars is not an obvious place to develop a market. Travel times and expenses are significant. What possible customer base could be found there? How will local trade function?

**External Trade**

The first and most important product will be propellant, although not directly for profit. SpaceX has indicated they would prefer to pay someone else to handle that work so they can focus on being a transportation business. Given our assumptions of 1000 tonnes fuel, 25 tonnes return payload and $200/kg cost, each return flight's price is $5 million. SpaceX is likely to run margins of ~30% and other costs are priced into the outbound flight, leaving about $3.5 million per return flight or $3.50 per kg of propellant. This is definitely **not** enough to bootstrap a colony since the ISRU spares alone would run to about $5 per kg of propellant.

The full suite of equipment and supplies for a single person on a one-synod tour is about 8 tonnes, which is $4 million in shipping costs alone. A self-contained Starship mission can take no more than 12 people along with their gear and a small amount of science instruments. There is no room for ISRU hardware in that budget, so even this self-contained flight is reliant on at least one cargo flight to make return fuel; the contracting agency will need to buy that flight and pay for both ships to sit on Mars for two years instead of making profits. Those twelve pairs of boots will cost $120 million in launch costs all-in. That's amazingly cheap by today's standards, but we can do better.

Another concern is equipment. If NASA wants to send astronauts to Mars then they need food, water, air and supporting equipment for exploration. Their habitat needs maintained and their ride home will need fuel. All of that takes time and money to build and operate. The time spent on infrastructure takes away from the exploration time of the astronauts, which is why pairing up with colonists is ideal.

Recall that NASA expects to pay approximately $108 million per astronaut for SpaceX flights to the ISS under CCtCap. They also planned to spend as much as $250 billion for three Mars missions under Constellation. The colony can start out at a very reasonable price point of $250 million per guest and drop the price over time to expand the customer base. This includes the cost of providing ISRU propellant for the return trip and colonists to run the base so guests can focus exclusively on their research. Guest fees account for some $18 billion of revenue in the first ten years alone, rising to $75 billion over ten flights.

Beyond the first flights, interest in Mars will grow as the cost of access falls. Martian industry will produce the equipment needed for new colonies and other expeditions. ISRU hardware, industrial equipment, food, life support systems and even rover/excavators will be offered for sale by the first colony. Since Earth has a hard cost floor due to shipping, the first colony can compete on price when selling to other sites. With excess propellant available a small number of cargo Starships could make suborbital flights to other outposts, which means the first colony can compete with Earth on shipping times as well. Hardware exports account for $1.6 billion in the first ten years, growing to $14.6 billion by flight ten as humanity explores Mars and the main belt in earnest.

Goods and services will be traded between outposts and eventually with asteroid belt mining missions. Propellant will be an important commodity for as long as reaction engines are in use. At many locations Mars may also be a cheaper source than Earth of buffer gases like argon and nitrogen and of structural materials like steel and aluminum. Investment in research by government, academia and industry will provide important sources of funds; data is one of the cheapest products to export back to Earth. Returned samples will net some $2.8 billion over ten flights; the earliest samples are comparable to Apollo samples in price but quickly become cheaper as their scientific value wanes.

These factors mean the colony's overall goal should be to reproduce the technological base of Earth as completely as is feasible given the lack of biosphere and fossil resources on Mars. Steel, obviously, but also microprocessors. Food, certainly, but also pharmaceuticals. Fuel, necessarily, but also polymers and ceramics. Mars may never outpace Earth in productivity or gross output, but industrial diversity can be competitive. The required hardware for survival is so broad that just meeting that objective will provide enormous potential.

In ten flights we spend just over $15 billion with SpaceX, just under $24 billion with equipment contractors and $4.9 billion in financing. As projected, the colony would hit a peak debt of $6.7 billion at flight four and turn positive at flight six. This is driven by increasing flight rates balanced against decreasing prices. We should end up with net profits of $49 billion when all flight 10 business closes. 81% of revenue comes from guest fees, which open at $250 million each on flight 1 and end up at just $3.2 million each by flight 10.

**Internal Trade**

How should work and profits be assigned? That's up to the colonists themselves, but I suggest a form of market. Let's assume for the sake of argument that the local currency is the Hour. There is a certain amount of work that needs to get done to keep the colony running, but not all work is desirable. Each shift is put up for auction; qualified people can bid some number of hours as payment and the lowest bids win. If nobody wants to clean toilets then the bid for that job can go quite high; janitor might be among the highest-paid positions. There may not be enough colony work to keep everyone busy; if someone doesn't win a civic shift bid they would be enrolled in training shifts (paying 1:1 which effectively provides a labor cost floor) to get qualified for additional jobs. These auctions would run fairly frequently, possibly with a bit of regulation to encourage comprehensive crosstraining.

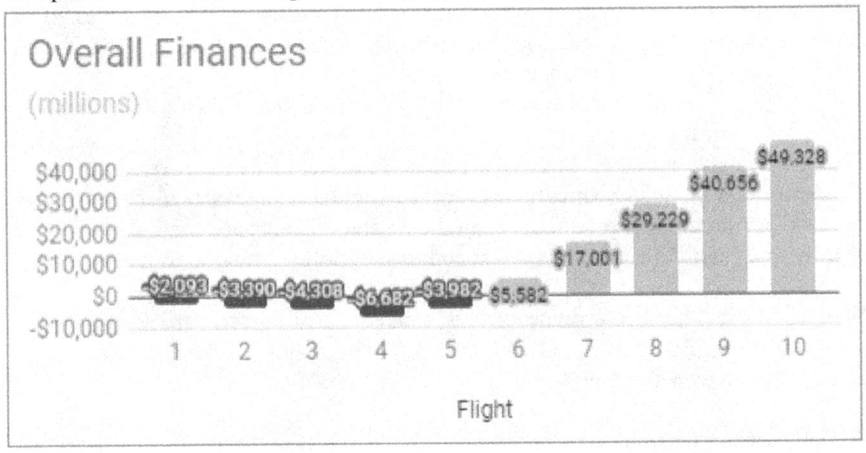


Who decides what work needs to be done? Just like a business, a management team controls each major function of the colony. Each team is led by an elected director, with directors forming a board to handle internal business. (A distributed Executive branch.) Individual managers have goals assigned by their leadership, and they post shifts for labor to achieve those goals. Unlike on Earth a management position would not be particularly lucrative; it does require a specific skillset but in the context of Mars that work would be spread across many more people to ensure continuity. That also means much less individual power, for better or worse. Don't forget that managers and directors still serve on civic shifts and are still expected to achieve competence in every critical system.

What about going beyond basic survival? We can loosely define three additional types of work to perform: outside contracts, private enterprise and colony expansion.

Outside contracts bring in cash, leveraging the colony's capabilities for resources that can be traded for goods from Earth. Colonists would not be paid for this work directly, it would be posted by a management team as shifts. This would be negotiated like a trade agreement, so there would be a wide range of possible terms and strong oversight to prevent abuse.

Private enterprise allows for innovation and personal expression to drive diversification. Anything requiring public resources would need approval from the relevant management team, but otherwise people would be free to start their own businesses. People would essentially be paying for the resources they use with their hours of credit (in lieu of dividends) and can succeed or fail on their own merits. Trade with outside entities would be handled as described above with an additional layer of negotiation between the business owner and the outside management team on distribution of profits.

Colony expansion applies any remaining resources towards building more capabilities. Because survival in space requires a conservative approach, supply will nearly always outstrip demand. Excess local materials will be invested in building more capacity and more variety in capabilities. Strong growth and steadily increasing independence are the goals.

**Using Martian Resources**

The ISRU plant involves large PV fields, semi-autonomous rover-excavators, water purification and chemical plants for high-temperature water electrolysis and the Sabatier process. It produces many thousands of tonnes of methalox fuel per year, with capacity increasing every synod to support the growing Starship fleet.

What is all that hardware made of? On a mass basis, mostly structural support which is either plastic or metal. We can make the majority of these components locally and save on shipping.

Initially we are dependent on replacement hardware from Earth. Very quickly we will develop capacity to make our own ISRU equipment and spares. We will build on that for several flights until our industrial base can grow with minimal outside mass. I've assumed we can hit a final productivity of 10% per month, which is about an eighth as productive as the ISRU gear is at making propellant.

**Bootstrapping Industry**

Our touchpoint of one thousand colonists is surpassed after six waves of flights. When that sixth wave arrives Mars will host about 9400 tonnes of industrial equipment, science gear and habitat hardware. We certainly won't be shipping all that mass; only about 3900 tonnes (40 cargo flights, $2 billion in shipping and $21.2 billion in material cost) will be manufactured on Earth. The rest comes from local manufacturing capacity which will be addressed in detail in the next section. By the tenth flight local hardware exceeds 60,000 tonnes as the Martian industrial base becomes more fully established; almost everything is made locally.

Modeling this kind of bootstrap process in detail requires a careful examination of specific technologies along with estimates of their likely productivity on Mars. An effort like that could keep a national space agency busy for years. I have approximated these details as geometric growth instead. My numbers[3] are not sufficient to plan a real colonization program, but they are a useful starting point for debate.

**Raw material production**

The heart of the chemical processes occur in stacked plate reactors or microchannels. The plates themselves don't need to be extremely strong, but they need to tolerate fairly high temperatures and aggressive chemical and electrical conditions. We can make them from iron or mild steel with a passivation layer such as nickel, and we can use a 3d printer (via SLM or thermal deposition of carbonyl) to do it. The catalyst is typically rhenium coated on a fine mesh of metal wires. The metallic supports can be printed locally, although the catalyst solution would probably be shipped in until a local source of PGMs is located. Seals are challenging; one promising option is based on silver foam, while the more common approach is glass. Material for this is likely to be shipped in for a while.

Some processes use a solid oxide ion-exchange membrane instead of a catalyst. For reactions at reasonable temperatures the ceramic membrane is sandwiched between plates with flow channels. Those requiring unreasonable conditions need more exotic materials and should only be considered if there are no alternatives.

The colossal amounts of power required (mostly for electrolysis) come from thin-film photovoltaic blankets, which are amorphous silicon deposited on a plastic sheet backing or a glass plate. Earth-made products would use Kapton, Mylar or similar. We don't have such stringent mass requirements for indigenous products, so simple-chain polymers like polyethylene or polypropylene will do fine as would local glass or even iron. Silicon is readily available and can be purified through zone refining to the necessary levels.

A few additions to this initial industrial base include several Fischer-Tropsch reactors with different product balances, a Haber-Bosch reactor, solid-oxide electrolysis cells and a water-gas shift reactor on the chemical side. A series of fractional distillation columns for the hydrocarbons are necessary as well. That gives access to several critical industrial chemicals including hydrogen, carbon monoxide and ammonia as well as hydrocarbons from C1 to aromatics to heavy waxes and alcohols if desired. Any undesired products can be fed back into a suitable reactor to continue growing or pyrolyzed to recover hydrogen and carbon.

For ore and water extraction the rovers are equipped with survey packages to map the terrain. This should include lidar for high-res surface and radar for subsurface, and could include more exotic tools like sampling arms, neutron backscatter sensors or IR spectrometers for composition testing. A variety of digging methods would be attempted, everything from a dozer blade to a bucket arm to a dragline and more. The water processing plant includes magnetic rakes to collect metallic meteorite fragments. The actual extraction is done by heating the soil to drive off adsorbed water and water of crystallization. Baked soils can be separated by several means to concentrate fractions of magnesia, alumina, silica and other minerals of interest.

If we are fortunate enough to find subsurface ice within a few tens of meters depth then rodwells can be bored with the core sampling gear; a downhole heat exchanger would melt ice to be pumped to the surface. This is vastly more efficient than extracting water from clays and would be a major boost.

On the metal refinery side we provide direct-reduction equipment using H2 and CO as well as Hall-Héroult cells (or more generally, oxide electrolysis cells both solid and molten depending on the product). This yields multiple routes to metals including iron, nickel, chromium, aluminum, titanium and magnesium as well as semiconductors such as silicon. This gear is composed largely of refractory insulation, resistance heating elements and electrodes.

High-temperature electric furnaces allow for the production of everything from steel to mineral wool using traditional processes. Carbon arc electrodes can be made from pyrolysis of methane or plant waste, while platinum electrodes would be shipped in until local sources are found. Electric heating elements use nichrome, barium titanate or molybdenum disilicide along with magnesia and either steel or quartz sheaths; the resistive elements themselves will be Earth-sourced at first, but the rest of the mass is available. Magnesia is abundant in Martian dust.

The Mond process is used both for purification and 3d printing or CVD of iron, nickel and their alloys. We require insulation and heating elements which can be largely local, but also high-pressure reaction vessels and pumps for carbonyl production which are more difficult to make. These will require quality steel.

Zone refining allows for the production of semiconductor-grade materials and the separation of traces like PGMs and rare earths from iron meteorites. This is done with an induction loop (iron) or a tungsten ring heating a charge inside a quartz tube; benchtop setups have been used since the 1930's. With plenty of silica available in local dust, the critical raw material for silicon PV can be produced in abundance. Trace dopants can be imported or extracted where available.

Plastics are accessible. Since there are no fossil petroleum reserves most synthetic plastics would start with syngas from the RWGS reactor. Bioplastics are a potential output of the hydroponics systems but are rate-limited. Serpentinite deposits could produce economically meaningful reserves of methane as well, but we can't bet on it.

From that humble and energy-intensive start the desired hydrocarbons are produced in F-T reactors, purified and then polymerized in a catalyst bed reactor. The polymerization reactors are typically steel, but conditions are mild so an iron cylinder with a nickel surface layer should work. The catalyst options are diverse but the most likely is a metal-oxide framework (related to zeolite/molecular sieve) which can be made locally through a bit of hostile chemistry.

The simplest of these products is polyethylene, which can be used for everything from thin films to plumbing to cutting boards. I expect this to be the #1 product because of its simplicity; fewer steps means less embodied energy and higher efficiency, plus it is a thermoplastic and can be recycled easily if unadulterated. Other plastics including aramids, urethanes, isocyanurates, ABS, PET, PP, etc., are producible in smaller quantities where needed.

**Component manufacturing**

Perhaps the most important component we want to make will be PV cells. Electrical energy is crucially important; almost every step of every process requires it in abundance. The first choice is whether to pursue high efficiency or ease of manufacture. Given the staggering cost of labor, simplicity is the clear winner. Amorphous silicon on glass will yield anywhere from 8-12% efficiency and requires no precise adjustments, unlike the sensitive labor-intensive processes in wafer cell production. The PVD processes will run at ambient pressure so no vacuum chamber is needed, only a buffer gas to flush the working area of CO2.

Fused silica glass is energy-intensive but requires fewer processing steps than plastic and is more durable against wide temperature swings and abrasion. Soda-lime glass would be easier to work with but we can't be certain of soda deposits; calcium for lime is present in the dust. The front contact would be indium tin oxide from imported indium and local tin (direct reduced from dust and electrochemically purified), applied with thermal deposition or sputtering. The active layer is a-Si:h deposited from silane. The back contact and support can be one layer, thermally deposited iron (from carbonyl). That part is somewhat complex due to the difference in thermal expansion between the materials, although thin films are more resistant to this than wafer cells.

We need resistive heating elements and refractory insulation for most of the material processing steps. Magnesia insulation is fairly easy to make once the raw material is purified; it can be cast in bricks or slabs and assembled into enclosures for plate stacks or reduction cells. Resistive elements are a bit more difficult as some of the ingredients are less common in surface dust.

This section could easily take another thousand pages and still be incomplete. Suffice to say that a team of skilled people will ultimately be able to reproduce the capabilities of Earth's industrial base in miniature. Many processes will require alternate routes due to the difference in local resources; this is a good thing since the Martian way may end up being useful on an Earth increasingly concerned about carbon balance, fresh water and pollution.

**Detailed History**

Let's explore exactly how the colony develops over the first six waves. Details are available in the reference sheet[3]. I've assumed Starship construction starts at one ship per six months and productivity grows by 0.5% per month.

**Wave 0 - Cargo**

The first flights are cargo only, a pair of ships each carrying 100 tonnes of supplies. The ships carry life support equipment for testing and to provide additional layers of redundancy for the future crew. Cargo per ship includes 60 tonnes ISRU, 10 tonnes production gear, 13 tonnes spares, 10 tonnes food/consumables and 7 tonnes scientific equipment. The ships might deploy small rovers from the aft cargo pods, but the bulk of the cargo is assumed to be waiting for people.

**Wave 1 - First Crew**

This flight marks the start of the reference sheet. A third cargo ship with the same payload as the precursors is joined by the first passenger flight. Six colonists and six astronauts make the trip. They spend their surface time setting up ISRU and production hardware, and exploring the immediate area. They live and work from the landed ships, with several hundred cubic meters of volume each. Limited hydroponics provide fresh greens and valuable data.

This crew runs critical tests on the surface hardware. We can be confident the ISRU gear will work as advertised thanks to decades of testing on Earth but other equipment for metal refining, chemical production, 3d printing and more will be put through their paces in the real environment. Feedback from this work will be very important for updating designs that ship in the next wave.

**Wave 2 - Making Tools**

This flight is a repeat of the previous two waves: 3 cargo ships and 1 passenger flight. The cargo flights carry more spares and consumables than the previous waves as personnel and equipment build up. Another six colonists and six astronauts fly out; everyone on the surface will continue to live in landed ships. Enough propellant is produced for three ships to return to Earth with the first astronaut team of six and up to 69 tonnes of samples. Six ships remain on the surface due to lack of propellant.

This crew uses the data collected by wave 1 and new equipment brought from Earth to start making ISRU hardware on Mars. This is early in the project so we're talking about structural supports, steel reaction vessels, rockwool insulation and possibly iron or aluminum wire. I've estimated that the 80 tonnes of industrial equipment can produce 3% of its mass in useful ISRU equipment each month.

**Wave 3 - Adding Complexity**

The growing fleet now has eight ships leaving Earth, six with cargo and two with passengers. The built-up supplies and hardware on Mars plus improvements to life support allow for twice as many passengers per ship; 24 colonists and 24 guests make the trip. Out of the 600 tonnes of cargo, nearly half is for industrial production. Food and spares take up about 100 tonnes with another 20 tonnes of science instruments. The rest is expandable habitat modules for the initial off-ship living quarters and larger scale hydroponics. Seven ships return to Earth carrying six guests and 169 tonnes of samples, leaving four ships on Mars.

With the colonist population doubled again, work can proceed on two key projects. First is setting up an interim surface base made of buried expandable modules. The hulls, airlocks and other critical parts are shipped in but as much as possible is made locally. Floor plates, furniture, hydroponics racks, insulation, even some of the plumbing and wiring is done with local goods.

Second is expanding the scope of industrial production, which means using the gear on hand to make more production lines. We scale up from simple metal and plastic forms to more complex products like motors and batteries. These are things we can make by hand with crude tools if necessary, but modern 3d printing and CNC equipment allows for much higher productivity and good quality. Motors enable pumps, fans, compressors and powered equipment like rovers or excavators to be built using almost all local mass; control, comms and power electronics would still be shipped in. We begin crediting the production mass into spares at 2% per month and more production mass at 2% per month.

**Wave 4 - Pushing Productivity**

Thirteen ships leave Earth this window. Once again we have two passenger ships, meaning eleven cargo ships. Further improvements to the ships and their destination allow another doubling of passengers to 48 each. When these ships arrive the permanent crew grows to 84 people, comparable to an offshore oil rig. Once again we receive a matching 48 guests, while the previous 24 guests return to Earth on 11 ships. For the first time we have a surplus of hardware, with 300 tonnes of locally produced ISRU, spares and production gear offered for sale.

Cargo is primarily industrial equipment, 860 tonnes of ISRU and production hardware. This is the last major Earth-sourced ISRU shipment; future expansion will need only small amounts of high-tech parts like microprocessors. If the colony will use nuclear power, this is the wave that delivers it. We also get the second batch of habitat hardware with improvements based on feedback from wave 3. It may sound odd to ship in the same category of hardware we're selling, but the imported ISRU mass is in parts we can't make locally yet while the exports are mostly local material by mass.

At this point the colony can start making silicon-based devices like PV cells, with a complexity scale-up over the next few waves through power transistors, sensors and eventually high-end parts like flash memory chips and microprocessors. The chemical industrial base is advanced enough for engineering fibers such as aramids, and has crossed with the metals refining section to use inorganic chemistry for making catalysts. We begin crediting the production mass into habitat mass at 2% per month. This is the last category of growth; our industrial base is now assumed to produce 10% of its mass every month in useful equipment of various kinds including more industrial capacity. The exact allocation of that output will be up to the colony; I've assumed static ratios for this analysis.

**Wave 5 - Full Turnover**

Nineteen ships leave Earth. Twelve carry cargo, seven carry the full complement of 80 people each. This dramatic increase in headcount will see us surpass 360 permanent residents. 48 guests return on 24 ships, leaving the colony with no fleet ships on the ground for the first time.

Cargo is primarily habitat parts and production equipment, eleven hundred tonnes of economy-of-scale systems plus a hundred tonnes of misc. items. This will pair up with local goods to greatly expand the available volume, leaving room to grow over the next few waves. There is a followup to the big production push that includes any remaining equipment needed to work with semiconductors and to enable full productivity.

Local industry has reached a solid surplus of propellant, spares and ISRU gear. This is where Martian-made goods start competing with Earth products for business at other colonies and space projects. Some twelve hundred tonnes of equipment representing $600 million in avoided shipping cost is up for sale along with four ships' worth of propellant.

**Wave 6 - Over a Thousand**

For the first time, the colony's demand for transportation falls behind the supply of ships. 27 leave for Mars while 5 remain behind; there is not enough habitable space to send more people and no particular need for more cargo. Only six of these ships carry cargo; the other 21 carry 840 each of colonists and guests. Our permanent residents now number 1,204.

At this point the colony is solidly established, with a significant industrial base and varied population. It will continue to expand, serving as a base of operations for further research and exploration. It also serves as a sort of frontier trading post for other colonies, mining expeditions, etc.

## ACKNOWLEDGEMENTS

NASA, for generating nearly every bit of information used here.
Winchell Chung – Project Rho
The r/SpaceX community

## REFERENCES

1: Mars Odyssey data, LANL/JPL
Hydrogen abundance map https://tinyurl.com/y5rn2tql
(Northeast edge of the orange blob center-right)
2: HiRISE images, NASA/JPL/Univeristy of Arizona
Tikhonravov Crater https://tinyurl.com/y2a8n8a8
3: Google spreadsheet, C. Wolfe 2019
https://tinyurl.com/yyem3krq

# 15: THE DVARAKA INITIATIVE
# FIRST SELF-SUSTAINING HUMAN SETTLEMENT ON MARS

**Bhardwaj Shastri, Arvind Mukundan, Alice Phen, Akash Patel, Heeral Bhatt**
Luleå University of Technology, Kiruna 981 92, Sweden
dvaraka.initiative@gmail.com

## ABSTRACT

The universe, from its farthest ends to our galaxy is made of numerous celestial bodies that although are so disparate in properties, hold the capability of appealing in their unique way. For as long as we can remember, human life has inhabited the most precisely positioned planet, Earth. Although we cannot be more humbled for all that Earth has provided us with, the expansion of the human race is necessary for it to not be a lone life supporting planet. As the saying goes what we have been searching for is right under our nose, the answer to our conundrum is our very neighbour, the red planet-Mars. This planet is subject with significant proximity and similarities with Earth which make it the closest candidate holding the potential to sustain human life. This paper will be dealing with the conceptual design for the first settlement on Mars. Considering 1000 people colony and the best settlement site on Mars, our team has proposed the technical, architectural, social and economic layout for the settlement. Aggregating assumptions, research, and estimations, the first settlement project suggested in this paper shall propose the best means to colonize, explore and inhabit our sister planet, Mars.

## 1. INTRODUCTION

### 1.1 About

This paper presents team Dvaraka's[1] response to the challenge posed by The Mars Society to design a working model for first self-sustaining human settlement on Mars. Team Dvaraka is comprised of students pursuing the SpaceMaster program at Luleå University of Technology, Kiruna, Sweden.

### 1.2 Mission

The dream of human exploration beyond the surface of the Earth is tied to the belief that new lands create new opportunities and prosperity. In the past, societies that have achieved unchallenged domination of their environments have ceased to develop and become stagnant. A new frontier is necessary if humanity is to continue its exponential progress. Fortunately a new frontier exists and awaits our arrival - Mars.

Among the celestial bodies near Earth, Mars certainly is the most habitable planet in our Solar System. The first serious study that moved the idea of settling Mars from the realm of fantasy into the realm of the possible was Wernher von Braun's 'The Mars Project' [von Braun 1952]. Since then a rich discourse has evolved concerning how to achieve this dream. Traditionally this discussion has been the monopoly of technical aspects with marginal participation from non-technical aspects. This paper will explore both, technical and non-technical aspects, involved in planning first self-sustaining human settlement on Mars.

## 2. PHASES OF THE DVARAKA INITIATIVE

With the challenges viewed in a historical context, and the developing motivations for technology, various literature were reviewed and assumptions were created around which The Dvaraka Initiative has been developed. First, it is assumed that multiple robotic, sample return, and human return missions shall be conducted before The Dvaraka Initiative is undertaken. The technology required to support humans for even a short duration mission is still under development and given the environmental dangers and difficulties in supporting the mission, it is here deemed necessary to test the required technology in-situ, through return missions. Second, it is assumed that the early stages of the mission will require regular resupply of equipment, raw materials from Earth, until Dvarakans[2] are able to provide resources for themselves. Third, it is assumed that technology, of carrying 85 tonnes of payload to surface of Mars safely with 1100 tonnes of propellant, is tested and ready to use by 2026[3]. Liquid methane and liquid oxygen ($CH_4/O_2$) will be used as propellant, operating in mixture ratios between 3:1 and 3.5:1 (oxygen: methane)[4]. Finally, considering the end goal of self-sustainable human settlement on Mars, full support of Dvarakans is assumed for the entirety of the mission. In order to better understand the challenges involved, the whole design for Dvaraka is conceptually divided into following phases: Pre-Initiative phase, Settlement phase, Self-Sustaining phase.

## 3. PRE-INITIATIVE PHASE

### 3.1 Martian Administration on Earth (MADE)

MADE will come into existence by 2022 when all around world's government will join hands and countries signing the Mars treaty will be the investors on this project. In return, the government with the highest investment shall have the highest representation in the training program as well as MADE authorities. MADE will look after prerequisite for Dvaraka, finances, recruiting and training of Dvarakans. MADE authorities will be selected by space organizations and nation's government based on the experience, managerial skills and plan proposal for Mars. In order to have more communication among departments, MADE will have flat hierarchy managerial structure. The authorities selected will be the board of directors for MADE. The board of directors will be responsible for managing the functionality of sub departments.

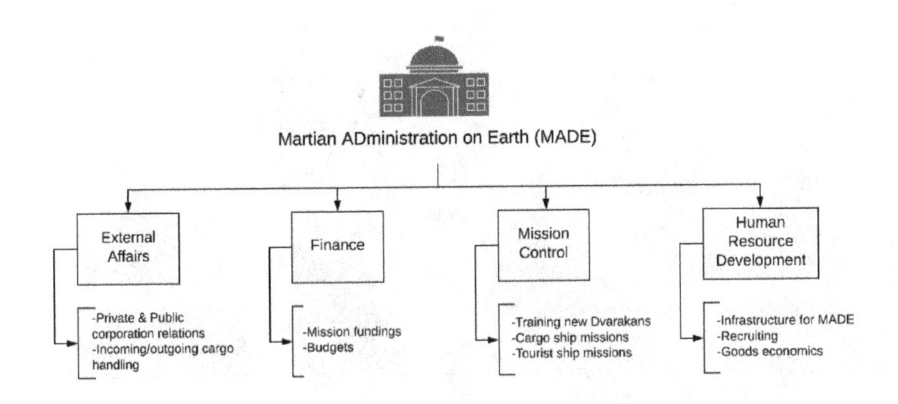


*Figure 1: Martian ADministration on Earth (MADE)*

Figure 1 shows managerial structure of MADE. Every mission will be under the surveillance of Mission Control of MADE. Goods arriving in future from Dvaraka, will be handed over to External Affairs department, for exchange of resources with government, private or research sectors. Finance Department, which will work with External Affairs department side-by-side, will hand out money and handle the budgets for funding of missions and research projects on Mars. Human Resource Development will look after recruitment process of MADE employees, trainee Dvarakans and infrastructure of MADE.

3.2 Settlement site

The various locations have been proposed for landing and sites suitable for habitats, however past proposals have predominantly been for the purposes of scientific settlements or research bases. The approach used for Dvaraka's site is to select with the long term goal of a self-sustainable human settlement. Four levels of criteria is used in the settlement site selection.

*3.2.1 Site selection criteria*

1st criteria for the identification of a suitable area - suitable atmospheric conditions such as temperature, pressure, sunlight, water and topography. Mars global climate zone map is taken into consideration for this criteria. Figure 2 shows climate zones based on temperature, modified by topography, albedo and actual solar radiation. A= Glacial (permanent ice cap); B = Polar (covered by frost during the winter which sublimates during the summer); C = North (mild) Transitional (Ca) and C South (extreme) Transitional (Cb); D = Tropical; E = Low albedo tropical; F = Subpolar Lowland (Basins); G = Tropical Lowland (Chasmata); H = Subtropical Highland (Mountain)[5]

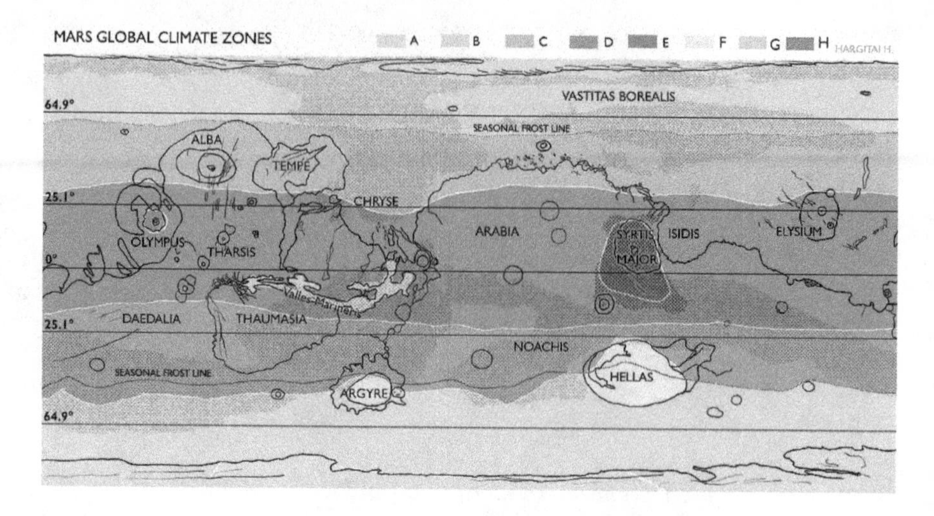


*Figure 2: Mars Global Climate Zones*

2nd criteria for the identification of a suitable area for growing plants. Map of the ideal sites showing different potentials for harvesting crops is taken into consideration. Figure 3 shows colour perspective as blue colours indicate high potentials, with the darkest blue as the best sites; red colours indicate less good sites with dark red as the worst.[6]

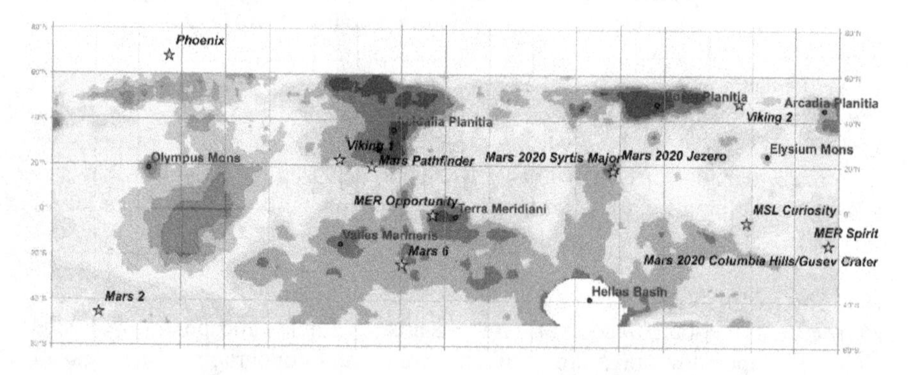


*Figure 3: Map of the ideal landing sites on Mars from a plant perspective*

3rd criteria for the identification of specific regions of interest – local magnetic field for radiation mitigation, minerals for in-situ resource utilization, and scientific importance. 4th criteria to justify the suitability of possible sites for habitation and under the local context - radiation protection based on type of habitat, asteroid and comet impacts, and potential for future expansion.

*3.2.2 Proposed settlement site*

The preferred region, which ranked highest in the combination of selection criteria, is close to western edge of Isidis basin which is located in Syrtis Major Quadrangle (Mars Chart-13). Jezero crater is a ~ 45km diameter impact crater located at 18.4°N, 77.7°E in Nili Fossae region of Mars as shown in Figure 4.

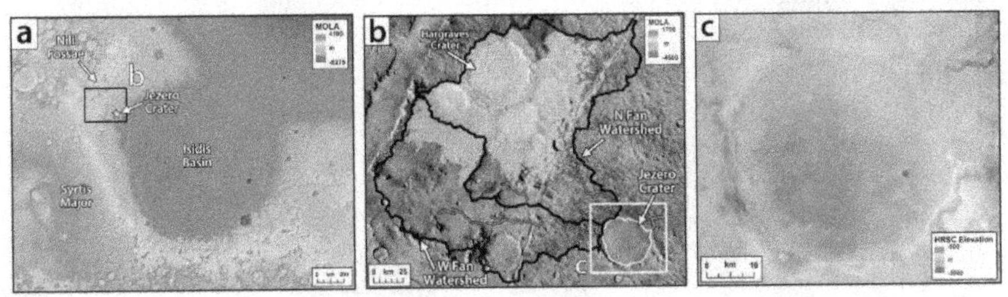


*Figure 4: Proposed settlement site - Jezero crater, Mars*

Figure 4(a) shows location map for Jezero crater (white star). Location of blow-up is indicated by labelled black box. Background is MOLA topography overlain on MOLA-derived hillshade. Figure 4(b) shows overview of watershed areas for the northern and western fan deposits within the Jezero crater paleolake basin. Watershed areas are indicated in labelled, thick black outlines and were derived from modern MOLA topography. Mapped valley networks are indicated in magenta. Regions within the watershed covered by material related to the Hargraves impact crater are indicated in crosshatched area. Location of blow-up is indicated by labelled white box, which is shown in Figure 4(c)[7].

*3.2.3 Characteristics of selected site*

Jezero is characterized by early deltaic sediment deposition into a low salinity (i.e. habitable) paleolake. The region is located close to the Martian equator, providing warmer temperatures than at the poles. There is also evidence of hydrated minerals. The availability of water is high and loss of water from the surface can be minimal due to low elevation the western delta is dominated by Fe/Mg smectites and exhibits well defined sedimentary layering, including bottom-set deposits while the northern delta is dominated by Mg-carbonates and associated olivine. A Volcanic Unit overlies most of the basin fill, embays the eroded delta scarps, and surrounds deltaic remnants which have been separated from the main delta bodies by aeolian deflation at some time prior to volcanism[8]. This region also contains comparatively less dust than others, providing a bit of protection from loose regolith from local dust storms, as well as provides flat ground surface for easy infrastructure development.

## 3.3 Dawn of The Dvaraka Initiative

2022, will be the dawn of The Dvaraka Initiative and by 2024, pre-initiative phase will begin, followed by cargo missions from 2026. Initial cargo missions will include transporting of pressurized and unpressurized rovers, nuclear generators, carrying equipments. Later, cargo missions containing robots, essential raw materials, equipments for air, water facilities will be launched. All the cargo spacecraft will be on round trip to Mars, by carrying few tonnes of hydrogen from

Earth and converting it into return trip propellant. After completion of pre-initiative phase, Dvarakans will be arriving.

*Dvarakans training phase*

Adjacently training program on Earth will be initiated around 2028, among which trainee Dvarakans selection will be done. Figure 5 shows the selection and training level layout for Dvarakans.

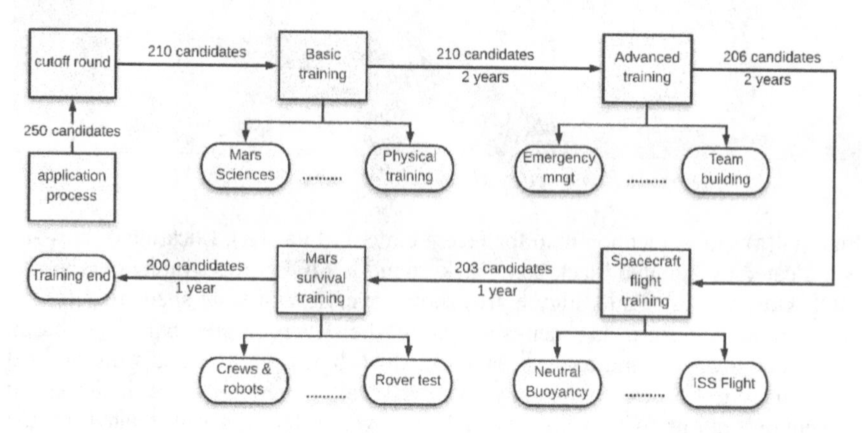


*Figure 5: Training phase for Dvarakans on Earth*

Selected space companies will have one year for application process and from that 250 trainees will be selected by MADE. The selection will be based on resilient, adaptability, flexibility, English language, medical, age limiting between 20-45 and no criminal background. After this, trainee will be directed for two weeks of cut off round. This round will test them on their Martian knowledge, checking their software skills and various group activities. At the end of cut-off round, trainee group will be left with 210 trainees. Every year 210 trainees will go through basic and advanced Martian survival training. At end, 200 trainees will be finishing the training and early batch trainees will be offered to join MADE, for better understanding of Martian governance. Basic level activities will include technical, physical and social training. Every candidate will be trained with multiple skills and the points will be awards to mark their progress. An average adult takes 66 days to adapt a new habit[9]. Therefore the whole training program has been designed for 6 years.

## 4. SETTLEMENT PHASE

### 4.1 Dvarakans' arrival

In beginning of settlement phase, 2036, first spacecraft will carry 100 people along with a food cargo for one year sustainability, since it is not feasible to send 1000 people together. Only engineers of various field and agronomists will be sent in first ship followed by 200 people in every two years. Therefore it will take approximately 10 years to have a 1000 people colony on Mars. In early phase, department representatives will be in contact with MADE, once Dvaraka's population reaches 60%, implementation of Martian government will be initiated. In order to avoid abnormal sex ratio in future, male to female ratio will be 505:495 respectively. Upon the arrival, living quarters will not be in full working conditions. The first 100 people will be spending approximately 6 more months inside spacecraft on Mars surface. During which they will make houses in habitable condition. After which they will move into respective living quarters to start their Dvarakan life.

## 4.2 Dvaraka's architectural concept

In developing the design for Dvaraka, a number of different issues were considered, each with a distinct set of precedents. Positive aspects were derived from each precedent and the results were synthesized into a new logic. Thus, the architectural layout of Dvaraka came into existence which is shown in Figure 6. Every settlement needs a vision and a plan for its future. The primary engineering requirements for the design are safety, efficiency and expandability. Safety requires that there be a number of interconnected and individually pressurized segments. In case of an accidental loss of pressure, a fire, or other failure, there must be at least two means of egress from every space.

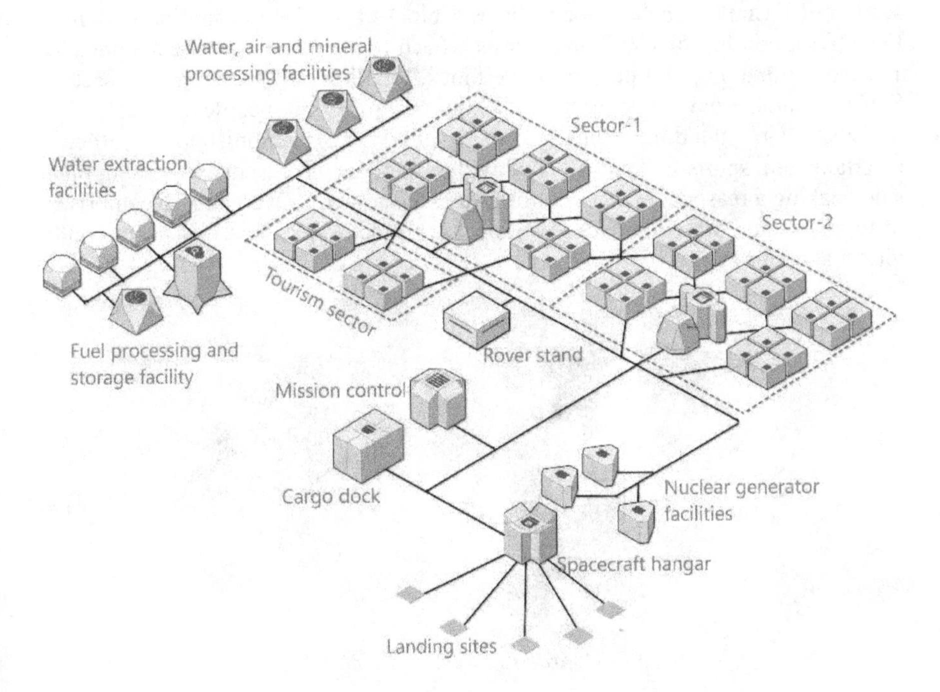


*Figure 6: Dvaraka colony concept*

Furthermore, the loss of one space must not cut off functioning portions of Dvaraka from each other. Efficiency is dictated by the large distance from Earth and the shortage of labour and energy on Mars. Expandability requires that the pattern of development be easily repeatable and expandable without compromising the qualities of the structures that have already been completed. All three parameters were kept in mind while developing Dvaraka's layout.

The whole Dvaraka is divided into sectors. Sector-1 and sector-2 consists of living quarters and greenhouses, while tourism sector consist of living space for tourists. Apart from that, water, air and mineral processing facilities are established adjacent to sector-1, making easy up on distributing the primary necessities. Along the linear line of water, air and mineral facilities, there are water extraction facilities located. On opposite, fuel processing and storage facility is built. Far at a distance on other side, connected to sector-2, mission control is established, near landing sites, along which cargo dock is built, making it comfortable to load and unload cargo on spacecraft. In same area, nuclear generator facilities are operating with the main aim of providing power to whole Dvaraka. A special area has been allocated to next to tourism sector, rover stand, where Martian Land Vehicles (MLV) can be built and repaired.

*Sector layout*

Figure 7 shows the top view and isometric view of conceptual layout of sector section of Dvaraka. Each sector includes 5 blocks and 1 administration building. Each block consists of 100 living quarters. Each living quarter will be assigned to 1 person so making 100 people in one block, totalling 500 people in one sector. Since, Dvaraka have two sectors, it can occupy 1000 people in total. The administration building consists of facility for administration offices, entertainment, sports-hall and other activities. Each sector is a square ~500 m on a side, making a reasonable distance for Dvarakans to walk. Dvaraka also comprises of tourist sector, which includes two blocks, providing a capacity of hosting 200 tourist at a time.

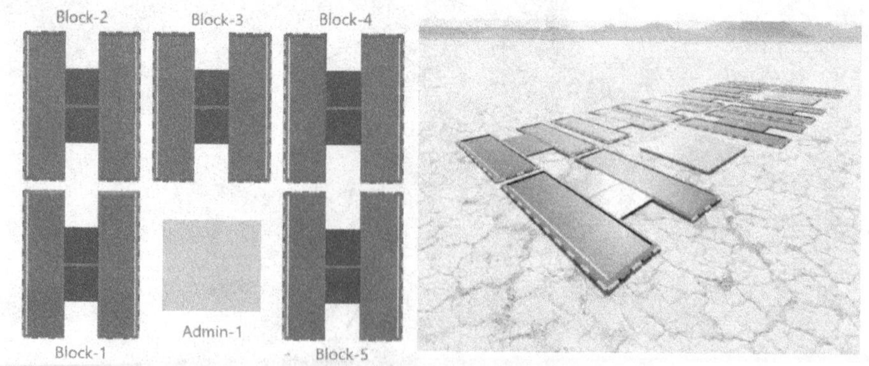


*Figure 7: Dvaraka's block and sector layout*

*Block Layout*

Each block covers an area of 31,488 $m^2$. Figure 8 shows block cross sectional viewed from top side. It consists of 2 greenhouses, 100 living quarters, 1 dining hall and 1 laboratory for conducting variety of research experiments.

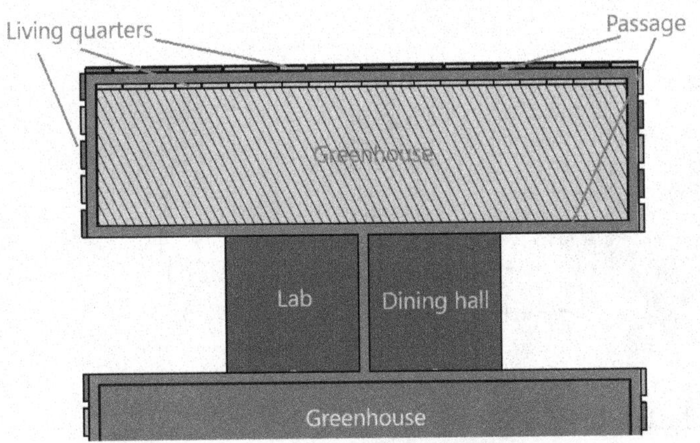


*Figure 8: Block's cross sectional view (top view)*

Considering the fact that, only about 15 $m^2$ plant growth area would be required to provide adequate nutrition to a single person if that person is willing to eat nothing but wheat. With addition of other crops, 50 $m^2$ should be sufficient[10]. It may be that the Martian soil is less fertile than Earth's soil, thus adding safety factor of 4, concluding 200 $m^2$ should be sufficient for 1 person. For 100 people in block, 20,000 $m^2$ plant growth area would be require. Each greenhouse of 50 m × 200 m, covers an area of 10,000 $m^2$. With 2 greenhouses, block has capability of providing enough food for 100 people. Dining hall and laboratory are of same size, 50 m × 50 m, covering an area of 2,500 $m^2$. There is a passage which is 4 m wide and 4 m high which provides a common connection to reach any part of the block.

*Living quarter*

The layout of each living quarter is shown in Figure 9. The size of one living quarter is 2 m × 10 m, providing a comfortable living space of 20 $m^2$ per person. The height of wall facing outside is 2.2 m, while the wall facing towards passage is 2.75 m. According to the requirement, it is possible to convert 2 single living quarter into 1 double living quarter. Outer living quarters will be provided with windows, while inner living quarters will have transparent roof, along with shades mechanism, to enjoy sunlight or night sky view, whenever require.

*Figure 9: Living quarter's layout*

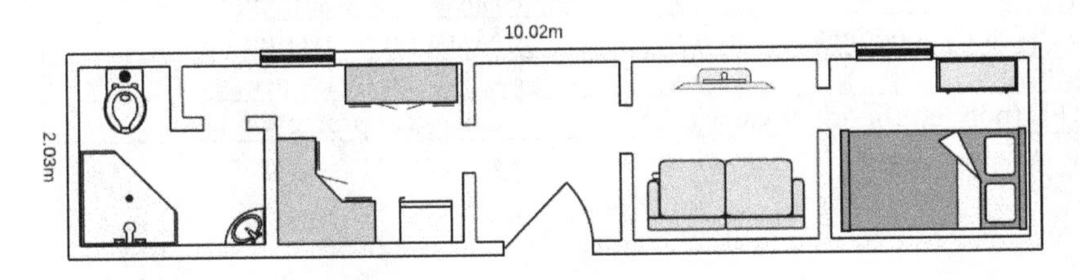

**4.3 Infrastructure materials**

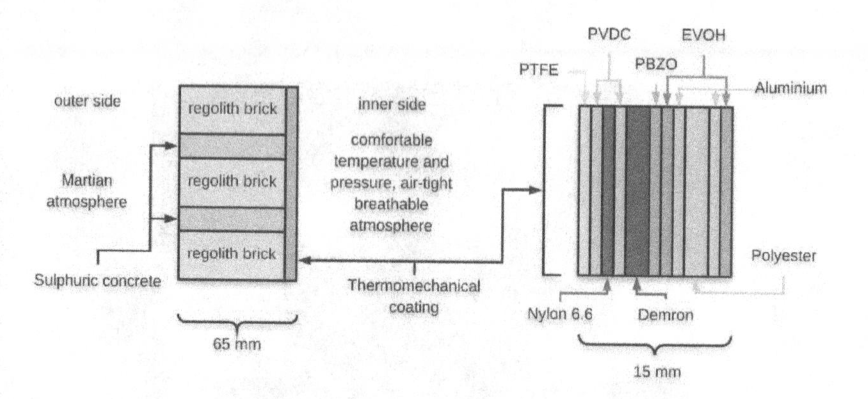


Figure 10: Materials for outer wall

Sending materials for building infrastructure from Earth is not economically practical idea. The Mars regolith is rich in oxides of iron, silicon, magnesium and aluminium. Based on the research and innovation done in material science, using different synthesizing processes, it is possible to extract essential materials for building infrastructure on Mars. Having abundance of $CO_2$ in Mars atmosphere, it is feasible to make materials like polyethylene, polypropylene and polycarbonate on Mars. For the main infrastructure of Dvaraka, bricks and sulfur based concrete is proposed to be used, along with thermomechanical coating (TMC), which can be extracted from Mars regolith[11]. The thickness of outside walls will be ~65 mm, of which 50 mm will be brick thickness and remaining will be TMC. Figure 10 shows the arrangement for outer wall along with configuration of TMC. TMC provides protection from excessively cold temperatures, impact, fire, heat, wear, abrasion, chemical degradation, thermal conduction and radiation heat losses and also flexible enough to allow an easy and repeatable folding and unfolding. The characteristics of TMC are listed in Table 1[12] [13] [14].

| Material | Layer |
|---|---|
| EVOH (ethylene vinyl alcohol copolymer) | Pressure layer |
| Aluminium | Insulation layer |
| Polyester | Structural layer |
| PBZO (polyphenylene benzobioxazole) | Impact/puncture, heat protection |
| Demron | Energetic radiation protection |
| PVDC (polyvinyldiene chloride) | Moisture protection layer |
| Nylon 6.6 | Abrasion resistance layer |
| PTFE (polytetrafluoroethylene) | Chemical protection layer |

*Table 1: Characteristics of thermomechanical coating*

Thus, TMC is in charge of creating a sealed environment that enables the human life, maintaining the pressure and breathable atmosphere. In addition to the needed structural resistance to withstand the difference of pressure between the environment and the interior and the possible impacts. For roof of greenhouse and inner living quarters, PMMA glass (polymethyl methacrylate) will be used. PMMA glass is transparent thermoplastic material which is perfectly suitable. The raw materials used for making PMMA glass are available in Martian regolith and can be extracted and synthesized to make PMMA glass on Mars[15]. Also PMMA glass provides protection from UV radiation for wavelength around 300 nm which is within the range in case of Mars. Apart from manufacturing of primary materials, ethylene can also be used for manufacturing of thermoplastic elastomers which is the primary material for space suits. Carbon, nickel, manganese, aluminium, steel and other metals required for manufacturing colony use equipments can also be extracted and processed.

## 4.4 Dvaraka's Life Support System

The schematic of Dvaraka's life support system (DLSS) for potable water production, breathable air production and propellant production system is shown in Figure 11. Atmospheric $CO_2$ and hydrogen electrolyzed from ground water are fed into a Sabatier reactor to produce oxygen and methane. The Sabatier reactor produces oxygen and methane at a ratio of 4:1. Since this is greater than the engine mixture ratio, excess oxygen will be produced. Therefore methane production is the driving requirement of DLSS. Considering, consumption of ~9.6 kg of water per person per day and ~2.8 kg of oxygen per person per day, thus 11,520 kg amount of water will be required for 1200 (including tourist) people per day and 3,360 kg amount of total oxygen per day.

*Figure 11: Dvaraka's life support system*

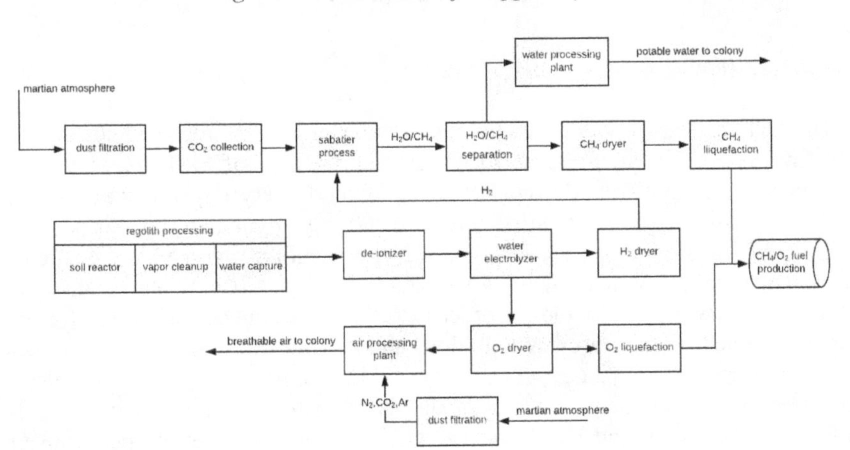


Considering, DLSS water extraction subsystem will be able to extract 68.2 kg/hr [4], hence 16 facilities will produce 26,188 kg/day. By doing electrolysis 23,372 kg/day amount of $O_2$ is produce, out of which 20,341 kg/day for propellant generation and remaining 3,032 kg/day is sent to air processing plant. From electrolysis, 2,922 kg/day amount of $H_2$ is sent to Sabatier reactor for producing

5,812 kg/day $CH_4$, which will be used for propellant generation and 13,051 kg/day water is produced, which is sent to water processing plant. Hence, $9.54\times10^6$ kg/year of propellant is produced, which satisfies the need for incoming and outgoing spaceships. The Figure 12 shows the path ways of breathable air within Dvaraka. Air processing plant will convert pure $O_2$ into a breathable air ratio. This breathable air will be supplied to living quarters, kitchen and labs. This air with more $CO_2$ proportion, released from mentioned areas, will be send to greenhouse chambers. The plants in greenhouses will produce $O_2$ using photosynthesis process. This air with more $O_2$ proportion, will be input for air processing plant.

### 4.5 Food production and waste management

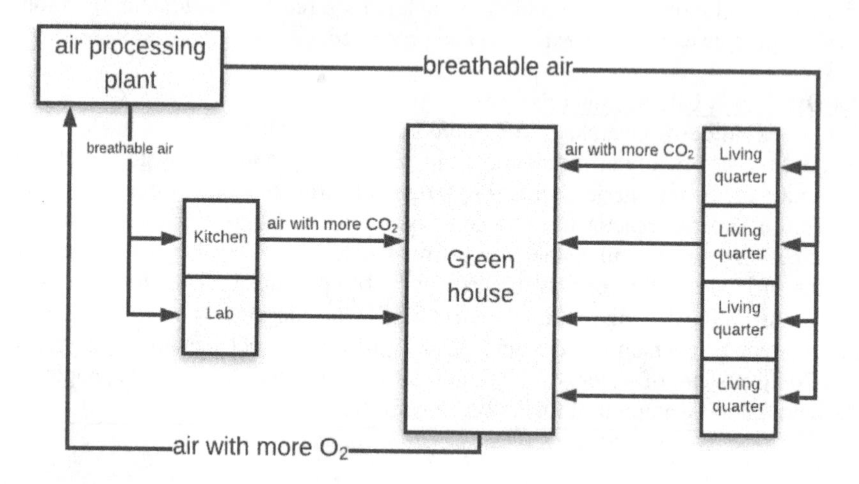


Figure 12: Breathable air circulation system

Each block consists of two green houses, which can provide adequate nutrition to 100 Dvarakans. The green house will be serving the need of food and oxygen for 100 people. A healthy body requires 2 kg of food every day for nutrition and energy[16]. Basic nutrients intake should be 1000 mg of calcium, 400 µg of folic acid, and 56 g of protein each day. But for iron quantity, it is double the amount for females, and vitamins intake varies with each type. Considering scenario for pregnancy, intake gets double. In order to fulfill these values, selection of crops are done based on their growth conditions and nutrition values. The crops grown will be: poppy seeds, winged and soya beans, lentils, potatoes, tomatoes, onions, green beans, spinach, beets, orange, lemons. All these crops require approximately 6-7 hours of sunlight and 12-14 hours of artificial light, manures, proper drainage system. These plants only require sprinkling of water from time to time. Soil is required to be rich in potassium, phosphorous and nitrogen nutrients.

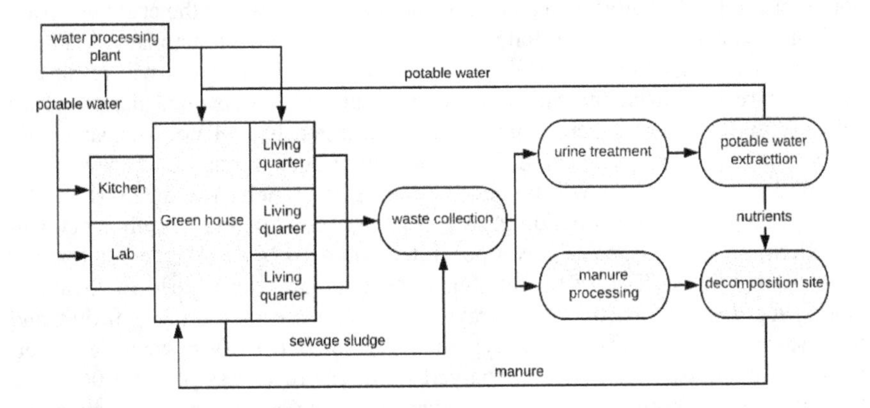


*Figure 13: Layout of manure and water recycle plant*

The soil quality will be attained by adding composites, lime and sulphur to maintain the pH balance for respective crops. Rhizobium bacteria's will be used to make nitrogenous rich soil. For first two years, 100 kg of manure's will be imported from earth, which cost $1000 for a ton and simultaneously the Martian soil will be prepared. Perchloride is excess in Martian soil, which can be reduced by adding gypsum. Some dry fruits, egg and milk powder will also be needed to import. During this time, fertilizers will be getting ready in the recycle plant, which is shown in Figure 13. Every lavatory will be connected to a plot, where waste products will be collected in separate tanks. Water waste will go for urine filtration process[17], which will separate 75% of water and 5% of other nutrients like nitrogen, phosphorous and potassium[18]. The solid waste products will be sent for decomposition process. The nutrients from filtration process additional mixed with animal manure's and sewage sludge to increase the process of decomposition. Manure's takes 6 months to be prepared. After which they will be sent to greenhouse along with water for the cultivation of crops.

**4.6 Electricity production**

In Dvaraka, the demand for electricity will come from the following functions: conditioning of buildings, domestic appliances, scientific and office equipments, processing facilities (water, air, and propellant), production of basic materials, water extraction facilities, and agriculture productions. Taking into account the various factors: scale effect, possibility of providing some of the needs in thermic mode and conservative assumption regarding the consumption of commercial activities, it is estimated as 3.5 kWe per person (4.2 MWe for 1200 people), the need for electric power, putting aside the lighting of greenhouses. Considering, power consumption of 17 kWe per processing facility [4]. Adding 58.3 kWe power consumption per greenhouse, with these values, total power requirement reach. ~ 6.1 MWe. Electric power generation and distribution will be a combination of uranium based nuclear reactors and a minimum system of nickel-hydrogen batteries. Appropriate nuclear reactors implementations will be independent of environmental conditions thereby providing consistent power for even severe conditions, e.g., dust storms that can last for several weeks and the ensuing period of digging out from resulting dunes. Currently, the baseline reactor is envisioned to be an evolution of NASA's Kilopower Reactor[19]. The evolution will result in a 100 kWe reactor from the maximum of the currently envisioned design of 10 kWe. With 30 nuclear reactors, total power generated is 3 MWe. Nuclear power generation is not only the option. The roof side of outer living quarters, passage, lab and kitchen buildings will be covered with solar panels. The area covered by solar arrays on 12 blocks buildings will be 126,336m$^2$, with 33% efficiency and considering 100 W/m$^2$, solar power generated will 4.16 MWe. Hence, total power generation will be 7.16 MWe satisfying the total electricity need of Dvaraka. During the dust storms, the solar arrays will be covered with a layer of dust and these storm last for few weeks to few months. To meet 6.1 MWe, we have nickel-hydrogen batteries which will be charged from part of excess power 1.06 MWe that we are generating from nuclear reactor during a Martian day when the weather is clear and there is no dust storm. Each living quarter will be equipped with nickel-hydrogen battery, will work independently of scenario like dust storm.

### 4.7 Thermal design

Assuming the thermal output at 95% efficiency from 30 nuclear generators, along with the electric power we are getting thermal output of 9.417 million btu per hour which puts us at 226.01 million btu per day. Mars' average temperature is 227.45 K (−50°F) and the greenhouse runs at 299.8K (80°F), so the temperature difference $\delta T$ is 72.35K (130°F). The total exposed area of one block is 60,000 m$^2$. This exposed area includes bottom, top and sides which are directly in contact with either mars land or mars atmosphere. Dvaraka infrastructure is made of bricks and concrete and the walls are covered with thermomechanical coating (TMC) which has total Thermal Resistance R value of 11.5. The total heat loss with reference to the above area is 195.32 million btu per day. This heat loss value is total heat loss calculated including every infrastructure building of Dvaraka. During the dust storms, the average temperature difference in 314.2K (106°F). Hence the total heat loss during the dust storm, 159.27 million btu per day. A better way to interpret this information is, 195.32 million btu is heat lost from Dvarka, and to keep it warm we are continuously supplying 195.32 million btu from 226.01 million btu thermal output. The difference between thermal output from nuclear generators and the heat loss is 30.69 million btu which will be supplied to keep the other facilities in Dvaraka, for example mission control, water/fuel extraction building, bricks and cement manufacturing machinery etc. Early placement of first few reactors will use atmospheric heat rejection system. After Dvarkans arrive, they will install the reactors and will utilize a more robust system of cooling by placing the heat rejection pipe system to Dvaraka, to make up the heat loss.

### 4.8 Against all odds of dust storm

Martian winds can frequently generate large dust storms. When Mars is at its perihelion, the southern hemisphere is suddenly heated, and a large temperature difference relative to the northern hemisphere is generated. This drives strong winds and dust from the southern hemisphere to the northern hemisphere. Most of the storms move with velocities of 14 m/s-32 m/s [20] and dissipate within a few days. An experiment is simulated in wind tunnel performing dust storm scenario with wind speed 50 m/s. Taking into consideration curves and contours of design, 2D analysis is done for the scenario mentioned, which is shown in Figure 14.

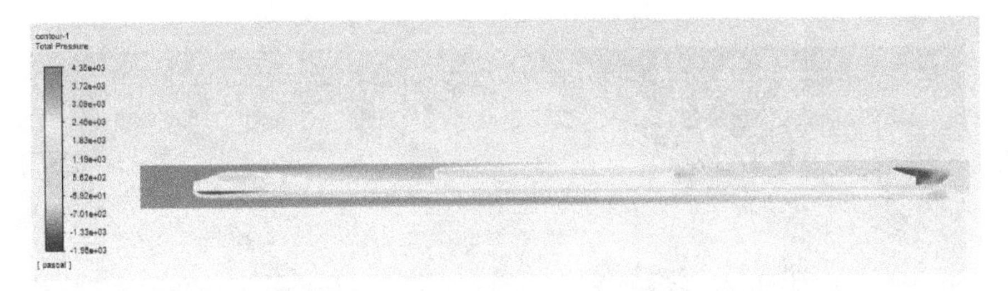


*Figure 14: 2D analysis of cross-section of Dvaraka's block design*

*Figure 15: 3D flow analysis of Dvaraka's block design*

Also 3D flow analysis is done on Dvaraka's block building which will hold living quarters for Dvarakans. Same scenario is performed as in 2D, but with three different orientation of block with respect to the direction of wind. Results obtained are almost same, stating maximum surface pressure of 4350 pa which is shown in Figure 15. Figure 16 and Figure 17 shows two other orientations, resulting in range of ~ 1300 pa – 2400 pa surface pressure.

*Figure 16: 3D flow analysis of Dvaraka's block design*

*Figure 17: 3D flow analysis of Dvaraka's block design*

Also, structural and thermal analysis is performed on living quarters with respect to 5000 pa surface pressure, considering the maximum obtained value of surface pressure. . Figure 18 shows structural analysis, showing displacement graph of maximum value 22.07 mm. Figure 19 shows thermal analysis, stating heat flux with maximum value of 168.9 W/m$^2$ on edges of living quarters.

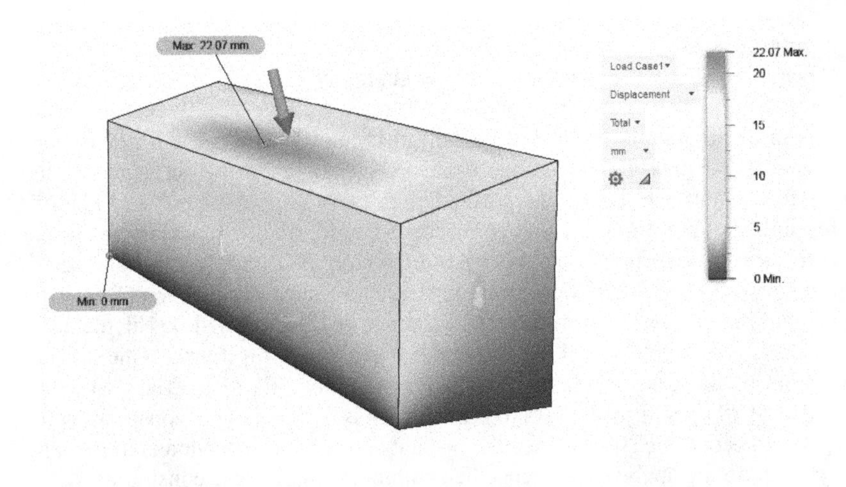


*Figure 18: Structural analysis of living quarter*

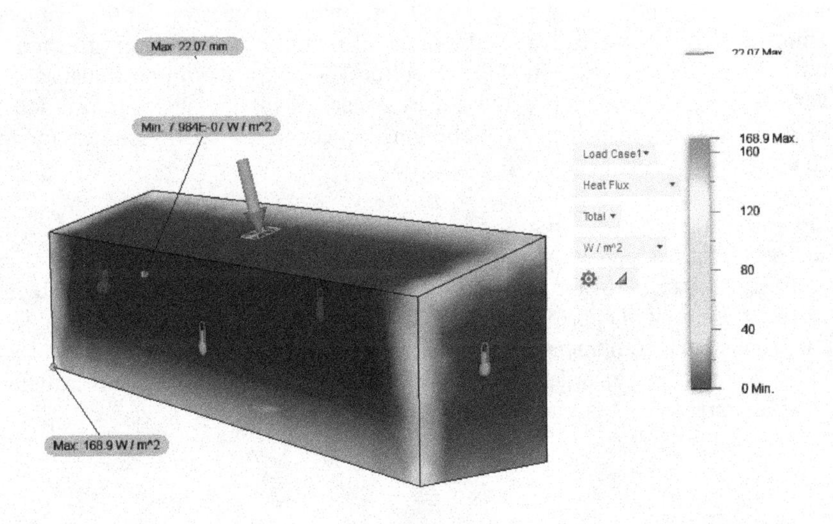


*Figure 19: Thermal analysis of living quarter*

From structural and thermal analysis it is concluded that Dvaraka's infrastructure is able to withstand the thermal displacement caused by pressure and temperature difference in case of dust storm. The inside pressure and temperature is maintained constant and the outside conditions vary time to time. The infrastructure is designed in such a way that it withstand against all the odds of dust storm and Martian temperature changes with a great safety factor of 2.33.

## 5. SELF-SUSTAINING PHASE

### 5.1 Economic system

The Dvaraka Initiative is a model of a new economy to introduce a trade system on Mars inspired by a similar pattern on Earth for several years. The reason for the same is to create a self-sustainable version of human life on another planet. Our experiences on Earth give us enough perspective to judge the influence of the suggested economic system on Dvaraka. The primary objective here is to identify the feasibility of the project which comprises several stages of analysis beginning with, the initial investment on the basis of which Dvaraka would be erected. Following which is the initiation of the settlement. The cash inflow-outflow in the project and the consumables in the establishment of Dvaraka will be calculated in US dollars, since the initial investment is generated on Earth. One of the constraints to the case is the transport of goods from Earth to Mars which is marked at $500/kg and from Mars to Earth which is $200/kg. This constraint helps us design an economic structure for the profitable operation of Dvaraka in the long run. The economic analysis can hence be commenced with these constraints.

#### *5.1.1 Initial Investment*

*Generation of Initial Investment* The foundation stage of Dvaraka will be laid through investments from across several government bodies on Earth. Dvarakan commerce will be influenced by the order of participation of these government bodies in the initial investment. $5 billion from the investors will be drawn as loan every two years. Furthermore, the money generated for training will also add to the initial account which is $3.15 billion. Hence the initial investment for the project is $65 billion.

*Outflow during the Pre-Initiative Phase*

The pre-initiative phase will involve five missions where in each of them comprises of transport of equipment, raw materials and other resources. The cost of every mission will approximately be $1 billion as can be inferred from Table 2. It also enlists the equipment transported, such as heavy machinery, raw materials and rovers to Mars in order to build the basic infrastructure of Dvaraka before Dvarakans arrive.

| Missions | Equipments | Material Cost estimation | Weight (kg) | Transport cost | Total cost |
|---|---|---|---|---|---|
| Mission-1 | Unpressurized rovers | $350,000,000 | 225,000 | $112,500,000 | $1,287,500,000 |
| | Load carrying trucks | $575,000,000 | | | |
| | Nuclear generator | $100,000,000 | | | |
| | Fuel processing | $150,000,000 | | | |
| Mission-2 | Pressurized rovers | $745,000,000 | 340,000 | $170,000,000 | $1,096,300,000 |
| | Cooling equipments | $65,000,000 | | | |
| | Heavy duty equipments | $116,300,000 | | | |
| Mission-3 | Water extraction | $575,000,000 | 340,000 | $170,000,000 | $1,037,000,000 |
| | 3D printing parts | $70,000,000 | | | |
| | Heating equipments | $222,000,000 | | | |
| Mission-4 | Air processing | $555,000,000 | 340,000 | $170,000,000 | $1,330,000,000 |
| | Water processing | $475,000,000 | | | |
| | Raw materials | $130,000,000 | | | |
| Mission-5 | Bricks making | $385,000,000 | 340,000 | $170,000,000 | $1,075,000,000 |
| | Infrastructure parts | $150,000,000 | | | |
| | Electronic parts | $370,000,000 | | | |

*Table 2: Mass - cost budget of Pre-Initiative phase*

*Outflow during the Settlement phase*

Settlement phase comprises of six missions as listed in Table 3. The first Dvarakans will be sent with cargo having food and various utilities, together costing $700 million. For the other five missions consisting of 200 Dvarakans and cargo, transportation costs will be around $1 billion. These values are rough estimations with reference to weight distribution.

| Missions | Flights | Material cost estimation | Weight (kg) | Total cost |
|---|---|---|---|---|
| Mission-6 | 100 people | $692,611,900 | 170,000 | $783,363,900 |
| | Food | $90,000 | | |
| | Space Suits | $37,500,000 | | |
| | Drugs and Sanitary items | $2,312,000 | | |
| | Medical equipments | $850,000 | | |
| | Fertilizers | $50,000,000 | | |
| Mission-7 | 200 people | $1,385,223,800 | 255,000 | $1,537,723,800 |
| | Cargo | $152,500,000 | | |

| | | | | |
|---|---|---|---|---|
| Mission-8 | 200 people | $1,385,223,800 | 255,000 | $1,537,723,800 |
| | Cargo | $152,500,000 | | |
| Mission-9 | 200 people | $1,385,223,800 | 255,000 | $1,537,723,800 |
| | Cargo | $152,500,000 | | |
| Mission-10 | 200 people | $1,385,223,800 | 255,000 | $1,537,723,800 |
| | Cargo | $152,500,000 | | |
| Mission-11 | 100 people | $692,611,900 | 170,000 | $845,111,900 |
| | Cargo | $152,500,000 | | |

*Table 3: Mass - cost budget of Settlement phase*

### *5.1.2 Profitable Operation*

In order to keep the economy of Dvaraka self-sustaining, this economy plan discusses various ventures and estimates the revenue generated upon their installation in a fully functioning settlement. The proposed methods of sustenance on Mars are as follows:

*Tourism*

Four schemes for tourism are proposed based on duration of stay, amount of propellant used and frequency of travel in a year. The aim behind having multiple schemes is to increase the frequency of travel to Mars. The amount of fuel expended for the different trajectories possible around the year determines the ticket prices. Hence the four schemes, A through D are named in order of the number of days that people will spend travelling. More fuel for lesser number of days demands a higher ticket cost as can be referred from Table 4.

| **Scheme** | **Total duration** | **Frequency (1 year)** | **Cost per tourist** | **Ticket price per tourist** | **Total revenue** |
|---|---|---|---|---|---|
| A | 95 days | 4 | $5,135,463 | $19 million | $5.5 billion |
| B | 160 days | 4 | $8,824,496 | $16 million | $2.87 billion |
| C | 320 days | 4 | $9,434,225 | $13 million | $2.85 billion |
| D | 450 days | 4 | $6,926,119 | $10 million | $1.23 billion |
| Total revenue generated in a year | | | | | $12.5 billion |

*Table 4: Revenue from Tourism*

As can be seen in Table 5, for an accommodation capacity of 200 tourists at a time in the colony and two spacecraft carrying 100 tourists each, the available windows of travel for a sample year were found[21]. The gradient in the colour of the data demarcates the journeys under different schemes.

| **2035 TRIPS** | | | | | | | | | |
|---|---|---|---|---|---|---|---|---|---|
| Earth_Departure | Dest_Arrival | Dest_Departure | Earth_Return | Stay time (days) | Duration (days) | Total DV (km/s) | Reentry speed (km/s) | Route | Mass Ratio |
| Jan-07-2035 | Feb-01-2035 | Feb-11-2035 | Apr-17-2035 | 10 | 100 | 9.72 | 11.21 | EA-AE | 28.55 |
| Jan-07-2035 | Jan-27-2035 | Feb-01-2035 | Apr-07-2035 | 5 | 90 | 9.86 | 11.09 | EA-AE | 29.96 |
| Jan-07-2035 | Feb-16-2035 | Feb-26-2035 | Jun-01-2035 | 10 | 145 | 8.15 | 11.18 | EA-AE | 16.62 |
| Dec-13-2035 | Apr-21-2036 | May-01-2036 | Jun-10-2036 | 10 | 180 | 9.36 | 12.56 | EA-AE | 25.22 |
| Feb-01-2035 | May-12-2035 | Jun-01-2035 | Nov-28-2035 | 20 | 300 | 7.92 | 11.87 | EA-AE | 15.35 |
| Feb-11-2035 | Jun-01-2035 | Jun-26-2035 | Nov-28-2035 | 25 | 290 | 8.26 | 12.1 | EA-AE | 17.26 |
| Mar-18-2035 | Aug-10-2035 | Aug-30-2035 | Jan-02-2036 | 20 | 290 | 7.93 | 11.55 | EA-AE | 15.40 |
| Mar-28-2035 | Feb-01-2036 | Mar-02-2036 | Sep-23-2036 | 30 | 545 | 4.76 | 11.7 | EA-AE | 5.16 |
| Oct-04-2035 | Mar-12-2036 | Apr-01-2036 | Sep-28-2036 | 20 | 360 | 5.65 | 11.63 | EA-AE | 7.02 |

*Table 5: Trajectory determination over one sample year (2035)*

*Deuterium Generation*

Deuterium, the heavy isotope of hydrogen is 166 ppm by composition on Earth, comprises of 833ppm on Mars. Deuterium is the key fuel for both the first and the second generation fusion reactors on Earth. On Earth, 1kg of deuterium priced between $10,000 - $16,000, depending on the purity. Therefore the estimated profit per kg is around $9500, even if the cost drops due to less demand. For one tonne the estimated profit is $9,500,000. As of now, on Earth, 452 nuclear power plants are either used or under construction which requires 400grams of Deuterium per day for 1500 MW production. Upon calculation, it can be inferred from Table 6, 43.8 tonnes of deuterium is required per year. On transporting 90 tonnes of deuterium once in two years, the profit is $746.58 million.

| Total D exported from Mars | 87,989.33 kg |
|---|---|
| Transportation costs | $133.32 million |
| Cost of D on Earth | $10,000 |
| Total cost of D on Earth | $879.89 million |
| Total revenue generated in 2 years | $746.58 million |

*Table 6: Revenue from Deuterium exports*

*Asteroid Mining*

Asteroids near Mars provide us with an abundant source of metals which are rich in quality. Our expeditions to some of these element rich asteroids shall start from 2042. Asteroids such as Aspacia, Rudra,[22] etc. help us estimate revenues that could possibly be generated from asteroid mining. Since the mass ratios in the calculation of fuel expended in asteroid mining is not determined for these individual asteroids yet, we took the mass ratio from Ceres which is further away from the asteroids in study. The total payload from the asteroid is estimated to be 7500 tonnes which can be extracted with an efficiency of 75%. The payload in every trip on the rocket is 85 tonnes which brings the total amount transported per year to 233.75 tonnes. Revenue generated through asteroid mining can be seen in Table 7. Assuming the composition by weight of platinum from the asteroid mined, to be 25%, the remaining ore is rich in elements such as nickel, iron, copper, etc. which can be utilized on Dvaraka for different purposes.

| Platinum extracted | 233.75 tonnes |
|---|---|
| Cost of Platinum on Earth | $26500 per kg |
| Cost of Transport per year | $386.37 million |
| Estimated cost of payload from asteroid per year | $6.19 billion |
| Total revenue generated per year | $5.81 billion |

*Table 7: Revenue from Asteroid mining*

*Broadcasting*

The trends in broadcasting on Earth for events of the magnanimity of the moon landings, the Olympics events[23] and the ISS launches were studied and a pattern for the broadcast of events such as the landings of the tourists and settlement dwellers on Mars and asteroid mining was designed. The revenue generated is listed in Table 8.

| Arrival of new Dvarakans | 5 |
|---|---|
| Tourism (in 6 years) | 66 |
| Asteroid Mining (in 6 years) | 4 |
| Total events broadcasted in first 6 years | 75 |
| Revenue generated per event (state of art) | $183 million |
| Revenue generated in 6 years | $13.75 billion |
| Total revenue generated in a year | $2.30 billion |

*Table 8: Revenue from Broadcasting*

### *5.1.3 Cost plan*

Figure 20, which shows the differentiation of the ventures across a span of 25 years, is studied and analysed to forecast the self-sustenance point of Dvaraka. The cost flow is laid out based on the timeline of the project taking into account mandatory ventures and those introduced to incur profits.

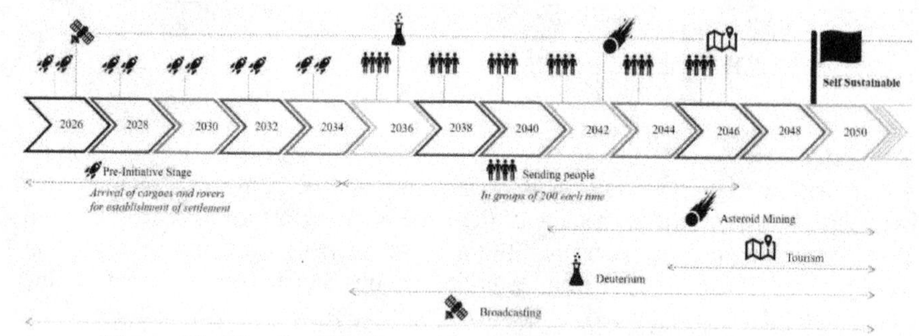


*Figure 20: Timeline for the order of events in making of settlement*

Table 9 shows the yearly expenditure and income of the Dvaraka initiative. The yearly planned average until the complete setup of Dvaraka can be studied as per the different phases we have divided the project in – the Pre-Initiative phase, the Settlement Phase and the full sprung self-sustaining phase.

| Phase | Year | Details | Outflow Cost estimation | Total Outflow | Outflow 5% margin | Inflow cost estimation | Total Inflow | Inflow 5% margin |
|---|---|---|---|---|---|---|---|---|
| Pre-Initiative | 2024 | MADE member's investments | | | | $5,000,000,000.00 | | |
| | | MADE infrastructure | $1,860,000,000.00 | | | | | |
| | | Spacecraft manufacturing | $1,500,000,000.00 | | | | | |
| | | MADE operating cost | $3,200,000,000.00 | | | | | |
| | 2026 | MADE member's investments | | | | $5,000,000,000.00 | | |
| | | MADE infrastructure | $1,860,000,000.00 | | | | | |
| | | MADE operating cost | $3,200,000,000.00 | | | | | |
| | | Spacecraft manufacturing | $1,500,000,000.00 | | | | | |
| | | Broadcasting | | | | $183,333,333.00 | | |
| | | Mission-1 | $1,287,500,000.00 | | | | | |
| | 2028 | MADE member's investments | | | | $5,000,000,000.00 | | |
| | | MADE operating cost | $3,200,000,000.00 | | | | | |
| | | Spacecraft manufacturing | $1,500,000,000.00 | | | | | |
| | | Mission-2 | $1,096,300,000.00 | | | | | |
| | | Broadcasting | | | | $183,333,333.00 | | |
| | 2030 | MADE member's investments | | | | $5,000,000,000.00 | | |
| | | MADE operating cost | $3,200,000,000.00 | $41,525,800,000.00 | $43,602,090,000.00 | | $40,366,666,665.0 | $42,384,999,998.25 |
| | | Training program at MADE | $1,260,000,000.00 | | | $3,150,000,000.00 | | |
| | | Spacecraft manufacturing | $1,500,000,000.00 | | | | | |
| | | Mission-3 | $1,037,000,000.00 | | | | | |
| | | Broadcasting | | | | $183,333,333.00 | | |
| | 2032 | MADE member's investments | | | | $5,000,000,000.00 | | |
| | | MADE operating cost | $3,200,000,000.00 | | | | | |
| | | Training program at MADE | $1,260,000,000.00 | | | $3,150,000,000.00 | | |
| | | Spacecraft manufacturing | $1,500,000,000.00 | | | | | |
| | | Mission-4 | $1,330,000,000.00 | | | | | |
| | | Broadcasting | | | | $183,333,333.00 | | |
| | 2034 | MADE member's investments | | | | $5,000,000,000.00 | | |
| | | MADE operating cost | $3,200,000,000.00 | | | | | |
| | | Training program at MADE | $1,260,000,000.00 | | | $3,150,000,000.00 | | |
| | | Spacecraft manufacturing | $1,500,000,000.00 | | | | | |
| | | Mission-5 | $1,076,000,000.00 | | | | | |
| | | Broadcasting | | | | $183,333,333.00 | | |
| End of Pre-Initiative phase | | | | | | | | -$1,217,090,001.75 |
| Settlement | 2036 | MADE member's investments | | | | $5,000,000,000.00 | | |
| | | MADE operating cost | $3,200,000,000.00 | | | | | |
| | | Training program at MADE | $1,260,000,000.00 | | | $3,150,000,000.00 | | |
| | | Mission-6 | $783,363,900.00 | | | | | |
| | | Mars mining | $133,318,094.00 | | | $746,575,239.00 | | |
| | | Broadcasting | | | | $366,666,666.60 | | |
| | 2038 | MADE member's investments | | | | $5,000,000,000.00 | | |
| | | MADE operating cost | $3,200,000,000.00 | | | | | |
| | | Training program at MADE | $1,260,000,000.00 | | | $3,150,000,000.00 | | |
| | | Mars mining | $133,318,094.00 | | | $746,575,239.00 | | |
| | | Mission-7 | $1,537,723,800.00 | | | | | |
| | | Asteriod mining | $5,500,000,000.00 | | | | | |
| | | Broadcasting | | | | $366,666,666.60 | | |
| | 2040 | MADE member's investments | | | | $5,000,000,000.00 | | |
| | | MADE operating cost | $3,200,000,000.00 | | | | | |
| | | Training program at MADE | $1,260,000,000.00 | | | $3,150,000,000.00 | | |
| | | Mars mining | $133,318,094.00 | | | $746,575,239.00 | | |
| | | Mission-8 | $1,537,723,800.00 | | | | | |
| | | Broadcasting | | | | $366,666,666.60 | | |
| | 2042 | MADE member's investments | | | | $5,000,000,000.00 | | |
| | | MADE operating cost | $3,200,000,000.00 | | | | | |
| | | Training program at MADE | $1,260,000,000.00 | | | $3,150,000,000.00 | | |
| | | Mars mining | $133,318,094.00 | | | $746,575,239.00 | | |
| | | Mission-9 | $1,537,723,800.00 | | | | | |
| | | Broadcasting | | | | $549,999,999.90 | | |
| | 2044 | MADE member's investments | | | | $5,000,000,000.00 | | |
| | | MADE operating cost | $3,200,000,000.00 | $130,114,068,922.00 | $136,619,772,368.10 | | $221,889,060,244. | $232,983,513,256.73 |
| | | Training program at MADE | $1,260,000,000.00 | | | $3,150,000,000.00 | | |
| | | Mars mining | $133,318,094.00 | | | $746,575,239.00 | | |
| | | Mission-10 | $1,537,723,800.00 | | | | | |
| | | Broadcasting | | | | $549,999,999.90 | | |
| | 2046 | MADE member's investments | | | | $5,000,000,000.00 | | |
| | | MADE operating cost | $3,200,000,000.00 | | | | | |
| | | Mars mining | $133,318,094.00 | | | $746,575,239.00 | | |
| | | Tourism | $15,901,810,848.00 | | | $28,400,000,000.00 | | |
| | | Mission-11 | $845,111,900.00 | | | | | |
| | | Broadcasting | | | | $366,666,666.60 | | |
| | 2047 | Toursim | $15,901,810,848.00 | | | $28,400,000,000.00 | | |
| | | Broadcasting | | | | $366,666,666.60 | | |
| | 2048 | MADE member's investments | | | | $5,000,000,000.00 | | |
| | | MADE operating cost | $3,200,000,000.00 | | | | | |
| | | Mars mining | $133,318,094.00 | | | $746,575,239.00 | | |
| | | Toursim | $15,901,810,848.00 | | | $28,400,000,000.00 | | |
| | | Asteriod mining | $386,366,310.00 | | | $6,194,375,000.00 | | |
| | | Broadcasting | | | | $733,333,333.20 | | |
| | 2048 | MADE member's investments | | | | $5,000,000,000.00 | | |
| | | MADE operating cost | $3,200,000,000.00 | | | | | |
| | | Mars mining | $133,318,094.00 | | | $746,575,239.00 | | |
| | | Toursim | $15,901,810,848.00 | | | $28,400,000,000.00 | | |
| | | Asteriod mining | $386,366,310.00 | | | $6,194,375,000.00 | | |
| | | Broadcasting | | | | $733,333,333.20 | | |
| | 2049 | Tourism | $15,901,810,848.00 | | | $28,400,000,000.00 | | |
| | | MADE operating cost | $3,200,000,000.00 | | | | | |
| | | Asteriod mining | $386,366,310.00 | | | $6,194,375,000.00 | | |
| | | Broadcasting | | | | $366,666,666.60 | | |
| | 2050 | Tourism | $15,901,810,848.00 | | | $28,400,000,000.00 | | |
| | | MADE operating cost | $3,200,000,000.00 | | | | | |
| | | Broadcasting | | | | $549,999,999.90 | | |
| | | Asteriod mining | $386,366,310.00 | | | $6,194,375,000.00 | | |
| | | Mars Mining | $133,318,094.00 | | | $746,575,239.00 | | |
| End of settlement phase | | | | | | | | $96,363,740,888.63 |
| Payback to MADE member's investment | | | | | | | $65,000,000,000.0 | $31,363,740,888.63 |
| Self-Sustaining phase | 2051 | Tourism | $15,901,810,848.00 | | | $28,400,000,000.00 | | |
| | | Made Operation cost | $5,000,000,000.00 | | | | | |
| | | Broadcasting | | $21,421,495,252.00 | $22,492,570,014.60 | $549,999,999.90 | $35,890,950,238.9 | $37,685,497,750.85 |
| | | Asteriod mining | $386,366,310.00 | | | $6,194,375,000.00 | | |
| | | Mars Mining | $133,318,094.00 | | | $746,575,239.00 | | |

*Table 9: Cost plan for Dvaraka*

All the three said phases comprise of the initiation of our profitable operations, which make it possible for us to see the economic trend of our project.
*2024 – 2025* Formation of MADE, Spacecraft manufacturing (2026 onward)
*2026 – 2035* MADE operations, multiple cargo missions, Broadcasting events, Training of Dvarakans
*2036 – 2045* Cargo and Manned missions, Mining on Mars-Deuterium, Broadcasting events, Asteroid mining mission (2042 onward), Tourism (2046 onward)
*2046 – 2050* Rounded functioning of profitable operations
*2050 – 2051* Payment of debt (initial investment), Self-Sustainable Settlement
From the data in Table 9, it is clear that the venture begins making profits from year 2046 onward, however the debt for the project is repaid in year 2050 keeping in mind the smooth operation of the profitable events. This credits further revenue generated directly into the future development of Dvaraka and the future settlements. Using the data from the cost plan, the graph in Figure 21 is plotted.

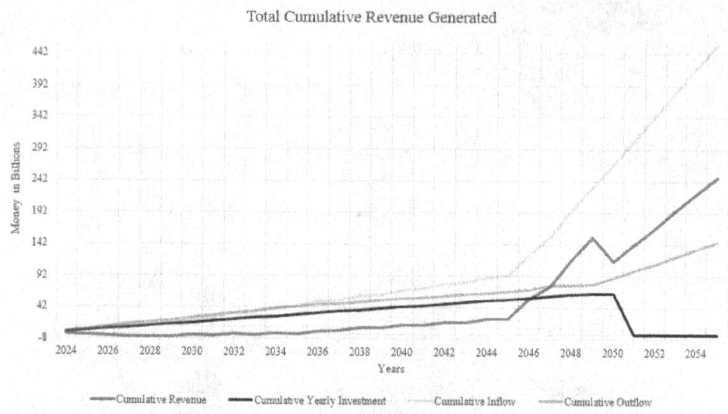


*Figure 21: Economic projection of Dvarakan settlement (2024-2055)*
We see the cumulative patterns of cost inflow, cost outflow, yearly investment and total revenue generated over a period of 32 years. The revenue curve defines the strength of the economic plan of the project. The nature of the revenue curve is as explained below:
*1. Gradual Increase until 2044:* Pre-initiative phase and development of the project
*2. Steep Increase until 2049:* Revenue generated from tourism and asteroid mining
*3. Steep reduction in 2050:* Repayment of the initial investment
*4. Increase in revenue after 2050:* Profitable ventures in a fully functioning society
Hence the graph concludes that the ventures enlisted above in the economy plan hold viable upon the installation in a fully functioning settlement.

*5.1.4 Economic viability and Future scope*
As is concluded from our analysis of the project, 2050 shall be the year Dvaraka becomes a self-sustaining society, having paid all their debt. This shows that the following economy plan is viable since it promises sustenance in less than 25

years. The project is well-rounded since it is symbiotic with the future developments that make Dvaraka progress. In the future, Dvaraka's economic independence from its contemporaries on Earth allows for MADE to govern as an independent body of its own. MADE shall become the Martian embassy on Earth comprising of official representatives among those fifty people who do not make through the training program. Eventually, TDC on Mars will be formed into a Martian government responsible for all political, social and economic operations on Mars. This enables the newly formed Martian Government to build more Martian colonies from the revenue generated through the profitable operations discussed above. Dvaraka can also act as a pit stop for refuelling in the future for missions like Human Outer Planet Exploration[24]. For research on Earth, Dvaraka shall work in providing samples and retrievable from Mars which will be priced as per the distance from Dvaraka, time and fuel required for the transport of the same. The same could not be estimated in our current plan due to the deficit of information. The Martian economy shall also flourish by aiding in bio-medical research which is primarily due to the variety in atmospheric conditions that make it possible to grow and treat certain microbes that cannot be on Earth. The semi-conductive material research which requires high vacuum like conditions[25] not available on Earth, can be smoothly regulated on Mars. This ground-breaking research would not only provide for Earthly consumption but also for future Martian developments.

## 5.2 Social and Political model

### *5.2.1 The Dvaraka Council (TDC)*

When the total population reaches 60% of Dvarakan population, The Dvarakan Council will come into action. Council members will be chosen by MADE from 1000 Dvarakans. The council will be the board of directors consisting of different departments' representatives as shown in Figure 22. Each representative will cover three departments, to maintain the discipline and check the functionality of the Martian settlement process. Labour and Commerce will be controlling mining, fuel management and entertainment. Education and Agriculture representatives will look into research programs, food and oxygen management while colony development will take care of water and electricity management and will improvise the infrastructure for future colonists.

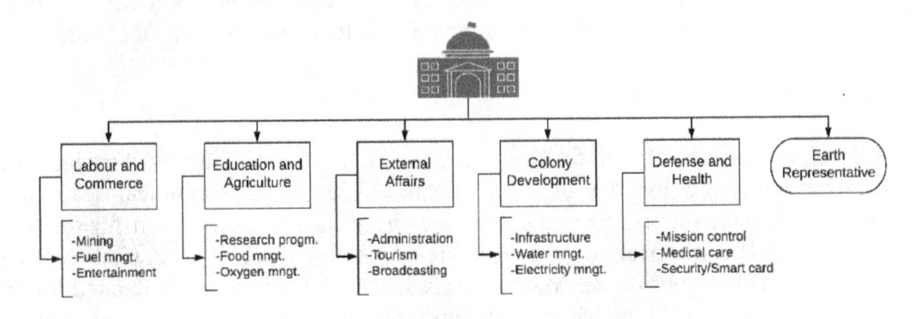


*Figure 22: Departments of The Dvaraka Council*

The entire mission will be monitored by the Defence and Health Department and will also be taking care of smart card for the credit distribution within the colony. The External Affairs department will look into tourism, broadcasting and other administrative work. Every department's main objective will be to make self-sustainable human settlement on Mars. TDC will also have an Earth representative, to check the functionality of the council but will not participate in Dvaraka's decision making process. The External Affairs department will be communicating with Mission Control on Earth. Once Dvaraka attains self-sustainability, TDC will become independent from Earth's finances.

*5.2.2 Smart card system*

Money flow within Dvaraka will be in the form of credits. Every Dvarakan will be given a handy card with their biometric data imprinted in a chip. This card will be loaded with credit points based on their working hours. During the training program on Earth, each candidate will be awarded with credits for their performance and progress. These credits will be loaded on the card and handed to them on their arrival to Mars so that they will have credit points to begin their new life. Every Dvarakan will be working 8 hour long shifts, at the end of which they will be given 250 credit points. These credit points will be used for buying food, water, drinks and enjoy some entertainment. Each meal will cost them up to 20 credit points. Water and drinks will cost between 1-5 points and entertainment will be costing 15 points per hour. With an average living standard, a Dvarakan will require 126 points each day. So at the end of the day approximately 124 points will be saved. The saved points can be used during days of casualty or in sickness.

*5.2.3 Worst case scenario*

The 1000 people community is quite a big population to manage. If ever there is a commotion or a riot that needs to be controlled, the responsible person will be punished with a deduction of 100 credit points by defence representatives. If mass riots take place then in order to avoid a biased consequence, the cutting of resources have been divided. If an entire block starts a turmoil then Education Department Representative in co-sign with Earth Representative will deduct oxygen supply while colony development person in co-sign with Earth Representative will hold back water supply for the block. If any of the representatives go rogue then Dvarakans can file a complaint on Human Resource Department on Earth Council and the Earth Representatives will be given instructions for actions. These precautions will help in smooth flow of Dvaraka.

## 6. CONCLUSION

In brief, considering all the factors and research, the Jezero crater is chosen as the most suitable site for the first settlement on Mars. The technical team has proposed a safe, sophisticated and eco-friendly infrastructural design for the first settlement. The materials used are environment friendly and can withstand the temperature variations on the Martian surface. The course has been planned for 26 years and includes 11 missions with three phases. The phases are described in detail with the economy layout and the budget comprising the money flow estimation. The economic flow between Earth and Mars has been discussed in detail in the respective section. The economic analysis has concluded that Dvaraka shall be debt free and independent of Earthly finances in the year 2050 thus rendering it a self-sustainable colony. In the social and aesthetic view of the colony, the team has pinned down the governance of the Martian society. The structure of MADE and The Dvaraka Council has also been sketched and its working procedure was discussed in detail. The currency used in Dvaraka has been replaced with smart card credits, the working of which was discussed. The worst case scenarios have been taken into consideration which make this plan risk proof as well. Finally the future of inhabiting a planet other than Earth, that seemed to be an illusion in the past, does not seem to be too estranged an idea anymore. With this idea, Earth will not be the only life supporting planet in the Solar System. With all the assumptions and estimations, our team has built a conceptual design for the first self-sustainable human settlement on mars - The Dvaraka Initiative.

**REFERENCES**

---

[1] The term 'Dvaraka' referring to the city with many gates in Hindu mythology, was chosen to symbolize the team's collective dream to usher in a new era of space exploration.

[2] Inhabitants of Dvaraka

[3] Making Life Multi-Planetary, SpaceX. https://www.spacex.com/sites/spacex/files/making_life_multiplanetary_transcript_2017. pdf.

[4] Paz Aaron Kleinhenz Julie E. "An ISRU Propellant Production System to Fully Fuel a Mars Ascent Vehicle". In: LUNAR AND PLANETARY SCIENCE AND EXPLORATION (Jan 09, 2017).

[5] Mars climate zone map based on TES data.Planetologia.elte.hu http://planetologia. elte.hu/mcdd/climatemaps.html.

[6] The ideal settlement site on Mars – hotspots if you asked a crop. WUR https://www. wur.nl/en/newsarticle/The-ideal-settlement-site-on-Marshotspots-if-you-asked-a-crop.htm.

[7] Goudge et al. "Assessing the mineralogy of the watershed and fan deposits of the Jezero crater paleolake system". In: Mars, J. Geophys. Res. Planets 120 (2015), pp. 775–808.

[8] Marsnext.jpl.nasa.gov. https://marsnext.jpl.nasa.gov/documents/LandingSiteWorksheet_Jezero_final.pdf.

[9] James Clear. Atomic Habits. Penguin Publishing Group, 2018.

[10] Frank B.Salisbury. "Bios-3: Siberian Experiments in Bio regenerative Life Support". In: BioScience 47 (1997), pp. 575–585.

[11] Lin Wan. "A novel material for in-situ construction on Mars: experiments and numerical simulations". In: Construction and Building Materials 120 (2016), pp. 222–231.

[12] Sushil N Dhoot, Benny D Freeman, and Mark E Stewart. "Barrier polymers". In: Encyclopaedia of polymer science and technology (2002).

[13] HF Mark. Encyclopaedia of Polymer Science and Technology, 15 Volume Set. Wiley, 2014.

[14] HW Friedman and MS Singh. "Radiation Transmission Measurements for Demron Fabric; Lawrence Livermore National Laboratory: Livermore". In: (2003).
[15] R. Heidmann. "Embarking for Mars". In: (May 11, 2017).
[16] WHO international news release health sheet. https://www.who.int/newsroom/fact-sheets/detail/healthy-diet.
[17] Steven Siceloff. NASA's John F. Kennedy Space Centre recycling of human waste. https: //www.nasa.gov/mission_pages/station/behindscenes/waterrecycler.html.
[18] Sirkka malkki. Human faeces as a resource in agriculture. TTS-Institute, P.O. Box. 13, FIN-05201 Rajamäki, 1999.
[19] Steven R. Oleson Mare A. Gibson. "NASA's Kilopower Reactor Development and the path to higher power missions".
[20] Descanso.jpl.nasa.gov. https://descanso.jpl.nasa.gov/propagation/mars/MarsPub_sec5.pdf.
[21] NASA Ames Research Center Trajectory Browser. https://trajbrowser.arc.nasa.gov/.
[22] Asterank. http://www.asterank.com/.
[23] International Olympic Committee. "Olympic Marketing Fact File". In: (2017).
[24] Pat Troutman. "Revolutionary Concepts for Human Outer Planet Exploration (HOPE)". In: (2003).
[25] Robert Zubrin. "The Case for Colonizing Mars". In: (1996).

# 16: "TEAM SPACESHIP" ENGINEERING REQUIREMENTS FOR A LARGE MARTIAN POPULATION AND THEIR IMPLICATIONS

**R. Mahoney, A. Bryant, M. Hayward, T. Mew, T. Green, J. Simmich**
robert.mahoney24@gmail.com, abryant143@hotmail.com, tmew91@gmail.com, tjg171@uowmail.edu.au

## 1. ABSTRACT

This report summarises the findings of Australia's 'Team Spaceship' from the Mars Society's competition which sought designs for a 1000-person colony on Mars. Using Mars Direct as inspiration, the best way to build a colony on Mars is to travel light and live off the land. With enough power, most modern construction materials can be synthesised using water, air and Martian regolith with only copper being difficult to find, and as a result, copper deposits will need to be specifically located. From these indigenous materials, steel habitats resembling terrestrial gas storage tanks can be constructed with existing automated practices. Inside, colonists will live and work, growing the majority of food necessary in intensive plant factories, while aesthetic secondary crops and spice gardens can be grown in spacious communal domes, adding visual and flavour variety to the local diet. Water can be extracted from the Martian cryosphere with glacier and near-surface ice being preferred sources. The proposed primary power system is a grid of three Sodium Pool Fast Breeder Reactors generating 30 MWe each. This design gives flexibility in the colony's siting, allowing the decision to be driven by important, and possibly exportable, secondary resources rather than survival considerations like geo-thermal wells or lava tubes. Based on the project brief's assumptions, an initial investment of approximately USD $6 billion would be needed. The initial colonisation would likely require government funding with a transition to a public-private partnership for ongoing operation, funded by Mars' main competitive advantage over Earth; asteroid prospecting and mining services. In time, this should provide enough return to self-fund further expansion. Survival and self-sufficiency will be the initial focus, but as the colony develops, the interaction of the settlers with their environment will develop the seeds of a unique culture.

## 2. SCOPE

The project brief laid out by the Mars Society provides several limitations and requirements for the design of the colony itself. These will help develop criteria for selecting different approaches and technologies for the colony to use and will broadly define the economics of the colony. These are summarised in Table 1.

*Table 1: Requirements as per the Mars Society Competition Brief*

| # | Requirement |
|---|---|
| 1 | The colony must support 1000 people on the surface of Mars |
| 2 | Imports to the surface of Mars cost $500 per kilogram |
| 3 | Exports back to Earth will cost $200 per kilogram |
| 4 | The colony must be as self-sufficient and automated as possible |
| 5 | The colony must be as automated as possible |

These requirements provide a basis for the proposed design, particularly from an economic standpoint. In addition to the project brief, there are a number of assumptions that need to be made in order to develop an appropriate basis of design. These are listed below in Table 2.

*Table 2: Assumptions*

| # | Assumption |
|---|---|
| 1 | Technologies used must exist or be in active development |
| 2 | Travel windows to/from Mars are not considered in the Export/Import Costs |
| 3 | There is a human presence with access to power (~10 MW) |

These requirements and assumptions provide appropriate framing to reasonably assess the practical and economic viability of the colony.

## 3. TECHNICAL DESIGN

### 3.1. Production

Data shows that there are abundant resources in the Martian soil and atmosphere. Sources of water and their extraction methods will be discussed further in Section 3.5. Based on the Viking landers and samples taken during the Pathfinder mission, it is reasonable to assume the average composition of the Martian Soil is as per below in Table 3 (1).

*Table 3: Average Martian Soil Composition by Weight Percent*

| | | | |
|---|---|---|---|
| **$SiO_2$** | 46.00 % | **$Na_2O$** | 2.20 % |
| **$Fe_2O_3$** | 17.85 % | **$TiO_2$** | 0.85 % |
| **$Al_2O_3$** | 7.75 % | **Cl** | 0.70 % |
| **MgO** | 6.88 % | **$K_2O$** | 0.23 % |
| **CaO** | 6.05 % | **Trace** | 5.09 % |
| **$SO_3$** | 6.40 % | | |

The above findings show that, the Martian soil is rich in useful metal oxides. These and the soil itself in the form of masonry will then serve as the main building blocks for the colony. The composition of the Martian Atmosphere has been summarised in Table 4 (2).

*Table 4: Martian Atmospheric Composition by Weight Percent*

| $CO_2$ | $N_2$ | Ar | Trace |
|---|---|---|---|
| 95.66 % | 2.30 % | 1.77 % | 0.27 % |

The colony will need two primary methods of gathering non-water resources. Both of these processes will be completely automated. Gathering and separating the main components of the Martian atmosphere is a simple process, summarised in Figure 1. The air will be filtered, compressed then cooled using liquid $CO_2$. The $CO_2$ is cooled by the ambient temperature on Mars, which averages 218 K or -55°C. The energy consumption of the pumps is based on a worst-case scenario for isentropic compression (4). Storage tanks will initially be lightweight composite overwrapped pressure vessels, with further storage being built locally from steel or aluminium.

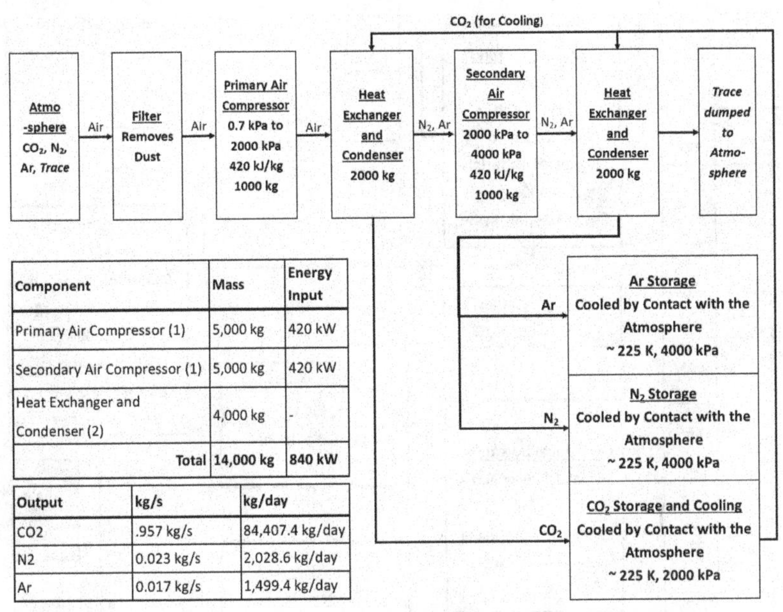


| Component | Mass | Energy Input |
|---|---|---|
| Primary Air Compressor (1) | 5,000 kg | 420 kW |
| Secondary Air Compressor (1) | 5,000 kg | 420 kW |
| Heat Exchanger and Condenser (2) | 4,000 kg | - |
| Total | 14,000 kg | 840 kW |

| Output | kg/s | kg/day |
|---|---|---|
| CO2 | .957 kg/s | 84,407.4 kg/day |
| N2 | 0.023 kg/s | 2,028.6 kg/day |
| Ar | 0.017 kg/s | 1,499.4 kg/day |

*Figure 1: Air Mining Process*

To gather the regolith needed, an automated collection drone based on the NASA RASSOR design (3) will be used. This will be an upsized version, referred to as the RASSOR XX, as described in Table 5.

At least 10 will be needed, with 4 being used at any one time for maximum production, 4 for duty cycling and 2 for backup. With a recharging base station system massing 5,000 kg, the total collection system will weigh 55,000 kg and consume 50 kW of power. Each drone will gather 2,500 kg of regolith per hour, allowing for 10,000 kg of regolith to be harvested per hour. The RASSOR XXs will also be capable of having their rotary buckets removed and replaced with other attachments for construction as necessary.

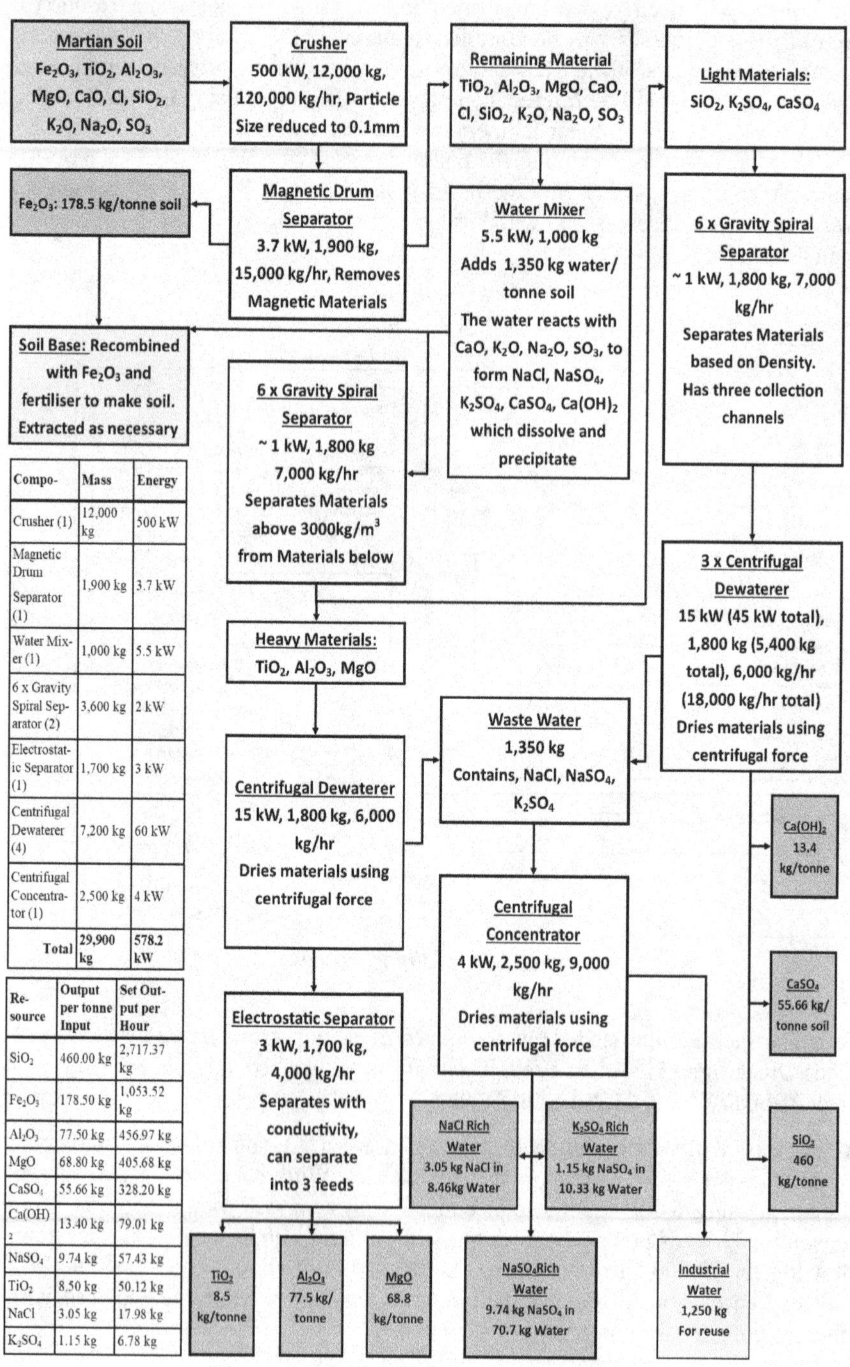


| Compo- | Mass | Energy |
|---|---|---|
| Crusher (1) | 12,000 kg | 500 kW |
| Magnetic Drum Separator (1) | 1,900 kg | 3.7 kW |
| Water Mixer (1) | 1,000 kg | 5.5 kW |
| 6 x Gravity Spiral Separator (2) | 3,600 kg | 2 kW |
| Electrostatic Separator (1) | 1,700 kg | 3 kW |
| Centrifugal Dewaterer (4) | 7,200 kg | 60 kW |
| Centrifugal Concentrator (1) | 2,500 kg | 4 kW |
| Total | 29,900 kg | 578.2 kW |

| Resource | Output per tonne Input | Set Output per Hour |
|---|---|---|
| $SiO_2$ | 460.00 kg | 2,717.37 kg |
| $Fe_2O_3$ | 178.50 kg | 1,053.52 kg |
| $Al_2O_3$ | 77.50 kg | 456.97 kg |
| MgO | 68.80 kg | 405.68 kg |
| $CaSO_4$ | 55.66 kg | 328.20 kg |
| $Ca(OH)_2$ | 13.40 kg | 79.01 kg |
| $NaSO_4$ | 9.74 kg | 57.43 kg |
| $TiO_2$ | 8.50 kg | 50.12 kg |
| NaCl | 3.05 kg | 17.98 kg |
| $K_2SO_4$ | 1.15 kg | 6.78 kg |

Figure 2: Martian Soil Separation Process

*Table 5: Technical Specifications for a RASSOR XX*

| Machine Type | Earth Analogue | Mass | Power | Earth |
|---|---|---|---|---|
| RASSOR XX | RASSOR | 5,000 kg | 10 kW | 2,500 kg/hr |

Once the soil has been collected, it will be crushed and separated using the process outlined in Figure 2. This process relies on existing technology utilised by the mining industry on Earth and has a total mass of 29,900 kg, energy consumption of 578.2 kW and uses 1,350 kg of water per hour. 92.5% of the water is directly recovered, while the remaining 7.5% is used to store salts, bases and acids that are formed during the separation processes and will be used for various industrial and production processes. Due to the use of Gravity Spiral Separators, this system will need to be located in a pressurised environment, however during the base setup phase, parts like the magnetic separator can be located outside. Once separated, the ores will be transferred to the 'Works' by drones for refining and manufacture.

3.1.1. The Manufactorum

The Manufactorum will be made up of five 'Works' that are as follows:

1. *The Alchemy Works:* will combine $CO_2$, water and energy to produce everything from oxygen for the colony, rocket fuel ($CH_4$ and $H_2$) and everyday plastics.
2. *The Steel Works:* will produce all the steel products used by the colony. Steel cannot be produced on Mars by the same method that is widely used on Earth. Instead, it will use the Direct-Reduction Iron (DRI) method, due to the availability of methane (4). Once the iron ore has been refined into DRI, it can be further refined in an Electric Arc Furnace, using calcium carbonate as a flux. The calcium carbonate will be manufactured from the calcium hydroxide separated from the regolith.
3. *The Volt Works:* So called due to its high electricity use, the Volt Works will produce aluminium and magnesium products. Aluminium will be processed using the familiar Hall–Héroult process, while Magnesium will be made using the Magnesium Electrolytic Process with indigenous hydrochloric acid.
4. *The Glass Works:* will be responsible for making glass, laminated glass, and fibreglass. Before manufacture, the $SiO_2$ will be heated with $H_2$ in a cyclonic mixer and then passed through a magnetic separator to remove any remaining $Fe_2O_3$.

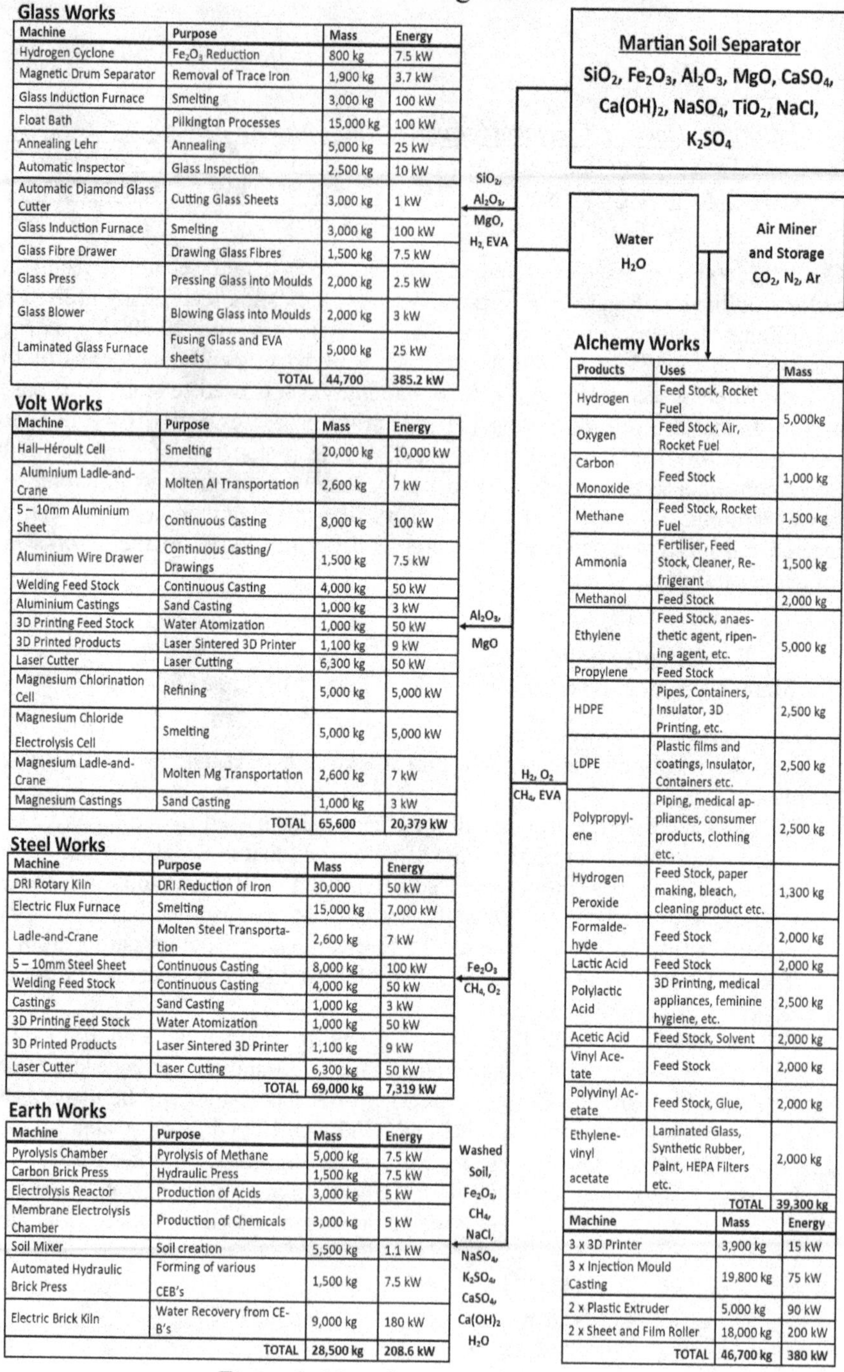

**Glass Works**

| Machine | Purpose | Mass | Energy |
|---|---|---|---|
| Hydrogen Cyclone | $Fe_2O_3$ Reduction | 800 kg | 7.5 kW |
| Magnetic Drum Separator | Removal of Trace Iron | 1,900 kg | 3.7 kW |
| Glass Induction Furnace | Smelting | 3,000 kg | 100 kW |
| Float Bath | Pilkington Processes | 15,000 kg | 100 kW |
| Annealing Lehr | Annealing | 5,000 kg | 25 kW |
| Automatic Inspector | Glass Inspection | 2,500 kg | 10 kW |
| Automatic Diamond Glass Cutter | Cutting Glass Sheets | 3,000 kg | 1 kW |
| Glass Induction Furnace | Smelting | 3,000 kg | 100 kW |
| Glass Fibre Drawer | Drawing Glass Fibres | 1,500 kg | 7.5 kW |
| Glass Press | Pressing Glass into Moulds | 2,000 kg | 2.5 kW |
| Glass Blower | Blowing Glass into Moulds | 2,000 kg | 3 kW |
| Laminated Glass Furnace | Fusing Glass and EVA sheets | 5,000 kg | 25 kW |
| | TOTAL | 44,700 | 385.2 kW |

**Volt Works**

| Machine | Purpose | Mass | Energy |
|---|---|---|---|
| Hall–Héroult Cell | Smelting | 20,000 kg | 10,000 kW |
| Aluminium Ladle-and-Crane | Molten Al Transportation | 2,600 kg | 7 kW |
| 5 – 10mm Aluminium Sheet | Continuous Casting | 8,000 kg | 100 kW |
| Aluminium Wire Drawer | Continuous Casting/ Drawings | 1,500 kg | 7.5 kW |
| Welding Feed Stock | Continuous Casting | 4,000 kg | 50 kW |
| Aluminium Castings | Sand Casting | 1,000 kg | 3 kW |
| 3D Printing Feed Stock | Water Atomization | 1,000 kg | 50 kW |
| 3D Printed Products | Laser Sintered 3D Printer | 1,100 kg | 9 kW |
| Laser Cutter | Laser Cutting | 6,300 kg | 50 kW |
| Magnesium Chlorination Cell | Refining | 5,000 kg | 5,000 kW |
| Magnesium Chloride Electrolysis Cell | Smelting | 5,000 kg | 5,000 kW |
| Magnesium Ladle-and-Crane | Molten Mg Transportation | 2,600 kg | 7 kW |
| Magnesium Castings | Sand Casting | 1,000 kg | 3 kW |
| | TOTAL | 65,600 | 20,379 kW |

**Steel Works**

| Machine | Purpose | Mass | Energy |
|---|---|---|---|
| DRI Rotary Kiln | DRI Reduction of Iron | 30,000 | 50 kW |
| Electric Flux Furnace | Smelting | 15,000 kg | 7,000 kW |
| Ladle-and-Crane | Molten Steel Transportation | 2,600 kg | 7 kW |
| 5 – 10mm Steel Sheet | Continuous Casting | 8,000 kg | 100 kW |
| Welding Feed Stock | Continuous Casting | 4,000 kg | 50 kW |
| Castings | Sand Casting | 1,000 kg | 3 kW |
| 3D Printing Feed Stock | Water Atomization | 1,000 kg | 50 kW |
| 3D Printed Products | Laser Sintered 3D Printer | 1,100 kg | 9 kW |
| Laser Cutter | Laser Cutting | 6,300 kg | 50 kW |
| | TOTAL | 69,000 kg | 7,319 kW |

**Earth Works**

| Machine | Purpose | Mass | Energy |
|---|---|---|---|
| Pyrolysis Chamber | Pyrolysis of Methane | 5,000 kg | 7.5 kW |
| Carbon Brick Press | Hydraulic Press | 1,500 kg | 7.5 kW |
| Electrolysis Reactor | Production of Acids | 3,000 kg | 5 kW |
| Membrane Electrolysis Chamber | Production of Chemicals | 3,000 kg | 5 kW |
| Soil Mixer | Soil creation | 5,500 kg | 1.1 kW |
| Automated Hydraulic Brick Press | Forming of various CEB's | 1,500 kg | 7.5 kW |
| Electric Brick Kiln | Water Recovery from CEB's | 9,000 kg | 180 kW |
| | TOTAL | 28,500 kg | 208.6 kW |

**Alchemy Works**

| Products | Uses | Mass |
|---|---|---|
| Hydrogen | Feed Stock, Rocket Fuel | 5,000kg |
| Oxygen | Feed Stock, Air, Rocket Fuel | |
| Carbon Monoxide | Feed Stock | 1,000 kg |
| Methane | Feed Stock, Rocket Fuel | 1,500 kg |
| Ammonia | Fertiliser, Feed Stock, Cleaner, Refrigerant | 1,500 kg |
| Methanol | Feed Stock | 2,000 kg |
| Ethylene | Feed Stock, anaesthetic agent, ripening agent, etc. | 5,000 kg |
| Propylene | Feed Stock | |
| HDPE | Pipes, Containers, Insulator, 3D Printing, etc. | 2,500 kg |
| LDPE | Plastic films and coatings, Insulator, Containers etc. | 2,500 kg |
| Polypropylene | Piping, medical appliances, consumer products, clothing etc. | 2,500 kg |
| Hydrogen Peroxide | Feed Stock, paper making, bleach, cleaning product etc. | 1,300 kg |
| Formaldehyde | Feed Stock | 2,000 kg |
| Lactic Acid | Feed Stock | 2,000 kg |
| Polylactic Acid | 3D Printing, medical appliances, feminine hygiene, etc. | 2,500 kg |
| Acetic Acid | Feed Stock, Solvent | 2,000 kg |
| Vinyl Acetate | Feed Stock | 2,000 kg |
| Polyvinyl Acetate | Feed Stock, Glue, | 2,000 kg |
| Ethylene-vinyl acetate | Laminated Glass, Synthetic Rubber, Paint, HEPA Filters etc. | 2,000 kg |
| | TOTAL | 39,300 kg |

| Machine | Mass | Energy |
|---|---|---|
| 3 x 3D Printer | 3,900 kg | 15 kW |
| 3 x Injection Mould Casting | 19,800 kg | 75 kW |
| 2 x Plastic Extruder | 5,000 kg | 90 kW |
| 2 x Sheet and Film Roller | 18,000 kg | 200 kW |
| TOTAL | 46,700 kg | 380 kW |



*Figure 3: Manufactorum Flowchart*

5. *The Earth Works:* will be responsible for producing the fertilisers, soils and the different Compressed Earth Bricks (CEBs) that will be used in the construction and operation of the colony. The Earth Works will also be responsible for producing solid carbon via the pyrolysis of methane for use in air filters and industrial electrodes. CEBs are preferred to conventional fired bricks even though they have a lower strength, due to the higher production rate, simpler machinery requirements as well as lower energy and water requirements for their production.

In addition to producing the raw feed stock for steel, plastic, etc., the Works will also be responsible for forming the feed stock into different products, be it metal sheeting, 3D printing feed stock, castings or 3D printed objects themselves.

A summary of the machines and process for the different Works can be found below in Figure 3, as well as their associated import masses and energy requirements. It should be noted that the energy and production requirements for the Alchemy Works has not been detailed, as many of the processes are exothermic, requiring energy to begin, but no energy to run.

### 3.2. Power

The power production for the colony must be considered in four sections. These are: existing power from early base infrastructure, base-load power required for safety and longevity, peak loading ability, and infrastructure to distribute the power. It is estimated that the colony will require 100 MW of sustained generation to remain operational.

#### 3.2.1. Existing Base Utilisation

It is assumed that an existing base will be established before the colony expands to the size of 1,000 people as is detailed in the scope. It is assumed that existing power resources, up to approximately 10 MWe, will be present and operational. A series of RTGs or nuclear fission reactors would be utilized to ensure the longevity of the colony's requirements and size (5). RTGs continue to be researched and prototyped to increase power output (6). The base outlined in this report will use this existing power supply as a backup power in the event of localised system failure or a large-scale cascade effect. It is believed that 10 MWe will be sufficient to sustain air, heating and vital water treatment necessary while essential systems are repaired and restarted.

#### 3.2.2. Base-load Power

The station will look to utilize Small Modular Reactors (SMRs) as part of the base-load power supply. SMRs are being developed as part of a push to replace coal fired stations with easily deployed, modular power plants (7). There are two distinct directions for which SMR technologies can be used in a Mars application. This will depend on supercritical $CO_2$ cycle reactors overcoming the many issues that still need to be solved, not least of which is material cavitation (8). Should supercritical $CO_2$ cycle reactors prove effective in delivering 10-15% efficiency gains, then these would be the logical choice for Mars, given its abundance of $CO_2$ (9). The current proposal is for three 30 MWe Sodium Pool Fast Breeder reactors to provide the base-load power for the base. These were chosen for the following four reasons:

1. Load Shedding & Maintenance: A grid setup of 3 by 30 MWe will allow the system to shed non-critical function in the event of downtime. Planned maintenance can be easily accounted for with this setup.
2. Reduced resupply: The fast breeder reactors were chosen as the fission reaction can be optimised to produce fuel that can subsequently be utilized in the reaction, decreasing the frequency required for re-fuelling missions from Earth.
3. Energy Security: The fast neutron reaction was chosen over molten salt reactors, another on the forefront of research, due to fuel salt requiring transport from Earth along with the fuel. By rendering the need for a supply-chain of salts obsolete for power production, the colony gains improved fuel security.
4. Improved redundancy against atmospheric disruption: The potential duration of dust storms may limit photovoltaic solutions (10) coupled with machinery constraints on tunnelling, and therefore reliable geothermal sources. This means that the three 30 MWe stations are required to maintain resource development and therefore the colony's economy and expansion.

Thermal management will be used to redirect waste heat from the condensation cycle to heating for habitation areas that will act as a heat sink. This is so that the load of the turbine does not affect the temperature in the residential areas. To utilise the waste heat, the geographical proximity of the reactors to the habitats will become a design factor. Additional heating for habitation areas will be supplied by electrical heaters that can fine-tune the temperature of these residential spaces where necessary.

Although ineffective for base loading, photovoltaic power may be considered for fuel saving initiatives. With a comprehensive photovoltaic array, and the robots to clean them, this power can be used to reduce the nuclear fuel required and also support the peak loading facilities described below.

#### 3.2.3. Peak Loading

The Mars Base will require peak loading to supplement the nuclear power provided. A fast response power solution is needed to supplement the relatively slow response of the nuclear generation. Rechargeable batteries are an obvious choice; however, these present a high weight-to-power ratio that must be transported from Earth. As such, gravity based 'Energy Vault' towers (11) are a solution that a Mars Base should consider.

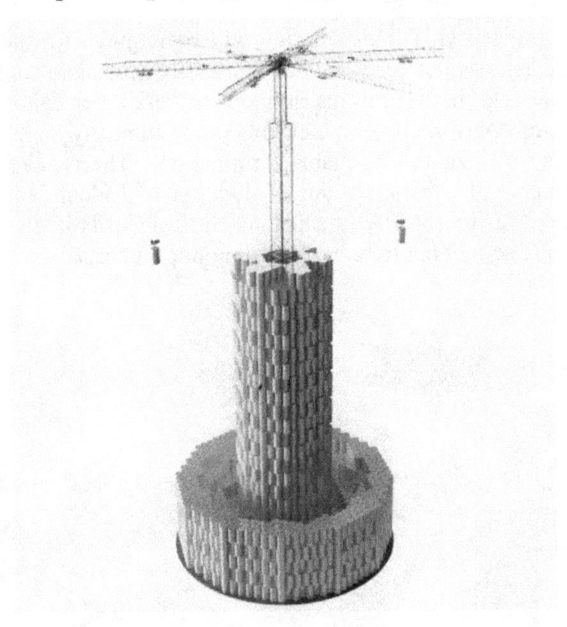

*Figure 4: 'Energy Vault' stacks blocks for peak energy (11)*

The blocks can be produced from waste incurred from the Manufactorum. Most components can be made from iron produced on Mars and only elements such as alternators would need to be sent from Earth. This would give the base the delta energy needed to restart equipment when required. Several of these towers could be required to support the system in the event of nuclear plant trips, however, as these expansions can mostly be constructed with resources on Mars it should be relatively easy to expand as the base grows.

#### 3.2.4. Infrastructure

Electrical infrastructure is one of the biggest challenges that a Mars base will face. Power reliability and no readily available electrical transmission mediums are two major discussion points. Power reliability is a major factor in any islanded power supply model, and presents a unique risk in an artificial environment.

A carefully designed grid system that allows power fluctuation response will be imperative for the colony's success. A grid system will incorporate a "contingency based and backup frequency-based load-shedding system to compensate for large generation and transmission disturbances" (12). This will need to be heavily prioritized for the power grid in the colony to keep critical systems available. Smart electrical protection devices will be required, with quick response to frequency disturbances. Due to their complex circuitry, these will need to be transported from Earth.

Distribution is another challenge that the colony will face. A modular solution is proposed for switch rooms. These should be used for the initial phase of the project where modules of various voltages (1.5t to 16t, 415 V to 66 kV) can be dropped-in and utilized (13). This solution will be required for the initial life of the colony as Mars does not currently have a readily accessible supply of copper. Aluminium cables will be used in distribution networks. For colony expansion it makes sense to transport in alloying elements (approximately 3 wt% of the cable) and copper lugs (14), for more reliable terminations. The lack of defined local sources of copper will necessitate the exploration and identification of copper deposits as soon as is reasonably practicable. Such deposits likely exist, but will need to be actively sought out to reduce ongoing import costs.

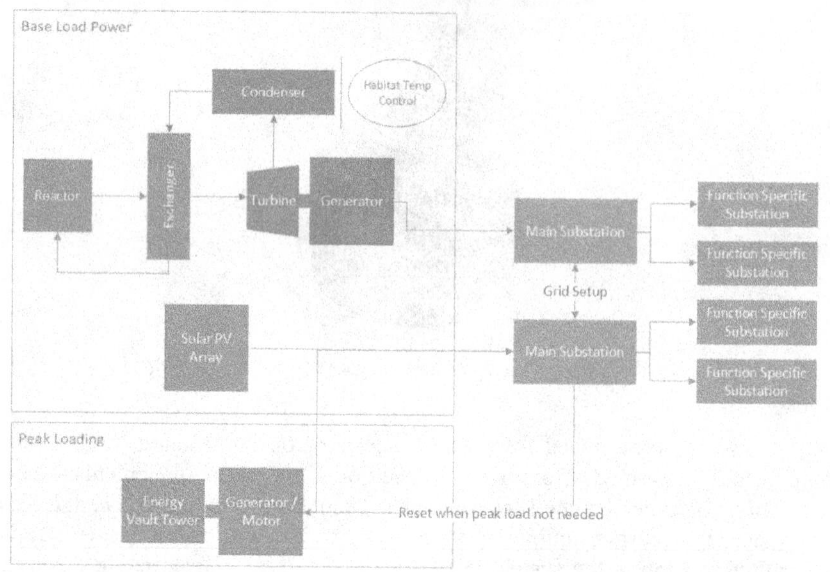


Figure 5: Flow chart of power system on Mars base

*Table 6: Approximate weights of components required for initial transportation.*

| Component | Weight |
|---|---|
| Nuclear Reactor | 100,000 kg |
| Water | 80,000 kg |
| Pipework/Vessels | 450,000 kg |
| Control Componentry | 120,000 kg |
| Main Substation | 16,000 kg |
| Minor Substations 1 | 5 x 1,500 kg |
| Minor Substations 1 | 2 x 8,000 kg |
| Total for Power | 789,500 kg |
| Initial Cable | 74 meters /1000 kg transported |
| Alloying elements after Al Available | 2456 meters /1000 kg transported |

3.2.5. Setup

The modularised nature of the power station and equipment means that the setup process will be very straightforward, with the components of the reactor being landed where it will be built.

The turbine will be able to be sent in segments from Earth and constructed on-site at Mars. Each segment will not exceed 20 tonnes. Once the system is sealed, the water and salt can be added, cabling will be laid from the generator to the main substation. As cabling will be shipped complete or manufactured on-site, connecting additional substations will be a simple process.

### 3.3. Habitats

3.3.1. Living Requirements

Designing a structure that can reliably contain a full atmospheric load is a difficult proposition. To this end, the atmospheric conditions inside the colony will be different to those on Earth, being at a lower pressure, but with a higher oxygen content. This will allow the occupants to still live in a pressurised environment, but allow for simpler habitat designs.

It has been demonstrated that flammability is chiefly governed by oxygen content, not pressure (15). To this end, the internal atmosphere will be 50 kPa and 30% oxygen content. This will provide a liveable, if thin, atmosphere but greatly reduces the stresses that the habitats have to withstand, while still keeping the internal flammability levels low. The internal temperature of the habitats will be kept at approximately 10-15°C in order to reduce heating requirements.

3.3.2. Habitat Design

Using the available materials outlined in Section 3.1, the most available construction materials are brick, glass and steel. It has been suggested that burying brick vaults on Mars would be the best way to make large dwellings, however this does not consider the excavation requirements. Such a vault would need to be buried at least 9 meters deep in order for the weight of the regolith, 1,520 kg/m$^3$ (16), to counteract the internal pressure of 50 kPa. This also ignores the depth of the vault itself, adding up to another 6 meters of depth for even a cramped dwelling. 15 meters of regolith would need to be moved, processed and then replaced.

Given the large quantities of steel that can be produced, a pressure vessel design was chosen. This design is based on gas storage tanks and is a steel silo 40 meters in diameter and 25 meters high with a domed roof made of either steel or geodesic laminated glass. All steel construction is 8 mm thick sheeting and all glass panels are 25 mm thick. A second floor is included to provide a recreational level with access to broad views if built with a glass dome, or large, vaulted ceilings if built with a steel dome. The second floor is reinforced with beams and rated to carry a mass of wet soil one meter thick to accommodate a possible park or farm. Both floors will have an area of 1256.64 m$^2$. The dome and second floor will be supported on brick columns and the tank side walls, with the side walls being welded onto a ground level steel sheet. This sheet is the first floor and extends under external brick walls that provide anchoring, radiation and physical protection. Below this base plate will be the foundation slab of Magnesium-OxyChloride or Sulphur-based 'Marscrete', 30 mm thick. The regolith under this will be levelled and compacted before construction, but the layer of 'Marscrete' will provide insulation and minimise settling differentials. Sacrificial anodes made of magnesium or aluminium will provide corrosion protection. This design will resemble vaulted structures common to classical architecture periods.

Once the construction of the outer shell is completed, airlocks and connectors can be installed and the habitat pressurised with oxygen produced from electrolysis or other industrial processes and nitrogen extracted from the atmosphere. This allows the second floor and up to four, three-story brick apartments to be built in a pressurised environment. Each three-story apartment will house occupants in a 25 $m^2$ studio style dwelling with 12 people per apartment, 48 per habitat. The total steel used is ~500,000 kg, with 480,000 kg being used for the bulk structure. In order to house the entire, 1000-person populace, 21 Habitats will need to be constructed, with and estimated other 6 for additional recreational or business uses. Each silo will lose ~270 kW of heat through radiative and conductive losses. This will be countered by solar radiation, waste heat from lighting and electronics, and providing heated water. Finite Element Analysis using FEMAP shows that the peak Von Mises stress seen by the silo is 125 MPa, which is below both the yield stress and fatigue limit of steel. A number of key sub-components require consideration and are discussed in Table 7.

*Figure 6: Gasometers in Vienna. An example of the proposed habitat design*

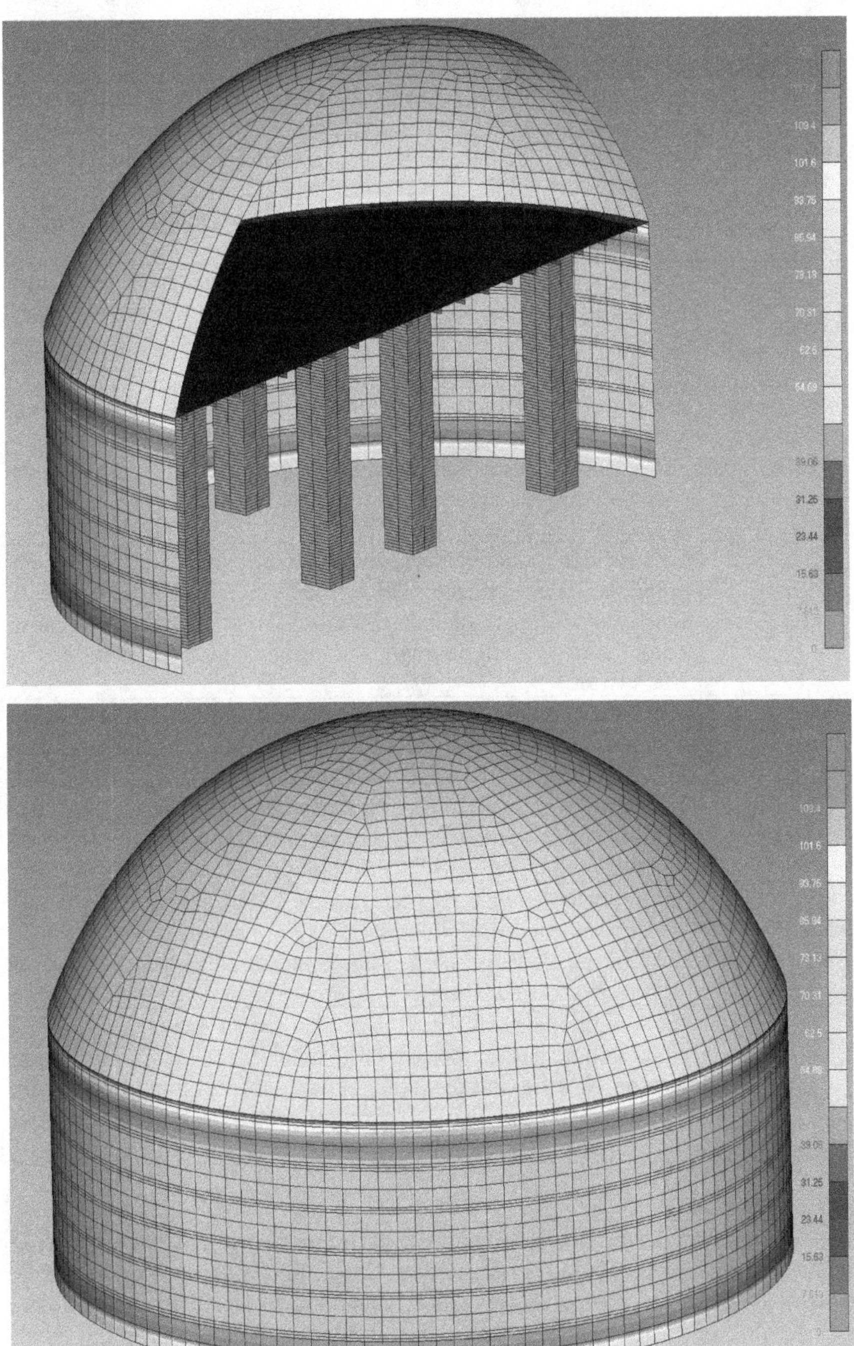


*Figure 7: Habitat FEA Analysis – Plate Top Von Mises Stress*

*Table 7: Additional Habitat Components*

| Component | Design Specifics |
|---|---|
| **Habitat Connection** | Above ground tubes 5 meters diameter and with a 3-meter-wide 'street' made from 8 mm steel sheet. Electricity cables, water pipes and air ducting will be suspended from the ceiling for easy access. Brick retaining walls can be built around the outside of the pipe and then waste material, such as ground slag, can be used to provide protection and insulation. |
| **Airlock** | Initial airlocks will be brought from Earth. Once the Steel Works is established, they will be constructed locally. These will vary in size from personnel airlocks to industrial airlocks the same size as the Habitat Connectors |
| **Dustlock** | Same design template as the airlocks, but will use high pressure $CO_2$ jets for dust suppression of personnel or equipment entering unpressurised areas that need to be protected from particulate material, such as storage silos, or before proceeding to an airlock to reduce dust entering the Habitats. |
| **Comms** | Communication equipment will be brought from Earth, although some heavy components will be made locally. There will be at least two long range communication hubs, but most communication will be small, and designed to access satellites in Mars Orbit. |
| **Computers** | All habitats will have Wi-Fi base stations and servers, connected to the other habitats. A local internet will allow access to stored information, local social media analogues and mirrors of frequented websites from Earth that will be periodically updated. Waste heat from the servers will be used to provide a portion of the heating requirements for the habitat. |
| **Fire and Safety** | All habitats will be equipped with fire control systems including mist sprinklers. All buildings inside the habitats will have their own independent fire suppression systems, as well as emergency oxygen masks and personal protective equipment. |
| **Air Conditioners** | Air conditioners will be in every habitat to recycle and purify air. It will be passed through HEPA and Activated Carbon filters to remove solid and gas pollutants. These air conditioners will then condense any excess water out of the air for recovery and use, minimising system losses. |

### 3.3.3. Construction Process

Once a construction site is selected, it will be prepared using a RASSOR XX equipped with Levellers and Compactors. Once this is done, the concrete foundation will be 3D printed. At this point, human labour is needed to layout the steel sheeting of the habitat base. Once this is done, a Continuous Spiral Steel Tank can be constructed. This will be done by building the first meter of the tank, constructing the domed roof and the frame work for the second level, and then 'growing' the habitat to its full height of approximately 25 meters.

Once completed, connectors, dustlocks, airlocks and air conditioners will be installed, with the habitat then being pressurised. Internal construction can continue in a pressurised environment, building the internal brick supports and completing the second floor while the outside brick walls are robotically constructed. Following this, the internals of the habitat will be constructed. It should be noted that all brick construction will be done in an automated fashion using 3D 'printing' or laying of interlocking, mortar-less CEBs. These machines will use the technology currently in development by Fast Brick Robotics for their Hadrian X robot. This process allows for the bulk of the construction to be automated or in a normal environment.

*Table 8: Habitat Construction Equipment*

| **Machine** | **Mass** | **Energy** |
|---|---|---|
| Leveller (Attachment for a RASSOR XX) | 1,000 kg | 1 kW |
| Compactor (Attachment for a RASSOR XX) | 1,000 kg | 1 kW |
| Marscrete Foundation Printer (Construction 3D) | 2,600 kg | 15 kW |
| Decoiler | 1,000 kg | 1 kW |
| Bender/Corrugator | 1,000 kg | 1 kW |
| Automated Welder | 1,000 kg | 1 kW |
| Adjustable Track Roller | 2,500 kg | - |
| Mobile Crane | 38,000 kg | 200 kW |
| Tower Crane | 48,000 kg | 300 kW |
| Auto-Brick Layer (Attachment for a RASSOR XX) | 1,500 kg | 7 kW |
| Transport Rack (Attachment for a RASSOR XX) | 500 kg | - |
| Scaffolding | 1,000 kg | - |
| Elevated Work Platform | 21,000 kg | 55 kW |
| **Total** | **120,100 kg** | **582 kW** |

## 3.4. Food & Agriculture

### 3.4.1. Diet

Developing and maintaining a self-sufficient food supply on Mars, is of fundamental importance to the colony. In order to sustain a population of people for the long-term, food must provide both energy and essential nutrients. Failure to provide sufficient energy will cause starvation and insufficient nutrients can cause a variety of nutritional deficiencies and diseases. These essential nutrients are listed in Table 9.

*Table 9: Essential Nutrients, Deficiencies and Sources*

| Nutrient | Deficiency Disease | Common Sources |
|---|---|---|
| | **Vitamins** | |
| **Vitamin A** | Hypovitaminosis A | Cod liver Oil, Liver, Kumara (Sweet Potato), Carrot |
| **Vitamin B1** | Thiamine Deficiency, Beri-beri | Whole grains, Seeds, Legumes, Pork |
| **Vitamin B2** | Riboflavin Deficiency, Ariboflavinosis | Eggs, Greens, Milk, Meat |
| **Vitamin B3** | Pellagra | Tuna, Meat, Capsicum, Mushrooms |
| **Vitamin B5** | Pantothenic Acid Deficiency | Egg yolk, Liver, Mushrooms |
| **Vitamin B6** | Pyridoxine Deficiency | Fish, Beef Organs, Starchy Vegetables, Fruit |
| **Vitamin B7** | Biotin Deficiency | Liver, Eggs, Yeast, Peanuts |
| **Vitamin B9** | Folate Deficiency | Yeast Extract, Legumes, Leafy Greens |
| **Vitamin B12** | Vitamin B12 Deficiency | Meat, Milk, Eggs, Fish |
| **Vitamin C** | Scurvy | Kakadu Plum, Blackcurrant, Capsicum, Kale |
| **Vitamin D** | Rickets, Osteomalacia | Sunlight, Mushrooms (UVB irradiated), Liver, Fish |
| **Vitamin E** | Vitamin E Deficiency | Wheat germ, Hazelnuts, Pine nuts |
| **Vitamin K** | Vitamin K Deficiency | Leafy Greens, Turnips, Eggs, Dairy |
| | **Fatty Acids** | |
| **Alpha Linolenic Acid** | Essential Fatty Acid Deficiency | Seeds, Nuts, Vegetable oils |
| **Linoleic Acid** | Essential Fatty Acid Deficiency | Nuts, Seeds |
| | **Minerals** | |
| **Calcium, Ca** | Hypocalcemia | Dairy, Leafy Greens, Soy |
| **Chloride, Cl** | Hypochloremia | Salt |
| **Chromium, Cr** | Chromium Deficiency | Broccoli, Potatoes, Whole grains |
| **Cobalt, Co** | Vitamin B12 Deficiency | Meat, Milk, Eggs, Fish |
| **Copper, Cu** | Copper Deficiency | Molluscs, Liver, Nuts |
| **Iodine, I** | Iodine Deficiency | Seaweed, Seafood, Dairy |
| **Iron, Fe** | Iron Deficiency Anemia | Beans, Eggs, Meat |
| **Magnesium, Mg** | Hypomagnesemia | Spices, Nuts, Cereals |
| **Manganese, Mn** | Manganese Deficiency | Nuts, Legumes, Seeds |

*Table 9: Essential Nutrients, Deficiencies and Sources*

| Nutrient | Deficiency Disease | Common Sources |
|---|---|---|
| **Molybdenum, Mo** | Molybdenum Deficiency | Legumes, Beans, Nuts |
| **Phosphorus, P** | Phosphorus Deficiency | Dairy, Eggs, Legumes |
| **Potassium, K** | Hypokalemia | Fruit, Vegetables |
| **Selenium, Se** | Selenium Deficiency | Nuts, Cereals, Mushrooms |
| **Sodium, Na** | Hyponatremia | Salt |
| **Zinc, Zn** | Zinc Deficiency | Meat, Seafood, Wheat germ |

Supplying nutrients commonly found in meat or other animal products on Mars poses difficulties for both growing and maintaining a stable breeding population of livestock. It may be possible to import some types of fish or insects to the colony, but such importation would only be feasible once the colony has been established, and excess food has been stockpiled as animal feed. For these reasons, a vegan diet is considered more achievable and will be considered for the remainder of this section. Vegan diets require awareness of the nutrient intakes that would usually be supplied by meat and other animal products and Mars itself will also cause changes to the nutritional requirements of the colonists. Both of these changes have been listed in Table 10.

*Table 10: Vegan and Martian Nutrients of Concern*

<table>
<tr><th>Diet</th><th colspan="6">Naturally Deficient Vitamin</th></tr>
<tr><td>Vegan</td><td>Vitamin B2</td><td>Vitamin B12</td><td>Iron</td><td rowspan="2">Vitamin D</td><td rowspan="2">Calcium</td><td rowspan="2">Iodine</td></tr>
<tr><td>Martian</td><td>-</td><td>-</td><td>-</td></tr>
</table>

It can be seen that the loss of nutrients from living on Mars are already covered in vegan diets. Essential nutrients and vitamins, are by definition substances that an organism cannot synthesize. This is not the case for Vitamin D, which humans can synthesize when exposed to ultra-violet B (UVB) radiation. However, UVB is blocked by glass and exiting the colony will require protective suits, therefore humans on Mars will not be exposed to UVB. Vitamin D is also found in food, but many of the common sources are animal-based or artificially fortified, excluding some fungi species. Fungi require UVB to synthesize Vitamin D2. An indoor UVB source will be required for either direct synthesis or irradiation of common mushroom varieties. Vegans also have lower calcium absorption, as plant derived calcium is less bio-available than animal derived calcium. This will be compounded on Mars by the lower gravity causing a reduction of bone mineralisation. This is currently experienced by astronauts, but will likely be less severe in the 0.38g gravity on Mars. No studies of sustained gravity between 0 and 1g have been performed, so early Martian visitors and colonists will need to be studied to ascertain the effects. It may be possible to get adequate levels of calcium from plant foods as grown, but if not the most practical options would be either direct supplementation or fortifying food with extra calcium. Calcium is an element and cannot be spontaneously created. As such, all plants will be drawing their calcium from their support media, so it is vital that the plants themselves get sufficient calcium. Fortunately, calcium minerals have been identified in the Martian regolith and can be extracted, as shown in Section 3.1, so indigenous sources of Calcium are a viable option.

Seafood is the most common dietary source of iodine. To bypass the capital works required to develop an aquatic facility to grow and harvest seafood for iodine, it is likely that food will need to be fortified to fulfil the iodine requirements. There is evidence of Martian iodine, which could be used for fortification, but due to the small amount needed, it may be more practical to import it. Vitamin B2 deficiency shouldn't become a problem provided that colonists eat whole grains and legumes. Vitamin B12 will be a problem if it is not planned for. Vitamin B12 is not synthesized by any plants or fungi, but only by some bacteria and archaea. The simplest option to ensure sufficient Vitamin B12 intake, would be to ferment these bacteria in a bio-reactor and take it as a supplement. Iron deficiency in vegans is mainly a result of the lower bio-availability of the non-heme iron found in plants. This can be countered by generally increasing intake, as well as combining iron rich foods with those containing Vitamin C, which aids in absorption. Iron is abundant on Mars, so while it may need to be processed to aid its uptake into the plants, it will not be hard to find.

### 3.4.2. Agricultural Techniques

Indoor volume, power and water are the primary design constraints for agriculture on Mars. Crops must be grown indoors to ensure they stay within their optimal growing temperature range, as well as to keep pressure at a viable level. However, plants do not need to be kept at Earth normal pressure with research (17) (18) showing that some plants are insensitive to total pressure down to 10 kPa provided that they are not $CO_2$ limited. However, for ease of construction, maintenance and operation, it is recommended that the agricultural areas are kept at the same pressure and atmospheric composition as the rest of the habitat. To determine the appropriate volume, power and water requirements, it is necessary to determine food requirements. Using recommended dietary data from the Australian Government (19), average energy and food mass requirements for men and women aged between 19 and 50 were calculated. It should be noted that due to restricting the colonist's diet to a vegan one; meat, poultry, eggs and fish are assumed to be replaced by legumes, seeds and beans; while dairy is assumed to be replaced by mushrooms. The total daily energy intake of the 1000-person colony (assumed 500 men and 500 women) is 7,825 MJ/day, with a daily food mass consumption of 1797.5 kg. To determine the growing area required, a safety margin of 30% and annual crop yield data from the United Nations (20) were used, resulting in a required crop area of ~800,000 $m^2$. This is based on the assumption of open field cultivation, and is highly conservative. Highly intensive agriculture in plant factories can increase per area yields by up to 75 times (21), depending on species. Taking a more conservative increase of 10 times, this gives a required crop area of 80,000 $m^2$, or 80 $m^2$/person. This is similar to the per person estimates done by the Mars Homestead Project (22), which specified 880 $m^2$ for 12 people or 73 $m^2$/person, but somewhat more than the 50 $m^2$ that NASA (23) calculated for meeting dietary energy requirements. These plant factories use a stacked hydroponic setup with fully artificial lighting and circulating nutrient solution as well as climate and humidity control, a set up that can be replicated on Mars.

The primary consumer of power for agriculture is lighting. Photosynthetic Photon Flux (PPF) requirements are species dependent, as are the photoperiods (light/dark cycle times), both of these impact the agricultural power requirements. Using the environmental set point data from the NASA's Kennedy Space Centre (24) for a variety of staple and supplemental crops, weighted average PPF requirements over time can be estimated to be 342 $\mu mol\ m^{-2}\ s^{-1}$ (3:1 staple to supplement ratio was used, with equal weightings within the categories). Light Emitting Diodes (LEDs) are the most efficient and long-lived light sources, while having the additional benefit of being able to focus their output into the most Photosynthetically Active Radiation (PAR) bands. The conversion efficiency of electrical power to PAR photons is about 1.7 μmol/J (25), which gives an average power per area as 201 $W/m^2$. This is crop area, not floor area, so the floor area previously calculated needs to be multiplied by the number of shelves, which is assumed to be three levels of 80,000 $m^2$, giving a lighting power consumption of 48.3 MW. This is almost half the base-load power generation capacity. Given the more variable loads of the other power consumers, it would make sense for the agricultural sector to act as a load leveller – dynamically changing its power consumption to keep the power plant running at its optimal efficiency – as providing excess power to the plants will improve their growth, until they become limited by another factor (26). Estimations for the mass of the LEDs vary greatly depending on the type. If based off a 2.32 μmol/s LED weighing 29 mg (27), masses as low as 1,026 kg will suffice for the planned crop area, though this includes none of the supporting equipment so this is an absolute lower bound. A 100 W LED with heat sink and built-in fan cooling weighing 360 g (28), by contrast gives a mass of 174,000 kg. This does not include all support equipment, but is not mass optimised so can be treated as an upper bound. The actual mass will be somewhere between these, but more than likely closer to the upper bound. Similarly, to power, the primary consumer of water will be agriculture. Irrigation requirements are species dependent and specific figures should be calculated based on the final composition of crops chosen. However, overall consumption was estimated by the method given in the United Nations Irrigation Water Management Training Manual (29) for the plants (mean daily temperature range of 15-25°C, humid atmosphere, crop influence estimated by sorting with similar crops), while through-life mushroom water consumption was taken as 28 times harvested mass (30). Combined these give a total daily water consumption of 2991 $m^3$ of water. This is a huge quantity of water, however the volume of water that leaves the plant by transpiration or guttation is estimated at 97% (31). The water lost by guttation, will evaporate and join the water vapour from the transpiration in the air. This can be condensed and recycled. Thus, effectively reducing the water consumption to ~90 $m^3$. It is also worth noting that almost all the water that is bound up in plants will eventually be recycled either by consumption or decomposition.

There will need to be some degree of food storage. While some crops will be harvested regularly, thereby ensuring a fresh supply, other crops will be more practical to harvest en-masse then store. This combined with the food surplus safety margin, means that there will need to be a substantial food storage capacity for the colony. Luckily, due to the absence of pests and pathogens, it will be relatively simple to store foods on Mars. It would be quite practical to have a silo with reduced climate control that acts as a mass cold store, possibly internally divided up by temperature into storage zones. Once a safety surplus has been reached, it is likely that other uses for food could be adopted, such as fermenting and distilling. For example, alcohol can be produced for cleaning, as a solvent, antiseptic and chemical feedstock.

#### 3.4.3. Supplemental and Ornamental Agriculture

Food crops will take precedence over other agriculture, however ornamental plants may also be grown. Plants such as poppies and belladonna could be primarily aesthetic, while providing a local source of the medicine's morphine, codeine and atropine. Similarly, spice plants such as mustard, pepper, fennel and saffron could be grown for beauty, but also add flavour to food. Flax or hemp could be grown for their fibres, allowing indigenous fabric production, but also supplementing the diet with their seeds and oils. Cellophane and Rayon can be made using the viscose process using non-edible plant matter (specifically the cellulose), with basic chemicals that would be easily manufacturable. This process is currently used for bamboo cellulose. Bamboo is also highly recommended as supplemental crop, due to its wide range of traditional and modern uses.

### 3.5. Water

Water is a critical requirement for long-term sustainability. While there has been significant research into the collection of water on Mars, few have considered the demand scale of a 1000-person colony. While short-term missions and small outposts of 12 people can be significantly rationed, such exceptional rationing is likely to be unfeasible in a larger colony. For a typical human, the minimum water use per day to maintain basic requirements for drinking, sanitation, bathing and food preparation is 50 L/person/day (32). European levels of water consumption vary, but have a minimum level of 120 L/person/day, which is above the level of basic subsistence and allows a level of freedom that would be required to attract citizens to the colony. By assuming this level of consumption per Equivalent Person (EP), a conservative estimate of civic water demands can be estimating using typical planning values, and is shown in Table 11.

*Table 11: Civic Water Demand*

| Demand Type | Demand | Unit | EP/Unit | Total EP | Water Use $m^3$ |
|---|---|---|---|---|---|
| **1 Bedroom Dwelling** | 500 | Each | 1.00 | 500.00 | 70.00 |
| **2 Bedroom Dwelling** | 250 | Each | 2.00 | 500.00 | 70.00 |
| **Cinema** | 400 | /100m$^2$ | 5.46 | 21.84 | 3.06 |
| **Hospital** | 100 | /bed | 1.37 | 1.37 | 0.19 |
| **Gym** | 400 | /100m$^2$ | 5.46 | 21.84 | 3.06 |
| **Manufacturing** | 1000 | /100m$^2$ | 0.98 | 9.80 | 2.94 |
| **Retail Area** | 400 | /100m$^2$ | 1.64 | 6.56 | 1.97 |
| **Rover Repairs** | 1000 | /100m$^2$ | 0.98 | 9.80 | 2.94 |
| **Meeting** | 200 | /100m$^2$ | 4.37 | 8.74 | 1.22 |
| **Office** | 400 | /100m$^2$ | 1.64 | 6.56 | 0.92 |
| **Restaurant** | 200 | /100m$^2$ | 5.11 | 10.22 | 1.43 |
| **Tavern/Bar etc.** | 200 | /100m$^2$ | 7.70 | 15.40 | 2.16 |
| **Warehouse** | 10000 | /100m$^2$ | 0.98 | 98.00 | 13.72 |
| | | | **Total** | **1,210.13** | **174.00** |

174 meters cubed per day (m$^3$/d) of water is required for Civic use. As per Sections 3.1 and 0, the water demand for Industrial and Agricultural use is 20 m$^3$/d and 3000 m$^3$/d. It has been assumed that 97% of the water requirements can be successfully recycled. This is an achievable recovery rate, as lost water will be either recovered by recondensation or is captured by the waste system.

The typical causes for leakage in municipal water and sewage reticulation are ground subsidence, construction damage and root infiltration. As the network will be laid in a pressurised steel environment, none of these risk factors will be present. Table 12 summarises the total demand and predicted recovery rates for the systems. Refer to Section 3.3.2 for the recovery of losses to the local habitat atmosphere. Water consumed by the Manufactorum will not be returned to the water system in order to minimise potential heavy metal contamination. As this load is small, this will not be a significant drain on resources.

*Table 12: Water Demand and Recovery*

| | Need (m$^3$/d) | Immediate Recovery (m$^3$/d) | Process Losses (m$^3$/d) | Net Intake (m$^3$/d) |
|---|---|---|---|---|
| **Civic** | 174 | 168.78 | 43.5 | 48.72 |
| **Agricultural** | 3000 | 2910 | -48.5 | 46.5 |
| **Industrial** | 20 | 0 | 0 | 20 |
| | | | **Total Intake** | **115.22** |

The net intake for the colony is 115 m$^3$/d. This flow rate should be sourced prior to demands being developed in order to provide a sufficient reservoir of water.

3.5.1. Water Mining on Mars

Numerous studies have investigated the possibility of water on Mars. In general, there are four targets for water collection. These are the Martian regolith, atmosphere, ground water, and ice. While the regolith and atmosphere are targets for short-term expeditions, they can be easily discarded as potential water sources for a large-scale settlement. The regolith is only 3-5% water by weight (33), and thus it would be required to process 1,533 $m^3$ of regolith to supply 1 day of water for the colony, assuming full recovery. The atmosphere has a water content of 0.03% by volume (34), and would require 3,840 $m^3$ to be processed per day. Both of these options are unfeasible to scale-up to the required demands, and will not be considered further. The water sources that potentially have the capacity to meet the scale of the colony are large ice deposits or drilling for the Martian hydrosphere.

The Martian hydrosphere has not yet been confirmed to exist, though modelling based on geological evidence has indicated a hydrosphere of hundreds of meters to kilometres potentially could exist (35). Ground Penetrating Radar (GPR) scans conducted by MARSIS and SHARAD have shown that the hydrosphere is unlikely to be within 300 m of the surface (36), however detection deeper than this is currently not possible without a stronger scan. While most of this is locked beneath the cryosphere, previous thermowell activity could have reduced the depth of the cryosphere to potential distance of 300-1000 m (35). This is well within the range of terrestrial drilling techniques. Additional missions to Mars will be required to locate and confirm the potential water sites, with stronger GPR satellites to augment the MARSIS and SHARAD data, followed by targeted seismic probe landers to confirm water content and volume. Once located, the water is well within terrestrial drilling technological capacity. Liquid $CO_2$ would be an adequate substitute for water for slim-hole drilling on Mars (35). The advantage of this system is that energy would not be required to melt ice. If the hydrosphere is not an artesian basin, the only energy cost would be pumping the water and operating a Reverse Osmosis (RO) unit to treat the water. Once in place the well would not be required to be removed or replaced unless the local water table drops. An estimated 40,000 kg would be required to be shipped to Mars to install the well, and an estimated 500 kW would be required to operate the extraction pump and RO unit.

Ice is a relatively common phenomena on the Martian surface, and potentially significant deposits have already been located from current observation. Significant potential deposits with a depth of hundreds of metres in Deuteronilus Mensae have been detected by SHARAD (37), with the potential for many more sites in the mid-latitudes of Mars. The proposed intake volume of water per day is not significant and could be managed with physical mining by using a Rodriguez well or 'Rodwell'. Physical mining would require a cutter tool for the RASSOR XX, and the target glacier would have to be cleared from surface debris. This could potentially require a large pit in order to gain access to sufficiently large volumes of ice. Approximately 45,000 $m^3$/year of ice will be required by the colony. This is feasible, but eventually the distance from the colony may pose logistical issues in the transportation of ice. Another extraction possibility would be a 'Rodwell'. The Rodwell can be formed by drilling or exposing the glacier then using a combination pump and heater to melt and extract the water from the glacier. Rodwell's are limited by extraction size before the well collapses. Piping hubs could be set-up with flexible tubing to link various Rodwell locations together as each is exploited and abandoned. While this is a proven technology, the scale of Rodwell used in terrestrial operations are significantly smaller than the proposed colony requirements. Testing can be conducted on Earth in Antarctic or Artic conditions to verify the most efficient design. The water can then be pumped out, with minimal manual handling. A significant proportion of the required piping could be manufactured on Mars. Due to the material requirements, it is likely that the down well pump, heater and casing and bit would be imported. This is estimated to be a small mass of ~1t. A continuous power draw of ~500 kW would be required to melt 115 $m^3$ of ice per day.

### 3.5.2. Water Storage

It has been assumed that a storage volume of 10 days for both potable and agricultural water will be required in case of potential mechanical failure and scheduled maintenance. In the advent of a significant failure and water cannot be mined further, agricultural water can be rationed to provide a sufficient volume of water for survival of the colonists to source new mechanical equipment or facilitate evacuation over an extended period. The total storage volume is 31,760 $m^3$, and can be stored in approximately 25 silos completely dedicated to water storage. The storage tanks can be lined with HDPE to minimise corrosion and contamination. The most significant challenge of storing water on Mars will be the loss of heat from the reservoirs, as maintaining the temperature in the silos would require 6.75 MW of heat. This storage volume could be reduced if a higher risk tolerance for water was adopted.

### 3.5.3. Wastewater Treatment and Recycling

The treatment processes of the colony's water and wastewater will be set up into two main streams. The first stream is the treatment of the agricultural water into drinking water. This can be achieved with a combination of Microfiltration and Reverse Osmosis (RO). Microfiltration is required to minimise the load on the RO units and prolong the lifetime of the membranes. RO processes do have a significant rejection rate, with 50-75% being typical. However, the concentration of the reject water would be approximately in-line with the sewage load for the Sewage treatment plant, and can be retreated through that process with no loss of water from the system.

As RO units are critical infrastructure, and made of high-grade materials, they will have to be imported. The advantage of using high-grade materials will be a longer operational life, and higher efficiency. These units will be sized to handle the treatment of 200 $m^3$/d, which is a relatively small flow rate for this technology. In addition, the lack of exposure to outside contamination means the system will be free of contaminants and the membranes will not be overloaded. The second treatment system is a typical Biological Nutrient Removal (BNR) plant. This plant includes a preliminary screening and de-gritting unit that removes major solids and grit from the waste stream. Manual bar screens will be sufficient for the flow, and a Vortex de-gritter can be used for the grit. The BNR unit will be small with a preliminary clarification tank, a biological treatment zone and final clarification tank. These units are capable of treating most civic loads. The initial clarification step eliminates suspended solids and fractional oils, with the biological zone removing the Chemical Oxygen demand, Nitrogen and Phosphorous. The process absorbs oxygen and emits Nitrogen and $CO_2$.

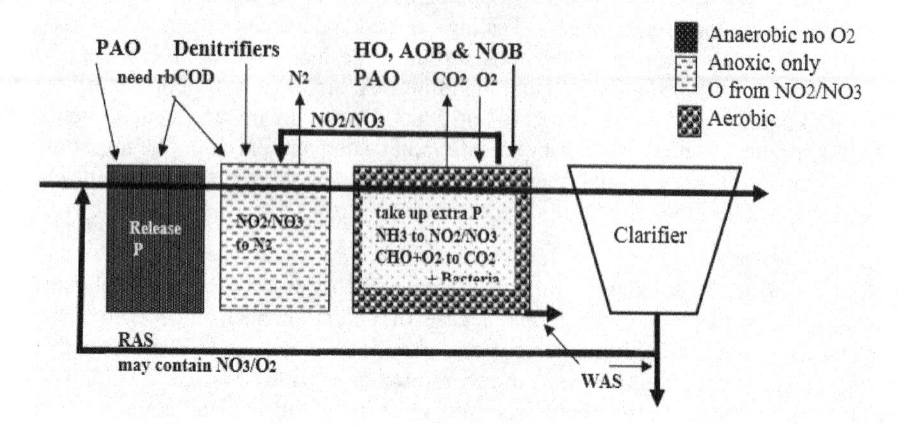


*Figure 8: BNR Treatment Plant Process (Grote, 2010)*

Typically, this unit will require chemical dosing of Methanol and Aluminium salts to maintain level of Nitrogen and Phosphorus removal rates. These have a very low dose rate, on the order of 10L/d and can be sourced locally. There are several concerns regarding a BNR plant that will have to be addressed on Mars, but are not considered to be significant roadblocks. These are:

- Low gravity will lead to extended sedimentation times, and likely increase the need for flocculants dosing.
- The gases given off by the plant and residual sewage waste ($N_2$, $CO_2$ and $H_2S$) will have to be vented or otherwise captured for reuse. Operations within the treatment facility will require constant gas monitoring and re-breathers for operators.
- Biological microorganisms will have to be brought to Mars, and their ability to survive importation to Mars is currently unknown.

The final stage of treatment will be a large UV unit. This unit will treat the effluent discharge and eliminate bacterial and virus contaminants. In addition to this, the agricultural recondensed water will be run through the UV unit to reduce long-term contamination. The combined treatment units will require approximately 300 kW of power to operate. Tanks and pipes can be made from Steel and HDPE which can be sourced locally. In addition, basic casting and 3D printing will provide the required pumps and non-critical mechanical equipment. It is estimated an additional 200 kW might be required for pumping; however, this will be heavily dependent on the arrangement of the colony. Specialised treatment units, including the UV, Microfiltration and Reverse Osmosis modules should be imported due to their criticality for the colony; and required material standards and operational lifetime.

### 3.6. Colony Summary

To summarise the entire colony design, it will require the initial importation of 3,069,100 kg of mass, costing $1,534,550,000. Of this, Power will make up 2,363,207 kg (77%) of the total import mass. The remaining 705,893 kg will be split between Production (10%), Agriculture (5%), Structures (6%) and Water (2%). Power usage will be approximately 86 MWe. The majority of this, 56%, will be used for food growth, with industrial manufacturing taking another 34%. With a total energy production of 100 MWe, there will be approximately 14 MWe for daily use by the colonists. This consumption rate is likely to vary, but 86 MWe represents a maximum value used at any one time, allowing energy to be stored in locally constructed Energy Vaults.

## 4. ECONOMICS

### 4.1. Colony Location

Given the flexibility of the colony design outlined in Section 3, the colony can be located anywhere there is a sufficient water supply. Recent evidence suggest that the Martian Cryosphere extends to a significantly low latitude and as a result, access to secondary resource can now be a driving factor for the location selection. This will primarily be copper deposits as copper is the one industrial metal that is not abundant in Martian soil, and it is a major source for Rare Earth Metals, as will be seen in Section 4.3.

### 4.2. Colony Deployment

It is assumed that there is already a supply of power and people on the surface of Mars as stated in Section 2. A crew of 50 people will be needed to deploy the first nuclear reactor, and set up the Manufactorum. Initially, only the Steel Works and Earth Works will be needed, and these will be located in pressured, lightweight 'tents' until steel and brick production is sufficient enough for the construction of initial habitats and the full Manufactorum. Setting up the initial power plant and manufacturing capability will take approximately one synod, or 26 months, with a second being needed for the other habitats for the full population. A full cost breakdown for the initial colony deployment can be found in Table 13.

*Table 13: Colony Deployment Costs*

| Item | Cost | Notes |
|---|---|---|
| **Labour** | $50,000,000 | 50 people, for 52 months at $200,000 per year wage |
| **Import Costs** | $1,534,550,000 | Total Import Costs for Power (77%), Production (10%), Agriculture (5%), Structures (6%) and Water (2%). Total import mass is 3,069,100.00 kg |
| **Procurement Costs** | $767,275,000 | Procurement Costs for initial construction. Due to the 'off-the-shelf' nature of the products, this is assumed to be 50% of import costs. |
| **Development Costs** | $383,637,500 | Assumed to be 25% of Import Costs due to the 'off-the-shelf' nature of the designs and mass savings of redesigns reducing mass. |
| **TOTAL** | $2,735,462,500 | Rounded to $3,000,000,000 for further growth / redundant parts. |

If a payback time of 10 years is assumed, then both wages and ongoing imports must be considered over this period. These are summarised below in Table 14.

*Table 14: Colony Deployment Costs*

| Item | Cost | Notes |
|---|---|---|
| **Labour per year** | $200,000,000 | 1000 people, at $200,000 per year wage |
| **Ongoing Imports** | $153,455,000 | ~10% of the initial import costs for maintenance and non-local production. ~300,000 kg per year |
| **TOTAL** | $353,455,000 | Per year ongoing cost |
| **TOTAL (10 Years)** | $3,534,550,000 | 10-year ongoing cost |

Based on the above costs, it can be seen that the colony, in its first 14 years, or 7 synods, will cost approximately $6,231,012,500. This is most likely beyond what a private company would be willing to invest without a clear return on investment. However, a Government would be able to afford the initial outlay and development of infrastructure to then provide facilities to private organisations that effectively lease the facilities for other purposes.

### 4.3. Goods

Given the export cost of $200/kg as per Section 2, any product exported from Mars would need to have a value of at least $600/kg to cover transport costs, manufacturing costs and profit. There are few products of value currently identified on the surface of Mars, or in the current production system as outlined in Section 3.1. However, the primary resource that will need to be specifically mined, copper, is also the source of many Rare Earth Metals. In fact, these are usually extracted from the 'anode sludge' that forms during the refining of copper.

*Table 15: Valuable Copper Production By-products*

| **Element** | **Rhodium** | **Iridium** | **Platinum** |
|---|---|---|---|
| **$/Kg** | $70,000 | $35,000 | $28,000 |
| **Element** | **Osmium** | **Palladium** | **Rutheniu** |
| **$/Kg** | $13,000 | $8,000 | $2,000 |

One drawback for relying on Rare Earth Metals is the need for detailed exploration of the surface to identify sources of copper. Additionally, importing the specialised processing equipment for Rare Earth Elements would be a cost above the $6 billion for the colony.

Another potential export is Heavy Water, which is far more abundant on Mars than it is on Earth, sells for approximately $778/kg and could be a secondary product from the current water extraction. The electrolysis process for producing Hydrogen could be easily modified to preferentially increase the concentration of Heavy Water in the remaining fluid. Given the estimated concentration of 1:1,284 Deuterium to Hydrogen, this represents a yearly production of 2,842.67 kg of Heavy Water, worth $2,211,604.36 yearly.

### 4.4. Services

Given the difficulties in reliably identify concentrations of valuable minerals before setting up the colony, an alternative to mining them directly would be to offer prospecting and support services. This would provide an additional income that does not require direct material exports, and could be linked to local demand to maximise efficiency. This service could also be extended to Asteroid Mining.

#### 4.4.1. Intellectual Property

Innovative designs and related inventions will arise throughout the course of a colony's establishment, development and maintenance. However, as such intellectual property revenue streams are *ad hoc* in nature, they cannot be relied upon for meaningful revenue streams for the colony. Additionally, their value is inherently tied to the robustness of patent and copyright laws. As a result, they cannot be relied upon as a source of revenue for the colony.

#### 4.4.2. Tourism

Space Tourism is a service ill-suited for Mars. Due to the long travel times and costs, Mars would not be an interesting tourist destination for anyone other than the super-rich. Even then, this would be an unreliable source of income, occurring too infrequently to provide a reliable source of revenue.

#### 4.4.3. Asteroid Mining

The nascent industry of Asteroid Mining could provide a 1000-person colony with the opportunity to provide services for asteroid mining initiatives instead. Briefly, it has been estimated that if the Asteroid Eros contains only 3% metals by weight, then it could be worth more than $20,000 billion USD (38). The asteroids 1028 Lydina (1923 PD), 414 Liriope (1869 CN) and 358 Apollonia (1893 K), which orbit beyond Earth, but relatively close to Mars through the asteroid belt, have an estimated value that exceeds $100 trillion each, and represent some of the most valuable asteroids identified. The extraction of their resources will require two key elements. The first is uninterrupted and near real-time telemetry to manually separate the metals via robotics; or, to oversee and sanity check such processes being conducted automatically. Second, the extraction of the resources may require a stable gravity well, as parts of the asteroids are separated or 'portioned' off the main body for further refinement. Presently, Earth would satisfy both elements as latency times would be short and the gravity well may serve as an effective 'capture' of 'off cuts' of an asteroid for further refinement, once the main body is harvested. However, Earth may not prove a desirable choice to stage an asteroid mining operation, given that such an operation may disrupt pre-existing space traffic management, and pose a collision risk with existing artificial satellites, thereby potentially initiating the Kessler syndrome. In addition to this, there may be a risk of re-entry and impact where a separation is uncontrolled. This, combined with the lower delta-V requirements to reach main belt asteroids, make Mars a better choice as a base for mining asteroids.

There are a number of ways Mars could be used to grow or capitalise on an Asteroid Mining Industry. Firstly, given Mars' closer proximity to the asteroid belt and the reduced atmospheric density and light pollution, better resolution of asteroids would allow Martian facilities to undertake more accurate spectral analysis of asteroids to allow for better mineral exploration. Such information could then be transmitted back to Earth, providing a valuable service. This may also potentially qualify as 'working' on an asteroid, or staking a claim on an asteroid, and therefore provide stronger proprietary rights for the organisation undertaking the spectral analysis. This exploration may open up a secondary market for trading of information on asteroids for future equitable interests.

Secondly, the presence of Phobos and Deimos, two captured, C-type asteroids, could be used as a testing ground for asteroid mining technologies or be mined themselves. Both of these would benefit from a local source of labour and resources, as well as a gravity well for astronauts to recuperate in, all of which a colony could provide.

Finally, the use of ground stations and gravity wells to secure 'orbital slots' for off-earth mining of asteroids on behalf of third parties such as individuals, corporations and nations would prove a reliable source of revenue for a colony based on Mars. Alternatively, where pre-existing contracts do not exist with Earth-based entities for mining, off-earth mining for speculative resources may be done independently by a prospective colony before being shipped back to Earth for bidding on the open market.

#### 4.4.1. Inter-Colony Trade

From Section 3, it can be seen that the key resource that will define the location of a colony is the availability of water. However, it can also be seen that secondary considerations will allow certain colonies to specialise in the production of certain resources. For example, access to larger quantities of purer water would allow more production of fuels and plastics, greater abundance of sunlight or geothermal power would allow more production of food and the right location may have access to copper. While such a situation would not cover the costs of a colony, it would still enable every colony on Mars to reduce its operating costs, and potentially pay off a portion of the deployment costs.

### 4.5. Economic Summary

Based on the above Costs, Goods and Services, an economic case for a Martian colony can be made. Continuing with the assumption that the Colonies set up cost is $6,231,012,500, including labour cost for 1000 people and additional imports over this period, and the payback period is 10 years. This necessitates a yearly revenue of $623,101,250 per year.

*Table 16: Economics Breakdown*

| Service | Teams | Est. Income | Total Income | Notes |
|---|---|---|---|---|
| **Mars Surface Miner Prospecting** | 4 | $35 million | $140 million | Prospecting companies on Earth can make $25,000,000 per year |
| **Mars Orbital Administration** | 1 | $20 million | $20 million | Administration and access to Orbital Slots around Mars |
| **Mars Asteroid Prospecting** | 4 | $35 million | $140 million | Prospecting companies on Earth can make $25,000,000 per year |
| **Mars Asteroid Mining Servicing** | 1 | $60 million | $60 million | Earth based Satellite servicing is estimated to cost $30,000,000 per satellite |
| **Inter-Colony Trade** | 1 | $75 million | $75 million | Income from exporting 750 tonnes of cargo to other colonies per year at $100/kg |
| **Deuterium Export** | - | $2.2 million | $2.2 million | - |
| **TOTAL** | | | $437.2 million | - |

The above revenue for the colony does not cover the required income needed to pay off the colony in 10 years. However, it could be paid off in approximately 15 years of its initial set up. This would not be an attractive proposition for a private company, however, if deployed and administered in a public-private partnership by a national government, the proposition becomes viable for a large number of national entities. As a result, it would make economic sense for a 1000-person colony to be run as a government industrial-park, set up and administered by a national government for use by private corporations.

## 5. POLITICAL, SOCIAL AND CULTURAL STRUCTURE

### 5.1. Political Structure

Article II of the Outer Space Treaty 1967 (UN) states that "Outer space, including the Moon and other celestial bodies, is not subject to national appropriation by claim of sovereignty, by means of use or occupation, or by any other means." It should be noted that the Treaty does not prevent state actors from establishing facilities, but from appropriation of celestial bodies through the establishment of facilities. Where a national government, be it the US or the Australian government, establishes a colony and attempts to claim sovereignty on the area either adjoining the colony, or below the colony, these claims would be voided pursuant to Article II. However, sovereignty within the facility would survive, as it does for a nation's satellites or ships in international waters. To this end, a potential colony would remain within the legislative control of the host nation, such as Australia, but no claims of sovereignty on the surface of Mars would be applicable. Essentially, even though a claim could not be staked on any 'land' on Mars, the Habitat itself would be legally subordinate to the laws of its host nation. This would extend to any facility or vessel, from a full habitat down to a space suit. Based on this, and given the suggestion in Section 4.5, the colony will be administered as a government territory, similar to the District of Columbia in the US or the Australian Capital Territory and Northern Territory in Australia. Territorial jurisdiction vested in facilities and infrastructure would provide capabilities for the territory to develop subordinate legislative powers (or regulations) as may be required from time-to-time. Such a scheme may also entitle the territory to parliamentary representation and may allow for delegated powers in a Mars-based Authority to make powers on behalf of the host nation. Local authority would come from a local 'Governor' appointed by the host nation, however, due to the isolated nature of a colony on Mars, it would be advisable that a local council be established, in order to provide a check on local authority – similar to the legislative councils of the Northern Territory and the Australian Capital Territory. New colonists would be required to undergo an accreditation scheme that includes a fit and proper persons test, as well as licencing requirements to operate within the external territory. Part of this licencing scheme would require being presented with a local Code of Conduct they would be required to sign, clearly stating the risks, restrictions and freedoms they are allowed at the colony, due to the hazardous nature of the environment. This would form a literal, 'Social Contract' they would be required to agree to before travelling to the colony. Local Workplace Health and Safety (WHS) Officers would form the basis of the local law enforcement due to the hostile nature of the outside environment. Penalties for breach of the code of conduct and territorial laws and regulations may vary depending on the infraction, escalating from restriction, suspension or revocation of a licence to work in the external territory, to being deemed an occupational hazard and having their access revoked to all operational areas of the base beyond their habitation quarters until extradited to Earth to face trial under the host nations laws. Using a WHS Officer as the delegated power to determine behavioural standard and their breaches rather than a constabulary force would be advisable to preserve due process and judicial review where extradition back to Earth is deemed necessary.

### 5.2. Society and Culture

Due to the small number of people, there will be little development of a local culture, however, the restriction of the Martian Environment will have a number of implications that will affect future cultures on the Red Planet. Living on Mars will be an intensely communal experience, with almost all spaces being shared except for one's personal apartment.

Most food will be prepared and consumed in a communal mess hall, and it would be advisable for hygiene facilities to be communal as well. Instead of these being a dull, sterile design, it would be advisable to make these communal experiences fun and enjoyable, designed more like a Mead Hall for a village and a Japanese Onsen, respectively. Food will be particularly interesting to develop, as it will start off very bland, but vary as people grow their own spices and ingredients. Due to the low quantities of cooking oils, most foods will be either baked, steamed or boiled, leading to each colony having its own unique style and flavour. Such variation and experimentation should be encouraged. Sports and fitness will be incredibly import, due to the high probability that living in reduced gravity will lead to wasting effects. Though not as pronounced as in zero gravity, they will likely cause issues and will need to be counteracted. Again, to capitalise on the communal environment, team sports should be encouraged, especially ones where the equipment can be made locally, such as Basketball or Tennis.

Both of these will prove to be interesting to watch in low gravity. Tennis in particular would be advisable as the shock-loading on bones would help reduce the loss of bone mass. Due to the restrictions on what materials can be produced locally, any musical instruments will either be small, electronic, or made of local plastic and metal. This in turn will cause early Martian Music to have a unique sound. It will also be a kaleidoscope of different cultures as people of different backgrounds will be forced into a communal environment. Finally, due to the incredible medical and ethical issues with raising a child on Mars, all colonists would be required to have a permanent form of birth control, be it an intrauterine device or vasectomy, in order to prevent conception until it has been determined that mammalian life can grow safely on Mars. It is most likely that this will be determined once colonies are above a population of 1000 people.

## 6. AESTHETIC

The exterior design of the Mars Base is driven by solid engineering and effective use of available resources. The internals of the colony, however, will need to be adapted to promote strong mental and physical health. Modern technology has made significant inroads into keeping collaborative spaces for work and play varied, while personal spaces can be open enough for healthy retreat while still being private. The importance of design of personal space is equally important to that of collaborative and working spaces for mental wellbeing (39). The social spaces, unlike in many science fiction settings, will need to incorporate natural environments. Some of the plants that are brought over from Earth will need to be sustained in these areas to help alleviate stress and anxiety (40). Piping the water filtration nearby to these collaborative spaces will promote calmness and create an ambiance that will make the colony a place where relaxation can be found sitting and reading in nature surrounded by running water, while looking out of an elevated dome across the Martian landscape. Collaborative working spaces will also be an important addition to the base. These spaces will define the colony with water capture and air recirculation built into the ceilings and large screens that can present data or even various outside views. This setup effectively uses space and will help in promoting a sense of community in the work place. With the ability to walk between nature filled areas meant for relaxation and socialising to collaborative work areas that focus on the seamless integration of technology, the colony will be a beautiful site that focuses on personal wellbeing and community in the harshest of environments. Outside, it would be possible to use excess material to adapt the exterior of the habitats to be more inspiring and architecturally interesting for new arrivals and those who work primarily on the surface, the vaulted domes and brick construction of the habitats recalling Roman, Islamic and Renaissance architecture.

## 7. SUMMARY

It can be seen that the best way to build a colony, capable of supporting 1000 people on the Red Planet is to travel light, and live off the land. The majority of modern construction materials can be synthesised using nothing but water, air and the soil, if enough power is present. This presents two challenges for a budding colony and as a result, technologies significantly reducing the weight of power components, and surveys looking for deposits of copper ores on the surface of Mars should be conducted. This will greatly reduce the costs of setting up a colony on the Red Planet, allow a colony to expand more rapidly, and begin exporting valuable minerals back to Earth. Before such a time, however, the initial investment for early colonies would need to be made by governments, with ongoing revenue and incentives for more colonists being provided by offering the colonies as a public-private partnership, geared to capitalising on Mars' competitive advantage over Earth when it comes to asteroid prospecting and mining. This will allow a colony on Mars to grow beyond its initial 1000-person population and while a new society and culture wouldn't be able to significantly develop with such a small number of people, the seeds can be planted.

## 8. REFERENCES

1. *Mineralogic and Compositional Properties of Matian Soil and Dust: Results from the Mars Pathfinder.* **Bell, J. F., et al.** E1, 2000, Journal of Geophysical Research, Vol. 105, pp. 1721-1755.

2. **Nasa.** Mars Fact Sheet. [Online] 2018. [Cited: 19 03 2019.] https://en.wikipedia.org/wiki/Atmosphere_of_Mars#Observations_and_measurement_from_Earth.

3. **Mueller, Robert P., et al.** *Regolith Advanced Surface System Operations Robot (RASSOR).* s.l. : NASA, 2013.

4. **Moss, Shaun.** *Steelmaking on Mars.* s.l. : The Mars Society, 2006.

5. *Systems engineering and design of a Mars Polar Research Base with a human.* **Rüede, Anne-Marlene, et al.** 2018, Acta Astronautica.

6. **World Nuclear Association.** Nuclear Reactors and Radioisotopes for Space. [Online] 2018. [Cited: 11 March 2019.] http://www.world-nuclear.org/information-library/non-power-nuclear-applications/transport/nuclear-reactors-for-space.aspx.

7. —. Small Nuclear Power Reactors. [Online] 2019. [Cited: 15 March 2019.] http://www.world-nuclear.org/information-library/nuclear-fuel-cycle/nuclear-power-reactors/small-nuclear-power-reactors.aspx.
8. *The application of supercritical CO2 in nuclear engineering: A Review.* **Qi, Houbo, et al.** 4, 2018, Sage Journals, Vol. 10.
9. *The Supercritical Carbon Dioxide Power Cycle: Comparison to Other Advanced Power Cycles.* **Dostal, Vaclav, Hejzlar, Pavel and Driscoll, Michael J.** 2006, Nuclear Technology, pp. 283-301.
10. **Mersmann, Kathryn.** *The Fact and Fiction of Martian Dust Storms.* Greenbelt, Maryland : NASA's Goddard Space Flight Center, 2015.
11. **McKevitt, Jennifer.** Can Stacking Concrete Store Energy? [Online] 2018. [Cited: 27 January 2019.] https://www.constructionequipmentguide.com/with-gravity-all-things-are-possible/41838.
12. *Distributed Generation Control in Islanded Industrial Facilities: A Case Study in Power Management Systems.* **Al-Mulla, Musaab M. and Seeley, Nicholas C.** Oslo, Norway : PCIC Europe Conference, 2010.
13. **PBE.** MEDIUM VOLTAGE CONTAINERIZED POWER SUBSTATIONS FOR MINING AND TUNNEL CONSTRUCTION. [Online] 2019. [Cited: 27 March 2019.] https://pbegrp.com/power/substations/containerised-substations-css/.
14. **Csanyi, Edvard.** Copper or aluminium? Which one to use and when? [Online] 2012. [Cited: 24 March 2019.] https://electrical-engineering-portal.com/copper-or-aluminium.
15. *Pressure Effects on Oxygen Concentration Flammability Thresholds of Polymeric Materials for Aerospace Applications.* **Hirsch, David, Williams, Jim and Beeson, Harold.** 1, 2008, Journal of Testing and Evaluation, Vol. 36, pp. 69-72.
16. *Martian Regolith Simulant JSC MARS-1.* **Allen, Carlton C., et al.** 1998, Lunar and Planetary Science, Vol. XXIX.
17. *Measurement of Net Photosynthetic and Transpiration Rates of Spinach and Maize Plants under Hypobaric Condition.* **Goto, Eiji, et al.** 2, 1996, Journal of Agricultural Meteorology, Vol. 52, pp. 117 - 123.
18. *Exposure of Arabidopsis thaliana to Hypobaric Environments: Implications for Low-Pressure Bioregenerative Life Support Systems for Human Exploration Missions and Terraforming on Mars.* **Richards, Jeffery T, et al.** 6, 2006, Astrobiology, Vol. 6, pp. 851-866.
19. **National Health and Medical Research Council.** Australian Dietary Guidelines Summary. Canberra : National Health and Medical Research Council, 2013.
20. **Food and Agriculture Organization of the UN.** FAOSTAT. [Online] 2019. [Cited: 3 March 2019.] http://www.fao.org/faostat/en/#data/QC.
21. **Brandon, Merrill F, et al.** Next Evolution of Agriculture: A Review of Innovations in Plant Factories. [ed.] Mohammad Pessarakli. *Handbook of Photosynthesis.* Boca Raton : CRC Press, 2016, pp. 723 - 740.
22. **Fisher, Gary C.** *Mars Homestead Project: Waste Recycling System Conceptual Design.* s.l. : The Mars Society, 2005.
23. **Wheeler, Raymond M.** *NASA's Controlled Environment Agriculture Testing for Space Habitats.* Kennedy Space Center : NASA Surface Systems Office, 2014.
24. **Wheeler, R M, et al.** *Crop Production for Advanced Life Support Systems - Observations from the Kennedy Space Center Breadboard Project.* Kennedy Space Center : NASA Biological Sciences Office, 2003.
25. *Economic Analysis of Greenhouse Lighting: Light Emitting Diodes vs. High Intensity Discharge Fixtures.* **Nelson, Jacob A and Bugbee, Bruce.** 6, 2014, PLoS One, Vol. 9, p. e99010.

26. **Zabel, Paul.** Designing a Closed Ecological Life Support System for Plants, Overview. [ed.] Erik Seedhouse and David J Shayler. *Handbook of Life Support Systems for Spacecraft and Extraterrestrial Habitats.* s.l. : Springer, Cham, 2018.
27. **Osram Opto Semiconductors GmbH.** OSLON® SSL 150, GH CSHPM1.24 . [Online] 19 February 2019. [Cited: 28 March 2019.] https://dammedia.osram.info/media/resource/hires/osram-dam-8841545/GH%20CSHPM1.24_EN.pdf.
28. **Stratus Productions LLC.** 100W LED Module - Version 3. [Online] 2019. [Cited: 28 March 2019.] https://www.stratusleds.com/module.
29. **Brouwer, C and Heibloem, M.** Crop Water Needs. *Irrigation Water Management: Irrigation Water Needs.* Rome : Food and Agriculture Organisation of the United Nations, 1986.
30. *A Sustainable Model for Rural Production of Edible Mushrooms in Mexico.* **Martinez-Carrera, D, et al.** 1998, Micologia Neotropical Aplicada, Vol. 11, pp. 77-96.
31. **Sinha, Rajiv Kumar.** *Modern Plant Physiology.* Pangbourne : Alpha Science International Ltd., 2004.
32. **Brown, M., Matlock, M,.** A Review of Water Scarcity Indices and Methodologies. s.l. : University of Arkansas, 2011.
33. *Water in the Martian regolith from OMEGA/Mars Express.* **Milliken, R., Jouglet, D.** 8, 2017, Journal of Geophysical Research: Planets, Vol. 119.
34. **Buckner, A., Adan=Plaza, S., Hoffman, C., Schneider, M., Carpenter, K., Elias, L., Hilstad, M., Grover, M.** Extraction of Atmospheric Water on Mars. 1998.
35. **Frankie, B., Frank, T., Lowther, S.** Drilling Operations to Support Human Mars Missions. 1998.
36. **Rummel, J.D., D.W. Beaty, M.A. Jones, C. Bakermans, N.G. Barlow, P.J. Boston, V.F.** A New Analysis of Mars "Special Regions": Findings of the Second MEPAG Special Regions Science Analysis Group (SR-SAG2). 2014.
37. **Plaut .J.J., J.W. Holt, J. W. Head,.** Thick Ice Deposits In Deuteronilus Mensae, Mars: Regional Distribution From Radar Sounding. 2010.
38. **Whitehouse, Dr. David.** 22 July 1999.
39. **de Botton, Alain.** *Architecture of Happiness.* s.l. : Penguin Books, 2006.
40. **Pearson, David G. and Craig, Tony.** The great outdoors? Exploring the mental health benefits of natural environments. [Online] 2014. [Cited: 27 March 2019.] https://www.frontiersin.org/articles/10.3389/fpsyg.2014.01178/full.
41. **Ellender, Damon.** Mars Foundation. [Online] 2005. [Cited: 20 03 2019.] http://www.marsfoundation.org/.

# 17: FOSSAE COLONY ONE
# A DESIGN FOR A VIBRANT MARS COLONY

**Stefanie Schur**
stefanie.schur@gmail.com

## ABSTRACT

Many proposals for colonizing Mars describe circular dome habitat structures and underground caves. While circular domes may be efficient for utilizing space, they are also very efficient for wiping out the entire colony with one small failure or accident. Caves offer lots of protection from radiation, but for long-term human habitation they are impractical since there is no natural sunlight and they have a very confining feel which can lead to low morale and depression.

This proposal instead uses a linear corridor as a principal design motif, which has the advantages of segmenting the colony construction in order to minimize the impacts of a catastrophic event on the colony, and to provide for incremental growth, so the colony can expand as it grows. It employs microterraforming to build an ecosystem in an enclosed environment envelope.

Fossaes on Mars are ancient linear fissures in the surface that extend for hundreds of kilometers in some cases. They vary in width from a few tens of meters at the beginning of the fracture to a few thousand meters, and depth from a few meters to 100 meters or more. These fracture features could provide protection from radiation, wind, and sandstorms, and are ready-made cross-sectional cuts into the Martian geologic strata—useful for resource exploration and providing a sloping landscape for terraced settlement layouts. An arching cover with radiation-resistant glass would keep a livable environment inside while providing natural sunlight to the colony and reducing cosmic radiation by reducing the ways it can reach the colony. The linear arrangement would also provide efficiencies in the transmission of energy, water, and other resources within the villages by building infrastructure along a line.

Human civilization has a long history of surviving in adverse environments by using compact linear settlement patterns and taking advantage of natural landscape features for protection and community organization. Ancient and current settlements such as Petra (Jordan), Canyon de Chelly (America), Wadi Rum (Saudi Arabia), Mykonos (Greece), Manarola (Italy), and Ajanta (India) are among many examples that can inform Mars colony design—in both community layout and beautiful landscapes to proudly call home.

## INTRODUCTION

While there may be regions that offer more dramatic settings, the geologic fractures known as fossae offer some of the best blending of features to be a good candidate for the first Martian colony. They provide shelter from sand storms, offer a cross-section cut through the geology for resource exploration, provide an efficient ratio of habitat volume to enclosure surface area, and as linear features they are ideal for laying out a settlement that can grow incrementally while segmenting the colony to contain damage in a catastrophic habitat failure.

Terraforming the entire planet and atmosphere is impractical in human time frames, but microterraforming in geographic features like fossaes can happen quickly and be an effective way to create permanent settlements with beautiful active ecosystems. While living in prefab sterile hab modules may be practical for temporary exploration missions, a permanent colony needs a living ecosystem in order to be self-sustaining and serve human physical, psychological, and cultural needs. People will need to find a sense of freedom, connection, beauty, and inspiration in their daily lives to thrive on Mars.

*figure 2.1*
*Earth examples can inform design for a glass dome to microterraform under.*

*figure 2.2*
*Internal air pressure supports a dome roof.*

**ARCH GLASS ROOF**

With sloped sides of a fossae providing the ground of the colony and partial radiation shielding, only a cover is needed to create a spacious microterraformed environment for habitation. Structures built on earth offer examples and insights into what could be constructed on Mars. An arching metal frame roof structure (figure 2.1) would be assembled robotically in advance. Standard-sized square transparent HDPE film window panels and lightweight metal framing could be transported from Earth for the initial colony roof assembly, then later additions and improvements to the glass quality could be manufactured on Mars. The simple square geometry would make replacement and manufacturing easy. Since the internal air pressure would be about 30 times greater (600 millibars) than the outside ambient atmospheric pressure, most or all of the weight of the roof structure would be supported by inside air pressure, the way an inflated dome stadium such as the Carrier Dome in Syracuse is (figure 2.2). It may even be necessary to add weight to the roof to balance the forces.

## COMPARTMENTALIZE WITH WALLS

Each wave of construction of the colony would be built as a 'cross-section' across the fossae fracture, at a length sufficient for the needs of the number of expected colonists for each growth phase (figure 6.2). A wall would be built at the "open" end of the colony (the end that would be added to in future expansion). This would ideally be a glass panel wall similar to the roof design, but could also be a kevlar material or even concrete form wall. With the colony built in this segmented fashion, a catastrophic failure in one location would be confined to that segment and not endanger the rest of the colony. A pressurized isolation evacuation corridor could be integrated into the linear colony layout to allow evacuees to escape to the next secure community segment (figure 6.4). In normal use, this tube corridor would be a linear thread along the entire length of the colony and could serve a transit system as well.

## ROBOTIC CONSTRUCTION AND MATERIALS MANUFACTURING

Ahead of the first humans arriving, it is well understood that a series of vehicles would be sent with stockpiles of survival supplies, tools, clothing, materials, and basic habitat modules to start. Autonomous robots will also be able to pre-assemble various pieces of equipment, organize stockpiles, and do preparatory construction. There is much discussion about 3D printers being widely relied upon to manufacture habitats. While 3D printers will be necessary and integral to the colony, it would be foolhardy to completely rely on such technology for all basic needs. Colony resiliency will depend on back-to-basics skills, construction methods, and resource collecting.

With that in mind, some of the key tasks that advanced robotic teams will have to do will include the mass production of in-situ regolith concrete masonry unit (CMU) blocks and structural beams, along with materials for mortar to use to build structures. These will be made from plentiful calcium, gypsum, sulphur, and clay minerals (phyllosilicates) found on Mars. Stronger structural units will be made of dense compressed material, while Autoclaved Aerated Concrete (AAC) units would provide lightweight insulating blocks for use as walls, floors, and paving.

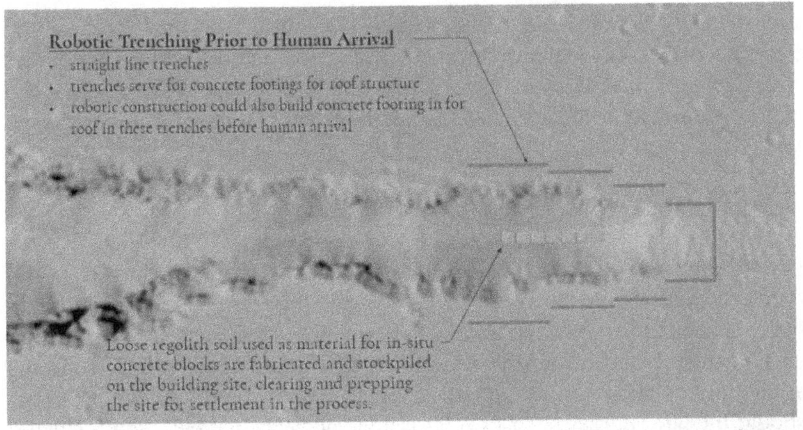


*figure 4.1*

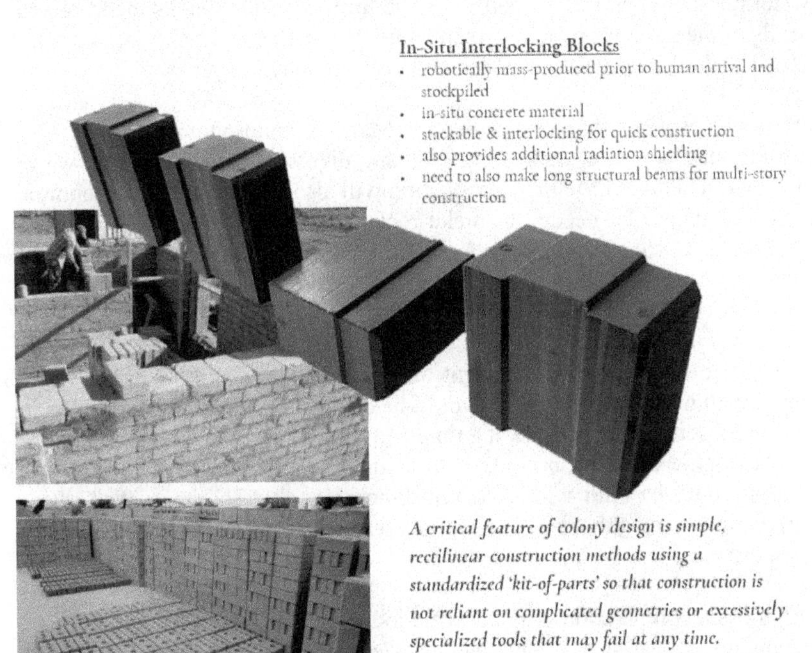


*figure 4.2*

The first colonists would then be able to build the structures that will become the village. At a bare minimum, stackable concrete blocks (roughly 600x300x200 mm) would be manufactured and stockpiled by robotic production, and structural beams (roughly 300x100 mm x 4 meters long) for floors/ceilings in multi-story structures for living and working space (figure 4.2). Techniques to mortar the stackable blocks together will have to be developed using in-situ regolith mortar. The goal in doing this is to make a standard "kit of parts" with which to build structures.

Autonomous construction equipment would dig linear trenches needed for concrete bases to support the roof structure (figure 4.1), construct the bases and metal framing, and install the square glass panels that would make the roof enclosure, in order to form a sealed habitable environment—the ***environment envelope.***

## ENERGY PRODUCTION

Necessary to the survival of any colony, the production of energy, mostly in the form of electricity, should be ***redundant, diverse, and dispersed***: many small-scale systems need to be deployed around the colony to create a power grid that is redundant, scattered, and using diverse sources of energy. These multiple micro-systems should be outside of, and surrounding, the colony environment envelope. They should include sources such as solar arrays, nuclear, fuel cell, and methane. Energy storage should also be dispersed, in the form of battery banks incorporated in the walls of the structures. Heating supplemented by thermal mass passive solar systems will reduce the total energy demand of the colony.

Methane can come from sources in the Martian atmosphere and geologic sequestration, and can also come from methane digesters used to process waste products from the colony. Methane production will also be an important economic commodity, as it will be needed to make rocket fuel for trips to Earth or other reaches of the inner solar system, especially the asteroid belt for resource mining.

## ENVIRONMENT ENVELOPE AND TERRAFORMING

In this concept, a series of adjacent enclosed *environment envelopes* are constructed within a fossae fissure canyon. The enclosures would be segmented in order to build discrete sections one at a time and to minimize the extent of damage from a catastrophic event by limiting it to a small section of the overall colony. Each section could be built with different dimensions, but sharing a wall (glass partition) with it's neighboring section to create a sense of continuity of the colony (figure 6.2).

Much of the site preparation and construction would be done by autonomous robotic equipment, assisted by people as needed when they arrive. Once an enclosure segment is completed and pressurized, people will be able to work inside it comfortably without space suits and build dwelling structures and all necessary infrastructure, as well as microterraforming the ecosystem and farms.

Along and above the rim of the fissure canyon, a zone of "***air gardens***" would be created that have a manipulated in-flow of Martian atmosphere, passing through a solar thermal mass diaphragm system that heats and pressurizes the air, which is then processed through plant photosynthesis into breathable air for the colony (figures 6.3, 6.4, 12.2). At night, cold denser atmosphere is pumped into diaphragm tanks, then during daytime sunlight heats and pressurizes the $CO_2$ in the tanks (supplemented as needed) to temperature and pressure that is higher than the colony environment envelope. It is then released into the air gardens to process into breathable air. The narrow linear layout of the colony maximizes the edge where this can happen, to help ensure there is sufficient production of heat and air for the colony. Energy production and heavy manufacturing would occur adjacent to the environment envelopes, while small cottage industries would operate within the community.

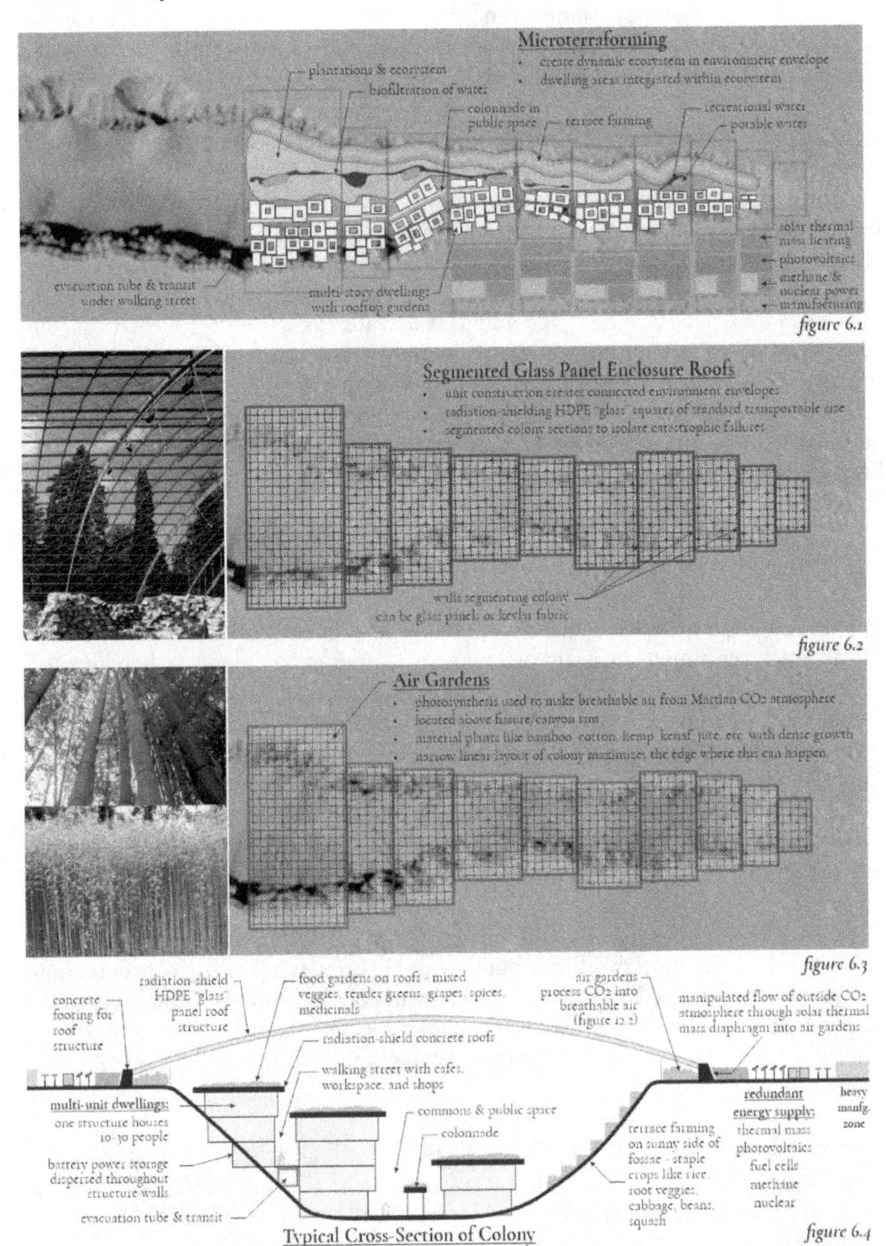


figure 6.1

figure 6.2

figure 6.3

figure 6.4

## INSIDE ATMOSPHERE

Atmospheric pressure and oxygen levels do not need to be maintained to match Earth sea level—it would require less energy to maintain lower levels. On Earth, cities and towns thrive at elevations over 4000m (La Rinconada, Peru; El Alto, Bolivia; Potosi, Bolivia), some even 4500m. At these elevations, the effective oxygen level and atmospheric pressure is about 60% of what it is at sea level. These should be the target levels for the settlements on Mars.

Heating and humidifying the inside atmosphere to a habitable range will help pressurize the space in the environment envelope. With ample solar energy coming through a glass dome, it will create a greenhouse effect that could also help support internal air temperature and pressure, which would help support the roof.

Surface water features and cascading waterfalls as an integral part of the ecosystem environment would help humidify the atmosphere, as well as oxygenate the water for aquatic ecology and water purification systems.
The internal atmosphere needs to be dynamic, and never stagnant. Manipulating air currents, and fluctuations in temperatures and humidity, will be necessary to keep the atmosphere cycling and purifying. This could be assisted by cycling air back through the air gardens.

## COMMUNITY DESIGN AND STRUCTURE

The underlying concept for this Mars colony is a linear settlement pattern. The fossae fractures found on Mars are relatively straight, running for hundreds of kilometers in many cases. They vary from tens to hundreds of meters in width and depth. They often narrow to a point at the end, and create a gentle slope from that point down into the canyon as you follow the line of the fracture (figure 6.1). Side slopes along the fracture canyon rim vary greatly from dramatic and beautiful rugged cliffs to a shallow even slope from rim to bottom. These evoke images of Sedona, Mykonos (figures 8.3, 8.4), and Cinque Terre (figures 8.1, 8.2) landscapes that make such beautiful settings for timeless villages on Earth.

The characteristics of those Earthly villages that can inform settlement patterns for a Mars colony include dwellings stair-stepping up steep V-shape cross-section topography, fragmented and piecemeal construction that follows natural contours of the land (a Mars colony would be built in segments), use of bright colors, narrow walking streets flowing down a slope, and live+work+play all blended together in a tight community.

**Courtyard buildings** will bring lots of natural sunlight into living and working areas (figure 8.5), improving the overall quality of life for residents. Building roofs will be thick concrete layered with metal/lead plates for additional radiation shielding.

With dimmer Mars sunlight and the colony roof glass panels reducing light more, bright vibrant colors and whites will be essential elements of streetscapes and interiors to maintaining healthy spirits in the community (figures 8.2, 8.3, 8.4). This will create a demand for color pigments, and skilled people will fill niches in developing techniques to create those pigments from native Martian materials.

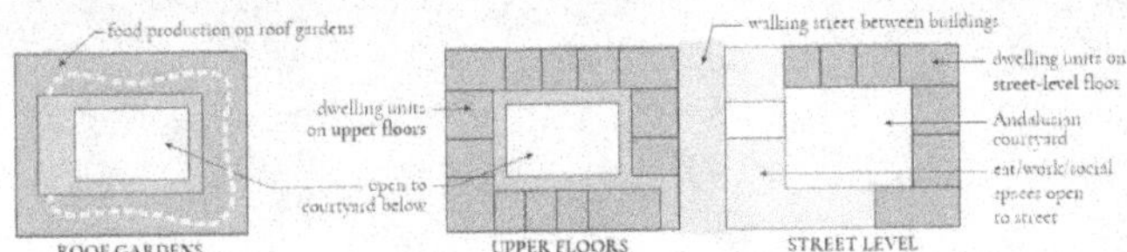


Typical Schematic Buildings with Andalusian Courtyard and Street for Mars Colony *figure 8.5*

## WASTE AND WATER

Things taken for granted on Earth, like private toilets and baths, will be impractical in a Mars colony that needs to aggressively conserve water resources.

**Public bathhouses and toilets** will be used to conserve water and provide efficiencies in waste processing. With a strong sense of community bond, a thoroughly modern and Martian interpretation of ancient Roman bathhouses would serve the needs of the community (figures 10.1, 10.7), and provide necessary efficiencies in resource use and recycling.

Water filtration can not completely rely on technology for the long run—there will also need to be biological cycles, analogous to what happens in healthy ecosystems on Earth, functioning throughout the colony. Simple sand and gravel based biofiltration systems could be constructed from basic Mars materials. A blending of technologies will be needed to initially cleanse perchlorates and other harmful chemicals from Martian rocks and sand, or, rock may be found in underlying layers that is free of those contaminates.

*Fig 10.1 - Ancient Roman baths provide a model for communal bath houses. Fig 10.2 - Organically shaped cascading waterfalls provide some recreation opportunity as well as aerating water and providing humidity in the settlement. Fig 10.3 - Big windows to provide lots of natural sunlight using radiation-shielding glass, and regolith concrete mass ceilings for radiation shielding. Fig 10.4 - Narrow walking streets with cafes and shops nestled in with dwelling units will create an inspirational charm for residents. Fig 10.5 - courtyards bring natural sunlight into dwelling areas. Fig 10.6 - Colonnade.*

The slope along the line of the fossae will provide a natural gravitational flow for water resources in the colony from storage at the higher end of the community down through the settlement to be used, and then into biofiltration and processing areas at the lower points where it would be recycled and conveyed back up to the storage areas.

**LIFE AND SOCIETY IN THE COLONY**

Daily life will be different from that commonly found on Earth, but with this design it would still provide plenty of opportunities to be unique—with your own interests, desires, and sense of identity. The linear nature of the settlement provides opportunities to take walks, socialize in public spaces, and feel some sense of being "outdoors". The community would have personal spaces for alone time, as well as space that feels like a street scene and a commons, to connect to community. With shared bath houses, those features of daily life will be more a social moment, rather than the very private moment that most on Earth tend to treat it as. Bathing will be much like using a group spa, or a public pool (figure 10.1).

On Mars, finding "shade" from radiation may be important, depending on the technological advancements made in radiation-shielding glass that can be used for the colony roof. On Earth we can easily get too much solar radiation, which leads to bad sunburns, and later-in-life skin cancers, cataracts, and premature aging, that conspire to shorten life. But we don't all live in underground caves to avoid it. The situation on Mars will be similar. Ceilings made of thick concrete, and radiation glass window walls (figure 10.3a, 10.3b), which all exist under the radiation-shielding glass cover, will provide layers of shielding (figure 12.1). When people go "outside" (still in the environment envelope), their clothing material could provide a layer of protection (as we've always done on Earth), along with stone/concrete covered concourses similar to a portico or **colonnade** (figure 10.6) that would all work together to make sure no one gets sustained doses of radiation that are unsafe, so they could wander freely under natural sunlight without risk of serious long-term radiation exposures.

Physical leisure activities would include walking (a sloped settlement means that walking involves hills), swimming in a communal pool/bath or aeration/humidification pools (figure 10.2), rock climbing, yoga, martial arts, group workout classes—all with a new interpretation in Point4G.

The settlement is planned to be walkable so that walking from place to place is a cultural norm supporting the need for daily physical exercise. The integral below-grade evacuation tube that runs the length of the settlements could double as a rapid transit tube when the community gets long enough to need it.

With an abundance of stone, concrete, and colorful ceramic tiles as primary building materials, it's easy to imagine the beauty of the colony to resemble Moorish, Mediterranean, or modern Arabic design motifs. Architectural styles like

Moroccan riads with Andalusian courtyards would work exceptionally well. With people from all over Earth settling Mars, the architectural styles would be eclectic, bringing beauty through diversity.

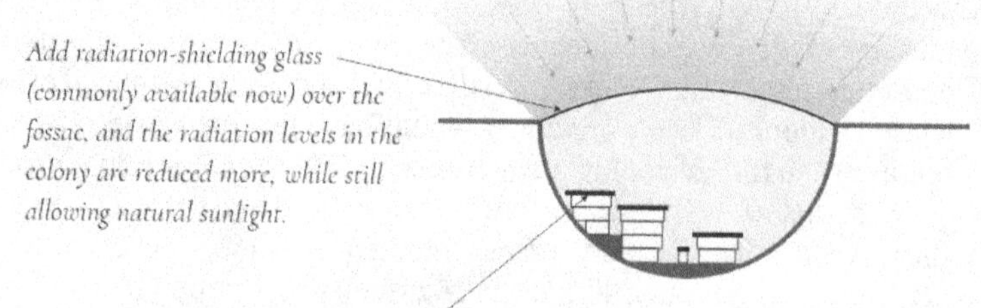


*figure 12.1* **Conceptualization of Levels of Cosmic Radiation Reaching the Colony**

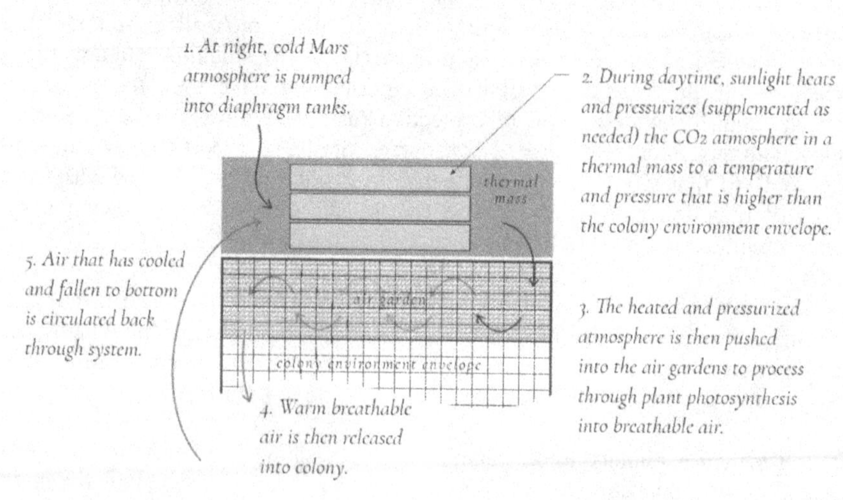


*figure 12.2* **Schematic Conceptualization of Air Garden Process**

While some may imagine the first 1000 settlers arriving to just set up computers and labs and go to work on scientific endeavors, the first settlers will have to have plenty of grit, be willing to do lots of hard physical work building and maintaining a community, be creative, and form a strong community identity that is unique to their situation.

The developing trend in coworking spaces offers a glimpse of what life on Mars might be. With manufacturing, mining, and production handled mostly by robotics, “work” would be done in courtyards and coworking spaces.

## FOOD, COOKING, AND LIFESTYLE SPACES

The typical culture we have on Earth of a kitchen in every home won’t exist in the first Mars colony. Personal private space, i.e., dwelling units, will consist of a bed, desk, dresser, and closet, akin to University dormitories. All other daily activities and functions will be in a shared community setting, including lounging, bath, toilets, recreation, and eating.

Without kitchens in individual dwelling units, shared cooking and eating will be the norm. With this in mind, a Martian version of street food culture will dominate meal habits in the colony. Sidewalk cafes, open air restaurants, and food carts will satisfy every meal. Community kitchens will provide opportunities for anyone who desires to cook and experiment with cuisine.

Architecture on Earth is founded on a basic principal: protect the inside from the outside. With a Mars colony built in an environment envelope, there is no real need to protect the interior from any kind of outside weather. This will create new architectural design paradigms that disregard tradition separating interior from exterior, and the two will blur and blend together in exciting new ways. Walls and windows will only be needed for privacy, sound control, and organizing spaces, rather than holding back a storm.

## ECOSYSTEM AND FOOD PRODUCTION

Ecosystem and food are two closely interconnected systems that need to be discussed together. Laboratory hydroponics will have a place, but it can't be the sole source of food, or even the primary source. Farming with Martian soil and water will have to be successfully developed. This can be a combination of sloped terrace farming, vertical gardens, orchards, and fields. Vulnerable monocultures need to be avoided, and diverse mixes that also serve scenic and recreational needs to enhance quality of life would be preferable.

A complex ecosystem integrated into the community will be required in order to be resilient and sustainable. Crops need soil insects for organic cycling, nitrogen fixing, and minerals availability. Soil insects need their populations controlled by higher trophic consumers, and those need small animals to control them. The small animals will need larger animals to prey on them, with humans at the top of the chain.

Ecosystems and climate could be varied in each colony segment for additional diversity, but *at a minimum*, **100 different species for each trophic level in the ecosystem** should be brought to establish in the colony.

Some Terrestrial Species

**Soil biota**: earthworms, nematodes, insects, silk worms, decomposing bacteria, algae, fungi, lichens, mosses.

**Plants for habitat and food**: root vegetables like potato, turmeric, cassava, carrots, beets, onion, garlic, etc. Dry climate plants like peppers, paprika, sage, rosemary, beans, tomatoes, squash, lentils, pinyon, etc. Berries including elderberry, grapes, etc. Pollinator plants like Ceanothus. Material plants like cotton, kenaf, hemp, jute, flax, etc.

**Insects**: bees, butterflies, ladybugs, praying mantises, hummingbirds, bats, other pollinators and birds.

**Small animals**: rodents, lizards, snakes, rabbits, chickens, cats, goats (probably nothing bigger than goats).

**Trees**: Drought-tolerant & poor soil species like jatropha, mesquite, acacia, etc. Durable resource-providers like tamarind, olive, bamboo, pomegranate, pecan, pistachio, juniper, cypress.

Some Aquatic Species

**Producers**: phytoplankton, cyanobacteria [algae], decomposers, etc.
Small primary consumers: small fish and amphibians, anchovies, tilapia, salmon, etc.

## ECOSYSTEM AND HUMANS

**Agriculture** for food production will be primarily terrace farming on the sunny side of the canyon slopes, and experimental agriculture will be an important scientific discipline. In a complex ecosystem, humans will be at the top of the food chain. Animals in the higher trophic levels of the ecosystem will need to be managed, which means they will become an important source of protein for colonists. Vegan diets will not work in a Mars colony–all colonists will need to eat small animals and insects to get nutrition and to contribute to the intensive management of the ecosystem. Sausages will be a common item on plates.

**Bamboo** will be an important plant since it is fast & dense growing in a variety of conditions, provides building and manufacturing materials, and is an aesthetically desirable tree.

## ECOSYSTEM AND WEATHER

There will be many reasons to create "weather" within the colony, and not let air be stagnant. No rainstorms or tornadoes, but with a dynamic ecosystem there will be times when certain populations in the ecosystem will have to be managed. Controlled weather like wind, heat, cold, or moisture level could be used to manipulate insects, plants, or animal populations. The added benefit of introducing such weather variations is breaking up the monotony of an unvarying climate condition for the colonists.

## POLITICAL SYSTEMS

The people on Mars who have the immediate needs of their own daily lives, will have little concern for what a government 60 million miles away thinks about how they should live. An independent government will need to be established quickly, identifying freedoms, responsibilities, elections, etc. Initially, a round-table council based on consensus governance will be the most effective, but as the population grows, a parliamentary democracy should be established. Representation would not be based on geographic location (as commonly done on Earth) but would be aligned with varying community interests. Systems of criminal courts and dispute resolution will need to be established quickly to prevent anarchy and chaos from taking hold, and to protect human rights. Incarceration will not be practical, so systems of accountability through community service and restitution will probably be used. With a cooperative community structure where resources are shared, crimes born of inequality would probably be extremely rare.

Land will need to be in the public domain, and private land rights will have no meaning in a confined environment; colony survival will always be paramount.

Water, air, land, and energy use will be critical issues, since they will be scarce resources that are essential to survival for every person. The use of these resources needs to be intensively managed by the political system to make sure no one is ever deprived of such basic necessities, and checks & balances must be in place so no one can contaminate, deplete, or have undue control of them.

Robotics and mechanization will need to have its rewards shared to support the people and colony, and not be allowed to concentrate power. There will be pressure for wild unfettered resource speculation and extraction, but the political system must establish the collective good of the people over raw profiteering, to prevent oligarchies and indentured servitude based on access to air, water, and a shuttle ride back to Earth.

The representative government will serve in the larger framework of a cooperative venture, similar to how **cooperative business** models operate. In this model, all workers are owners and share in the responsibilities and rewards.

Systems of community Merit may be established, where service to the community can be rewarded. There will still need to be a currency system so people can feel that hard work and creative innovation can improve their own lives.

**LOCAL ECONOMY**

Scientist and engineers will be the first settlers, but will ultimately not be the backbone of the colony. A complex community and economy must quickly emerge to keep the settlement healthy and viable. Children will be born and grow up with their own unique sense of self and life dreams. There must be room for all of that to happen and thrive.

Design innovation and creativity will be important—"*Martian Ingenuity*" will become a mantra.

Some cottage industries will include:

**Synthetic fibers** - for clothing and materials; from raw materials found on Mars.

**Natural fibers** - silks and natural fibers from plants that prove to be viable in the growing conditions there.

**Clothing and fabric making** - skills and methods for making clothing.

**Pigments** - by Earth standards, Mars is not a colorful place. Pigments for paints and dyes will be a hot commodity, so the creation and manufacture of color pigments from Martian material will be a demanded skill.

**Construction materials and methods** - will be in high demand, such as stone work, masonry, metal work, etc.

**Stone artisans** - elegant stone cutting by both high-tech and traditional methods from in-situ materials will be an important trade skill.

**Ceramic tiles** - making tiles from Martian clays and pigments will be a high-demand craft.

**Food preparation** - A simple cooking vessel—a wok—will probably dominate cooking. Colonists will eat in shared kitchens and dining, so chefs and nutritionists will be in demand.

**Farming** - while factory hydroponics will have its place, it would be folly to rely on that for food. Terraced farms and rooftop gardens will need to be created. Crops suited to Mars need to be developed. Animals will need to be tended.

**Ecologists & biologists** - complex self-sustaining ecosystems will need to be developed to maintain air for breathing, create living systems that can survive lapses in technology, bring natural beauty to the community, and provide natural biofiltration of toxins in the air and water.

**Architects, artists, creative thinkers** - people to help create typologies, vernaculars, and styles that bring a unique pride and identity to Martian culture. People who can help prevent a feeling of a bland and sterile life.

**Hospitality** - tourism from Earth and maybe elsewhere will require lots of people to service the needs and desires of those visitors, who would be staying for 12-24 months at a time.

**Exercise& Recreation** - health coaches will be needed, as well as recreation support infrastructure. Eventually, ski & snowboard slopes could be built—imagine the aerials in Point4G!

**Infrastructure management** - water systems maintenance, air systems, energy systems, public toilet & bath house management, etc.

Other critical skills needed in the community include: **teachers, tech innovators, tech manufacturing, medical professionals, medicine manufacturing**, to name a few.

## PRIORITIES IN MANUFACTURING

While there will be many materials and production that will have to be perfected for the needs of the community, a list of priority items would include:

- Regolith concrete
- Concrete masonry unit (CMU) blocks & structural beams
- Stone masonry craft for structures, tile, and flooring
- Metals manufacturing - structural, electrical, mechanical, ornamental
- Steel for rebar in reinforced concrete
- Steel wire for cables used for construction and colony roof support
- Ceramic tiles for flooring, walls, decoration
- Color pigments for dyes, paints, glazes, etc.
- Radiation-shielding glass technologies
- Alcohol and methane from raw materials and from methane digesters - for community use and rocket fuel
- Water & air filtration systems
- Cookery, utensils, ceramics, and glass vessels
- Perfection of materials and resource recycling: zero waste necessity.

## INTERPLANETARY ECONOMY

On a most basic level, "Point4G" tourism will be important: LowGrav retreats, exotic sports, and sightseeing. Typical retreats would last 18-24 months, given the travel time from Earth, and would cater to people taking a gap year (or two) from school, or those like backpack travelers on Earth who take a year or two off of work to travel and gain life experience, often while working locally in their destinations. Those people could be an important part of the work force, and may resemble *Peace Corps* or *Doctors Without Borders* volunteers who work under multi-year commitments.

Refined exotic metals and specialty materials for electronics and quantum devices could be an important export. Unique silks, pigments, medical technology that arises out of necessity of life on Mars, air/water filtration technology (which may become increasing in demand on Earth) could be developed and exported back to Earth or to other outposts.

On the grander scale, Mars could become the leader in manufacturing of exotic space materials used in the construction of space stations and vehicles. It could also become the lead manufacturer of space stations, space vehicles, and construction equipment. A manufacturing industry, much like the history of the auto industry, could arise where standardized space module components are mass produced, being able to be launched into space at much lower cost due to lower gravity and thin atmosphere. A Mars colony would also likely be a staging and jump-off point for exploration and resource extraction in the asteroid belt and beyond.

**CHILDREN AND EDUCATION**

With plants, animals, colors, textures, and complexity in the community, children can thrive and be stimulated. Earth-like schools won't exist, instead the entire colony will be the classroom. With technology and innovation being important traits, education will be constant and blended into all aspects of daily life, for children and adults.

**RESOURCES**

Construction Materials:

Wan et. al., 2015. *A Novel Material for In Situ Construction on Mars: Experiments and Numerical Simulations*. Dept. of Civil and Environmental Engineering, Northwestern University.

*M7MI Hydraform block making machine brochure*. Hydraform.com website.

Microterraforming:

Interview with Omar Pensado Diaz, 2004. *Mars: Goldilocks' Oasis? Thinking Locally, Before Acting Globally*. Astrobio.net website - Interview, March 10 2004.

Radiation Shielding and Glass:

Thibeault et. al., 2012. *Radiation Shielding Materials Containing Hydrogen, Boron, and Nitrogen: Systematic Computational and Experimental Study - Phase I*. National Institute of Aerospace, Hampton Virginia.

Lin, et. al., 2019. *Glass-like transparent high strength polyethylene films by tuning drawing temperature*. Polymer, Vol. 171, 2019-05-08: 180-191.

Hurley [ed.], 2019. *Higher-Strength Plastics: How to Turn Polyethylene into 'Transparent Aluminum'*. TechBriefs.com website - Billy's Blog, 2019-04-17.
*Corning Radiation Shielding Glass* - Corning.com website.

Filtration:

Matthews, 2019. *How to Make a DIY Water Filtration System Using Sand or Gravel*. Mother Earth News website - DIY section, 2019-03-12.

*Biosand Filter Manual*. Center for Affordable Water and Sanitation Technology website - cawst.org; biosandfilters.info. 2008-01.

# 18: STAR CITY, MARS

**George Lordos**
Massachusetts Institute of Technology
glordos@mit.edu
**Dr. Alexandros Lordos**
University of Cyprus
Alexandros.Lordos@gmail.com

## PROLOGUE

*The Mars Treaty was finally signed at the United Nations in Geneva on December 1st, 2024, two months after the launch of the Starship* Heart of Gold *on the first-ever journey to Mars by a private company. The Treaty, signed by member states from every continent, enabled 74 private companies to join forces with large and small states to fund the Mars Consortium, which established Star City on Mars.*

*For the new generation of far-sighted leaders who signed the Treaty, the establishment of a multicultural city on Mars was above all a political project. They came together to support the opening up of Mars for humanity with a singular goal: to send powerful messages of peaceful collaboration, shining the bright light of hope back to every corner of the Earth. These messages from our fellow humans on Mars helped to consolidate the long healing process that had just begun after a decade of rising extremism and polarization which ended in a wave of the peaceful revolutions of the '20's by moderate majorities in one country after another.*

*For the Consortium privateer members and their investor backers, the Mars vision was far more prosaic: companies such as SpaceX and Blue Origin had started transporting people and equipment to the Moon and Mars at low cost, opening up new opportunities for commercial activity on other worlds. Star City would go on to integrate and deploy technologies for advanced subtractive construction, in-situ resource utilization and additive manufacturing, thereby powering its own sustainable expansion. In a short period of time, with its accumulated and hard-earned know-how, the Star City Development Guild on Mars naturally emerged as the lowest-cost provider for the construction of new outposts and cities on Mars.*

*And so it was that in the month of September of 2029, the first 1,700 tons of nuclear reactors, roadheaders and other heavy equipment was delivered to the chosen Star City site in Utopia Planitia, followed by the first sixty humans in November of 2031. By July 2042, Star City's population had surpassed 1000 persons, and the Guild already had the first orders for new outposts: one from the United Arab Emirates, one from Switzerland, and one from Israel. Star City was open for business, and the order book for new outposts started growing.*

## PART I – THE STAR CITY CONCEPT

**If We Want to Put Humans on Mars, We'll Need Humans on Mars.**
Star City exports Hope, Promise and Know-How. The Star City Development Guild builds new cities on Mars, inspires millions of young people who watch Star City's blinking lasers from their backyard telescopes and trains leaders who return to Earth to rebuild devastated communities. If humans from all around the world can work together to build cities on Mars, where even the air we breathe must be made in a machine, they can inspire the rest of us to work together to make our Planet Earth a better place for all.

The New Space companies are in the transportation business, moving people and equipment between Earth and Mars. The Star City Development Guild (also known as "the Guild") are the experts who can build anything on Mars, from a small dome to a complete town for thousands of people. With their fleet of mobile nuclear reactors and heavy roadheaders, their extensive industrial base, and above all their know-how and expertise, the Guild earns enough from the Mars construction and outfitting business to pay for the necessary imports of advanced high-tech parts and spares.

The Star City University manages exploration and scientific research on Mars, on behalf of sponsors from Earth. It also offers the ultimate leadership training programs, pairs researchers and visitors from Earth on expeditions, and produces art, poetry and Martian musical instruments sold to Earth. The Star City Spaceport is investing in an ever-growing network of refueling nodes at the Sun-Mars Lagrange points, on Phobos and in the asteroid belt. Their strategy is to become the go-to supplier for propellant beyond Earth orbit, opening up deep space to their customers.

## PART II – BUILDING STAR CITY

**Site Selection**

We have reviewed the USGS maps of Mars [01] NASA's Human Landing Sites workshops [02] and also its ongoing project for Subsurface Water Ice Mapping (SWIM) [03]. In producing a shortlist of regions suitable for Star City, the criteria were: in Northern Hemisphere; low altitude for safer landings; abundance of shallow subsurface water ice deposits; rock quality suitable for tunneling and supporting tunnels; suitable landforms for tunneling inside mesas or crater walls; availability of other critical resources such as metals, silicates, sulfates etc. and proximity to areas of scientific interest and natural beauty. Tunneling in competent rock formations is used for the construction of the majority of the habitable volume. Accordingly, the first landings in the 2024 - 2026 timeframe investigate one or more of the candidate sites to validate the geomorphological assumptions and provide data to refine the tunneling concept of operations for the Star City campaign which follows from 2028 - 2042.

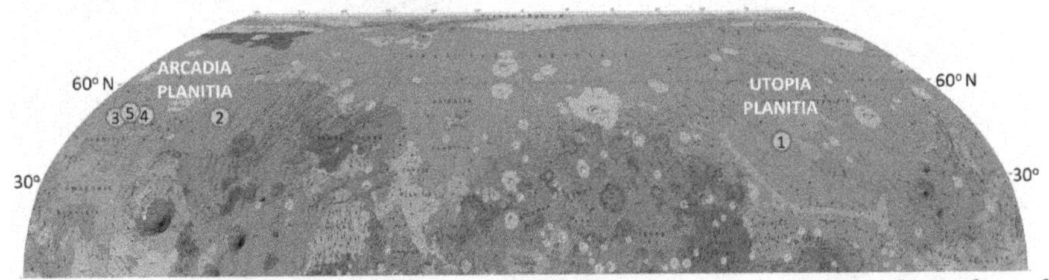


**Figure 1**: *Candidate locations for Star City along mid-northern latitudes of Utopia and Arcadia Planitia. (Image credit: USGS, https://doi.org/10.3133/i1286)*

**Subtractive Construction**

Habitable volume at Star City is created primarily by *subtractive construction.* Each of the five founding villages around the crater uses crawler-mounted roadheaders, two 130-ton Sandvik MT720 and two 60-ton MT361's. These can be operated economically in rock formations with uniaxial compressive strength up to 140 MPa. Each roadheader is all-electric and can be remote-controlled via PLC to create up to 480 $m^3$ of new habitable volume per day. The Guild uses these to construct secure, long-lived tunnels for transportation, storage, residences, aquaponics, industry, services, and pumped hydro storage.

**Figure 2:** *a Sandvik MT720 Roadheader (Image credit: Sandvik)*

**Colony Urban Concept**

Star City can be constructed in craters of diameter from 2km - 10km. The scheme presented in Fig. 3 below has been fitted to a 10km crater located in Utopia Planitia at 40.8N, 99.5E, altitude -4,000 meters [04], but can be adapted for other craters. The site first receives power, ISRU and heavy equipment in the 2029 window. In November 2031, two Starships by SpaceX with 30 passengers each land near the site. Each of the two Starships carries five teams of six persons, one from each of the five continental groupings. Thus, there are 12 founders of the "red" sector 1 village in Figure 3 below, 12 founders of the "blue" sector 2 village, and so on.

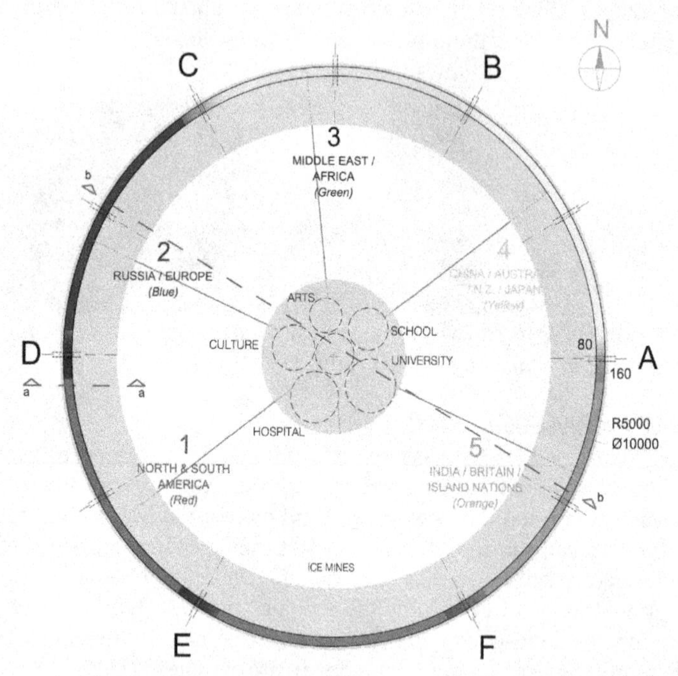


**Figure 3:** *Urban plan of Star City. Each of the five founding villages grows perimetrically along the crater wall. The villages are connected by transportation and services tunnels so they can share people and resources and come to each other's aid. As Star City grows, the five villages merge into a city with five culturally distinct neighborhoods. At some point, once they can afford it, they pool their resources to build facilities for arts, culture, school, a university and a hospital at the center of the crater. (Image credit: Delta Architects)*

In all, the first 60 persons on Mars start five villages simultaneously. Between 2031 and 2042, more people and equipment arrive and the villages continue to expand their habitable volume perimetrically along the crater wall, until they merge to form Star City. All tunnel tailings are mined for water ice and useful resources such as metals and silicates, while the north-facing inside wall of the crater is reserved for supplemental water ice mining.

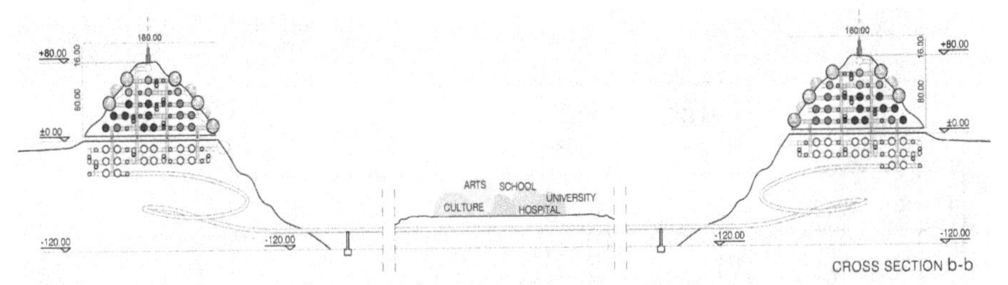


**Figure 4:** *Cross sectional view b-b. The five villages have panoramic views of each other across the crater, and of the Martian horizon. Each village is self-sufficient, with its own power generation and backup power system, and its own aquaponics and industrial zones. Each of the five villages is tackling the same survival challenges in their own way, and in a contingency or emergency, each can rely on help from the other four. (Image credit: Delta Architects)*

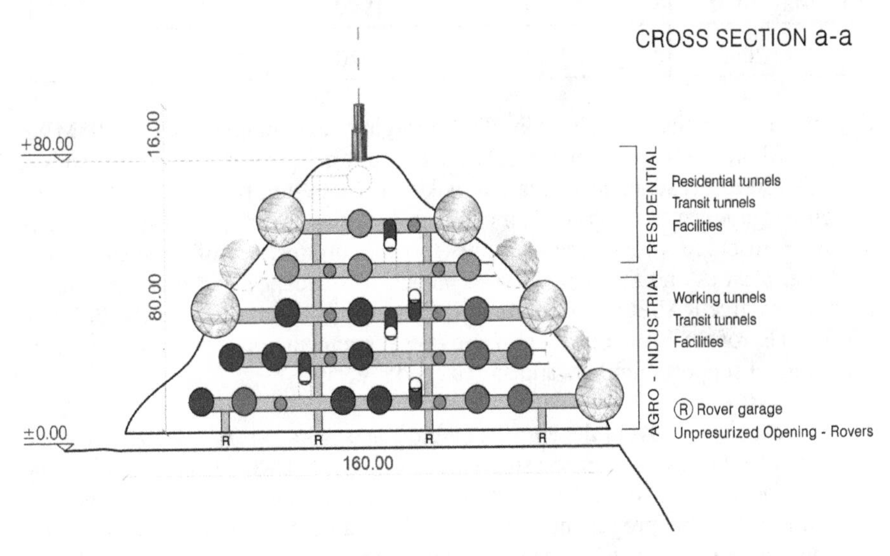


**Figure 5:** *Cross sectional view a-a. The residential tunnels and domes are located along the top of the crater wall, with panoramic views from the domes to the other four villages or to the distant horizon. Rover access is at ground level of the surrounding plains, with unpressurized rover garages situated directly below the agro-industrial tunnels. (Image credit: Delta Architects)*

## Concept of Operations

The logistical plan is counterintuitive: start big, with 17 cargo Starships in September 2029, followed by 16 cargo and 2 passenger starships in 2031. The full schedule of cargo and passenger arrivals is shown in Table 1 below.

**Table 1:** *Schedule of Starship arrivals per synod, 2029 – 2042*

| | Landing | Cargo ships | Passenger ships | Cargo (tons) | Cumulative passengers |
|---|---|---|---|---|---|
| 1 | Sept 2029 | 17 | 0 | 1700 | 0 |
| 2 | Nov 2031 | 16 | 2 | 1720 | 60 |
| 3 | Dec 2033 | 10 | 2 | 1060 | 180 |
| 4 | Feb 2036 | 8 | 4 | 920 | 420 |
| 5 | April 2038 | 8 | 4 | 920 | 660 |
| 6 | May 2040 | 8 | 4 | 920 | 900 |
| 7 | July 2042 | 8 | 4 | 920 | 1140 |

Seventeen cargo Starships land in 2029, bringing two complete 85-ton, 95MWe Integral Molten Salt Reactor (IMSR) nuclear power systems [05], five Sandvik MT720 120-ton roadheaders, five Sandvik MT361 60-ton roadheaders [06], five 35-ton Megapower 2 MWe reactors [07] to be paired with five 40-ton crawler-mounted resource processing systems for the production of oxygen, water, methane, plastics and the separation of metals. The balance of the cargo space is taken up by five RASSOR rovers for surface trenching / leveling / ISRU [08], five ATHLETE rovers [09], other earth moving equipment, power harnesses, spare parts, tunnel support members and stored food.

The nuclear power rovers deploy at a safe distance away from the crater and are connected to the resource processing rovers. RASSOR rovers harvest ice-bearing regolith for delivery to the resource processors, which produce methane and oxygen to refuel the propellant tanks of the landed cargo Starships. The regolith-moving equipment is used to shape ramps, berms and landing pads. The heavy equipment is gradually, remotely positioned at the site of each village as required.

Another sixteen cargo Starships land behind berms on the prepared landing pads in 2031, along with the first 60 humans in the two passenger Starships. Additional heavy equipment this synod consists of one more 85-ton, 95MWe IMSR reactor, another five MT720 & five MT361 roadheaders, five 80-ton additive manufacturing systems, five sets of life support systems for the first tunnel, tunnel support members, more spare parts and more stored food. The first colonists commission the heavy equipment and commence tunneling operations while living in the cargo and passenger Starships.

**Figure 6:** *a Sandvik roadheader cutting a tunnel (Image credit: Sandvik)*

The extraction rate is expected to be between 20m$^3$ - 40m$^3$ of rock per net cutting hour. With 10 out of 20 roadheaders working at any one time, and with 12 contact hours per day (Sol), each village can create up to nearly 1,000 m$^3$ of new habitable volume per day, and the Guild can complete up to 60 km of 6m-diameter tunnels per Earth year. The priority for each village is to cut a short transversal tunnel through the crater wall and to excavate, support and finish the first pressurized transit tunnel which will connect all villages. Both of these tasks can be completed within 9 months of the crews arriving on Mars. Each roadheader team requires four people to remotely operate or monitor all systems including cuttings removal and installation of tunnel support in the form of reinforced Martian concrete. The completed tunnel is pressurized with warm, humid air. Any air leaks into the cold rock/regolith will seal themselves as water vapor freezes and expands into the pores. The aquaponics farm, residences and additive manufacturing systems are temporarily deployed inside the single transit tunnel which connects all five villages, for easy access and mutual support. At each village, an off-ramp leads to the rover garage airlocks at ground level.

Passenger flights from 2033 onwards bring 60 persons per Starship, twelve for each village. The population expands to 180 in 2033 and while tunneling continues, the priority shifts to creating and outfitting permanent spaces for aquaponics and industry in their own dedicated tunnels and tunnel segments. From 2036 onwards, arrivals increase to four passenger ships of 60 persons each. As the population expands, the emphasis shifts once again from tunneling and setting up working spaces to optimizing production output, to stockpiling water, and to the personalized outfitting of the residential spaces at the top of the crater rim. Net of those who will choose to return to Earth, the population is expected to surpass 1,000 by 2042.

**Energy Rich Colony**

An integrated power grid stepped down to 1000V, 240V and 110V AC is supplied by the three 95MWe IMSR load-following nuclear reactors by Terrestrial Energy, which operate at a useful 600°C providing high quality process heat for industry, and are walk-away safe in the event of an accident (Fig. 7 below).

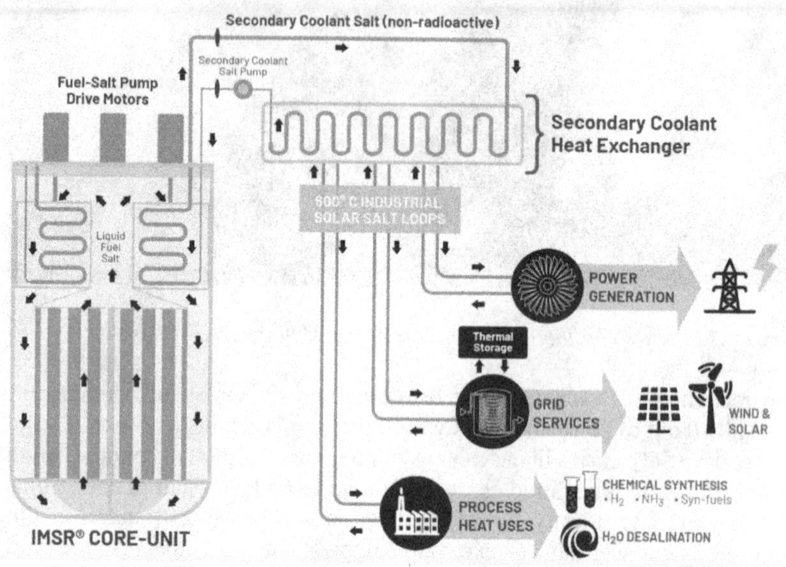


**Figure 7:** *Replaceable sealed IMSR core unit (Image credit: Terrestrial Energy)*

All cooling takes place in the molten salt loop, with a secondary salt circuit transferring heat to the water loop and driving the steam turbine. The lifetime is 7 years, after which the core must be replaced. Each of the three IMSR reactors can carry the entire load if required. Exploiting the height differences, five pumped hydro systems holding ~22,000 tons of water each are installed in the mid 2030's, providing up to 30 minutes of backup @ 20MWe. Five additional MegaPower 2MWe reactors [07] serve for ISRU propellant and also as redundant emergency backup for the five villages.

**Resource Processing**

Each of the five resource processing system-of-systems consists of the following elements: RASSOR, a robot designed at NASA's KSC to collect regolith and deliver it for processing [08], a milling and flotation separation system for up to three concentrates, similar to lab desktop milling systems used to prove up and qualify mining deposits on Earth, a scaled-up version of MOXIE, a Sabatier reactor and small scale aluminum, plastic and steel- making equipment to produce pure resource feedstock suitable for additive manufacturing. The tailings extracted every day from tunneling operations are swept up by automated RASSOR units and transferred to the nearest processing unit for separation of the water and other valuable resources. Clays and other regolith is baked into Mars bricks and the remaining regolith is removed from the tunnels and shaped into berms and ramps.

The most important resource is water, on account of the diversity of its uses and its criticality in life support and the manufacturing of return propellant. Accordingly, the principle of a "water rich" city has been adopted at the highest design level and a ten-year water budget has been created for Star City. Conservative assumptions made are that at least 8% by volume of the tunnel cuttings is water ice or the water content from hydrated gypsum, and that food and life support water recycling loop closure is only 80%. All water uses have been accounted for and a margin of over 40% in excess water production capability is targeted, as shown below in Table 2:

**Table 2:** *Star City's 10-year water budget is based on the design principle of a water-rich city. The budget shows unallocated water reserves of 460,000 tons plus pumped hydro reservoirs of 113,000 tons.*

| Water allocation | Pumped Hydro | Propellant | Food | Life support | Industry | Radiation Protection |
|---|---|---|---|---|---|---|
| Recycle % | 0% | 0% | 80% | 80% | 0% | 0% |
| Units req'd | 5 | 95 | 1776000 | 1776000 | 6600 | 100 |
| Tons / unit | 22,723 | 1,000 | 1.0 | 0.20 | 0.25 | 226 |
| Total allocation (t) | 113,614 | 95,000 | 355,200 | 71,040 | 1,650 | 22,619 |
| Total water allocated or consumed in 10 years: | | | 659,123 | tons | | |
| Total water unallocated (margin): | | | 462,157 | tons | 41% | |

**Additive Manufacturing**

A small, flexible industrial base is automated and replicated across the five villages. These systems produce the majority of life support and habitation systems, primarily by employing the additive manufacturing of plastic and metal components whose feedstock was itself also produced on Mars from in-situ resources. Robotic, automated assembly systems integrate the additively manufactured parts with complex parts imported from Earth and carry out quality control. In addition, there is a specialized steel casting unit for the mass production of replacement roadheader picks, as the pick consumption rate can be as high as 1 pick per $10m^3$ in rock with high uniaxial compressive strength (>140MPa).

**Food Production**

The food production system requirements have been modeled on the following per-person assumptions for the primary production only, i.e. plant matter: 10KW of power for LED's per person per year, 60 $m^2$ of growing area per person, and 57 kg / person / year of NPK fertilizer [10]. If the growing racks are stacked in double height, the maximum aquaponics tunnel length for 200 persons (i.e. per village in 2042) is 1,000m. The power requirement for 200 persons would be 2MWe, which can be met by the backup generators. In addition to edible plants, the system produces fish, chicken and eggs. The fish are in symbiosis with the plant roots, helping to recycle nutrients, and the chicken are fed with waste biomass from the aquaponic greenhouses. Each village of 200 would require 11 tons of nutrients per year, elevating the importance of recycling of nutrients as we do not yet know enough regarding the abundance of phosphorus and potassium at candidate landing sites.

**Habitats and Private Homes**

Once the second passenger flights arrive in 2033, the villages deploy their heavy equipment towards creating dedicated, separate tunnels for aquaponics, resource processing, industry and private residences. As shown in Fig. 5 above, private residences are near the top of the crater rim, benefiting from easy access to small, shared neighborhood domes which offer real sunlight and views across the crater rim or out to the horizon. Cross-level transit tunnels, available only within villages and shown in purple in Fig. 8 and Fig. 9 below, interconnect the residential, working and rover garage levels. There is shirtsleeve access from every residence, to every workplace and to every pressurized rover. Spacesuits, accessible via suitports on rovers, are only required for field expeditions and for limited external work.

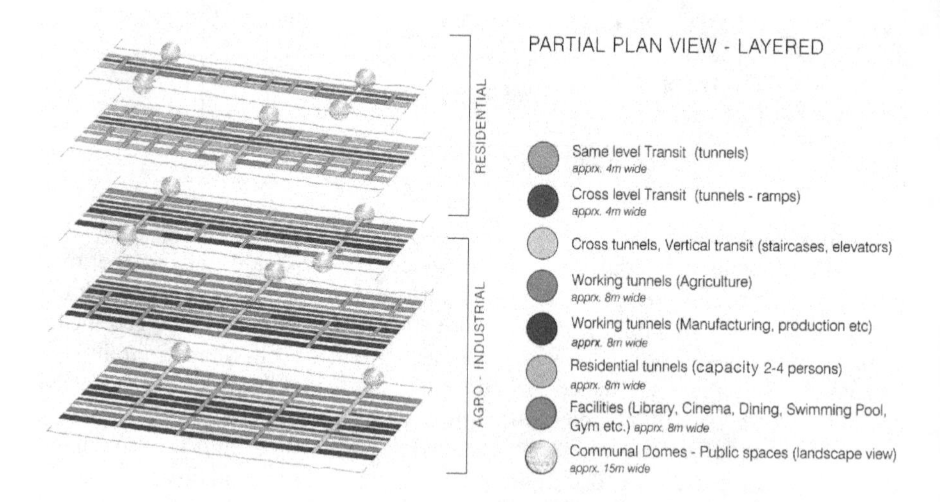


**Figure 8:** *Residential tunnel segments (brown) are interspersed with facilities segments (red: libraries, cinema, dining, swimming pools, sports, etc.), and have easy access to a nearby neighborhood dome with spectacular panoramic views of Star City and the surrounding plains. (Image credit: Delta Architects)*

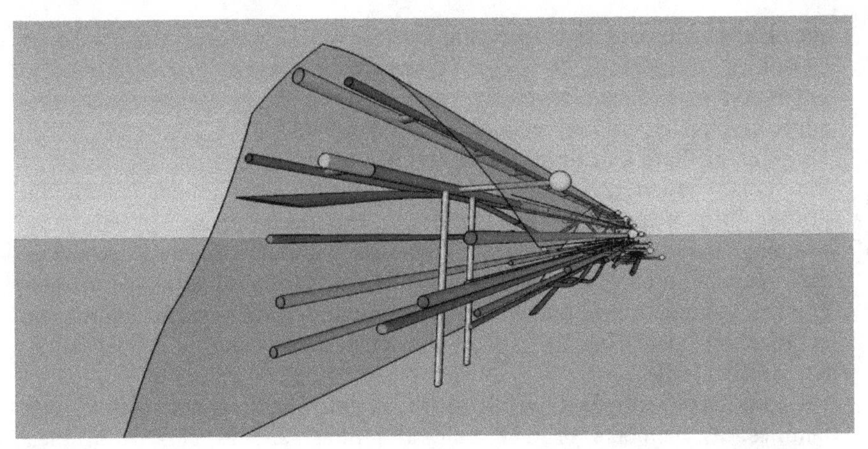

**Figure 9:** *Every tunnel in the village has roadway access to every other tunnel at all levels, enabling the transportation of large volumes of equipment and products. This 3D cutaway perspective from an early phase of construction shows the transverse interconnections between cross level transit tunnels (purple) and intra-level cross tunnels (green) (Image credit: Delta Architects)*

## PART III – A HUMAN COMMUNITY ON MARS

### Principles of Human Development

Society, culture and governance systems in Star City have been designed to bring out the best in human nature - but at the same time, staying within the realities of human nature. Specifically, the following axioms and insights, well established in human historical experience and through scholarly research in the fields of sociology, psychology and social anthropology, have been factored into Star City social system design, to foster the development of a peaceful and resilient human community on Mars:

1. **Humans are not perfect, but can learn and grow** through education, mentorship, reflective practice, and innovation. However, rigidly defined hierarchies, a fixed mindset, lack of opportunities to reflect, lack of intellectual humility, absence of mentors, or norms that are hostile to change might undermine capacity for learning and growth [11]
2. **Humans enjoy psychological and physical well-being** when they experience intimacy and authenticity in their relationships, accept and address daily life challenges, pursue objectives that are meaningful to them, and are mindfully present in the here-and-now. If however these needs are not met, mental health might deteriorate leading to multiple disturbances in the individual, household and community systems [12]
3. **Humans can be resilient in the face of various types of adversities**, if equipped with appropriate cross-cutting skills (e.g. collaborative skills, distress tolerance), task-specific competencies (e.g. conflict mediation competence, parenting competence), supportive networks of relationships (e.g. with spouses, with peers or with professionals) and adaptive norms (e.g.

social tolerance, inclusive leadership). However, absence of such resilience capacities can magnify the negative impact of stressors and contribute to utilization of maladaptive coping strategies (e.g. apathy, dominance) that adversely impact various levels of the social system [13]

4. **Humans affiliate and organize themselves in groups**, at multiple levels of the social system (e.g. household group, professional group, community group) as a means to derive livelihood, experience connectedness, draw meaning, and protect against threats. However, various maladaptive processes can creep into group dynamics, such as marginalization of individuals or sub-groups, in-group confirmation bias, out-group stereotyping and inter-group polarization, increasing the risk of disrupted social cohesion, tribalization, and conflict [14]
5. **Humans have multi-layered identities**, as individuals, as members of their families, as members of their cultural groups, and as citizens of their community or country. Simultaneously embracing several such layers of identity underpins personal coherence and social cohesion while enabling adaptive functioning across multiple contexts. However, when whole layers of identity are neglected, suppressed or otherwise undermined, negative consequences can include loss of direction, apathy and social fragmentation [15]
6. **Humans establish institutions to govern their collective affairs**, particularly to systematize decision making processes, establish rules and norms, and regulate the flow of resources. Institutions that are effective, reflective, inclusive and adaptive contribute significantly to human development. However, institutions can become dysfunctional and a burden to the community if they become rigid and bureaucratic, or lose their service orientation, or become corrupted by self-interest [16]
7. **Humans have aesthetic and / or spiritual needs**, which, while diverse in scope and means of expression, tend to revolve around a desire for self-transcendence. Meeting such needs enhances capacity to, adopt a wise perspective in daily challenges; resist the corrosive influence of narcissistic self-absorption; derive strength to persist through adversities; and generally, enjoy an enhanced sense of meaning and purpose. However, spiritual self-expression can occasionally mutate to sectarian hostility of other cultures or reduced willingness to engage with societal challenges through science and civics [17]

All aspects of Star City's design, including physical configuration, societal norms, education, economy, culture, institutions and governance, have been designed in accordance with the above understanding of human development.

**Urban Planning for Social Cohesion**

The city's physical configuration, comprised of five separate villages - where distinct professions and cultures can flourish -- and a central hub that houses the university, the sports centre, the cultural centre and the general hospital - encourages the development of multi-layered identities, where citizens simultaneously experience a sense of belonging and connection at the level of their village and at the level of the city.

**Figure 10:** *View from the central hub to the red sector village (Delta Architects)*

The distinct culture of each village is expressed through 'soft' elements, such as distinct traditions in clothes weaving, interior design and aesthetic or spiritual practices, as well as 'hard' elements, such as the unique contribution which each village is making to the city's infrastructure and economy (e.g. by specializing in tunneling, water extraction, or rover construction). While each village is unique in all these ways, it is at the same time connected to the other four villages, both for the necessities of the city's functioning and for cultural - aesthetic enrichment. Citizens of each village routinely visit the other villages, both for work-related reasons (e.g. to coordinate mining or life support operations) and for recreation (e.g. to enjoy a meal at a restaurant from a different cuisine, to purchase clothes or artifacts with a different cultural aesthetic, to procure interior design services of a distinct style, to enjoy local artistic productions, and to converse with philosophers or spiritual guides from the specific cultural perspective). All such recreational facilities, including restaurants, clothing shops, artisan workshops, small theatres and places of spiritual / philosophical contemplation, are located in woodland domes, with at least one such dome in the administrative boundaries of each village. Other institutions that are located in each village - for the benefit of its own residents - include, a local nursery and primary school, for children up to age 12; a health center, where basic medical services are provided; and the village's local administration, which is responsible for permits, management of village public spaces, and facilitation of inclusive governance processes at the village level.

In contrast to the five villages, the central hub is home to institutions that transcend culture and serve all of Star City. These include the University of Mars at Star City, where secondary, tertiary and life-long education is provided, and where basic and applied research in both the physical and social sciences is taking place; the General Hospital, where all specialized medical services are provided; the Central Auditorium - with seats for all 1,000 city residents - where artistic productions and other public events take place; the Central Park, which is a domed, richly wooded terrarium that hosts a large variety of freely circulating land animals and birds; a multi-purpose Sports Centre, where gravity-adapted versions of sports that are popular across different cultures (e.g. football, soccer, basketball, baseball, rugby) can be played as friendly games or as competitive tournaments; the Mars Museum, where younger and older citizens are able to interactively engage with the history of Martian colonization but also of the history of diverse Earth cultures; and finally the city's central administration, which is further described in the section on governance below. Overall, the layout of Star City, with its five distinct villages and the central hub, has been designed to promote a healthy, harmonious, balanced and fulfilling human experience, where citizens can combine the intimacy of small-group identities with the creative openness of a diverse society; and the calming, mindfulness-promoting influence of a natural, aesthetically pleasing environment with the physical and intellectual stimulation of a forward-looking technological civilization.

**Life Skills for Societal Resilience**

Similarly to the physical layout, the educational policies, programs and institutions of Star City have been designed to nurture well-rounded, competent, engaged and versatile citizens, who can contribute to a vibrant economy while adapting as circumstances change and new challenges emerge. Specifically, Star City's education places a lot of emphasis on the acquisition of a broad repertoire of cross-cutting life skills, which then form the basis for the acquisition and effective utilization of task specific competencies, in accordance with the professional roles of each citizen. Life skills are neurally encoded personal assets, which become activated in the course of daily life to resiliently meet emerging challenges [18]. Specific skills that every citizen is encouraged to acquire include: skills for emotional well-being (acceptance, mindfulness, emotion regulation, distress tolerance, self-respect, a hopeful outlook); skills for learning (intellectual humility, intellectual curiosity, growth mindset, critical thinking, creative thinking, and capacity for knowledge application); skills for self-direction (future orientation, decision making, planning, persistence, flexibility, and a capacity for course correction); skills for collaboration (empathy, kindness, communication, cooperation, negotiation, and leadership); skills for co-existence (sense of fairness, gratitude, forgiveness, tolerance of diversity, gender equality mindset, and capacity for inter-generational partnership); and skills for effective participation into the affairs of Star City as a whole (civic awareness, civic responsibility, civic adherence, sense of agency, inclusive mindset, and civic judgment). To ensure acquisition of life skills, AI-enhanced assessment systems are routinely deployed to all residents of Star City, using cutting-edge psychometric batteries which include self-report rating scales, situational judgment tests and behavioural observation. These ongoing assessments result in determinations, for each resident, of which skills each resident possesses to an outstanding extent; which skills he or she is adequately competent in, but without excelling; and for which skills the resident appears to be currently in deficit. This results in a personalized curriculum for life skills acquisition, which assigns strugglers to ongoing training for needed skills while assigning those who excel to provide mentorship and supervision. As for the youngest residents of Star City - those children born on Mars - acquisition of a comprehensive suite of life skills has been built into the primary education curriculum for the more basic skills, and in secondary or tertiary education for the more complex and advanced skills.

**Dual Professional Specialization**

On the foundation of cross-cutting life skills, as described above, robust professional identities and specializations have been established for each resident of Star City. Specifically, each resident has received training and acquired expertise in two distinct professions with the balance of time devoted to each fluctuating according to city needs and their own preferences. One of the professions is related to city basic infrastructure and services (e.g. life support, tunnel construction) while the other is related to culture, society or self-expression (e.g. artist, educator, clothing designer). The dual specialization system has multiple benefits: Firstly, it reduces the risk of any citizen finding themselves without a useful function as economic needs shift, given that they would be able to fall back on their other profession during the transition period. Secondly, it ensures that all citizens remain personally flexible, adequately stimulated and acquire a broad spectrum of life skills, as they learn to operate across 'hard' survival-oriented professions and 'soft' expression-oriented professions. And thirdly, as a corollary to the above, it prevents the evolution of polarized socio-economic tribes, of "conservative miners" Vs "liberal artistically oriented urbanites". Specific "hard" professions (i.e. related to basic infrastructure or services) for residents of Star City are available within various economic sectors, as follows:

- **Energy:** Operating and maintaining the energy grid
- **Habitation:** Constructing tunnels, maintaining heavy equipment, processing resources
- **Life Support:** Producing water, operating city-wide life support systems
- **Industry:** Manufacturing parts for domes, manufacturing life support systems, manufacturing spares for heavy equipment
- **Food:** Working at aquaponics & chicken farms, packaging and distributing food
- **Medical:** Staffing the medical centers in the five villages and the general hospital in the central hub, supported by AI diagnostic systems and robotic surgery assistants

While each resident is contributing to one of the above essential sectors related to the physical infrastructure, they have each also freely chosen one ancillary profession, pursuing their own sense of calling and personal mission, which contributes to the social, cultural and civic life of Star City. Such ancillary professions include, but are not limited to, the following –

- **Educators:** Primary educators, working out of local village schools, are responsible for the acquisition of cross-cutting life skills, as well as literacy, numeracy, and basic principles of science and engineering, in children up to age 12. Secondary and tertiary educators are housed at the University, and have specialist professional experience in the specific fields they are teaching, which in many cases is their other profession.

- **Artists:** Residents who pursue the calling of Artist have been encouraged to be versatile across prose, poetry, visual arts, and acting, so that they can provide holistic expression to the experience of life in Star City. Artists are based in the villages, where they have their workshops, but hold regular events and exhibitions both in the villages and in the central hub. Their work typically represents a blend between the unique culture of their own village and the emerging culture of the city as a whole, and of Mars-based humanity more broadly. Beyond providing a local medium for cultural self-reflection, their literary and artistic productions are also licensed for distribution on Earth, which contributes to a regular stream of external revenue for Star City.
- **Scientists:** They are housed at the University of Mars, conducting both basic and applied research in the physical and social sciences. Their work, however, is thoroughly connected with relevant Star City institutions (e.g. of industry, governance, social support) to ensure that the gap between science and policy is bridged at all levels. Furthermore, they take on research grants and consultancies on behalf of clients and institutions on Earth, and in this way constitute an intellectual export industry of Star City. Such external grants and consultancies can fall into one of two broad categories: Firstly, support for the development of other Martian cities, or even cities elsewhere in the solar system, based on lessons learnt from the technical and societal development of Star City; and secondly, transferring lessons learnt in Star City for more peaceful and sustainable communities on Earth.
- **Cultural Elders:** Specific roles and titles for cultural elders are contingent on each village's unique cultural heritage, but can include citizens who take on the role of priest, rabbi, imam, philosopher, shaman or whatever else each culture calls the citizens who serve as repositories of the community's spiritual or philosophical heritage. Cultural elders contribute to the fulfilment of spiritual or philosophic needs, of Star City citizens who opt to seek their guidance.
- **Coaches:** Citizens can opt, as their ancillary profession, to take on the responsibility to train others on sports that are popular in Star City (e.g. football, soccer, basketball, rugby etc.) while also providing general guidance and advice on enhancing and maintaining physical fitness in Star City.
- **Chefs:** Several citizens of Star City have opted to take on the role of Chef as their ancillary profession, opening restaurants of diverse cuisines in their respective villages, supported by robotic cooking and serving staff.
- **Interior Designers:** Citizens who chose to become interior designers as their ancillary profession advise other residents on how to create a unique aesthetic for their otherwise standard accommodations, then proceed to 3D print or otherwise manufacture the decorative features which their clients have requested. Interior designers receive inspiration from their own culture of origin for their creative work, thus each village's designers tend to provide a distinct decorative style.

- **Gardeners:** Gardeners work with village authorities and central authorities to enhance the biodiversity and aesthetic value of woodland domes, but also support individual residents who wish to use plants and floral arrangements in their private residence.
- **Fashion Designers:** Finally, some citizens opt for the ancillary profession of clothier. Their work blends traditional aesthetics of their culture of origin with the evolving aesthetics and functional needs of life in Star City. Once they have designed their creations, either as standardized designs or for unique customers, they then rely on the support of 3D printing and other robotic assistance to craft each item of clothing.

The acquisition of task-specific competencies, whether for the main, infrastructure-oriented profession or for the ancillary, self-expressive profession, begins with formal education at secondary and tertiary level, as needed, then continues with an apprenticeship under supervision of an experienced professional. Individuals who decide that they wish to change a profession, or who need to change a profession due to changing economic needs, have the opportunity to do so by following the appropriate training and apprenticeship programme.

**Participatory Democracy**

Just as the professional life in Star City resembles professional life on Earth, in terms of the breadth and type of professions that are available, but also represents a significant divergence, in that all citizens will be expected to combine a technical and a self-expressive profession, so the governance system of Star City includes elements of continuity but also discontinuity when compared with contemporary governance systems of Earth. Specifically, the element of tripartite division of power between the legislative, executive and judicial branches of government has been imported into Star City governance. What has been left out, however, is the inclusion of any type of professional political class or career politician. As a small city with highly qualified, engaged and civically aware residents, it became clear early on that the ancient Athenian model of direct democracy [19], which does not make room for any permanent distinction between citizens and governors, was more appropriate for Star City than the Roman model of representative democracy which, while efficient when dealing with large populations, is also vulnerable to gradual degradation through the manifestation of elite corruption and public apathy. Specifically, the governance system of Star City includes the following features:

1. A governing council, exercising all executive authority in Star City, which is selected by random ballot of all eligible citizens to serve a non-renewable two-year term. Eligible to participate in the random ballot are all citizens over the age of 30 who are in good standing as to their cross-cutting life skills and have not been reprimanded or convicted of any wrongdoing in Star City court. During the period of their service, members of the governing council are excused from their infrastructure-related profession, but are still permitted and encouraged to practice their ancillary, self-expressive profession, to the extent that they have available time.

2. Whenever legislative or regulatory decisions need to be made, the governing council is responsible to facilitate open consultations, then administer a vote through a secure online platform where all relevant information is made available and multiple options are given to those casting their votes. While some votes are universal, others are restricted to specific groups or professional classes that would be solely affected by the decision.
3. The judiciary is also randomly selected, for a fixed term of four years, but only amongst citizens over the age of 40 who have previously served in the governing council. Legal counsel for citizens who have cases before the court is generally provided by AI, thus keeping legal costs in check and removing financial barriers to accessing the court system. In cases where citizens are found guilty of a penal violation, the preferred sentence is an enhanced regimen of community service and direct restitution of any harm done to victims, though in rare occasions secure confinement is decided, but only to prevent further crime and to prepare the ground for rehabilitation by addressing root causes of criminality. The concept of retributive justice (i.e. imprisonment as a deprivational punishment that is proportional to the offence) has not been incorporated into Star City's legal framework.

## PART IV – THE ECONOMY OF STAR CITY

### Overview of Economic Concept

The central economic motivation for the consortium of investors is to establish Star City Development Guild as the lowest cost provider for construction of habitable outposts on Mars and to thereby reap an economic benefit by selling the establishment of other outposts to second-movers. This space is already being actively monitored, as evidenced by the space interests of decidedly non-space companies like Caterpillar, Obayashi and Kajima and of professional firms like Foster + Partners and SOM. This would not normally be a viable business model, except that a new space race is under way. Major powers and/or well-funded private corporations have clearly and repeatedly declared their intent to return humans to other worlds and to create permanent settlements there.

The second and more subtle pillar underpinning the economic concept is the drive to maximize in-situ resource utilization (ISRU) from the earliest crewed flights. In doing so, the colony minimizes its dependence on the logistical supply train from Earth to contain primarily spare parts for the energy system, tunneling machines, resource processing and additive manufacturing systems. Crucially, these systems do not scale with the size of the colony, because it is not the absolute size but the *rate of growth* of the colony which depends on the constant output rates of these primary ISRU systems. The longevity of the colony's habitation systems will drive the fraction of ISRU output available for growth. Therefore ISRU and longevity of locally manufactured habitation systems will be the keys to minimizing dependence on Earth resupply, leading to organic economic growth, self-sufficiency and the economic independence of Mars.

### Background: Space Race 2.0

A handful of catalysts are simultaneously driving the new space race. These include perceptions that we may be at the precipice of new, multi-trillion dollar industries, national pride and competition, a new approach by NASA to acquire services instead of acquiring hardware, and the romantic motivations of billionaires with the financial and technical wherewithal to take the risky first steps and develop critical capabilities before there is certainty of profit or economic return [20].

**Mars Treaty and Constitution**

The adoption of a Mars Treaty which further clarifies the legal environment using the Law of the Sea as a basis will facilitate commercial enterprise and public-private risk-sharing. Specifically, infrastructure on other worlds that is inhabited or regularly inhabited should be treated as a cross between a stationary ship and an island, conferring upon its owner the right to its peaceful use without interference by third parties, and also the right to access planetary resources within a certain radius from the infrastructure depending on the number of human lives that will depend on those resources. Given that the Treaty is expected to facilitate the settlement of Mars, it would be desirable for the signatories to also execute a brief Mars Constitution providing for the most fundamental rights and obligations of persons who settle on Mars, with the objective of promoting their peaceful collaboration and coexistence. A single Constitution for Mars signed by major countries from all continents, which allows for independent, self-governed city-states, would provide a solid foundation for the future political evolution of Mars. This foundation will in turn support private risk-taking, investment and economic development by spurring demand for city-states on Mars by state actors such as the UAE, Norway, Switzerland, China, Luxembourg, Israel, India and many others.

**Lowest Cost Construction Capability**

As Star City progresses towards completion, it will accumulate mobile heavy infrastructure including nuclear reactors and roadheaders, as well as the know-how to operate it. The dominant strategy for second-moving organizations wishing to establish an outpost on Mars will be to do so in the same region as Star City and to employ the Star City Development Guild to carry out the construction. The same applies to corporations or nation states that want to construct their own buildings within existing cities such as Star City. Thus, the Guild will be the lowest-cost supplier of construction services on Mars. To maintain its edge, it is likely that the Guild will expand into new cities and/or partner with space transportation companies to move its heavy equipment point-to-point on Mars.

**Secondary Income Sources**

Supplementary / secondary sources of income for Star City include the provision of exploration and research services, the provision of leadership training services, the sale of propellant to NASA and other space agencies and corporations, and the production and export back to Earth of Mars art, Mars poetry and Mars musical instruments. Leadership training services in particular will consist of hosting community leaders sent from Earth who will spend two or four years working with Star City personnel to start up new cities at other sites, while the blinking color lasers of peace from Star City's five villages will be visible from backyard telescopes on Earth. These community leaders will later return to Earth to lead the reconstruction of communities destroyed by war, civil strife or economic dislocation using the technologies, their fame and reputation and the "aura of peace" from Star City, Mars.

**Balance of Payments and Investment Appraisal**
Based on the Star City concept described above, Star City can be valued by the Mars Consortium as a single enterprise for the purpose of estimating a Net Present Value (NPV). An order-of-magnitude NPV calculation is presented in Table 3:

**Table 3:** *The NPV is -$5.7bn, indicating necessity of public-private partnership*

| Amounts in millions USD, 2019 | **Sept 2029** | **Nov 2031** | **Dec 2033** | **Feb 2036** | **April 2038** | **May 2040** | **July 2042** |
|---|---|---|---|---|---|---|---|
| Investment | 5290 | 4360 | 410 | 2405 | 1390 | 450 | 1885 |
| Transport cost | 751 | 803 | 440 | 450 | 433 | 432 | 453 |
| Salaries @ $500/day | 44 | 132 | 307 | 483 | 659 | 834 | 834 |
| Total outflows | **6085** | **5294** | **1157** | **3338** | **2481** | **1716** | **3172** |
| Construction income | 0 | 0 | 0 | 1000 | 7000 | 9000 | 13000 |
| Leadership training | 0 | 0 | 20 | 50 | 350 | 400 | 600 |
| Research grants | 0 | 0 | 50 | 60 | 80 | 100 | 120 |
| Propellant resupply | 0 | 0 | 0 | 0 | 400 | 500 | 600 |
| Exports @ $1k/kg | 0 | 0 | 40 | 40 | 40 | 40 | 40 |
| Shipping @ $200/kg | 0 | 0 | -8 | -8 | -8 | -8 | -8 |
| Total Inflows | **0** | **0** | **102** | **1142** | **7862** | **10032** | **14352** |
| Net Inflows / Outflows | -6085 | -5294 | -1055 | -2196 | 5381 | 8316 | 11180 |
| Discounted @ 15% rate | -6085 | -4003 | -603 | -949 | 1759 | 2056 | 2090 |
| NPV | **-5737** | | | | | | |

**PART V – NEXT STEPS**

The balance of payments and investment appraisal, calculated from the detailed equipment manifest, using the given values for transportation costs and using a conservative 15% discount rate results in a negative NPV of about $6 billion. This indicates the necessity of proceeding with a public-private partnership as described above.

The fundamental principles for the design of Star City all drove towards the same objective: the creation of a secure and self-sustaining community on a hostile planet. Star City's security lies in its social fabric of five culturally distinct villages who join their strengths to create one interdependent city, and in its wealth of habitable volume, energy and water supplies. The physical and societal fabrics strengthen each other via the mediation of the deep sense of personal security they engender in Star City's citizens. The wealth accumulation which makes all this possible is critically enabled by the energy and labor efficient technique of subtractive construction, which can create up to 1,000 $m^3$ of new habitable volume per village per day while at the same time providing easily manageable cuttings for processing and separation into pure water and other useful resources.

Back to the present day, in 2019, the next step is the engagement of the international multi-stakeholder group to work towards the adoption of a Mars Treaty and the public-private funding of a Mars Consortium. A funded Mars Consortium would invest in the procurement of an extended human campaign to Mars, along the lines described in Parts I, II and III of this proposal, and its investors could reasonably look forward to an economic return over a decadal horizon, as described in Part IV above.

*Per aspera, ad astra!*

**STAR CITY WEBSITE**

Read more about the authors' vision for Star City and download drawings, figures, renders and tables from the Star City website: https://stellaurbis.com/starcity

**ACKNOWLEDGMENTS**



**Special Thanks**

Prof Olivier de Weck, Prof. Jeffrey Hoffman and Prof. Herbert Einstein (advisors / mentors to GL), Alexander Guest (astronomy), Andreas G. Lordos and Marianna A. Lordou (readers), to the MIT Redwood Forest and MIT BEAVER teams and Jeff Greenblatt (for collaborations in other projects regarding living on Mars).

**Cover Art and Drawings of Star City**

Original sketches by George Lordos. Drawings by Delta Architects (www.deltaarchitects.gr) - Nikos Papapanousis, Tatiana Kouppa, Efi Koutsaftaki, and Aliki Noula. 3D renders by NOX3D (www.nox3d.com) - Aris Michailidis.

**Historical note**

The original *Star City* is near Moscow. It started as a single building, with the first occupant being Yuri Gagarin. Today, all astronauts who travel to the ISS aboard Soyuz rockets receive training at Star City.

**REFERENCES**

[01] USGS Maps of Mars, retrieved from https://astrogeology.usgs.gov/solar-system/mars
[02] First Landing Site / Exploration Zone Workshop for Human Missions to the Surface of Mars, Lunar and Planetary Institute, Oct 27-30 2015, Houston, TX https://www.hou.usra.edu/meetings/explorationzone2015/
[03] G. A. Morgan, N. E. Putzig, M. R. Perry, A. M. Bramson, E. I. Petersen, Z. M. Bain, M. Mastrogiuseppe, D. M. H. Baker, R. H. Hoover, H. G. Sizemore, I. B. Smith, B. A. Campbell, A. Pathare, C. Dundas, (2018) "Mid-term Review for The Mars Subsurface Water Ice Mapping (SWIM) Project", presented at NASA HQ https://swim.psi.edu/resources/media/presentations/Morgan_SWIMoverviewLPSC_Small.pdf
[04] Digital Terrain Models from HiRISE photography, NASA/JPL/University of Arizona. Retrieved from https://www.uahirise.org/dtm/dtm.php?ID=PSP_002175_2210
[05] Terrestrial Energy website: https://www.terrestrialenergy.com/technology/
[06] Sandvik Roadheaders website: https://www.rocktechnology.sandvik/en/products/mechanical-cutting-equipment/roadheaders-for-mining/
[07] Megapower patent: Mobile heat pipe cooled fast reactor system US 20160027536 A1
[08] Mueller, R.P., Rachel Cox, Tom Ebert, Johnathan Smith, Jason Schuler and Andrew Nick, (2013) Regolith Advanced Surface Systems Operations Robot (RASSOR), IEEE Aerospace
[09] JPL page on ATHLETE Rover https://www-robotics.jpl.nasa.gov/systems/system.cfm?System=11
[10] Hinterman, E., Zhuchang Zhan, Natasha Stamler, Sheila Baber, Tajana Schneiderman, Joe Kusters, Hans Nowak and Sam Seaman, "MIT Beaver: Biosphere Engineered Architecture for Extraterrestrial Residence", to be submitted to NASA BIG Idea Competition, April 2019
[11] Dweck, Carol S. "Mindsets and human nature: Promoting change in the Middle East, the schoolyard, the racial divide, and willpower." American Psychologist 67, no. 8 (2012): 614.
[12] Brown, Kirk Warren, and Richard M. Ryan. "The benefits of being present: mindfulness and its role in psychological well-being." Journal of personality and social psychology 84, no. 4 (2003): 822.
[13] Ungar, Michael. "The social ecology of resilience: addressing contextual and cultural ambiguity of a nascent construct." American Journal of Orthopsychiatry 81, no. 1 (2011): 1.
[14] Cox, Fletcher D., and Timothy D. Sisk, eds. Peacebuilding in Deeply Divided Societies: Toward Social Cohesion? Springer, 2017.
[15] Thoits, Peggy A. "Multiple identities and psychological well-being: A reformulation and test of the social isolation hypothesis." American sociological review (1983): 174-187.
[16] Armitage, Derek R., Ryan Plummer, Fikret Berkes, Robert I. Arthur, Anthony T. Charles, Iain J. Davidson-Hunt, Alan P. Diduck et al. "Adaptive co-management for social–ecological Complexity." Frontiers in Ecology and the Environment 7, no. 2 (2009): 95-102
[17] Frankl, Viktor E. "Self-transcendence as a human phenomenon." Journal of Humanistic Psychology 6, no. 2 (1966): 97-106.
[18] Trilling, Bernie, and Charles Fadel. 21st century skills: Learning for life in our times. John Wiley & Sons, 2009.
[19] Sinclair, Robert K. Democracy and participation in Athens. Cambridge University Press, 1991.
[20] Special Report: The New Global Race to Space, Axios, Jan 1, 2019 https://www.axios.com/global-space-race-bezos-musk-spacex-e1ee8fa7-63a8-4f41-94c8-10860e59365f.html

# 19: PROJECT MENEGROTH

**Wesley Stine**
Purdue University
wstine@purdue.edu; wesleystine@gmail.com
Illustrations by Rebekah Stine

## INTRODUCTION

The design of a suitable habitat for the first thousand Martian colonists presents a multitude of technical and human challenges. Air, water, food, and warmth must all be extracted with great effort from a hostile environment. Indeed, the full force and ingenuity of modern technology will be needed to eke out the barest level of subsistence.

This paper will make the case for Mars' first settlers to build an underground city state, called *Menegroth,* while describing its physical layout, power supply, industrial capabilities, agriculture, economics, political organization, society, and culture. Menegroth will be built into a natural lava tube cave beneath Mars' surface, which will be sealed off with cement, filled with breathable air, and warmed by a lakeside nuclear reactor.

Within the cave, Menegroth's inhabitants will be at liberty to engage in most activities with no extra protection from the outside environment. Residences, laboratories, foundries, machine shops, body shops, spacesuit manufactories, schools, parks, restaurants, chapels, and assembly halls will fill its expanse. Outside the cave, colonists will cultivate food in greenhouses and mine water ice, iron, aluminium, uranium, and other necessary minerals, and will also conduct scientific research and mine precious metals for export.

Menegroth's economy will begin in a state of heavy dependency on imports. In course of time, Menegroth will balance its trade with Earth, through the export of precious metals, scientific research, and entertainment media. If the population continues to grow, research and media will dwindle in importance, losing ground to local industry. However, while Mars' population numbers only a thousand, it will be important to exploit these sources of income for all they are worth.

The society and culture of Menegroth will be inspired by successful pioneer societies on Earth, and the labour and ingenuity of the colonists will be devoted toward achieving as much economic independence as possible. Their ultimate goal will be the establishment of a strong and distinct Martian culture, supported by a robust mineral trade that could eventually link Earth, Mars, and the asteroids.

I will assume that human astronauts have already explored the surface of Mars and located sufficient mineral deposits to support a colony, and that experiments have confirmed the ability of mammals raised in Martian gravity to not only grow up healthy, but to survive a return to Earth as well. Menegroth will be a self-perpetuating colony, and thus it should not be established unless children born there would have a place to go in the event that it ends up being abandoned.

## THE CAVE HABITAT

### Overview of Lava Tubes

Martian lava tubes, like lava tubes on the Moon and Earth, are formed when flowing lava recedes and leaves a tunnel several kilometres long in the surrounding rock. Portions of the roof sometimes collapse, leaving skylights by which the tube's presence can be discerned from orbit.[1] Cave diameters as great as 252 meters have been observed.[2] Mars' low gravity allows these caves to become larger than lava tubes on Earth.

Project Menegroth would be located in a smooth-floored lava tube about 120 meters in diameter, with a length of about 2 kilometres used for habitation. The site of settlement would need to be searched out in advance by ground-based explorers, as the collapsed caves visible from orbit are unusable.

The benefits of an underground habitation are numerous. Foremost is the light work required to create a habitable volume: as soon as a proper entrance is constructed, and any leaks are sealed with cement, enough space can be filled with air to support thousands of inhabitants. Temperature stability and protection from radiation are also important benefits.

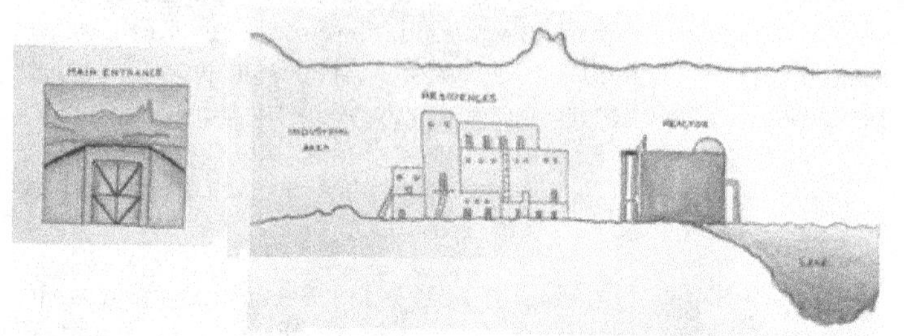


Figure 1 – Outline of Menegroth (not to scale)

### Outline of the City

The principle structures which Project Menegroth will require are shown in the cutaway in Figure 1. An airlocked main entrance, large enough for vehicles, will handle most traffic with the exterior. As one goes further into the cave, one would encounter the industrial districts, with body-shops placed closest to the entrance to service the mining machines and pressurized rovers. Beyond the industrial districts are the residences, schools, and other non-industrial buildings.

The nuclear reactor will stand on the lakeshore at the lowest point of the cave; if a suitably deep lakebed does not already exist, one can be created with explosives. The reactor and lake are, in a way, the 'heart' of the colony; they provide it with power, water, and thermal stability. The water will cool the reactor, which will in turn keep the lake warm enough to create a liveable environment in the surrounding air. The lake can be strewn with agricultural waste from greenhouses in order to feed an ecosystem of fish.

**Residential Areas**

Menegroth's residences will be built near the lake, on both sides of the colony's main street, which will run all the way from the lake and reactor at one end of the lava tube to the great entrance at the other end. Smaller, footpath streets would extend out on both its sides.

The residences present no major engineering challenges. Allowing 30 $m^2$ per inhabitant in condominia three stories high, a total ground area of 10,000 $m^2$ is needed; in reality the buildings would be spread out with streets and parkspace in between, over an area of 36,000 $m^2$, covering 300 m of the cave's length, in between the lake and the industrial areas.

These rooms will be supplied with electricity and water, and furnishings will be sparse, with colonists expected to spend little time at home. Nevertheless, they will be encouraged to adorn their quarters with paintings of landscapes on Earth.

Any architectural tradition would be suitable, so long as it uses little timber or other organic materials. My personal preference is for the Santa Fe Style, whose mild minimalism, clusters of cubelike rooms, and pueblo-inspired tendency to blend smoothly into the surrounding rock-faces make it a good fit for Menegroth.

Figure 2 – A Residential Street

The apartments' surroundings will be decorated with rock gardens, shrubberies, ornamental trees, and children's playsets. Schools, restaurants, chapels, laboratories, assembly-halls, and athletic parks will be placed closer to the residences than to the industrial districts, along with any other installations for which a quiet environment is desirable.

Menegroth's whole lava tube will be lit in the daytime, never to noon intensity, but enough to see as well as at dawn on Earth. The ceiling of the cave will be painted in hues of blue, gold, and white, like a terrestrial sky, with figures of gods and heroes soaring above.

**Atmosphere and Life Support**

Maintaining the chemical balance of Menegroth's air is a vital process. The cave would be filled with a three-to-one mix of oxygen and nitrogen, with traces of carbon dioxide and water similar to Earth's. Total pressure is 20 kPa, so partial pressure of oxygen is 15 kPa, the same as on Earth at an elevation of 2800 m (the altitude of Quito). Air temperature is 25°C, and air density is thus 261 g/m$^3$. Assuming a cave length of 2 km and a semi-circular cross-section with a d = 120 m, the air mass will be 2,950 metric tons, but because the exact cave geometry is yet unknown, this can be regarded as an order-of-magnitude estimate only.

Unlike in present-day spacecraft, there is no need to match Earth's atmospheric composition for the sake of easier launch and landing procedures. The compromise chosen for Menegroth is similar to Skylab's atmosphere, which was 74% $O_2$ and 26% $N_2$. Menegroth's nitrogen will benefit bacteria and plant life by giving nitrogen-fixers like legumes something to work with. Gasses will need to be produced in-situ. Initially, the cave will be filled only with oxygen, as this gas is easier to produce and more vital. Curiosity has recently confirmed the existence of nitrates on Mars, which can be mined for nitrogen, though little is known of their distribution or abundance.

The population of one thousand will consume about 750 kg of oxygen per day. Replenishing this oxygen will require electrolysis of 844 kg of water with about 4.7 MWh of electrical power. It is also necessary to replenish both nitrogen and oxygen lost through the airlocks during normal operations, and to release excess carbon dioxide into the Martian atmosphere. Air must also be carefully monitored and filtered for industrial pollutants in order to avoid a health disaster.

**Demographics and Professions of Menegroth**

Menegroth is intended to subsist as a demographically complete frontier community. Thus, men and women will be present in equal numbers, and the population will be replenished primarily through natural growth. I estimate that out of Menegroth's 1000 inhabitants, about 300 will be under the age of sixteen.

In most modern terrestrial societies, including the United States, children generally perform no economically valuable labour until after they have come of age; in some cases they remain full-time students until their mid-twenties. This is a historical anomaly that need not be repeated on Mars. In any pioneer society, Menegroth included, everybody works.

Children's tasks will vary with age; young children will generally be occupied in greenhouses. It would, perhaps, be easier to have robots perform some of the agricultural tasks, but others mechanize very poorly. In any case, the benefits of teaching children to work at a young age, and surrounding them by the world of green things on a planet where life is otherwise scarce, are the primary reasons to employ them in the greenhouses.

In their early teens, children could begin apprenticeships in the various forms of manufacturing. The most dangerous jobs, including work in mining or other trades that requiring venturing into the outside environment in spacesuits, will be reserved for adults.

These assumptions regarding the number of people available for various lines of work will be relied upon heavily in the remainder of this paper. Henceforth, I will assume that of Menegroth's 700 adults, 20 will be employed at the nuclear reactor, 50 at foundries, 200 at the machine shops, 200 in mining, and 30 in agriculture, with the remaining 200 to be divided between scientific research and the various smaller occupations that Menegroth will require.

These smaller applications are too numerous to list in full, but they will include, to begin with, educators, physicians, safety officers, diplomats, laboratory technicians, merchants, shopkeepers, artisans, cooks, and computer specialists.

Among scientists, the field which I believe will be held in the highest esteem is that of the geologists, whose knowledge is required not only for building Menegroth, but for discovering sources of ice water and every other mineral resource on which the colonists must rely.

## POWER AND INTERNAL CLIMATE

### Menegroth's Power Requirements

I have estimated Menegroth's power needs at 33 MW. Residential power requirements are negligible; industry will dominate. This conclusion is reached by comparison to Earth's highest per-capita electricity user, Iceland, which consumes 5,777 W per capita, largely due to its aluminium foundries. All industrial sectors combined employ 16.6% of Iceland's workforce. If Menegroth employed 90% of its workers in energy-intensive industries, with the same power consumption each, the per-capita need would be $(90/16.6) \times 5.77$ kW = 31.3 kW, or 31.3 MW for the full thousand, increased to 33 to utilize the chosen reactor design.

To make full use of Menegroth's reactor, industries must run day and night, though the number of labourers on shift need not remain constant. Robotic smelters and computerized lathes would allow the full power to be used even when few human workers are present.

Solar power is unsuitable in the long term. The solar constant is only 586 W/m$^2$ at Mars' distance; 5 ¾ hectares of solar panels would be needed to provide 33 MW at perfect efficiency, so the true area requirement would be three or four times greater, and Menegroth would be vulnerable to destruction by dust storms.

Nevertheless, initial settlers will need solar energy to perform the minimal tasks required to keep themselves alive, seal off the cave, mine enough water for a rudimentary lake, and assemble the nuclear reactor. After power-up, construction and industry can begin in earnest.

**The Nuclear Reactor**

Menegroth will use an *Integral Molten Salt Reactor* (IMSR), a design currently under development by the Canadian firm *Terrestrial Energy* (T.E.). The smallest variant of this reactor has a power of 80 MW-thermal and 33 MW-electric, enough for Menegroth's needs.

The IMSR is a continuation of molten salt technology originally demonstrated at Oak Ridge National Laboratory. Its advantages over a conventional, solid-core reactor are numerous. The higher working temperature allows a greater efficiency then in ordinary water-cooled reactors, and operating at atmospheric pressure eliminates the risk of explosions. There is also no chance of a meltdown, as the fuel is already liquid.

Figure 3 – Cutaway of the IMSR core, courtesy of Terrestrial Energy

The IMSR uses a thermal-spectrum, graphite-moderated, molten-fluoride-salt reactor system with standard-assay, low-enriched uranium (less than five percent $^{235}U$) fuel. The sealed and replaceable reactor core unit has a lifetime of seven years, and can be installed in a power plant using ordinary electrical turbines with a secondary, non-radioactive salt as a heat transfer fluid. The result is, in Terrestrial Energy's words, "a small modular reactor that delivers a combination of high energy output, simplicity of operation."

Estimating the reactor's mass is difficult, as Terrestrial Energy does not, itself, design the non-nuclear components, such as turbines and secondary heat exchangers, which account for most of the mass. Nevertheless, any such components which fly to Mars would be optimized for lightness. I therefore assume a power-to-mass ratio similar to that of U.S. Naval reactors, such as the 12.5 metric tons per $MW_{th}$ of the S8G reactor. By that estimation, the reactor will weigh about 1000 metric tons, and be delivered to Mars in pieces at a shipping cost of $500 million. The intrinsic price is estimated by Terrestrial Energy as "less than US $1 billion." [3]

**Climate and Temperature Control**

Climate inside Menegroth's lava tube will be driven by a heat cycle in which air is circulated throughout the colony to stabilize the temperature and return all industrial heat to the lake. Most of the 33 MW of electricity will be converted to waste heat inside the cave. Along with the 47 MW of thermal heat, this will present a serious climate control challenge.

These problems can be solved as follows: The lake will be kept at 25°C. A steady wind will blow from the lake outward through the remainder of the lava tube, carrying cool air into the industrial regions, from whence warmer air will be piped back toward the lake and bubbled up through the bottom, returning the 33 MW of waste heat to the water.

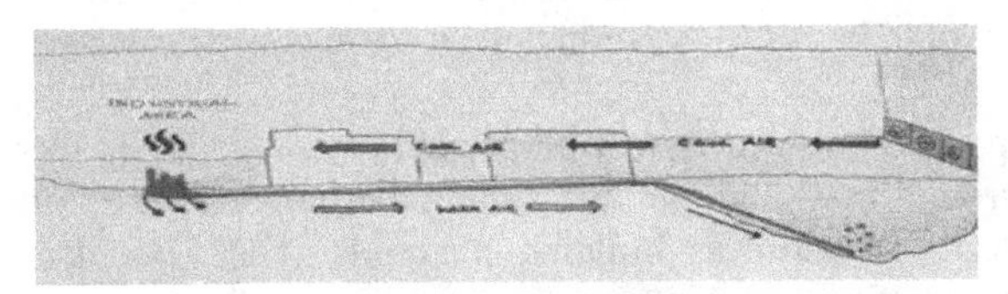

Figure 4 – Air Circulation in Menegroth

Between the 47 MW of reactor waste heat and the 33 MW industrial heat, the lake will experience 80 MW heat gain. Assuming a circular lake 200 m in diameter and 40 m deep, the temperature will increase by 1°C each 18 hours if nothing is done to cool it. (This is an order-of-magnitude estimate only, actual size depends on the geometry of a yet unexplored cave).

I considered three methods of cooling the lake. First, a coolant near lake-temperature could be pumped through radiating surfaces exposed to the outside sky. However, with a blackbody flux of 447 W/m$^2$ at 25°C, at least 18 hectares of complicated hydraulic fins would be needed. Second, a heat pump could be used to operate the radiators at a higher temperature than 25°C, but the heat pump would drive up power needs and create yet more thermal waste.

The third method, which I believe to be best, consists of simply adding new ice to the lake each day and allowing an equal amount of water to evaporate out into the Martian atmosphere, through a system of valved pipes, in order to carry away heat. Melting a ton of ice consumes 334 MJ, heating the water from 0°C to 25°C consumes 105 MJ, and vaporizing it at 25°C consumes 2442 MJ; thus 2881 MJ is shed for each ton of water, and 2500 tons must be processed daily in order to operate the reactor at full power.

The United States' 50,000 coal miners produce 739 million metric tons of coal annually, for 36½ tons of coal per miner per day. If Menegroth's ice miners produced ice at the same rate, 68 miners would be needed to meet the colony's needs. A short-term interruption in ice production would not be a threat, because a lake with the previously-estimated dimensions would last 432 Martian days, without replenishment, before boiling away.

## FOUNDRIES AND MACHINE SHOPS

### Manufacturing Requirements

Menegroth will require a great deal of diverse manufacturing capabilities in order to sustain life on Mars with a minimum of terrestrial imports. Ore will be smelted in foundries to produce iron, aluminium, copper, and other metals. Machine shops will rely heavily on a variety of computerized technologies, including CNC milling, PLC controllers, and additive manufacturing, in order to allow a technologically advanced economy to be supported with only a small number of highly skilled workers.

Any attempt to enumerate, in such a brief paper, the full range of necessary manufacturies would be futile. Nevertheless, a basic outline can be commenced.

Metals, once purified from ore, will be needed as raw material to produce other implements through both additive and subtractive manufacturing methods. Clay, brick, and glass will serve as building materials. The vehicular industry will mainly concern itself with mining machines and pressurized rovers. An advanced textile craft will be necessary for space suits. Machines both inside and outside the cave will require the continuous, rapid fabrication of spare parts. Another line of industry must be devoted to synthesizing industrial and agricultural chemicals from the raw minerals delivered by miners, and so forth.

### Smelters and Refineries

As on Earth, iron and aluminium will be the metals produced in the largest quantities. Other important metals will include copper, needed for electronics, nickel, needed for high-strength alloys, zinc, needed for cast-metal parts, and uranium, needed for nuclear fuel. Besides these, almost every other metal will find some use, and will need to be produced or imported in small quantities. The mining of metals will be treated in a later section; the present one will deal with the challenges of refining iron, aluminium, and other metals on Mars.

Making steel on Mars presents a unique problem. Terrestrial steelmaking relies on two major resources, iron oxide and coal, of which the first is plentiful on Mars and the second is non-existent. An alternative to coking coal is needed in order to heat and smelt iron ore. Hydrogen gas, which can be made through the simple but energy-intensive method of electrolysis, can supply that alternative; indeed, steelmaking will be only the first of many areas in which fossil fuel use must be replaced by the hydrogen economy.

Hydrogen Breakthrough Iron-Making Technology (HYBRIT) is a subject of extensive research in Europe, where decarbonisation is the driving concern. HYBRIT relies on hydrogen produced by electrolysis of water to heat blast furnaces. According to the Stockholm Environment Institute, HYBRIT faces no serious technical obstacles and is substantially closer to commercial deployment than its main alternative, electrowinning. [4]

Electrowinning, though unsuitable for steelmaking, is easily used to refine bauxite into aluminium (whose melting point is only 660 °C compared to iron's 1538 °C), and this process can be continued on Mars with little modification. For industrial processes requiring even lower temperatures, the molten salt reactor can provide heat directly; one of its stated capabilities is the transportation of heat to industrial sites up to 5 km away at up to 660 °C. [3]

Other metals, used in smaller quantities than iron and aluminium, will require their own production methods, but these are beyond the scope of this paper.

Menegroth's foundries are estimated to employ fifty laborers. The furnaces used to smelt iron and aluminium would probably be the heaviest of all of the colony's industrial equipment. Shipping them from Earth would be foolhardy, so they will need to be manufactured mainly from ceramics produced on Mars.

**Subtractive and Additive Manufacturing**

Menegroth's metalworking craft will use two increasingly important technologies – CNC Milling and Additive Manufacturing (3D printing) – to enable a small number of workers to staff a heavily automated manufactury and produce all necessary custom metal parts.

Three-axis and five-axis vertical CNC lathes and milling machines will form the top line of subtractive manufacturing equipment. These will be similar to the machines used in a high-tech shop on Earth, with some modifications to minimize their weight. In addition, a number of other machines such as cutters, grinders, welders, saws, drills, presses, water-cutters, laser-cutters, non-destructive testing equipment, and heat-treating equipment will also be needed.

Menegroth's machine shops will also rely heavily upon additive manufacturing to produce metal and plastic parts. The technology of 3D printing is uniquely suitable for Mars because, in addition to allowing a high degree of customization and specialization, this mode of manufacturing is also fairly autonomous and wastes far less metal than traditional methods. Materials to be printed would include ceramics, plastics, composites such as alumide, and metals such as stainless steel, aluminium, titanium, and Inconel, among others.

Figure 5 – The SuperDraco thrusters, which are 3D-printed in Inconel

Because of the importance of CNC, 3D printing, and other automated manufacturing technologies, the skills of CAD/CAM programming will be held in high regard in Menegroth and be taught at a young age during the schooling or apprenticeship of the town's children.

Any attempt to estimate the weight of all this equipment is, of necessity, very imprecise. Nevertheless, if one assumes a workforce of two hundred machinists, a ratio of five major machines per worker, and a mean mass of two metric tons per machine (the machines being optimized for lightness), the total equipment mass will be somewhere near 2000 metric tons.

However, this full weight need not be transported from Earth. A smaller machine shop could start with only a quarter of the machines operational, then produce most of the heavy parts for the others, so that only the fine electronics, tool heads, and motors of some 75% of the equipment need be brought from Earth. Assuming that 80% of the machines' weight consists of easily producible components, the mass to be sent from Earth will only be 25% + 0.2 × 75% = 40% of the final mass, or 800 metric tons. Transport cost is then $800 million. Since only the highest-quality machines will be chosen, an intrinsic price of $1.5 million per machine is assumed, so the total cost comes to $2.3 billion.

Once fully-established, Menegroth's system of machine shops would be capable of producing spare parts for itself, as well as more of any of its principal machines, except for circuit boards and other fine electronics and tool heads, which must still be imported from Earth.

**Glass and Ceramics**

The raw materials required to create glass and other ceramics are inorganic and will exist in abundance on Mars. Clay and sand can be easily produced and processed to create a variety of ceramics. Brick, concrete, and high-strength glass will all be essential to Menegroth's economy; glass will be needed both for the many surface greenhouses, and for the face-plates of space helmets and the windows of pressurized rovers.

Other uses for ceramics stem from their thermal properties. Any spacecraft built on Mars would need ceramic heat shields, and the bodies of pressurized rovers could be constructed from a lightweight cermet to provide strength and thermal insulation. Furthermore, the aforementioned hydrogen-burning foundries would rely heavily on brick and ceramics.

The kilns, ovens, mixers, extruders, presses, water-cutters, laser-cutters, milling machines, and chemical processing equipment needed to support Menegroth's ceramics industry are included in the same machine cost and weight estimates mentioned earlier.

**Organic Materials**
In contrast to metals and ceramics, organic materials will be scarce on Mars and colonists will need to adapt to use them sparingly. Nevertheless, for those applications where plastics or organic textiles are indispensable, an organic chemical industry must be developed.

Petroleum will be unavailable as a feedstock for synthesizing organic chemicals. Samples taken by Curiosity have demonstrated the existence of the hydrocarbons benzene and propane in some quantities on Mars; however, too little is known about the abundance or distribution of native organic materials for Menegroth's colonists to safely rely on their presence.

When synthesizing organic chemicals, biomass could be used as a feedstock in place of oil; however, this would require the sacrifice of organic matter that could be composted and used to renew the colony's tenuous soil supply. A better alternative is to create methane via the Sabatier reaction, then use the Fischer-Tropsch process to convert the gas into the long hydrocarbon chains needed for Menegroth's chemical industry. These processes, though demonstrated on Earth, are far too energy intensive for widespread use while fossil fuels are still present.

While plastics will rarely be used in applications for which a ceramic or metal part can substitute, they will be indispensable in the making of textiles, where inorganic fibres are simply too brittle. Spacesuits will require nylon and other synthetic polymers; synthetic rubber and explosives will also find many uses.

## AGRICULTURE AND GREENHOUSES

**Greenhouse Labour and Architecture**
Menegroth's agricultural land use is estimated at 45 hectares, by calculations to be described shortly. A total of thirty greenhouses, each 50 m wide and 300 m long, will be needed. Each of the thirty adult farmers will command a greenhouse. They will be assisted in their labours by about 110 children ranging in age from seven to their early teens, after which the children would become eligible for apprenticeships in industries other than agriculture. Children will work in the greenhouses for only half the daylight hours, spending the rest in their schooling. Children need not be divided evenly among the greenhouses; rather, their numbers and assignments will vary according to the nature of the day's work.

Greenhouses will be constructed from concrete, brick, steel, and high-strength glass. Soil will be made by washing Martian regolith to leach out toxic perchlorates, and adding enough nitrate fertilizer, dung, microorganisms, and worms to sustain plant life. Each greenhouse will have its own climate control system, and its own airlock at which small pressurized rovers can dock. Greenhouses will also have numerous of safeholds or small sealable capsules into which the workers can flee in case of a depressurization, and children who work in the greenhouses will be frequently drilled in their use.

Figure 6 – One of Menegroth's Greenhouses

Because of the amount of work required to build greenhouses, and the need for so much glass, their construction will not begin until after Menegroth's mines and machine shops are fully operational. Before then, colonists will eat imported food; by bringing dehydrated foods such as flour and beans, they can live on 200 kg per person per year. This will require a total of 1000 tons, imported at a cost of $500 million, to feed Menegroth for the first five Earth-years.

**Martian Staple Crops**

The regolith of Mars, once purged of perchlorates and mixed with sufficient fertilizer, will produce an iron-rich soil suitable for growing terrestrial crops. Normal growth of several crops in simulated Martian soil, under Martian lighting conditions, has been experimentally demonstrated.

Students at Villanova University successfully grew many vegetables including kale, carrots, lettuce, onions, dandelions, and hops, and sweet potatoes. Ordinary potatoes did poorly in the Villanova experiment, but this was attributed to a clayrich soil; a sandier soil could probably support them. A separate experiment successfully grew sorghum.[5]

Figure 7 – Plants in Simulated Martian Soil at Villanova University

Nevertheless, the majority of terrestrial staple crops have simply never been tested. The best option, at this point in planning a colony, is to assume that any crop can be grown if the soil is created from regolith of the proper consistency, and the right fertilizers are added. Thus, Menegroth's residence can plan on eating a plant-based diet whose staple foods will include wheat, sorghum, amaranth, potatoes, maize, and beans. Essential data for these crops are given in Table 1.

**Table 1 – Comparison of Staple Crop Properties**

| Crop | Season | Yield | Spec. Energy | Caloric Flux |
|---|---|---|---|---|
| Summer Wheat | 90 days | 9 ton / ha | 3400 Cal / kg | 340,000 Cal / ha-day |
| Sorghum | 120 days | 11 ton / ha | 3400 Cal / kg | 310,000 Cal / ha-day |
| Amaranth | 90 days | 4 ton / ha | 3700 Cal / kg | 165,000 Cal / ha-day |
| Potatoes | 120 days | 50 ton / ha | 930 Cal / kg | 390,000 Cal / ha-day |
| Maize | 100 days | 15 ton / ha | 3700 Cal / kg | 555,000 Cal / ha-day |
| Common Bean | 85 days | 9 ton / ha | 3500 Cal / kg | 370,000 Cal / ha-day |

I collected the data in Table 1 from a variety of online sources, choosing the most optimistic results reported for each crop under the assumption that the controlled climate and intensity of labour in the greenhouses will make them equivalent to the best farms on Earth. The six staples in Table 1 produce a mean flux of 355,000 calories per hectare per day, enough to provide Menegroth's population with a 3000 Cal/day diet on only 8.4 hectares.

However, growing crops on Mars will bring its own challenges, not least among them is the weaker sunlight, which will likely reduce yields to some degree. In light of this, I have chosen to allot greenhouse space using a safety factor of 2.5, thereby devoting 21 hectares, rather than 8.5, to staple crops. Assuming that another 21 hectares will be used to grow lower-calorie crops such as fruits and vegetables, and non-food crops like cotton, and reserving 3 hectares for research, gives the earlier-mentioned total of 45 hectares.

**Non-Staple Crops**

In addition to the major crops, which will meet the bulk of their energy needs, the colonists, the colonists can also grow a variety of fruits, vegetables, and spices. The iron-rich soils of Mars will be especially good for growing cranberries, blueberries, strawberries, raspberries, and grapes, as well as turnips, tomatoes, sweet potatoes, and squash. Mars' first winery could be made into a profitable export enterprise by virtue of its novelty.

**Domestic Animals**

From a purely economic standpoint, veganism would be optimal; however, scientific and other benefits justify raising a small number of animals, which can also be used for food. These will consist mainly of goats, chickens, and rabbits, whose meat would be eaten only at feasts. In addition, cats, dogs, and birds will be kept as pets, and mice for research.

Figure 8 – Animals of Menegroth

A single nanny goat is sufficient to establish an entire population, though two or three should be brought from Earth for redundancy. By also bringing frozen sperm from hundreds of males, each new goat can be conceived with a different father in order to establish a genetically diverse population. The goats will be kept in a dairy inside the lava tube, where males will be slaughtered for meat, and females will enable the colonists to enjoy luxuries such as ice cream and pizza.

## MINING

### Martian Mineral Resources

Menegroth's most important mineral resource will be water ice, whose production will occupy about a third of the mine workforce of 200 laborers. Other major minerals will include bauxite, iron core, copper, and uranium. Dozens of other base metals must be mined in smaller quantities, and precious metals will be mined for export.

Specialized sand and clay will be needed for Menegroth's ceramics industry. Nitrates and phosphates will be needed to make fertilizers and nitrogen gas; they are known to exist on Mars, though probably in smaller quantities than on Earth, and finding them will be tricky.

Very little is currently known about the actual distribution of minable ore deposits on Mars. The planet has been imaged from orbit, but never explored for any kind of mineral resource. It can be assumed that metal ores will occur in formations similar to those found on Earth, but as for water, nitrates, and abiogenic hydrocarbons, far less is known. It is for this reason that I believe geology will become the most valuable of the sciences, as finding a diverse array of mineral deposits on a previously unexplored planet will be essential for the colony's survival.

**Pressurized Rovers and Mine Convoys**

Menegroth's location will be selected only after a thorough scouting operation has located nearby sources of the principle minerals, such as water ice, iron ore, bauxite, and nitrates. Nevertheless, rarer minerals will require travel to minesites hundreds of kilometres away.

Miners will travel in convoys with perhaps a dozen autonomous mining machines and open-topped ore carriers accompanied by a single pressurized rover. Mineral exploration will be conducted by smaller, solitary rovers. These rovers will all be masterpieces of technical design, each with a life support system capable of sustaining six or eight occupants for months at a time, with full redundancy at every level.

Figure 9 – A Pressurized Rover

Powering these rovers, and the other mining machines, will be a challenge. Radioisotope thermal generators are too inefficient to support the size, quantity, or speed of vehicles that Menegroth will need. For example, Curiosity, with a specific power of 125 mW/kg, can only make 0.14 km/hr. Scaling up for an eight-ton rover going 20 km/hr suggests a power requirement of 140 kW.

Laying out 440 $m^2$ of solar panels for eight hours during the middle of the day (assuming $\eta = 0.2$ and a mean sun angle of 45°) will produce enough power to travel 40 km in the evening. Each vehicle in the convoy will require a similar amount of power, and an entire truck will be required to transport the solar panels.

For common, nearby resources, this method of travel can be dispensed with as soon as an electric railroad is built and the mine is supplied with power from Menegroth's reactor. However, monthlong journeys will continue to be required for reaching deposits of rare minerals that are only to be found far from home.

As for the most basic mineral, water ice, the location of the colony will have to be chosen so that extensive deposits of high-grade ice are within walking distance of its doors – there is simply no other way to mine the 2500 tons per day that will be needed to cool the nuclear reactor during Menegroth's construction.

Assuming that the average Martian mining machine weighs eight tons and costs $10 million, that machines outnumber miners two to one, so that there are 400 machines in total, and that half of them must be brought from Earth before Menegroth gains enough self-sufficiency to produce the others, the setup cost of the mining industry will be 1600 tons × $500,000 / ton + $2 billion = $2.8 billion.

**Minesites**

At the minesite itself, whether it is reached by rail or convoy, most of the work will be autonomous. Miners will keep their distance from the active machines and instead concern themselves with site inspection, formulating instruction sets, performing repairs, and keeping the power flowing. Dozens of truckloads of minerals, mostly water ice by weight, will be returned through Menegroth's great doors every day.

Water ice will never be mined through the convoy method, though in the beginning all other minerals will. Iron ore and bauxite will probably be the first to make the jump to rail-mining, and nitrates will similarly be produced on a large scale once Menegroth begins growing its own food.

Most other metals will never be mined in large enough quantities to justify a railroad. Menegroth will produce copper, nickel, tin, lead, zinc, molybdenum, and tungsten through the convoy method, along with cobalt and lithium, which will be needed to make batteries and uranium for the nuclear reactor.

The same will hold for the precious metals which will account for most of Menegroth's material exports. Gold, silver, platinum, palladium, and beryllium will be searched out and mined at sites far from Menegroth, as will gemstones.

With time, cultural differences will develop between rail and convoy miners. Rail miners will sleep in their houses and ride the train to work each day; their job, while more dangerous than working inside the cave, will still be much easier and less lonesome than that of the convoy miners. Teenage miners learning the trade will always begin at the rail minesites.

Convoy miners and their accompanying scientists, mechanics, and physicians will be chosen from among the bravest and most skilful colonists, who have volunteered to venture out into the harsh Martian wilderness in a single rover heavily stocked with food, water, oxygen, spare parts, and extra spacesuits. The captains of these mining parties will be second only to the President himself in the honor and gravity of their office.

## IMPORTS AND EXPORTS

### Initial Imports and their Costs

In the early years, Menegroth will be entirely dependent upon settlers' levies and corporate sponsorship to defray the cost of its initial equipment, which will come to about $15 billion. The costs previously calculated, namely $1.5 billion for the nuclear reactor, $500 million for food, $2.3 billion for manufacturing equipment, and $2.8 billion for mining machines, come to a total of $7.1 billion. However, it is reasonable to assume that just as much will be spent on other supplies too numerous to list in such a brief paper, and even more when including costs of exploration and site selection, hence my estimate of $15 billion.

The considerable research and development costs of interplanetary spacecraft are not included, as the entire paper is written under the assumption that there is a mature technology that can transport freight from Earth to Mars for $500/kg, and from Mars to Earth at $200/kg.

### Overview of Earth-Mars Trade

Even after the full establishment of Menegroth as the burgeoning city of a thousand inhabitants described in this paper, it will nonetheless be reliant on many imports from Earth. Chief among these will be precision electronics, rare metals, and intellectual property.

While Menegroth's machinists will be able to manufacture at least 90% of the components, by weight, of nearly any piece of equipment they need, they will still be dependent on integrated circuits and other fine components imported from Earth. Furthermore, the modern technology they use will be dependent upon nearly a hundred rare metals in addition to the common ones mentioned previously in this paper, and a city of one thousand will never achieve the specialization needed to produce them all. Finally, the colonists will not be intellectually self-sufficient, and in manufacturing, they will usually follow designs already in use on Earth, which will, in many cases, be proprietary.

In order to pay for these imports, as well as to pay down the initial settlement cost, Menegroth will need to develop three principle export sectors, which are:

1. **Precious Metals**. Gold, silver, platinum, and other metals valued at more than $200/kg can be sold on Earth for more than their shipping costs. Martian miners, beginning with Menegroth, will have the advantage of working on a land area equal to Earth's, on which even the easiest mineral deposits have yet to be exploited.
2. **Research**. Researchers at universities, as well as government and private laboratories all over the world, will take an interest in all manner of research performable only on Mars. By hiring out their services to earthbound scientists in need of experimenters, Menegroth's technicians can bring in a considerable amount of income.

3. **Media**. Entertainment media will be a significant revenue source only to Mars' earliest colonists, but it will be important while it lasts. Broadcasting Menegroth's progress on television has the potential to break viewership records, most of the colonists will win book deals, and even mundane events can be exploited for advertising revenue.

In addition to the trade in goods and intellectual property, there will also be an exchange of people between Earth and Mars. At first, immigrants will account for nearly all interplanetary passengers, but in time there will also be emigrants who have tired of life in Menegroth, and medical patients returning to Earth for treatment. Scientists and tourists will often come to Mars for a single synod, and be charged a heavy tax for the privilege of temporary residence.

**Precious Metals**

Once the novelty of Martian rocks and Martian wine has worn off, precious metals and gemstones will be Menegroth's only significant physical exports, with metals being the more important of the two. Some metals which can be shipped off-planet for less than their current market prices are shown in Table 2.

**Table 2 – Precious Metal Prices**

| Metal | Price |
|---|---|
| Gold | $47,900 / kg |
| Silver | $580 / kg |
| Platinum | $31,900 / kg |
| Palladium | $50,300 / kg |
| Beryllium | $550 / kg |

There are a few other metals such as rhodium, rhenium, iridium, and germanium, which are expensive enough to be worth exporting, but these are extremely rare, and they are usually obtained as a by-product of other industries.

Of the metals included in Table 2, gold will probably be the most important. Gold, though rare on earth, can be mined in every part of the world. It is therefore reasonable to assume that gold will be discovered near Menegroth as well. Beryllium and the platinum group metals are much more geographically restricted. The discovery of deposits near Menegroth would be a matter of good fortune that would be taken advantage of, but not counted upon. Silver has most of the same advantages as gold, but with the downside that shipping costs would eat up about a third of its price.

Figure 11 – Possible gold coinage of Menegroth, depicting the god Mars

**Research and Media**

Scientific research will be a major component of Menegroth's economy. More than a billion dollars annually is already spent on the robotic exploration of Mars, and Menegroth's establishment will lead to a surge in scientific interest. Governmental and private researchers from all over the world will hire Martians to perform all manner of research into subjects as diverse as geology, meteorology, astrobiology, and the effects of low gravity on living things.

Television will dominate Menegroth's media exports. The colony's output will include an overarching show focused on the President and chronicling Menegroth's general progress, more technical shows focused on specific industries, natural wonder shows in the Martian wilderness, educational programming aimed at children, sporting events such as low-gravity basketball and the solar-system-record long-jump and high-jump, and even a cooking show.

Further opportunities to profit from media include product endorsements, as well as books and magazine articles written by nearly every citizen.

**Long Term Sustainability and Growth**

The resources bases upon which Menegroth will rely, namely, intellectual property and previously-untouched mineral deposits, are inadequate to support much further expansion of the human presence on Mars. Ore formations of the kind that would have been mined by pre-industrial societies on Earth are the only kind that will be economically competitive at all when mined on Mars. Growth beyond the small, early colony which I have heretofore described will require access to a major resource that can be utilized only on Mars.

I believe that metal-rich asteroids are the best candidate for that resource. Asteroids can contain precious metals and rare industrial metals at hundreds or even a thousand times the concentrations at which they are found in the crusts of differentiated planets like Earth and Mars. Trying to refine asteroidal material *in situ*, without air, water, or gravity, is a fool's errand. But if asteroids are dislodged from their orbits and collided with Mars, their material could be refined in city-states like Menegroth. The planet Mars is well-suited for this future economic role, on account of both its proximity to the asteroids, and the much lower risk of damage in colliding them with a sparsely populated planet with a thin atmosphere.

## MARTIAN POLITICS AND SOCIETY

### The Founders' Dilemma

Any attempt to centrally plan human society and culture is perilous. The normal course of events is for one generation to receive its culture and traditions from its elders and pass them down, with only small changes, to the next generation. It is beyond the rights of a single generation to endeavour to reinvent everything, and efforts to do so have a very disappointing history.

Nevertheless, setting up a city state in a radically new environment will demand a certain degree of social engineering, whether or not anybody want to do it. To avoid the worst of the perils, Menegroth's initial leaders should avoid concentrating, in themselves, too much of the power to create new culture and traditions; they should rather allow Menegroth's thousand inhabitants, along with all their future descendants, to create that culture organically.

These founders will do well to recall Thomas Paine's words: "There never did, there never will, and there never can, exist a parliament, or any description of men… possessed of the right or the power of binding and controlling posterity to the 'end of time,' or of commanding for ever how the world shall be governed, or who shall govern it.... Every age and generation must be as free to act for itself in all cases as the age and generations which preceded it. The vanity and presumption of governing beyond the grave is the most ridiculous and insolent of all tyrannies. Man has no property in man; neither has any generation a property in the generations which are to follow." [6]

### The Republic of Menegroth

The ships which bring Menegroth's settlers to Mars will be flagged to terrestrial nations and subject to their laws, but upon disembarking and commencing life on Mars, the settlers ought to declare immediate political independence. Justification for this can be made by appeal to the Outer Space Treaty, according to which its signers – including all terrestrial spacefaring nations – cannot claim sovereignty over a celestial body beyond the Earth.

The Republic of Menegroth itself would not be signatory to the Outer Space Treaty, though in the interests of fairness, the settlers would offer to sign a reciprocal treaty in which they renounce the right to make or enforce territorial claims on *Earth.* The new government will not be so presumptuous as to claim authority over all of Mars. Menegroth will merely be a city state, willing to coexist as equals with any other Martian city states that may arise.

In order to encourage recognition by as many nations as possible, Menegroth should offer free transit to Mars for anyone appointed to serve as his or her country's ambassador to the new state. This will, I believe, result in friendly relations with a hundred or so small countries that otherwise might never send anyone to Mars, even if the ambassadors themselves are, in many cases, actually scientists whose diplomatic duties take up only a little of their time.

Menegroth's government will function through an Assembly of all citizens, called weekly or whenever else the President thinks necessary. Suffrage will extend to everyone over the age of sixteen. The Assembly will vote on any changes to Menegroth's laws, and will also elect the President, through as many rounds of balloting as are needed for a single candidate to gain a majority.

The President will have authority to appoint any other officers he thinks are necessary, provided the Assembly confirms them. The citizens may call for a vote to remove any officer, including the President, and the vote of three-fifths will suffice for removal.

Each President will serve until resignation, or removal in such a vote of no confidence. It would be good to encourage a tradition of honourable resignation, so that statesman who feel that their that work is done or that their best days are past do not keep hanging on to office for as long as they can. There will be no political parties, and without fixed-term elections, Menegroth will be spared the drama of cyclical politics, as well as the problems attendant with egotists running for office merely because there is a regularly scheduled time to do so.

All trials will be by jury. Menegroth will have little tolerance for crime, and felonies will be punished with confinement in the brig until the next launch window, then exile to Earth.

**The Martian Calendar**

Proposals for a Martian calendar are numerous; nonetheless, most have at least one of two major defects: first, not setting the year at a whole number of weeks, and second, creating confusion by using day and month names that are the same as those used on Earth. In Menegroth's calendar, there will be no such thing as a "Thursday" or the month of "October," because those terms refer to elements of Earth's temporal cycle, not Mars'.

Menegroth's name is of Elvish etymology; it means "Thousand Caves" and is the name of a subterranean city in Tolkien's backstory to The Lord of the Rings. In keeping with this theme, Mars' weekdays will take their names from the six-day Elvish cycle, being called *Minë*, *Atta*, *Neldë*, *Canta*, *Lempë*, and *Enquë*, after which comes the Sabbath. Because the Martian Sabbath bears no particular relation to either Friday, Saturday, or Sunday on Earth, religious controversy over which, if any, of those three days is the correct Sabbath can be avoided.

There will be 665 Martian days, or 95 weeks, in a common year, and 672 days, or 96 weeks, in a leap year. Out of every thousand years, 514 will be leap years. The twelve months will be named for Zodiac signs, but rather than the constellation the Sun is in, which is never visible, each month will be named for the sign currently opposite the Sun, which appears in the sky all night long, and in which Mars' moons will, on most nights, undergo their eclipses.

Because of Mars' eccentric orbit, month lengths will be unequal. Northern spring and summer will be longer than autumn and winter, and the months will range in length from 49 to 70 days, each having, like the year, a whole number of weeks. The year will begin shortly after the northern vernal equinox, the first day being *Minë, 1 Tauri*. The last day of the year will be a Sabbath, either *49* or *56 Arietis*, depending on whether or not there is a leap weak.

Nonetheless, the Martian year will be only one of *three* cycles around which Menegroth's inhabitants will plan their affairs. The second cycle will be the Earth year, as the colonists will be reliant on trade with Earth, and will continue to reckon their ages in Earth years for the sake of cultural continuity. In addition to both of these annual cycles, Martian colonists will also closely follow the Earth-Mars Synodic Period of 25 ½ Earth months. Each Synod begins and ends at conjunction; launches from both Earth and Mars must occur during the first half of the Synod, and arrivals during the second. Synods will be numbered beginning with Synod I, during which Mariner 4 made the first flyby of Mars in 1965. At the time of publication, we are in Synod XXVII, which began on 2 September 2019.

**Education in Menegroth**

The scientific and technical education of Menegroth's children will be of the utmost importance. However, there will be no concept of a division between STEM subjects and other fields, nor a prioritization of the former. Rather, children will be encouraged to study, in history and literature, the stories of explorers, adventurers, and innovators throughout the ages.

Children will be taught the alphabet and numerals in the home by their parents. They will start grade school at age seven, advance to the lyceum at twelve, and graduate at sixteen. Children at each level will attend school either only in the morning or only in the afternoon, spending the other period at agricultural labour or, for older children, an apprenticeship.

Education in grade schools will consist mainly of instruction in reading, writing, and mathematics, with a smaller amount of art, music, and basic science. History will effectively be merged with reading, as it will be taught by having children read old books and report on what they learned. The school day will last only four hours, with one period devoted to physical education in subjects such as swimming, gymnastics, and martial arts, so that children are never required to sit still for unnatural lengths of time. There will be no academic records, no passing or failing grades, and no chance of a child being held back. How to respond to poor marks on individual assignments will be a question for the parents.

The four years in the lyceum will be much more demanding, as Menegroth's youth are required to take a more active and rigorous role in their own education. They will study history and literature intensely as they hone their skills at rhetoric through debate and essay writing. They will also learn higher mathematics and all the intricacies of computer coding, and gain a much deeper understanding of the sciences through relentless laboratory work.

Children should be taught that, if they are doing things right, they will get more of their education from the library than from the schools. And while many high-tech societies are tempted to dispense with printed books, Menegroth will retain them, even at the cost of greenhouse space for growing fibre. The ill effects of computer screens on youngsters are well known, and even adults can develop a love for their favourite volumes that would not be possible with mere digital files. Furthermore, having a book that tells a specific story or describes a specific topic helps to focus the mind in a way that a sea of computerized information cannot.

Menegroth will have no universities and no formal system of post-secondary education, with all professionals trained by apprenticeship and hands-on experience. Education will not, by any means, cease with graduation from the lyceum; academic discourse will continue among all of the citizens. Rather, there will be no need for a university because there is no division between scientific pursuit and everyday life, and there will be no concept of a college degree because there is no distinction between the learned and the unlearned.

**Marriage Customs**

Menegroth's colonists will need to adopt a more disciplined attitude toward marriage than is the norm in modern Europe and America. In Menegroth, as among pioneers in general, marriage will be seen as a practical means to build families, care for children, and meet the need for lifelong companionship. Entering and leaving relationships on the basis of fleeting passions will not be romanticized, and adultery and no-fault divorce will not be tolerated. Weddings will be celebrated, as in most cultures, with feasting and dancing. The custom in which the bride wears white to symbolize purity, while the groom wears black, will be dispensed with. In Menegroth, both bride and groom will wear white.

**Private Property**

Menegroth will begin as a commune, but rapidly move toward operating many of its industries on free-market structures, in the expectation that entrepreneurship will foster innovation and lead to more efficient use of resources. Thus, after beginning with all equipment and industries state-owned, Menegroth's leaders will encourage private enterprise as soon as possible. Individuals will earn money in wages from their employers (initially, the state) as well as in media deals, and they will be free to purchase equipment and go into business for themselves. The state will not sell off anything that would create a monopoly – the nuclear reactor, for instance, is off limits – but manufacturing and mining equipment, as well as greenhouses, could be bought, so in time most machine shops, farms, and mining convoys would be private.

Menegroth's currency will be the *Stater*, defined as 10 *g* Au held in a bank on Earth. (Σ1.000 = $479.32 as of 15 Sep 2019). Staters would be readily accepted on Earth; on Mars itself, tokens would be exchanged in their place. The Stater is subdivided into 1000 mites. (1μ = 47.9¢)

**Sports and Recreation**

Menegroth's inhabitants will, from the earliest age, be taught to value physical activity. Until seven, they will be deemed better off at play, chasing the chickens and the dogs, than sitting in a classroom. After starting school, children will get daily, rigorous lessons in physical training, and frequent reminders that, being adapted to a world with higher gravity, they are stronger than they think, but without hard work, they will never unlock their true potential.

Athleticism will continue throughout life. Racquetball courts will be ubiquitous, as racquetball, though played in a small, enclosed space, gives a very intense workout. Wrestling will be encouraged, as will swimming, for on entering the water, one loses the advantages of low gravity and must work just as hard as on Earth. Popular team sports will likely include basketball and water polo.

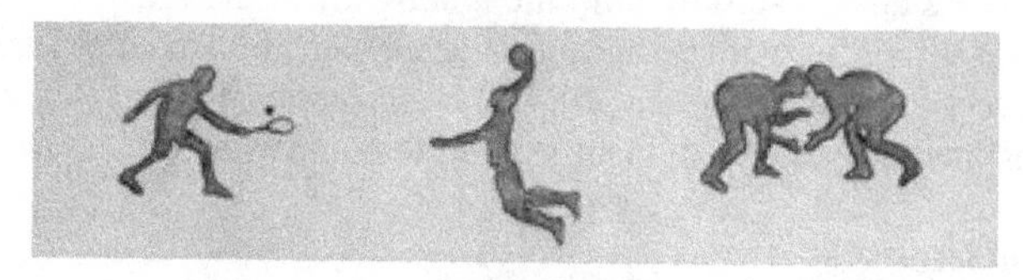

Figure 12 – Athletics of Menegroth

Nearly all sports will be informal, as Menegroth will be too busy with the challenges of survival to support any kind of organized leagues. The exception will be during festivals, when dozens of diverse matches, races, and tournaments are held as everyone, from the President down to the youngest grade schooler, tests his or her strength by striving to become the best at *something*. Sporting festivals will become media spectacles, broadcast on Earth, especially where low gravity is a factor, as in the solar-system record track and field events.

**Art and Music**

Menegroth's aesthetic will be minimalist, with a few exceptions. First, the chapels will break the trend of plain buildings, as the colonists put their utmost effort into beautifying their houses of worship. Second, art reflecting the beauty of Earth will be highly esteemed. No building, public or private, will be complete without a magnificent landscape painting, and even the very roof of the cave will be painted like a sky, to preserve not only the memory of wild nature, but also the old world's history and legends. A colonist might look up and see Thor hurling his hammer, or the Montgolfier brothers in their balloon.

Music will also be important. Each child will learn to play an instrument in school, while at the same time, in the greenhouses, they all acquire the habit of singing or whistling while they work. The youngsters' love for songs of any shape and size will grow as they discover the works of Bach, Mozart, and Beethoven, men who could, in the course of an hour, build theme upon theme to tell a soul-stirring story in the language of pure music. As adults, many will join Menegroth's orchestra and chorus, or learn to play the organ in the Great Hall.

**CONCLUSION**

My design for Menegroth imagines Martian settlers living in a natural, subterranean cave. Once regular interplanetary flights with Earth have been established, and sufficient scientific exploration of Mars has been conducted, an initial group of a thousand dedicated settlers, working with a $15 billion starting investment, could build a successful city state. Nevertheless, their task will not be easy, as Mars is a harsh environment where the full force and ingenuity of modern technology is needed to eke out the barest level of subsistence.

By sealing off a long, flat-bottomed lava tube, the colonists will immediately gain a spacious habitat in which to safely reside and perform most of their labour. Nuclear power, supplied by a molten salt reactor, can meet their heavy electricity needs. Their industries, once operational, will be able to smelt iron and other metals and produce, through additive manufacturing and CNC milling, a diverse array of parts with little human involvement.

Menegroth's children will begin their labours by working in the greenhouses and growing wheat, sorghum, amaranth, potatoes, maize, and beans as staples. Meanwhile, miners will direct robots as they mine ice and other needful minerals. Some minesites can be reached by rail, but others will require convoys to make slow and arduous journeys into the wilderness.

Menegroth will be hard-pressed to recover its initial costs, and will remain dependent on Earth for physical imports and intellectual property. But by exporting precious metals, performing research for paying institutions, and making a media phenomenon out of its own rise and progress, Menegroth can sustain its economy and balance its trade with Earth.

Figure 13 – Martian Settler and Martian Flag

Political independence will be declared immediately, but in the end, only a disciplined population, willing to build the best possible education system and pass on a strong culture to the next generation, can become the nucleus of a lasting human presence on Mars.

## REFERENCES

[1] Candidate Cave Entrances on Mars -
www.caves.org/pub/journal/PDF/V74/cave-74-01-33.pdf

[2] Themis Observes Possible Cave Skylights on Mars
https://agupubs.onlinelibrary.wiley.com/doi/full/10.1029/2007GL030709

[3] Terrestrial Energy, *Integrated Molten Salt Reactor* -
www.terrestrialenergy.com/technology
The online descriptions of this technology are for the 195 MWe reactor. Work on a 33 MWe / 80 MWt variant was confirmed in a personal communication with the company.

[4] Stockholm Environment Institute, *Hydrogen Steelmaking for a Low-Carbon Economy*
www.sei.org/wp-content/uploads/2018/09/hydrogen-steelmaking-for-a-low-carbon-economy.pdf

[5] Villanova's Experiment on Crops in Martian Soil Simulant
https://eos.org/articles/tests-indicate-which-edible-plants-could-thrive-on-mars

[6] Thomas Paine, *The Rights of Man*
www.let.rug.nl/usa/documents/1786-1800/thomas-paine-the-rights-of-man/text.php

# 20: THE BUBOLAQ
# A DESIGN SKETCH OF THE FUNCTIONS AND CAPABILITY OF AN EARLY MARTIAN COLONY.

**Stellie J. Ford, Tessa Young, Julianna Ricco**
Bubolaqmars@gmail.com

## INTRODUCTION

Humanity owes much of the success of our species to curiosity. Building upon millennia of refined discoveries and catalyzed by our ephemeral lifespan, humanity has pushed ever closer to the boundaries of our world. The colonization of Mars is inevitable. Like the hostile environments colonized by our ancestors, Mars presents us with a literal world of challenges. To forge a new life on a dead planet, the potential of every grain of sand must be considered. The early years of colonization will fundamentally shape the course of humanity in a way that we have never been known on home world. We can launch this project with a collective investment of resources and will power, but it is human curiosity and the drive to define ourselves anew that will enable a colony on Mars will be built.

In the Fall 2037, during the 7th Hohmann transfer window since human arrival, 1000 residents of the Bubolaq colony will be taking their firsts steps towards autonomy on the red planet. With 1000 colonists, economies of scale begin to aggregate and sheer numbers allow for specialization. Ubiquitous automation and sensing in environmental control and life support systems (ECLSS) This makes the colony robust in the face of accidents, and leaves room for the cultural and creature comforts necessary for humans to live fulfilled lives. To this same end colonists will receive a universal basic income (UBI) to ensure they will be able to focus on colonization over subsistence.

The excess of wealth and ability that is present on Earth will be instrumental in founding the Bubolaq colony. Consequently, a significant portion of the administrative body of the colony will be located on Earth. The remote nature of this group precludes their involvement in inherently personal parts of the colony like short term resource allocation and domestic policy, but they will undoubtedly control much of the long term contracting, and logistics. The local government on Mars will be directly beholden to colonists and meeting operational goals. The various departments that will facilitate this will coordinate with both local and Earth governments. These departments will be: Power Generation, subdivided into nuclear, solar, thermal, and grid management, Agriculture, Areology, Habitation and Utilities, Resource Extraction, Healthcare and Bioproduction, Fabrication, Community Services, Martian Discovery Institute, Finance, and Administration.

In contribution towards its funding, the colony will capitalize on high value export goods: data, intellectual property, luxury goods, and access to the Martian surface. Moreover, individuals who are willing to pay for access to visit the Martian colony will be welcome. The data and feedback will come for the daily activities of the colony and its inhabitants, is valuable as an export, but more so for the naturally arising IP that it represents. The discoveries that enable Martian life will be secured by the colony's legal systems. To support this innovation, and to serve as a focal point for investment a post-graduate research institute will be established where Martian ingenuity can take root[eu]. There is a near limitless potential for growth on the Red Planet and investment in that potential can be compounded for centuries.

**THE BUBOLAQ**

At its core, the Bubolaq is a collection of lightweight habitation domes within excavated ice caverns of Deuteronilus Mensae. Farming domes outside the caverns and aquaponic systems will grow nearly all of the colonist's food, supplemented by local $CO_2$. Mars itself will supply the raw materials needed for glass, metals, fertilizer, and even the gasses we will breathe. While the colony will import and export a significant amount of material with colonist turnover and build out, bulk infrastructural and simple biological necessities like antibiotics, food, and water, will be produced within the colony. The colony's power supply will be based nuclear energy, with demand variability managed by small modular reactors, solar, and extensive power storage systems.

The modular design of domes and caverns allows for satellite colony clusters to be supported in the immediate vicinity, and for the resupply of remote surface and orbital outposts. Construction of new caverns and outposts will be a unique service that the Bubolaq can provide, but either affiliated or incidental to this colony, collaboration will benefit all residents of Mars.

The caverns will be protected by two primary airlocks in the primary egress tunnel, from which a series of channels provides access to habitation and industrial caverns. These caverns will provide a secondary environmental shell that is central to the life support and heat retention strategy. The model cavern is 30m high by 100m, supported by free standing structural pillars, and filled with 50% O2 air at 30kPa. Though lined with, airtight sealant [ch], aerogel-impregnated composite fabrics, and reinforced concrete, it will be kept well below 0C . Despite harsh conditions, caverns offer a much safer working environment than the open surface for infrastructure, maintenance, and repair operations. Colonist habitation domes will, of course, be kept at a reasonable temperature for daily life by tying into the central heat pipe network.

**The Caves Of Deuteronilus Mensae**

Competing interests make selecting an ideal location difficult. Avoiding cosmic radiation, toxic perchlorates, and meteor impacts are the primary concerns, but must be weighed against the practicality of system implementation for life support and sustainable in situ resource utilization (ISRU). Though the plentiful lava tubes [cd],[g],[cf],[e],[f],[bq],[cc] throughout Mars contain subsurface water [bq] and appear to be a promising location at first, there are numerous drawbacks. First, the tubes can have several-meter-thick basalt walls, which though prized for their impact resistance, is incredibly tough and difficult to drill or blast through, restricting access to the local resources [f] [cd]. Second, though some tubes may be accessible via collapse features called skylights [f], in many locations, the tubes are mere meters in diameter [bq] and follow irregular organic outflow features that would limit growth and be harder to seal. The In addition, as many of these lava tubes are in regions that are not well studied, they are areas of limited scientific interest [eu], which be a major setback to the academic aims of the colony.

To maximize access to resources, ease of habitat construction, and expansion potential, the colony will be located Deuteronilus Mensae, a high mid-latitude region along the dichotomy boundary [eu] centered near 44N 337W. The lobate debris aprons, infilled channels, and craters that dominate the region [ed],[er],[eu] provide necessary resources: plentiful water ice, easily mineable regolith, and diverse geological features. The debris aprons are not only impact and radiation resistant, but they are immense, offering ample space for colony design and expansion: they can be tens of kilometers (kms) deep, tens to hundreds of kms wide and over 300m high [ed]. Inside these debris aprons are naturally occurring water ice deposits, the larger ones containing cubic kms of water ice [ee],[ef],[eg]. The immense quantity of water and regolith in the region and the relative ease of their extraction indicates that extraction will be a rate limiting factors on the ability of the colony to grow or respond to acute demand. Moreover, modeling combinations of temperature, debris covering, and formation time has given a definitive picture of apron evolution [ee]. Before excavation, a drone survey characterizing the regolith and ice composition [gm] will help establish design parameters [en], akin to terrestrial high ice permafrost tunnel construction [ek]. The glacier surrounding the colony will be stabilized [ee][eh] by embedding cryocooling loops[gx-gz] into the surrounding ice and permafrost [ha][hb]. This is invaluable data in planning for both the fabrication requirements and strategic regional resource placement. The plentiful local resources and scientific areas of interest have made this region a popular candidate for human exploration [ev] and will make it an ideal colony location.

**Power**

Energy consumption is tied to productivity in island economies [fq] and will be an even greater factor in the cold, inhospitable Martian landscape. Baseload heating supplemented by the tremendous power requirements for environmental control and life support systems pose a challenge to the need for surplus power. Though they do reduce variability in power demand. A combined 390MWe of nuclear and solar power supported by a 250MWh battery system, solar thermal [hj] and chemical energy storage [hs] systems are the foundation of the colony and provide the needed capacity for grid stability. Optimization of demand timing and a DC grid further reduce variability as does strategic use of thermal power in manufacturing and distillation. Careful management of the colony's power resources creates the capacity for growth.

The nuclear components of the colony power systems consist of three modular elements: (1) the B&W mPower [s] or and equivalent Small Modular Reactor (SMR) [t],[he], such as the Holtec SMR-160 [hd], (2) the Pylon reactor design from the Ultra Safe Nuclear Corporation[ap], and (3) NASA's Kilopower reactor[w]. Molten salt heat exchangers coupled to the cooling loops and insulated with aerogel composite allows the heat from these reactors to be transported where it is needed in the colony. The primary reactor, here represented by mPower, is a pressurized, water cooled small modular reactor able to produce 180MWe, and 530 MWt per core, and with a footprint of less than one hectare. Two mPower reactors produce the baseload power supply for the Bubolaq. The upfront cost of these reactors is $2.65 billion [s] for transportation and installation. Site prep will begin immediately with reactor delivery and installation targeted for the 3rd Hohman window. Numerous specialists are needed during their 3-year installation and startup window, and 105 colonists will be tasked with operating the turbine and reactors' combined management system. This itself is a significant advancement over present day reactor staffing [hf] but is within the scoping [hg] for remotely monitored and managed SMRs [hh]. Additionally, a small contingent will be required every 4+ years for refueling [s], heavily assisted by robots [jl].

Reactor safety is of upmost concern. The 4ha cavern housing the SMR-turbine complex will be a brick and concrete installation consisting of impact resistant housing, access controls, and . The complex will be physically separated from the habitation caverns by airlocks and hundreds of meters of glacier and will maintain 72-hr minimum walk-away stability [hd]. For security personnel, a 15-person team will maintain 5-person shifts to monitor the nuclear and secure systems, focused on access-controlled areas, particularly the mPower reactors and, to a lesser extent, the Pylon reactors.

The Pylon reactor is a near-term gas-cooled Micro-Modular Reactor (MMR) [ap], that will be used for load balancing, remote habitats, and as a source of dedicated process heat. Each Pylon, per present day design, produces 150kWe – 1MWe and 800kWt – 4MWt [ap][bk], with the development of 5MWe, 15MWt [ht] anticipated by its usage during the 7th Hohmann window. An estimated 12 Pylons will be necessary to establish the colony, and an additional two will be installed during the 7th Hohman window to offset capacity loss during the first mPower refueling. With approximately 40MWe and 160MWt, collectively, resource accumulation will still be limited during SMR refueling. Each pylon will be contained within an access restricted, hardened structure, and no less than 50m from any inhabitable dome.

NASA's Kilopower reactors are heat pipe cooled, low enriched uranium reactors massing 400kg and 1500kg for 1-3kWe and 4-10kWe [w] variants, respectively. Though NASA's design couples to Stirling engines and a radiative cooling fan, the reactor will be modified to link with heat exchangers for the majority of colony installations [d], allowing for smaller industrial systems and domestic temperature control. The reactor is ideal for transport to and maintenance of construction sites and remote scientific outposts as it has a16:1 turndown ration and can even be turned off and restarted [jm]. As with the central system, the remote reactors' heat too can be captured, making outposts' habitability and energy requirements much more feasible than with solar power alone. Particularly in groups and when paired with mobile battery systems these reactors provide versatile support for daily operations. An estimated 50 will be necessary to sustain in the colony during the 7th Hohman window.

**Solar**

Flexibility and maximization of available resources govern the colony design. While solar flux at 44N will be low [jm], a solar panel farm will provide grid stability and an additional source of power during high demand hours. By the mid 2030's terrestrial solar power will perform at 250W/kg [bu] or 107W/kg on Mars given the 56% reduction of solar irradiance [il]. To yield a peak 30MWe, a 0.5 square kilometer array at 50% panels will be built adjacent to the colony on the Martian surface. Panels will be equipped with an electrostatic dust clearing system [im] that has no moving parts, reducing maintenance requirements. Panels will initially be imported but by H7 solar cells may be fabricated locally from regolith sourced materials [n]. While the fabrication of panels can be complex [jp], once established, mass fabrication is possible using vapor deposition [jo]. Additional solar capacity is a long term investment that will stabilize the colony.

Heliostat mirrors installed on terraces will also maximize local resource capture by focusing sunlight onto the agricultural domes. This system could also be used to improve the energy production of the solar panels or the solar thermal system, especially during dark winter months when flux will drop to nearly 25% of its summer levels. While these mirrors enhance solar capture, they are considered a secondary system are not critical to colony survival.

**Batteries**

Two grid-level battery systems will be in operation throughout the colony. The Tesla Powerpack [bo][bp] will provide load balancing and surplus supply storage, and an aqueous flow battery [u][x] will provide a large storage capacity offering durability and capacity to the grid. The entire grid will be actively modeled and compared against grid data. Over time, this data will allow for better automated management of these systems.

The Tesla powerpack is an electric battery cell power storage system capable of storing 200kWh per unit at 8.11kg/kWh [bo]. It is modular and scalable beyond the 15MW/50MWh in the Bubolaq. Its built-in battery management system ensures stable operation and pack longevity with minimal user input [io]. This system will be the primary means for load balancing due to its ability to rapidly adjust output, providing time for output to be modified.

A 200MWh, air breathing sodium-polysulfide redox flow battery will provide a massive buffer to the colony's power grid. At 29Wh/L, 7,000m3 of 10M polysulfide and NASO4 solutions will provide an emergency power reservoir and absorb output as needed by the grid. These are significantly safer than traditional NaS flow batteries [io] as they do not use metallic sodium and can operate at lower temperature. Sodium and sulfur are both readily obtainable from regolith[n].

Extensive fixed and mobile small battery units will also be in use locally throughout the colony. These will be for vehicles, sensors, and stand-alone equipment support. Fabrication of batteries often involves harsh solvents that can be difficult to produce. Optimization of battery fabrication will be a valuable research target.

**Thermal**

The life support systems of the colony will produce, manage, and store thermal energy in concert with the electrical energy storage systems just discussed. Thermal systems will rely on thermal energy from nuclear power generation, captured and converted into chemical energy where it can be stored in bulk for extended periods, and supplemented by solar thermal energy capture to maximize passive resource accumulation.

Surplus heat will be stored as chemical energy in the form of reforming gasses and liquified natural gas (LNG). The CO2/CH4 reforming reaction (Fig. 1) [hu] can be reversibly catalyzed to store heat at a rate of up to 30MWt/ton catalyst [hs], and density of 2.2 m3 of stored gas per MWt. The reaction can occur at pressures up to 69 bar [hs] allowing for a significant volume of gas to be reacted. A 1million m3 storage vault (100m x 200m x 50m) could hold 126 MWht at atmospheric pressure. LNG, at 5.6 MWh/m3, offers grid scale backup. Importantly this reaction can be artificially pushed via the Sabatier process for methane production. Due to boil off, LNG and stored thermal power will be available in the colony for preheating systems, stand-alone power generation, plastics, chemical synthesis, and local transportation.

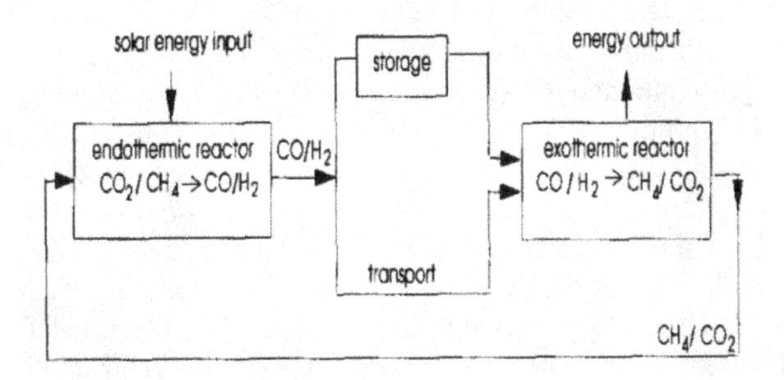


While conservation of energy and strategic timing of energy will lower energy demands, fundamental process optimization will yield more robust results. The use of low CO-2 cements [hi][hq] can help offset resource demands and make use of process heat, as can the manufacture of aluminum [hl], hydrogen [ho],[hp],[hq] chemical and plastic synthesis [hn],[hq], and desalination [hr], which will occur in the colony (see below).

The Bubolaq's solar thermal system will be parabolic trough installations that concentrate incident radiation onto molten salt heat pipes to transfer into colony heat management system. Solar thermal systems have 40% efficient heat capture[hu] The option to divert heat directly to the agricultural surface domes will be made available, particularly as the positioning and relatively low operating temperature [js] of this solar thermal system limit its impact on higher temperature applications. While concentrated solar thermal towers would solve this issue, they require significantly more setup [jn].

**Domes**

As with the Arakeen Conservatory, there is a verdancy to contained ecosystems. This will be taken to heart when landscaping the inhabited portions of domes. However, minimal time requirements for all horticulture is necessary. Combined with the availability of surplus farming technology, automated care and monitoring equipment will be abundantly applied to . This is a strong motivator for modular and accessible system selection.

The dozens of industrial and habitation domes within the colony caverns will be made of PVC [gl][gq] and polyester [gs][gt] composite films, a significant portion of these to be built through ISRU. Kevlar will be used in outdoor domes and flooring, and can be manufactured by the Bubolaq's polymer chemists, though nitrogen will likely need to be imported for this purpose. Widely used in subarctic climates on Earth, inflatable domes have been well studied [gl][gm][go-gt] and proven in the closest terrestrial analog to the Red Planet. High frequency welding [gu] will be used to join individual sheets while retaining their original strength [gt]. Further, cuts do not propagate well in domes [gj]. HF welding is an easily automated process[gm], ideal for the colony, with simple and primarily static components. Pressure regulation within the domes, will compensate for thermal variability in internal gas volume. The caverns themselves will also have pressure regulated both systems will be linked to active control systems should issues arise

[gi]. While Polyethylene is good for radiation protection, it is structurally limited and transparent to infrared, necessitating additional insulation, [gk] though this is a common design challenge. For example, the TransHab design relied on a series of redundant Polyurethane/Saran bladders [gp]. This design poses an additional barrier, as the synthesis of polyurethane consumes nitrogen, which will be limited on Mars.

The habitation domes are designed to be airtight structures. To diminish the chance of air leakage into the cavern, the number of utility ports for plumbing, utilities, and egress should be minimized. Preparation/preinstallation of utility ports into dome structures will minimize the chance of contamination during setup and remove the need for on-site sealing. The foundation will be covered with a layer of Kevlar, contiguous with the wall and roof portions, allowing for the domes to be airtight if all the ports are sealed. Despite the higher rate of air loss to the environment, agricultural domes will not use a full envelope as tiling [be] and other stress from agriculture risks breaching dome material. Instead, the walls with be joined to the floor with an anchor and seal system. Kapton tape is a widely used in aerospace [gm] as a vacuum compatible emergency repair material and will be used in the colony; unfortunately, it must be imported due to its complicated and nitrogen-consuming synthesis[gw].

Habitat foundations will be composed primarily of compressed regolith bricks, aircrete [gk], and reinforced concrete for structural stability and low thermal conductivity [ew]. Bricks can be made with automated systems like the Reva RBME [bc] and then handled by general purpose robots. Aircrete will be mixed in specialized rovers and form an exposed crawlspace containing the utility piping above habitat. Aircrete's high insulating capacity ensures its compatibility with thermal piping and is easier to manufacture at scale than aerogel.

## COLONY INFORMATION/TECHNOLOGY

There will be a fiber optic cable network between the Bubolaq domes with traditional wireless access points in the domes. WiMAX will connect the Bubolaq's network to remote outposts and ground relay stations that further link up to an areostationary satellite network.

WiMAX minimizes the labor, mass, expense[eb], and hazards of laying fiber optic cable for outposts allowing faster and more flexible deployment. Submarine fiber optic cable has faced greater challenges than those expected for a Martian fiber network [dy],[dz]. Robotic cable laying systems have been in existence for over a decade [dp][dr] and known copolymer buffer systems [ds] would be within the colony's capacity to manufacture. However, the emergence of satellite constellations may supersede this development pathway. The base stations will be connected to a direct broadcast satellite network in three-tier mobile network[hk] uplink centers, allowing for the majority of network operations to be maintained by the colony.

To remain serviceable and maximize useable floor space within smaller domes, the low-profile cable must be laid no closer than 0.5m from the base of a dome wall, allowing common access near the curved based of the dome's walls. Moreover, this configuration should be common for dome infrastructure to promote standardization of design conditions, access protocols, monitoring, and environmental control.

**Internet of Things**

There will be diverse and pervasive implementation of sensors and simple automated systems in the colony. The Bubolaq's DC grid reduces mass requirements for small electronics by up to 40% [da] and eliminates the need for phase synchronization. As a general standard, maintenance of most components of these systems should be able to be performed by any end user, guided by readily-available technical specifications and the local network of community and professional support. In this way, the time and focus of experts can be reserved for situations demanding escalated assistance. This is highlighted as small automated systems are so pervasive, they represent a sub-category within most other domains. Having colonists feel comfortable and competent in the manipulation of these as a matter of course is a part of the ethos of self-reliance that must dominate any complex remote settlement.

There will be levels of complexity and functional scope for colony IoT systems. A 3125-node network based on the LoRa[dw] and NRF24L01[dv] boards is an example of a standard implementation platform for long term projects, like monitoring growing conditions or controlling infrastructural subsystems. Despite the limited nature of the system, it is domestically producible, low power, modular, and widely customizable. There is already a robust library [dx] for interfacing with functional and control elements. While LoRa boards are lighter, transmit much farther, and consume less power, they are not 5V compatible and have lower bandwidth than NRF24L01 boards [ds]. Deciding between these two systems will be on a case-by-case basis with LoRa favored for hubs and more remote scenarios and NRF boards for ad hoc projects and sensors. These networks will be one of the core elements supporting the automation of daily tasks [dt] and environmental monitoring [du][ep].

## RESOURCES

**Water**

Local water extraction is a part of the life-sustaining work of the colony. The lobate debris apron will supply millions of cubic meters of potable water during the Bubolaq excavation phase and additional potable water is easily excavated nearby. The water will be purified by vapor distillation due to the abundance of nuclear heat and the robust simplicity of the process. Fortunately, the excess eliminates much of the concern over this resource. Water will be stored in abundance and will be available for industrial and personal use.

**Regolith Mining**

Along with extensive water resources, lobate debris aprons will provide the colony with tons of mineable material. During the initial tunneling, significant quantities of shield regolith will be broken up and collected by heavy electric rovers with front loading scoops. The collected regolith will first be fed into a rotary trammel [iw] for size-based separation. Larger pieces will then be characterized and sorted by optical and x-ray luminescent properties [ix] and stockpiled for eventual batch processing as ore. This sorting allows for filtering of unwanted ores like excess silica and can reduce the energy requirement for processing by 60% [iy].

This material will be available from the very start of colonization. The rover operators will be some of the first Martian residents and will remain a ubiquitous feature. They will operate on an event [ir] and task list [is] framework with significant machine autonomy dramatically multiplying one person's impact. The excavation of the Bubolaq caverns will provide a substantial amount of material and additional regolith will be available as the colony expands. Less common resources like nitrogen may require mining of nearby deposits, but the landscape offers diverse deposits.

In general, the ability to extract minerals from regolith is well demonstrated [iy-jf] so focus will be placed on process optimization to maximize overall value chain [jg]. An example of one such practice is the use of oxalic acid to leach low concentrations of iron out of high silica sands, increasing yield [ik]. The mobilization of soluble minerals under varying pH will be used widely to exploit the optimal solubility windows [jc]. A similar mechanism will be used for selective precipitation of minerals [je] from extract solutions. When the limits of this system are reached, other factors, like redox potential will be exploited. Copper will be selectively separated from Zn and Co via electrowinning [jf] and wash with acid for purification. This will take about 1.5 MWh/ton Cu [bz] plus chemical production costs which is very reasonable when compared to the $750,000 it would cost to import an equal mass of copper. The smelting of aluminum is very energy intensive when starting from ore waste streams; instead low value tailings can be consolidated and returned to feed streams or used as soil filler [ff][iz].

Some regolith determined to have suboptimal characteristics for refinement will be minimally processed and used as a bulk component of soil; however, toxic perchlorates must be removed before the regolith can be safely used. One method for achieve this is by capitalizing on perchlorate's high water solubility [cq], though the wastewater must also be remediated. Perchlorate can be oxidized via high temperature reactions [q], but this is energy intensive. Instead, efficient remediation will be achieved via a fluidized bed bioreactor. The reactor will be stocked with perchlorate metabolizing bacteria, supplemented with acetic acid and inorganic nutrients to promote activity [cp]. This process breaks perchlorate into chloride and O2 which can be captured for colony use. Moreover, the enzymes required can be purified and stored when not actively in use, according to the process described by Davila and Wilson (2013). The large amount of stored oxygen in perchlorates yields up to 91 O2 per kg regolith, so perchlorate contaminated wastewater becomes as a mineable source of oxygen for the colony.

## CONSUMER GOODS AND SERVICES

### Chemical synthesis

In order to facilitate dome fabrication, complex research aims, and myriad vital purposes, a 34-person crew will operate a chemical synthesis plant. The simplest products may prove the most impactful on Mars. A network of 3d printed reaction vessels [bz] will be used for batch production of vital and contracted chemicals. The production of these vessels is a long-term investment that is necessary for true autonomy as it enables ISRU, biotechnology, and complex manufacturing. Much of the cost of these vessel is their design and fabrication [ju], but this has been greatly aided by 3d printing [jy] which can now achieve suitable quality for the fabrication of such vessels [ka]. Surplus printer time will be suitably allocated towards the production of these tanks.

The chemistry department will supply over 100 purified compounds and base reagents with the ability to produce a few thousand compounds on demand. For example, polyethylene will be formed via vertical gas phase reactor [ji] with reaction properties digitally monitored and tuned [jk] to yield the desired density and melt index. PVC and polyester will be synthesized for use in new domes and ongoing maintenance. Kevlar will be synthesized [jx] for outdoor domes and various other soft goods, like suits and rovers. Though Kevlar is more difficult to synthesis and requires nitrogen import, the value of local durable goods production is greater. Plastics will be extruded and cut as pellets for bulk storage.

### Heavy Construction

Heavy construction is a generalized department that will work in tandem with many other departments on major construction projects. This includes coordinating excavation, building generator housings, industrial manufacturing support, cavern systems, dome installation, and to a minor extent interior build out. With a permanent staff of 35, the department will be supplemented by other relevant experts on a project basis. For larger projects including the reactor installation, a temporary project team will be formed separate from the Heavy Construction department, to minimize the project's impact on operations.

Excavation, observation, and versatile heavy machinery can be shared with the mining, survey, and repair groups; additional equipment like rebar tying machines, heavy cranes, hydraulic systems, and concrete pumps will be maintained by Heavy Construction. Modular prefabrication is faster and allows for controlled conditions and would share resources with the fabrication team. Scaling up prefabrication facilities has powerful commercial benefits. Reactor containment structure frames can be assembled on site with insulation and utility connections pre-installed. Following a concrete pour, plug and play remote base deployment could occur within months.

Metal-working will be a large part of this department's responsibility in coordination with the Fabrication department, everything beyond the capability of 3d printers should be considered heavy construction. Rebar and structural trusses are likely to be common. The Colony's metal will come from regolith extract and waste stream recovery will be significant. A full machine and fabrication shop where CNC plasma cutting, MIG, TIG and arc welding capability will be maintained.

Industrial and process engineering knowledge will be critical to the fabrication facilities design. The need for a more comprehensive analysis of safety hazards, materials sourcing, modes of failure, operational demands, and maintenance put a significant pressure on this subset of the Heavy Construction team. Earth based design firms will be contracted for design and planning support to mitigate this risk and maximize the effectiveness of those on Mars.

**Fabrication**

It is essential to produce goods that are conceived of or designed after leaving Earth. The fabrication department will consist of 77 highly skilled individuals who will convert concepts into physical goods. Through plastic and metal 3d printing, vacuum forming, spinning, weaving, painting, ceramics, and a host of other prototyping and light fabrication techniques, the fabrication department will support the demand for niche goods within the colony.

The Metal-X Mark Forged printer will be used for small parts [cl], and the equivalent of Relativity Aerospace's Stargate laser-wire printer [cn] for large and specialized parts. Glass, small electronics, ceramics, silk, and other fabrics will also be crafted from of domestically produced raw materials through polymer spinning, and the growth and processing of organic fibers. This department must also educate others as the need arises to maintain the ethos and reality self-sufficiency. If possible, the fabrication department should allocate time to assisting the colony's theater and arts initiatives as part of its good-will mission. As colonists will interact with members of this department more than most others, the fabrication department will need to maintain positive relationships with other departments and individuals.

Fabrics will be an important material that can be produced at industrial scale in the colony. Linen, hemp, and synthetic fibers can be spun and woven on common equipment and blended as needed. Fabrics cover many of the supply gaps that 3d printing cannot, like canvas for painting and soft structures that do not need to be air tight.

Ceramics serve a similar role to fabrics, but for rigid structures. Their unique heat resistance, physical properties, and medical applications are compelling features of ceramics. The addition of enamel coatings for durability on vehicles and structures exposed to surface conditions will reduce maintenance and benefit many departments. Liquid nanoclay will need to be manufactured from mineral rich regolith slurry as a soil binder for agriculture and will share a feed stream with the clays needed for ceramics.

**Domestic Services**

While the agricultural facilities of Mars will be the property of and managed by the colony authority, colony universal basic income (UBI) will contain an allocation proportional to each colonist's equal share of production less overhead and commercial agreements. Colonists will be able to digitally order produce from the available crop, and have it delivered via light cargo drone, or purchase goods from a central marketplace.

Commercial dinning will exist in several forms in the colony, all of which will be regulated by permit and subject to ballot approval before open if disputed. These permits will include specified quantities of food to be produced by colonial agriculture and any commitments for import or export. This may limit commercial dining opportunities somewhat, but the communal cafeterias and food services will offer a more egalitarian environment and is a more efficient way of distributing communal resources.

Light transport services will be available in the colony as both battery electric vehicles like scooters and shared colony vehicles that can double as a reserve for industrial vehicles. The availability of batteries, LNG, and hydrogen offer a range of fuel sources for vehicles. This flexibility offers further resilience in Martian transport, which is a critical capability for the colony.

**Commercial Services**

The success and stability of this colony will be symbolic, validating the expansionist ambitions of humanity. 1000 individuals are enough to spawn edge cases and begin amassing institutional knowledge. The 1000-person scale and the collective management of resources will allow specialization to thrive even in an extreme environment. At this stage of colonization, there will be a small portfolio of commercial services available both on Mars and for delivery elsewhere. These will be motivational resources for future colonists and a life line for the present day.

The sharing of the accumulated knowledge of the colony will itself be a valuable service, with much of this data being collected anyway to build the colony. One example is the export of deidentified health data [dn] which could provide minor but stable revenue. The survey and initial planning of the Bubolaq will provide detailed characterization [fm] of the mineral [fl] and water resources in Deuteronilus Mensae available for future colony sites[fk], not to mention invaluable experiential knowledge. High spatial resolution mapping can cost about $7/km2 for terrestrial projects [fm] with price increasing with detail and spectral diversity. This is a small way to reap profit from a necessary capital investment in surveying equipment, with revenue estimated at over $100,000 per survey. High resolution imagery can also be used as valuable marketing material to promote emigration and generate tourist spending at extremely favorable ratios. Moreover, local weather data will reveal behavior vital to the planning of surface operations and maintenance requirements. While these represent relatively small revenue stream, they can also be largely automated [gn] [fn] and increase the value of existing capability. In addition to the colony's food supply commercial crops such as wine grapes, cannabis and flowers [dd], and utility crops like hemp, flax will be grown. While a portion of these will be consumed in the colony, the export of these luxury crops will yield an estimated $25million/y[db][dc] [de] from sale on Earth.

At this stage available tourism capacity and transit restrictions will keep the colony's return on marketing far below the incredible 1:65 ratio of Puerto Rico [fp], but the future colony may yet capture this value. Focusing on the uniqueness of the Martian experience and promoting sustainable tourism maximizes value relative to demand [fo]. In general, the identification of premium goods for market customers will help to develop domestic infrastructure and increase the real wage of colonists [fr].

The ability for the colony to host and entertain a cohort of 10 tourists will be a definitive milestone of its success. Some may be adventure-seekers coming to hike the Elysium Mons, camp in Valles Marineras, or tryout a pair of gliding wings in one of the larger domes. Some may be passionate explorers wanting to witness humanity's foot hold on another planet. The curiosity and determination of humanity is what has led us to the cusp of colonization today, and it is that curiosity that will invite a deluge of visitors to the Red Planet.

In keeping with the core function of the colony, launch and logistics services will be provided for inbound and outbound colonists as well as other contracting parties. This department will consist of 55 individuals and be a supporting source of third party revenue.

## SCIENTIFIC RESOURCES

The colony will maintain baseline facilities to support biological research, Martian geology and atmospheric science, novel fabrication, aerospace and astronomy for its own development and security. These facilities will work in conjunction with the discovery institute but are separate entities will always maintain some operational independence.

Cellular biology and genetic engineering will be prominent in the Bubolaq's biological sciences. A wide variety of bioreactors, basic media, nutrients, and supplemental factors are required for the diverse growing conditions of different cell lines. 50 roles will be allotted for the production of media, small batches of proteins, and cell derived products. Scanning electron microscopy (SEM) and diffraction microscopy services, gene sequencing, and healthcare products production. This last will be focused on antibiotic and phage synthesis [cu]. This is especially critical to colony security as a large phage library offers a resistance free measure to mitigate bacterial outbreaks. In addition to the benefits to agriculture, genetic engineering for bioproduction of recombinant proteins is a commercial service and improves the healthcare system.

To support astronomical study, an observatory will be built and maintained by a staff of 10. It will house a large aperture Schmidt-Cassegrain telescope [kg] with dedicated data storage and computational resources. This system is flexible [kc] being suitable for UV [kf], spectrographic[kh], and CMB studies. [ke]. This system can also act as an emergency communications relay [kd].

**The Martian Discovery Institute**

Most of the Bubolaq's research will be carried out by the Martian Discovery Institute, which will be owned and operated by the colony. Founded on the unifying goal of building a new and better world, the Institute's broad focus encourages researchers to innovate and be focused on the collective benefits of progress. Beyond the baseline work done for the establishment of the colony, there will be numerous research objectives available on the surface of Mars. Supported by the colony, researchers will be able to go to Mars with a focus solely on discovery.

The Institute will consist of 25 principle investigators, with spots for 3 accompanying researchers with each investigator. Sources of research funding whether public, private, or industry, will be considered with all applications, but not necessarily provided by the colony itself. The potential benefits to the colony will be an important factor in the selection of investigators and the funding of the projects. While the colony will contribute scientific resources: an observatory, bioreactors, isolated domes, reaction vessels, in kind knowledge, and materials, researchers may contribute equipment or additional infrastructure that enhance the mutual value proposition. Other factors considered will be the resource requirements, the likelihood and value of generating IP, and the significance of the work. In the long run, many individuals who work with the institute will return to Earth, as will residents in other roles, inspiring others to pursue the journey to Mars through their stories. Should the experience prove profound enough, some may be invited on as permanent colonists.

## HEALTHCARE

Despite the small size of the colony, a robust suite of medical services and capabilities is essential to the safety and endurance of colonists. Beyond easing colonists' fears for classical health concerns, this system must also mitigate both the known and to-be-discovered health hazards presented by living on Mars.

**Preventive Healthcare**

Given the inherent limits to available space, trained personnel, perishable goods, and complex biologics, the aim of the healthcare system will be to preempt the need for medical attention whenever possible. This will be achieved through screening and preventative care. Biometric monitors on colonists will continuously collect data as part of an ongoing health record. While robust physical state monitoring is valuable, data collection devices cannot require regular calibration and maintenance, as this would require too much time and high-demand resources. Simplified urine analysis systems could help uncovering potential health hazards [dl] such as urinary tract infections. Heart rate, skin temperature, and respiration [dj] can all be monitored with existing passive accessories, like smart watches and it would be simple enough to integrate sweat sensors into these devices [dk]. Sweat analysis presents a non-invasive way to monitor glucose, lactate, Na+, and K+ in a self-calibrating multiplexed array during sustained physical activity [di] and offers a more complete picture of physical state during exertion. Additionally, an integrated AI personal trainer will be used to manage exercise routines, prevent progression of dehydration and electrolyte imbalances, and monitor for emerging trends in personal health status [df].

The regular collection of these data points has the secondary benefit of standardizing the baseline metrics that will be available for observational biological studies and clinical trials. For example, toxin exposure is directly linked to consumption and bodyweight [hw], so computer monitoring of patterns in food consumption recognize systemic contamination before becoming apparent to colonists.

The healthcare program will do their upmost to limit the ability for parasites to enter into and persist within the colony. All applicants will be screened for gastrointestinal parasites before being accepted into the colony program. Human feces will be degraded via contained combustion to produce syngas [at] and will not be returned directly to the food production cycle.

**Responsive Healthcare**

A theme in the Martian healthcare system as elsewhere will be cross disciplinary collaboration. Internal medicine and family medicine physicians will share responsibility of primary care needs, hospital medicine, gynecologic conditions, and urgent care needs.

As there are so many unknowns about the Mars colony, the colony will employ a wide range of doctors to ensure overall health. As our understanding improves, the makeup of this department should be altered to meet the colonists needs. The following design is only a starting point.

Acute trauma is one of the most demanding case scenarios anticipated and can include multiple patients arriving at once which may stress the system. It may be necessary for health care practitioners to be trained in some additional aspects of trauma care so there is an infrastructure of support if needed. Some of the specialties included, such as urology and ophthalmology, have expertise in handling very specific aspects of trauma care. Physicians will likely have additional training for sub-specialties within their fields, such as electrophysiology or interventional cardiology, with the expectation that they will also offer general care in their field. Surgeons, too, must have a general understanding of trauma as well as individual specialties to treat emergent or life-threatening conditions: 1 general, 1 trauma, 1 colorectal, 1 cardiothoracic, 1 orthopedic, and 1 ENT.

| Martian Healthcare Team | | | |
|---|---|---|---|
| Dietician | 2 | Oncologist | 1 |
| Family Medicine | 6 | Gynecologist | 2 |
| Internal Medicine | 10 | Urologist | 1 |
| Nurse | 12 | Psychiatrist | 2 |
| Surgeon | 5 | Psychologist | 8 |
| Neurosurgeon | 1 | Dentist | 2 |
| Neurologist | 1 | Ophthalmologist | 1 |
| Cardiologist | 4 | Optometry | 1 |
| Pulmonologist | 4 | Radiology | 2 |
| Respiratory Therapist | 2 | Diagnostic/Lab Services | 4 |
| Anesthesiologist | 3 | Tech Support | 8 |
| Total | 82 | | |

Some conditions are foreseen and should be monitored from the outset to see if they develop into issues. First, the response to space and ultrafine dust may result in cardiopulmonary problems, so the colony should employ cardiologists and pulmonologists to respond to and closely monitor colonists' issues. Second, if the incidence of neurological diseases and cancers are comparable to those on Earth, a large staff of neurologists and oncologists will not be necessary. However, there should be at least one of each available for rare or emergent conditions. (If nuclear thermal rockets are available, direct transport back to Earth for specialist care will be the standard course of action in cases where only severely inferior care can be offered on Mars.) Third, while the colony should be prepared for radiation and toxin infiltration, if the barriers prove sufficient, the monitoring doctors may be relieved of their roles. Finally, the fields of pediatrics and obstetrics has been purposefully excluded due to lack of knowledge regarding pregnancy and child development on Mars and the expectation that no children will be reared in the colony. Correspondingly, long term birth control measures will be available to all colonists.

**Revenue**

There are two revenue streams available to the healthcare department: sell bio-informational data and clinical trials. The data harvested from the personalized sensors can be de-identified and sold for analysis. However, the relatively low price of this aggregated data [dn] makes it an only minor contributor. Additionally, the sale and use of such data, even when de-identified, has been contentious [dm] and so will require colonist consent.

The more significant healthcare revenue stream will be the coordination and management of clinical trials on Mars. This will be carried out by an independent contract research organization (CRO) in conjunction with the Martian health network. Onsite CRO staff will be minimal, which is consistent with decentralized nature of clinical trials on Earth [et]: two project managers and four clinical research associates will be on site to ensure proper training of subjects and clinical staff, drug storage, IP accountability/destruction, subject safety, and data acquisition. They will be supported by terrestrial statisticians, drug supply managers, data managers, compliance and quality officers, regulators, and technical specialists; there is already a wide variety of software [fv][fw][fx][fy] available to minimize the labor and error rate in data collection. These will be used with software-based study planning systems [fv] to maximize the capacity of the CRO. Direct-to-patient drug shipping is a present-day example [fu] of a trial model that requires an absolute minimum of personnel. Median research grants exceeding $13,000/patient [do] combines with industry investment to an average spending of $44,0000/patient for a phase 3 trial [ft]. Assuming an average of 1 trial per colonist per year and modeled off of the US market the net impact of clinical trials on the Martian economy is estimated at $20 million/year [fs]. Colonists will be incentivized to participate in trials through adequate financial compensation, and, hopefully, their innate interest in not only their own health, but also their interest in being part of scientific progress.

## DIET AND AGRICULTURE

The ability to produce sufficient food on Mars will be a definitive marker of stability and autonomy for the colony. The ability to grow, harvest, and enjoy fresh food provides a foundational stability to the colony. Not only would it be financially infeasible to import nearly 1 ton of food per year per colonist [ib][hx], it would be a major source of vulnerability. A diversity of crops will help to build culture, ensure nourishment, and offer a motivational case for highly productive engineered crops that could be exported back to Earth. At a grander scale, agriculture will be the primary means of converting the Martian atmosphere into the Martian biosphere.

Fertilization with vermicompost can increase plant yields by 20%-35% [fe] and frequency optimized led lights will be used to supplement the light received by plants to >600umol/m2/s [ax]. This will ensure that the concentration of CO2 is the rate limiting factor for plants. Supplemental lighting and CO2 let Martian plants fix carbon at double the rate than with natural light [ax]. With adequate nutrient supply and at least two crops per terrestrial year, yields of greater than 25t/ha are achievable [bj]. By producing crops with approximately 80% traditional farming, 15% aquaponics and 5% additional sources, including terrestrial imports, there will be ample sustenance for the colony.

### Diet

Individual calorie intake is about 2700kcal/day and an additional 1200 kcal/day is wasted giving a net food demand of 3900kcal/d/person [hx]. While much of the value that is lost in this wastage is from meat and fish, which will be limited on Mars, the majority by mass is wasted grain, vegetable and dairy [hy]. This 750 kg/y [hz] individual need on Earth is assumed here as 950kg/y to account for harvest variability, waste, and buffer. Distribution of food will be managed primarily digitally. Harvests will be recorded, and resource managers will release goods into a common digital store. Colonists will then be able to log on and buy produce with Zubles, the colonial currency, derived from their income and UBI. These goods can then be picked up or delivered via drone [km]. Additional food can be purchased from the limited shops and dining options in the colony (see domestic services).

A wide variety of vegetables will be grown in the terrestrial farms of Mars. Carrots, sweet potatoes, onions, beets, garlic [bj], grapes, wheat, beans [ac], peppers [ab], and others will all be part of the Martian diet along with copious fresh produce from aquaponics. Oyster mushrooms are a notable source of protein and are highly efficient at converting mulberry stems, and straw into mushrooms [by]. Though they are largely water, 34kg of protein is produced per ton of otherwise inedible biomass[cs].

Supplementing the local food production, some popular food items will be imported to provide a sense of comfort and dietary stability to colonists. However due to the high import cost, a reduction in cheese consumption to half (7kg/y/person) of US norms [ib], will be promoted via domestic price controls. This will still require seven tons of cheese imported annually at a cost $3,500/person. Dried dairy products, while shelf stable enough for import, [ie] fall into a similar category: 10kg/y of dried dairy products, instead of the 25kg/y on Earth, will be provided. This equates to 100L milk/y/person once rehydrated, with consumption unevenly distributed. 65% of adults on earth are lactose intolerant [if] and may wish the bulk of their dairy allotment to go to other culinary delicacies. Interestingly, egg consumption need not be reduced thanks to the stability of powdered egg whites [ic] and 88% mass reduction in drying [ig]. At just 1.8kg, or $900/y/person imported, eggs will be an important dietary protein for the colony. They will also enhance the culinary offerings to include traditional baked goods. Whole egg powders are especially suitable for baked goods and dressings [ii] but, like dried whole milks [ij], require attention to proper fat emulsion before drying and have a limited max shelf life [ic].

**Agricultural Domes**

Thirty-five hectares of multilayered composite agricultural domes will be on the Martian surface, adjacent to the colony. Along with access tubes between domes and to the colony airlock, surface accessible airlocks will be located on some agricultural domes to facilitate access by rover. They will use processed regolith as soil with locally sourced chemical fertilizer and compost for nourishment. The desire for direct exposure to incident solar radiation, prevents the stacking of crop beds, though mushroom beds will be plentiful. Low toroidal domes [bw] are the best design for low-mass agricultural domes. To further minimize daytime energy input to domes, mirrors will be used to concentrate sunlight onto domes, including the heliostats mentioned earlier, if feasible. As mentioned previously, these domes will differ from the habitation ones as they will not be Kevlar-lined, and instead secured to the ground by an anchoring system.

Drones will be the cornerstone of Martian agriculture [fk]. Multispectral [fl][fm] drone fly overs will provide detailed characterization to validate embedded sensors. Networks of wheeled and walking drones will be able to coordinate around dictated[is] tasks [it] and remove the physical labor burden from farmers. Projects like the Hands Free Hectare [kj] demonstrate this in present day and systems like Sweeper bot [y], Agrobot [z] Agribot[aa], and Agribotix [as], along with seed sowing [kk] and fertilizing [kl] robots will be needed by the dozens to create digital management for Martian farming.

Sensors located throughout the agricultural domes will allow for extensive analysis and contextualization of agricultural nuance on Mars. Light level, light composition, air composition, humidity, soil moisture, temperature, nutrient conditions, and biological indicators will all be part of the typical passive monitoring node in an agricultural dome. However, chemical fertilizers will be used to supplement any measured deficiencies. Ideally, there will be hundreds of nodes per hectare linked to local fixed hardware and mobile drones. The role of the farmer will be focused on coordinating and monitoring the crops and robots that tend them.

**Vermiculture Compost (Vermicompost)**

A core part of farming on Mars will be the recycling of crop waste biomass into sources of nutrition for future harvests. Raising worms on this biomass yields two essential products with minimal labor and without heavy equipment [ca]: high quality fertilizer [ca][bx] and protein for fish feed. The long-term goal of the Martian farming network will be to produce hectare quantities of topsoil capable of supporting plant growth without the use of fertilizer. This goal will help unify and direct colonists and ensure the prosperity of future colonists. This endeavor sets humanity on the path to creating a new ecosystem by providing in trust the sustenance it will need, whatever its form.

Vermiculture is preferred over chemical fertilizers for its sustainability and superior nutritional impact on crops [fe]. It also removes the risk of leachate, the nutrient rich but hazardous byproduct of chemical fertilizers[ff], affecting nearby systems or harming workers. Crop [bj] and silk worm waste will be pre-composted with urine [bh],[cr] and serve as feedstock for vermiculture, returning these waste products to the agricultural cycle. This system has also been shown to provide sufficient heat to kill pathogens in biosolids [fd], which will add to the stability of the ecosystem, though no human or fish biowaste will be vermicomposted. It will also yield over 400kg of worms per hectare of biomass, which can be fed to the fish and crustaceans living within the aquaponics system (see below). This is important for keeping carbon in the biosphere and generating complex protein and nutrients in situ.

Some biomass poses such a threat that they will not be composted, primarily human and fish wastes. For human waste, the risk of opportunistic pathogen cycling [bh], and added complication of biohazard handling, outweigh the gain in compost mass. The fish waste poses a different issue: the need for an additional filter species before being fed to worms who will eventually be fed back to fish. In both these cases, the risk outweighs the biomass reward, and, instead these waste products will be heat dried and sterilized for use as organic filler in topsoil.

To ensure the safety of the vermiculture, the compost will be routinely sampled and tested for composition and the presence of bio contaminants. Should harmful species be found, the compost will be subject to thermal oxidation as the primary disposal method in a syngas reactor modeled on the OSCAR system[at]. Chemical pesticides are undesirable in their potential bioaccumulation and long-term danger to resident insect populations[fb]. Fortunately, the closed nature of Martian biosystems eliminates the environmental reservoirs where pests may live. All effort must be made to ensure the constancy of this barrier to ensure ecosystem stability. The first worms brought to Mars will be screened to prevent the introduction of parasites into newly formed soil. As an emergency measure, shelf stable [dh] albendazole will be kept for acute treatment [dg] preceding a larger decontamination effort targeting local reservoirs.

The target blend for soil will be a 60:40 ratio of processed sandy regolith/tailings to composted biomass[ff]. New soil will be inoculated with fungi and worms for aeration and liquid nanoclay will be used as a binder and to improve water retention. Making high quality soil will require 5200 tons of composted biomass per hectare for a .5m thick layer of soil. Crop biomass will provide roughly 10 tons of compost/ha/planting after composting, estimating a 40-60% reduction during composting[fe]. Moreover, crop fertilization demands an annual 5-6 tons of compost/ha [bx][fe] in good soil and will require supplementation before that point. At peak efficiency, using continuous auger-driven systems to reduce the work of compost bed turnover, worm feeding, and harvesting [fg] and estimating two plantings per year, it will take dozens to hundreds of years to generate both the soil and a self-sustaining amount of compost for this system. To assist in reaching the goal within a realistic timeframe, partially composted biomass can serve to improve soil [ff], diminishing the work hours on composting while still reaping many benefits.

**Aquaponics**

Supplementing the traditional agriculture, aquaponic installations will be used to produce a substantial portion of the colony's nutritional requirements. There will be 10 installations following the UVI model[cx][cy] that will collectively occupy less than 0.75ha and produce annually 100 tons of vegetables, a full culinary spread of herbs, medicinal crops, 30 tons of fish and additional crustaceans and catfish. Aquaponics can be rapidly expanded within a small footprint should demand increase but is more equipment and labor intensive than the machine grown soil crops.

Aquaponics works by allowing plants to grow in floating beds and on plastic towers with their roots partially submerged in nutrient-rich running water derived from fish waste. Many standard vegetable crops can be grown this way more effectively than in soil because of the high nutrient availability, though larger plants require media beds for physical support. The substantial fish and crustacean production inherent in the process allows for two food sources within one footprint.

Fish are one of the most efficient forms of livestock: being cold blooded and aquatic, they need not waste energy on heat generation or bones for fighting gravity [h]. Tilapia have a >30% efficiency in converting food to fish mass [i][ad]. The 30 tons of fish require 72 tons of food to survive and nourish the gardens. As tilapia (and other omnivorous finfish) can subsist on a dietary protein content as low as 5% [i], duckweed will be the primary bulk fish food. Only 0.5ha of floor space is required for the 65.5 tons needed annually: it can be produced in stacked shallow growth troughs of nutrient solution. 20% of the total crop can be harvested daily [ag] via simple mesh strainers on articulating arms and transported by drone [ar] to a countertop oven for drying. As an added benefit, excess duckweed can be converted into oxalic acid [ad], a useful chemical precursor and cleaner. At minimum an additional 6.5 tons of high protein feed, comprising of silk worm pupae, red wigglers, and recovered fishmeal will be needed. This is blended and supplemented based on biometric data from in tank sensors and sampling, before being fed to the fish. Roughly 15% of fish mass will go to non-food purposes: 10% fishmeal for fish feed and 5% fish oil for explosives manufacturing, industrial, and medical use. The remaining 25 tons of fish mass will be given to the central food distribution system, mentioned earlier.

The scavenging crustaceans and catfish will control algae buildup and remove uneaten food in fish tanks. They will also serve as a luxury culinary item and help diversify the diet of colonists. Notably, the Bubolaq with at least 35 ha of cropland could produce over 10 tons of worms from vermicomposting which could be used to expand the hydroponic systems or grow more protein intensive fish.

**Silkworms**

The colony agricultural cycle will rely on silkworms and their cocoons from which silk strands are harvested. Silk has been a luxury fabric for most of human history [gb][gd], transitioning to major domestic fabric and industrial good in the modern era. Silk will be a core component of the Martian biotechnology and will be critical step of the nitrogen and carbon cycles on Mars, contributing to the creation of a Martian biosphere.

Annually, the colony will produce 2.6 million silkworms over three worm farms, fed with domestically produced mulberry leaves. This could be used to create 1800 garments, 200m of antibacterial bandages, suture thread and tons of protein and compostable material. More broadly, the sericin of silkworms can be used as a platform for recombinant protein production at meaningful scale.

Mulberry is the silkworm's preferred food, and mulberry shoots also grow in synchronicity with the silkworms' two month cycle [aj]. To produce the necessary 80 t/y [an] of fresh mulberry leaves, a dense shoot farming system will robotically plant the necessary 1.67ha required to produce 8tons/ha/harvest [gh]. Three full time roles will be dedicated to the care and harvesting of the Mulberry shoots and associated robot management.

Three central staff members will be assigned to the population-level care and breeding of silkworms. Of the remaining nine silkworm focused roles (3 per tower farm), one colonist at each farm will be dedicated to juvenile worm care, one to adult bed care, and one will serve in a harvest and flex resource capacity. For the first 3-7 days after hatching, the worms require more manual manipulation and feeding. At that age, they will be transferred to one of the three worm farms and grown for an additional week [aj] before being transferred to stacked beds of layered fresh mulberry leaves. Each farm will consist of 33 towers containing eight 1m2 beds, separated by a 30cm gap, stocked at a density of 800 adult silkworms/m2 [ai]. Fresh mulberry leaves will be added daily, but no leaf removal is required until harvest, minimizing labor required for adult silkworms. The adult worm beds will be further monitored by cameras mounted to underside of the bed above the one being monitored. With continuous automated analysis, illness and suboptimal conditions can be identified and handled before it can affect the bed or tower. The physical separation between beds and towers will also acts as a barrier to the spread of disease should monitoring fail to detect it.

Once the silkworms have formed cocoons, they will be frozen, pupae will be extracted, and the cocoons will be spun into silk by the fabrication department. This involves a wash in a simple soap solution and spinning on common equipment. In the case of bioproduction using silkworm vectors, harvest can occur before or after cocooning and processing will be the responsibility of the coordinating party.

The bio-waste from silk production is recoverable through composting, yielding over 10m3 of compost per annual crop of 2.6 million worms. More than 2 tons of silkworm pupae[b] will be harvested annually along with the cocoons and used to fuel aquaculture. These will be combined with worms harvested from composting to fully satisfy fish-feed needs.

**Trees**

Fruit trees, like mulberry and apple, will be a common feature in the Mars colony, both inside the agricultural domes and within the habitat domes. From a commercial lens, trees are easily maintained through community labor input and almost passively produce food and useable biomass. For example, grafted mulberry trees will begin fruiting within 2 years [p] at up to 5kg/y [ao]. After 20 years, a single tree will produce 300kg of fruit/y [ao] and is tons of useable biomass. Pear and Apple trees can yield 28 tons/ha greatly supplementing dietary needs [ju]. Some non-fruiting tress like curry and tea will also be grown to provide diverse colony resources. Moreover, trees function as a habitat enhancement for colonists, serving as a reminder of home.

**COMMUNITY**

**Media**

The 14-member media department will enliven the colony. While three members will report on local interest pieces, the other 11 will be tasked with this historically important role of documenting the colonization of Mars. Not only will this footage be relayed back to Earth as promotional material, but it will also be archived for future generations. The department will be open to documenting specific features of the Red Planet and aspects of colony life per specific requests. Moreover, recordings can be licensed for human-interest projects. Commercial projects will also be welcomed and supported by the Martian Media Department to a negotiated degree.

**Arts**

The arts program aims both to serve the community and to produce high value exportable items. With a group of 30 professionals to design and manage projects, local performance and visual arts can flourish.

The habitation domes will feature a thoroughly landscaped central square for public gathering. This location can serve as the community hub. By setting up a small stage, this location will be ideal for watching informational speeches and comedic performances alike. As with many major cities, this space can be converted for free performances. By creating a projector screen out of the locally produced silk and metal rails, the community can gather to watch movies, the local news reports or other videos. This encourages people to get out of their dwellings, form meaningful relationships with their neighbors and get a taste of "home." The screen can also be used for satirical shadow puppet performances.

The performance arts of theater and dance will provide creative and physical outlets. With a domestic workforce made up of mainly technical professionals, a community theater is an invaluable asset for relaxing and promoting community. As on Earth, the performances can be as amateur or professional as the skill and time allows and the subject matter will not be regulated, except as absolutely necessary. Run by engineers and supported by the Fabrication Department, the 3d printers will be put to good use, creating innovative scenic elements and props. A community dance troupe provides the same benefits, while also promoting flexibility and fitness for the dancers. Ideally, several levels and styles of coursework will be available, from true beginner ballet to a modern dance performances troupe.

In addition, the visual arts will allow colonists to design and decorate their living and work spaces. Though standard furniture and clothing will be created for each person's dwelling, it can easily be modified. Assisted by currently available online historic and modern pigment fabrication databases, researchers can easily look into synthesizing a wide variety of colors from regolith, basic chemical agents, and other locally available products including fruit and vegetables. Once ground into locally-produced linseed oil, or another drying binder, these pigments are easily transformed into standard paints. Paint brushes can be made domestically, allowing colonists to paint murals or other artworks on locally-produced silk or hemp canvases. Carving tools will allow for furniture modification, sculpture and wood-block printing. Weaving looms can be 3d printed, as well as many other basic artistic tools. Creating personalized clothing and interior textiles (rugs, blankets, upholstery, etc.) will give colonists a sense of self and a pride of place. These visual arts can adorn work spaces as well as dwellings, customizable and supportive environment for the colonists. It is also possible that these artworks will be exported back to Earth as a luxury good.

**Community Benefit Jobs**

Some goods and services desired by colonists will be beyond the capacity of this design or any colonization government to coordinate. Many are commonplace on Earth, but difficult to plan for with such a limited number of colonists: examples include barber and cosmetology services, soap manufacture, canning, a crawfish tank, small electronics fabrication and repair, domestic horticulture, commissioned or public art, automation installments, and process improvement projects. These positions are any tasks or goods production and distribution that is performed by colonists and sanctioned as "beneficial" by a government resources board or departmental manager.

To respond to and facilitate these tasks, the Community Benefit Job Board will publish Requests for Proposals in identified underserved community functions. Colonists may perform more than one function, but these should ideally occupy no more than a cumulative 10 h/week. Colonists who choose to serve in one of these roles can apply for additional monetary compensation even if the role itself generates a profit. The scope of each position in terms of time and material resource requirements, breadth of impact, uniqueness, and potential hazards will be evaluated carefully. It will also serve as a way of enticing flex support should unforeseen factors shift labor demands. Colonists are free to accept payment from others for their services, and more generally restrictions on the use of currency will be limited, though this will be taken into account when assessing colony contributions to a project.

## GOVERNMENT

The Bubolaq Colony Charter will be a minimal guideline in defense of social rights and liberties and will be based on extant democratic-socialist nations. The government of Mars will be operated by a dense web of colonists and vested parties that will heavily color the nature the government and colony as a whole. The core functions of the colonial government will be to ensure the infrastructure of the colony through stewardship and to coordinate the equitable management of resources including intellectual property. This extends beyond the physical resources of the colony to include and the physical and emotional health of colonists. There will be rotating citizen involvement in the operation of the government to promote civic attentiveness. Similarly, direct referendum will be available and accessible through colony IT systems.

In order to ensure the continued function of all life sustaining systems, regular meetings will be convened between departments. These meetings are in addition to community meetings where the general public can voice their concerns. The executive council meetings will serve as a hub for board organization for incidental and recurring administrative needs. For example, a 9-person resource board in a 7 departmental + 2 Administrative set up to monitor and analyze the consumption, productivity, and projections of all departments. A separate eight-person Community and Housing Board of four government and four rotating citizen seats will be responsible for communicating the needs and concerns of the colonists at large to departmental leads outside of community meetings. There will be two executive administrators and two regulatory administrators who will act as leadership for the Administrative Department and consequently the colony at large.

These legal services will collectively consist of a three-person judicial arbitration board, and nine lawyers, serving overlapping roles in government counsel, business, and personal law. There will be no capital punishment, and incarceration will result in deportation, unless deemed unsuitable by physicians.

The colony government will work with the legal and financial departments more directly than others due to the need to utilize their functions for colony management. Particularly for management of the economy. The Zuble will be for domestic use only and primarily managed by two colonists and an Earth based board. Physical tender will not be issued at this stage but is not barred in the future.
Four roles will be allocated to personal banking services. While most domestic financial functions will be available digitally, services like personal and government loan management require direct oversight. Assistance with the more complex Earth-associated finances will also be necessary, though heavily supplemented by Earth based personnel. Colony legal services will be available to support domestic banking.

The referendum process will be available to colonists for the regulation of domestic policy. Given the Colony's incontrovertible financial obligations, this will primarily be for social policy, though can be used to voice colonial opinions on interplanetary policy. Engagement and satisfaction with Bubolaq Administration is highly important. Colonists are wholly dependent on these governing bodies in a far less flexible manner than on Earth where countries and numbers can dilute the cognizance of authority. Ideally, this need for collaboration will build more effective forms of government, hammered into shape by living and collaborative administration.

**Intellectual Property**

Given the fundamentally unique nature of Martian design requirements and dedicated study at the University, advancements will be made in manufacturing, processing, and systems. These can then be licensed on Earth, Mars, and the wider solar system. Three roles will be held for Intellectual Property Managers, who will be supported by an Earth-based team. These individuals will work with the government counselors to identify and secure IP. A diverse portfolio will be instrumental in bringing long term investment into the colony but is dependent on establishing a strong brand associated with superior discoveries [jr]. Consequently, the foreign trade board will be briefed on relevant technology to serve as a better feeder channel [it] for licensing.

In conjunction, the colony will support a four-person venture capital (VC) board, backed by Earth based analysis and legal teams, as an investment arm for the resources of the colony. This is a key step in ensuring technology transfer [jq] and value capture [jr] and is a potential expansion point from community benefit jobs. VC board members individuals will also serve in an advisory capacity for the IP and Financial contingents.

## CONCLUSION

The colonization of Mars is a clear milestone in human history. Today, it would mark the first life off ever known to take root off of Planet Earth. We are still living in the echoes of the colonization of the Americas. Only in hindsight will we see the full impact of adding a planet to the range of humanity. In a more immediate sense, Mars will be a testing ground for the plethora of ideas that are being cultivated today. Nowhere else will colonists be so thoroughly isolated from outside systems and so thoroughly committed to their attempts at forming a society. As they develop the ability to grow their own food, produce their own goods, and govern their own societies, a new branch on the tree of humanity will grow. The Bubolaq must bear in mind this awesome responsibility for the longevity, durability, and future societies of Mars.

## REFERENCES

[a] Hiroyuki TOMOTAKE, Mitsuaki KATAGIRI, Masayuki YAMATO, Silkworm Pupae (Bombyx mori) Are New Sources of High Quality Protein and Lipid, Journal of Nutritional Science and Vitaminology, 2010, Volume 56, Issue 6, Pages 446-448,
[b] http://www.wildfibres.co.uk/html/silk_cocoons.html
[c] Engineering ToolBox, (2010). Dirt and Mud - Densities. Available at: https://www.engineeringtoolbox.com/dirt-mud-densities-d_1727.html [Accessed Day Mo. Year].
[d] Lee, Kuan-Lin, Calin Tarau, and William G. Anderson. "Titanium Water Heat Pipes for Space Fission Power Cooling." Nuclear and Emerging Technologies for Space (NETS) (2018).
[e] Frederick, R.D. MODIFIED MARTIAN LAVA TUBES REVISITED
[f]. J. Léveillé, Richard & Datta, Saugata. (2010). Lava tubes and basaltic caves as astrobiological targets on Earth and Mars: A review. Planetary and Space Science. 58. 592-598. 10.1016/j.pss.2009.06.004.
[g] Dupuis and Mueller 2015. NASA. https://ntrs.nasa.gov/archive/nasa/casi.ntrs.nasa.gov/20150016608.pdf
[h] Bene, Feeding 9 billion by 2050–Putting fish back on the menu
[i] Hasan, M. R., & Halwart, M. (Eds.). (2009). Fish as feed inputs for aquaculture; practices sustainability and implications. FAO Fisheries and Aquaculture Technical Paper. No. 518. Rome: FAO. *********430pg technical guide************
[j] McAdam, Amy, et al., Studies of Young Hawai'ian Lava Tubes: Implications for Planetary Habitability and Human Exploration https://ntrs.nasa.gov/search.jsp?R=20170002432
[k] Aquaponic system

[l] van Huis, Arnold, 2011 Edible insects contributing to food security?
https://doi.org/10.1186/s40066-015-0041-5
[m] Irons 2016
[n] Geiszler 2015
[o] Saidi and Abardeh 2010
[p] http://www.twisted-tree.net/propagating-mulberry-trees/
[q] M. Ralphs, B. Franz, T. Baker, S. Howe, Water extraction on Mars for an expanding human colony
[r] https://www.nuscalepower.com/technology/technology-overview
[s] https://www.nrc.gov/reactors/new-reactors/smr/mpower.html
[t] Maisler, J., Hawkinson, J., (2015) Small Modular Reactors. Enercon
hpschapters.org/florida/6spring.pdf
[u] https://spectrum.ieee.org/energywise/energy/renewables/new-sulfur-flow-battery-could-provide-affordable-longterm-grid-storage
[v] http://www.arcraftplasma.com/welding/weldingdata/powersources.htm
[w]
https://www.nasa.gov/sites/default/files/atoms/files/kilop180ower_media_event_charts_16x9_final.pdf
[x] https://www.energy-storage.news/news/chinas-biggest-flow-battery-project-so-far-is-underway-with-hundreds-more-m
[y] http://www.sweeper-robot.eu/
[z]http://www.agrobot.com/
[aa] http://agribot.eu/?lang=en
[ab] https://dengarden.com/gardening/indoor-hydroponic-garden
[ac]http://www.aquasol.org/uploads/7/0/7/0/70700977/guidelines_for_12_common_aquaponic_plants.pdf
[ad] Sascha Iqbal, March 1999. Duckweed Aquaculture. SANDEC Report No. 6/99.
https://pdfs.semanticscholar.org/8a68/3ea9a35a234945515519c9e2952f1254d8b6.pdf
[ae] FMI, Oxalic Acid Market: Global Industry Analysis and Opportunity Assessment 2016-2026, March 2019 https://www.futuremarketinsights.com/reports/oxalic-acid-market
[af] http://edepot.wur.nl/408365
[ag] Hasan, M. R. ; Chakrabarti, R., 2009. Use of algae and aquatic macrophytes as feed in small-scale aquaculture: A review. FAO Fisheries and Aquaculture technical paper, 531. FAO, Rome, Italy
[ah] Horvat, Tea & Vidaković-Cifrek, Zeljka & Orescanin, Visnja & Tkalec, Mirta & Pevalek-Kozlina, Branka. (2007). Toxicity assessment of heavy metal mixtures by Lemna minor L. The Science of the total environment. 384. 229-38. 10.1016/j.scitotenv.2007.06.007.
[ai] Chen et al. 2018. Transgenic Silkworm-Based Silk Gland Bioreactor for Large Scale Production of Bioactive Human Platelet-Derived Growth Factor (PDGF-BB) in Silk Cocoons
file:///Users/sford/Downloads/ijms-19-02533.pdf
[aj] https://everythingsilkworms.com.au/silkworms/life-cycle-of-a-silkworm/
[ak] Chen, F., Porter, D., Hallrath, F., May 2012. Structure and physical properties of silkworm cocoons. J R Soc Interface. 2012 Sep 7; 9(74): 2299–2308. 10.1098/rsif.2011.0887
[al] http://www.fao.org/livestock/agap/frg/mulberry/Papers/HTML/Mulbwar2.htm
[am] http://www.suekayton.com/Silkworms/whole.htm
[an]http://e-krishiuasb.karnataka.gov.in/ItemDetails.aspx?depID=3&subDepID=%203&cropID=0
[ao] Paul Alfrey, Oct 2017. Mo' Mulberry—The Essential Guide to all you need to know about Mulberry https://medium.com/@balkanecologyproject/mo-mulberry-the-essential-guide-to-all-you-need-to-know-about-mulberry-28a0c11b611
[ap] Morrison, C.G., Deason, W., Eades, M. J., Judd, S., Patel, V., Reed, M., Venneri, P., The Pylon: Near-Term Commercial LEU Nuclear Fission Power for Lunar Applications. Survive the Lunar Night Workshop 2018 (LPI Contrib. No. 2106)
[aq] https://nssdc.gsfc.nasa.gov/planetary/lunar/apollo_lrv.html
[ar] https://robots.ieee.org/robots/spotmini/
[as] Baden et al A Lightweight, Modular Robotic Vehicle for the Sustainable Intensification of Agriculture https://eprints.qut.edu.au/82219/1/Published%20paper_Perez.pdf
[at] Meijer and Hintze, 2018. Closing the Loop on Space Waste
[au] Davila and Wilson 2013 Perchlorate on Mars: A Chemical Hazard and a resource for humans
[av] RABINOVITCH, A., FISHOV, I., HADAS, H., EINAV, M., & ZARITSKY, A. (2002). Bacteriophage T4 Development in Escherichia coli is Growth Rate Dependent. Journal of Theoretical Biology, 216(1), 1–4. doi:10.1006/jtbi.2002.2543

[aw] Ivan Zajic, Tomasz Larkowski, Dean Hill, Keith J. Burnham, 2011.Energy consumption analy sis of HVAC system with respect to zone temperature and humidity set-point,IFAC Proceedings Volumes, 44(1)
[ax] http://fluence.science/wp-content/uploads/2016/09/High-PPFD-Cultivation-Guide-9.27.16.pdf
[ay] https://www.lumigrow.com/toplight/
[az] Jennifer C. Stern, Brad Sutter, Caroline Freissinet, Rafael Navarro-González, Christopher P. McKay, P. Douglas Archer, Arnaud Buch Evidence for indigenous nitrogen on Mars. Proceedings of the National Academy of Sciences Apr 2015, 112 (14) 4245-4250; DOI:10.1073/pnas.1420932112
[ba] Wang, Fan & Bowen, Brenda & Seo, Ji-Hye & Michalski, Greg. (2018). Laboratory and field characterization of visible to near-infrared spectral reflectance of nitrate minerals from the Atacama Desert, Chile, and implications for Mars. American Mineralogist. 103. 10.2138/am-2018-6141.
[bb] http://clowder.net/hop/railroad/EMa.htm
[bc] https://www.revaengineers.com/fully-automatic-egg-laying-block-making-machine-rbme-07.htm
[bd] https://www.alibaba.com/product-detail/ECO-2700-soil-interlocking-brick-machine_2005045010.html?spm=a2700.7724857.normalList.24.e0f43fc0v0AZAk&s=p
[be] Isam I. Bashour, DESERT RECLAMATION AND MANAGEMENT OF DRY LANDS: FERTILITY ASPECTS LAND USE, LAND COVER AND SOIL SCIENCES – Vol. V – Desert Reclamation and Management of Dry Lands: Fertility Aspects –
[bf] Yadav, K. D., Tare, V., & Ahammed, M. M. (2011). Vermicomposting of source-separated human faeces by Eisenia fetida: Effect of stocking density on feed consumption rate, growth characteristics and vermicompost production. Waste Management, 31(6), 1162–1168. doi:10.1016/j.wasman.2011.02.008
[bg] Rose, C., Parker, A., Jefferson, B., & Cartmell, E. (2015). The Characterization of Feces and Urine: A Review of the Literature to Inform Advanced Treatment Technology. Critical reviews in environmental science and technology, 45(17), 1827-1879.
[bh] Heinonen-Tanski, H., & van Wijk-Sijbesma, C. (2005). Human excreta for plant production. Bioresource Technology, 96(4), 403–411. doi:10.1016/j.biortech.2003.10.036
[bi] Christina Moulogianni and Thomas Bournaris, 2017. Biomass Production from Crops Residues: Ranking of Agro-Energy Regions. Energies
[bj] Yasir Iqbal, Iris Lewandowski, Axel Weinreich, Bernd Wippel, Barbara Pforte et al. 2016. Maximising the yield of biomass from residues of agricultural crops and biomass from forestry. Ecofys, Project number: BIENL15082
[bk] Morrison, C. UNSC (2018). 21st Annual International Mars Society Convention
[bl] P. Hanley, Brian. (2017). Financing Mars settlement: Default Insurance Notes to Implement Venture Banking.
[bm] Patel, V., Eades, M., Venneri, P., & Joyner, C. R. (2016). Comparing Low Enriched Fuel to Highly Enriched Fuel for use in Nuclear Thermal Propulsion Systems. 52nd AIAA/SAE/ASEE Joint Propulsion Conference. doi:10.2514/6.2016-4887
[bn] Thangavelautham, J., Robinson, M. S., Taits, A., McKinney, T., Amidan, S., & Polak, A. (2017). Flying, hopping Pit-Bots for cave and lava tube exploration on the Moon and Mars. arXiv preprint arXiv:1701.07799.
[bo] https://instylesolar.com/blog/2018/05/28/tesla-powerpack-2-review/
[bp] https://www.tesla.com/powerpack
[bq] Ramsdale, J. D., Balme, M. R., Conway, S. J., & Gallagher, C. (2015). Ponding, draining and tilting of the Cerberus Plains; a cryolacustrine origin for the sinuous ridge and channel networks in Rahway Vallis, Mars. Icarus, 253, 256-270.
[br] http://goldengatebridge.org/research/factsGGBDesign.php
[bs] Hermansson, V., & Holma, J. (2015). Analysis of suspended bridges for isolated communities.
[bt] United States Department of the Army, (1964) Technical Manual Series, (Headquarters, DOA, Washington, D.C.) Cableways, Tramways, and Suspension Bridges. http://www.tramway.net/US%20Army.pdf
[bu] Rao Surampudi, Julian Blosiu, Ed Gaddy et al. (Dec 2017). Solar Power Technologies for Future Planetary Science Missions, Strategic Missions and Advanced Concepts Office. JPL D-101316
[bv] Destefanis, R., Amerio, E., Briccarello, M., Belluco, M., Faraud, M., Tracino, E., & Lobascio, C. (2014). Space environment characterisation of Kevlar®: good for bullets, debris and radiation too. Universal Journal of Aeronautical & Aerospace Sciences, 2, 80-113.
[bw] Gary C. Fisher And members of the Independence Chapter of The Mars Society, (1999) Torus Or Dome: Which Makes The Better Martian Home [http://www.marshome.org/files2/Fisher.pdf]

[bx] Vermicompost - Production and Practices. ICAR Research Complex for NEH Region, Umiam – 793 103, Meghalaya
[by] Madan, M., Vasudevan, P., & Sharma, S. (1987). Cultivation of Pleurotus sajor-caju on different wastes. Biological Wastes, 22(4), 241-250.
[bz] Panda, Bijaya & C Das, S. (2001). Electrowinning of copper from sulfate electrolyte in presence of sulfurous acid. Hydrometallurgy. 59. 55-67. 10.1016/S0304-386X(00)00140-7.
[ca] Suthar, S. (2007). Vermicomposting potential of Perionyx sansibaricus (Perrier) in different waste materials. Bioresource Technology, 98(6), 1231–1237. doi:10.1016/j.biortech.2006.05.008
[cb] Algahtani, Ali. (2006). Manufacturing Of High Strength Kevlar Fibers. Journal of King Kalid University - Science. 2.
[cc] Kronberg, Peter & Hauber, E & Grott, Matthias & Werner, S.C. & Schäfer, Tanja & Gwinner, Klaus & Giese, Bernd & Masson, Philippe & Neukum, Gerhard. (2007). Acheron Fossae, Mars: Tectonic rifting, volcanism, and implications for lithospheric thickness. Journal of Geophysical Research. 112. E04005. 10.1029/2006JE002780.
[cd] Rodriguez, J. A. P., Bourke, M., Tanaka, K. L., Miyamoto, H., Kargel, J., Baker, V., ... & Hernández, M. Z. (2012). Infiltration of Martian outflow channel floodwaters into lowland cavernous systems. Geophysical Research Letters, 39(22).
[ce] Boston, P. J., Spilde, M. N., Northup, D. E., Melim, L. A., Soroka, D. S., Kleina, L. G., ... Schelble, R. T. (2001). Cave Biosignature Suites: Microbes, Minerals, and Mars. Astrobiology, 1(1), 25–55. doi:10.1089/153110701750137413
[cf] Hauber, E., Bleacher, J., Gwinner, K., Williams, D., & Greeley, R. (2009). The topography and morphology of low shields and associated landforms of plains volcanism in the Tharsis region of Mars. Journal of Volcanology and Geothermal Research, 185(1-2), 69–95 doi:10.1016/j.jvolgeores.2009.04.015
[cg] https://www.greenbuildingadvisor.com/article/aeroseal-rolls-out-air-sealing-technology-for-houses
[ch] Tang, Z. B., Zhao, Y. L., Kong, D. S., & Kan, D. (2013). Study and application of a new type of foamed concrete wall in coal mines. Journal of Coal Science and Engineering (China), 19(3), 345-352.
[ci]https://www.extension.iastate.edu/forestry/tri_state/tristate_2014/talks/PDFs/Aquaponic_System_Design_and_Management.pdf
[cj] Shi, L., Yan, B., Shao, D., Jiang, F., Wang, D., & Lu, A.-H. (2017). Selective oxidative dehydrogenation of ethane to ethylene over a hydroxylated boron nitride catalyst. Chinese Journal of Catalysis, 38(2), 389–395. doi:10.1016/s1872-2067(17)62786-4
[ck] https://www.relativityspace.com/stargate/
[cl] https://3d.markforged.com/metal-3d-printer.html
[cm] Sinter II data sheet https://s3.amazonaws.com/markforged-marketing-assets/Hosted+Assets/F-SR-0002-S.pdf
[cn] https://3dprinting.com/news/relativity-space-developing-new-technologies-space-exploration-using-stargate-3d-metal-printer/
[co] Liu, C., Ye, J., Jiang, J., & Pan, Y. (2011). Progresses in the Preparation of Coke Resistant Ni-based Catalyst for Steam and CO2 Reforming of Methane. ChemCatChem, 3(3), 529–541. doi:10.1002/cctc.201000358
[cp] Polk, J.; Murray, C.; Onewokae, C.; Tolbert, D.E.; Togna, A.P.; Guarini, W.J.; Frisch, S.; Del Vecchio, M. Case study of ex-situ biological treatment of perchlorate-contaminated groundwater. In Proceedings of 4th Tri-Services Environmental Technology Symposium, San Diego, CA, USA, June 18-20, 2001.
[cq] Srinivasan, A., Viraraghavan, T., 2009. Perchlorate: Health Effects and Technologies for Its Removal from Water Resources. ISSN 1660-4601 [www.mdpi.com/journal/ijerph]
[cr] Karak, T., & Bhattacharyya, P. (2011). Human urine as a source of alternative natural fertilizer in agriculture: A flight of fancy or an achievable reality. Resources, Conservation and Recycling, 55(4), 400–408. doi:10.1016/j.resconrec.2010.12.008
[cs] Royse, D. J. (1996). Yield stimulation of shiitake by millet supplementation of wood chip substrate. Mushroom Biology and Mushroom Products, Penn State University, Pensnylvania
[ct] Soria-Salinas, Álvaro & Zorzano, María-Paz & Martín-Torres, F. J.. (2015). Convective Heat Transfer Measurements at the Martian Surface.
[cu] Gill, J. J., & Hyman, P. (2010). Phage choice, isolation, and preparation for phage therapy. Current pharmaceutical biotechnology, 11(1), 2-14.
[cv] Leonard, W. R. (2010). Measuring human energy expenditure and metabolic function: basic principles and methods. J Anthropol Sci, 88, 221-230.

[cw] https://www.agrilyst.com/stateofindoorfarming2017/
[cx] Bailey, D. S., & Ferrarezi, R. S. (2017). Valuation of vegetable crops produced in the UVI Commercial Aquaponic System. Aquaculture Reports, 7, 77-82.
[cy] Gruetze, S., Oxley, K., & Rivard, C. L. 2016 Evaluation of Standard Pickling Cucumber Varieties in Kansas. Midwest Vegetable Trial Report for 2016, 33.
[cz] Li, H., Robinson, M. S., & Jurdy, D. M. (2005). Origin of martian northern hemisphere mid-latitude lobate debris aprons. Icarus, 176(2), 382-394.
[da] Waffenschmidt, E. (2015). Direct Current (DC) Supply Grids for LED Lighting. LED Professional N48, Mar.
[db] Gerling, C. (2011, December). Conversion Factors: From Vineyard to Bottle GRAPES 101. Apellation Cornell.
[dc] Dal Bianco, A., Boatto, V., & Caracciolo, F. (2013). Cultural convergences in world wine consumption. Revista de la Facultad de Ciencias Agrarias, 45(2).
[dd] Caulkins, J. P. (2010). Estimated cost of production for legalized cannabis. RAND, Drug Policy Research Center. http://www. rand. org/content/dam/rand/pubs/working_papers/2010/RAND_WR764. pdf.
[de] MacCoun, R. J. (2011). What can we learn from the Dutch cannabis coffeeshop system?. Addiction, 106(11), 1899-1910.
[df] Hunn, N. (2016). The market for hearable devices 2016–2020. Wifore Consulting.
[dg] Albonico, M., Smith, P. G., Hall, A., Chwaya, H. M., Alawi, K. S., & Savioli, L. (1994). A randomized controlled trial comparing mebendazole and albendazole against Ascaris, Trichuris and hookworm infections. Transactions of the Royal Society of Tropical Medicine and Hygiene, 88(5), 585-589.
[dh] GlaxoSmithKline (2009) ZENTEL® (albendazole) PRODUCT INFORMATION. Boronia, Australia Retrived from:
http://gsk.com.au/resources.ashx/prescriptionmedicinesproductschilddataproinfo/565/FileName/32A72A20722E06315E6A1BB5EB300FF3/PI_ZENTEL_Issue2.pdf
[di] Gao, W., Emaminejad, S., Nyein, H. Y. Y., Challa, S., Chen, K., Peck, A., ... & Lien, D. H. (2016). Fully integrated wearable sensor arrays for multiplexed in situ perspiration analysis. Nature, 529(7587), 509.
[dj] Bandodkar, A. J., & Wang, J. (2014). Non-invasive wearable electrochemical sensors: a review. Trends in biotechnology, 32(7), 363-371.
[dk] Kim, J., Campbell, A. S., & Wang, J. (2018). Wearable non-invasive epidermal glucose sensors: A review. Talanta, 177, 163-170.
[dl] Gronowski, A. M., Haymond, S., Harris, B., Peng, C. K., & Kohli, S. S. S. (2018). A Q&A with the qualcomm tricorder XPRIZE winners. Clinical chemistry, 64(4), 631-635.
[dm] Kaplan, B. (2015). Selling health data: de-identification, privacy, and speech. Cambridge Quarterly of Healthcare Ethics, 24(3), 256-271.
[dn] Vezyridis, P., & Timmons, S. (2015). On the adoption of personal health records: some problematic issues for patient empowerment. Ethics and Information Technology, 17(2), 113-124.
[do] Ernst R. Berndt and Iain M. Cockburn, "Price indexes for clinical trial research: a feasibility study," Monthly Labor Review, U.S. Bureau of Labor Statistics, June 2014, https://doi.org/10.21916/mlr.2014.22.
[dp] Atalah, A., Chang-Jin, C., & Osburn, K. (2002). Comparison Study of Installing Fiber Optic Cable in University Campuses Using Trenchless Techniques Relative to Open Cut. Pipelines 2002. doi:10.1061/40641(2002)70
[dq] Yang, H. C. M., Holder, J. D., & McNutt, C. W. (1998). U.S. Patent No. 5,761,362. Washington, DC: U.S. Patent and Trademark Office.
[dr] Edmundson, G. W. (1992). U.S. Patent No. 5,125,060. Washington, DC: U.S. Patent and Trademark Office.
[ds] Sonavane, S. S., Kumar, V., & Patil, B. P. (2008, December). MSP430 and nRF24L01 based wireless sensor network design with adaptive power control. In The International Congress for global Science and Technology (Vol. 25, No. 29, p. 11).
[dt] Chen, M., & Chang, T. (2011, June). A parking guidance and information system based on wireless sensor network. In 2011 IEEE International Conference on Information and Automation (pp. 601-605). IEEE.
[du] Rui, P. Y. G. X. Z. (2010). Design of smart wireless temperature measurement system based on NRF24L01 [J]. Electronic Measurement Technology, 2.

[dv] Romer, K., & Mattern, F. (2004). The design space of wireless sensor networks. IEEE wireless communications, 11(6), 54-61.
[dw] Bor, M., Vidler, J. E., & Roedig, U. (2016). LoRa for the Internet of Things.
[dx] Ramanathan, N., Balzano, L., Estrin, D., Hansen, M., Harmon, T., Jay, J., ... & Sukhatme, G. (2006, May). Designing wireless sensor networks as a shared resource for sustainable development. In 2006 International Conference on Information and Communication Technologies and Development (pp. 256-265). IEEE.
[dy] Fisher, W. C., & Bevis, M. E. (1991). U.S. Patent No. 5,054,881. Washington, DC: U.S. Patent and Trademark Office.
[dz] Hope, T. S. (1983). U.S. Patent No. 4,401,366. Washington, DC: U.S. Patent and Trademark Office.
[ea] Dunn, J. M., Stern, E. H., & Willner, B. E. (2004). U.S. Patent No. 6,836,476. Washington, DC: U.S. Patent and Trademark Office.
[eb] Mingliu Zhang, & Wolff, R. S. (2004). Crossing the digital divide: cost-effective broadband wireless access for rural and remote areas. IEEE Communications Magazine, 42(2), 99–105. doi:10.1109/mcom.2003.1267107
[ec] Markendahl, J., & Makitalo, O. (2007). Analysis of Business Models and Market Players for Local Wireless Internet Access. 2007 6th Conference on Telecommunication Techno-Economics. doi:10.1109/ctte.2007.4389894
[ed] Baker, D. M., & Head, J. W. (2015). Extensive Middle Amazonian mantling of debris aprons and plains in Deuteronilus Mensae, Mars: Implications for the record of mid-latitude glaciation. Icarus, 260, 269-288.
[ee] Fastook, J. L., Head, J. W., & Marchant, D. R. (2014). Formation of lobate debris aprons on Mars: Assessment of regional ice sheet collapse and debris-cover armoring. Icarus, 228, 54-63.
[ef] Petersen, E. I., Holt, J. W., & Levy, J. S. (2018, March). All Our Aprons are Icy: No Evidence for Debris-Rich" Lobate Debris Aprons" in Deuteronilus Mensae. In Lunar and Planetary Science Conference (Vol. 49).
[eg] Petersen, E. I., Holt, J. W., & Levy, J. S. (2018). High Ice Purity of Martian Lobate Debris Aprons at the Regional Scale: Evidence From an Orbital Radar Sounding Survey in Deuteronilus and Protonilus Mensae. Geophysical Research Letters, 45(21), 11-595
[eh] Gupta, P., & Atrey, M. . (2000). Performance evaluation of counter flow heat exchangers considering the effect of heat in leak and longitudinal conduction for low-temperature applications. Cryogenics, 40(7), 469–474. doi:10.1016/s0011-2275(00)00069-2
[ei] Petitpas, G., Bénard, P., Klebanoff, L. E., Xiao, J., & Aceves, S. (2014). A comparative analysis of the cryo-compression and cryo-adsorption hydrogen storage methods. International Journal of Hydrogen Energy, 39(20), 10564–10584. doi:10.1016/j.ijhydene.2014.04.200
[ej] Kleinböhl, A., Schofield, J. T., Kass, D. M., Abdou, W. A., Backus, C. R., Sen, B., ... & Teanby, N. A. (2009). Mars Climate Sounder limb profile retrieval of atmospheric temperature, pressure, and dust and water ice opacity. Journal of Geophysical Research: Planets, 114(E10).
[ek] Aceves, S. M., Espinosa-Loza, F., Ledesma-Orozco, E., Ross, T. O., Weisberg, A. H., Brunner, T. C., & Kircher, O. (2010). High-density automotive hydrogen storage with cryogenic capable pressure vessels. International Journal of Hydrogen Energy, 35(3), 1219-1226
[el] Christensen, P. R., Anderson, D. L., Chase, S. C., Clancy, R. T., Clark, R. N., Conrath, B. J., ... & Roush, T. L. (1998). Results from the Mars global surveyor thermal emission spectrometer. Science, 279(5357), 1692-1698.
[em] Kanevskiy, M., Shur, Y., Connor, B., Dillon, M., Stephani, E., & O'Donnell, J. (2012, June). Study of the ice-rich syngenetic permafrost for road design (Interior Alaska). In Proceedings of the Tenth International Conference on Permafrost (Vol. 1, pp. 25-29). Salekhard, Russia: The Northern Publisher.
[en] Kanevskiy, M., Fortier, D., Shur, Y., Bray, M., & Jorgenson, T. (2008, June). Detailed cryostratigraphic studies of syngenetic permafrost in the winze of the CRREL permafrost tunnel, Fox, Alaska. In Proceedings of the Ninth International Conference on Permafrost (Vol. 1, pp. 889-894). Fairbanks, Alaska: Institute of Northern Engineering, University of Alaska Fairbanks.
[eo] Shock, D. A. A., & Sudbury, J. D. (1969). U.S. Patent No. 3,436,919. Washington, DC: U.S. Patent and Trademark Office.
[ep] Nordin, V., & Murov, V. (2016). Logistical approach to monitoring operation of underground gas storages (UGS). Zeszyty Naukowe. Organizacja i Zarządzanie/Politechnika Śląska, (91), 241-249.

[eq] YAMASHITA, M., ISHIKAWA, Y., KITAYA, Y., GOTO, E., ARAI, M., HASHIMOTO, H., ... FUJITA, O. (2006). An Overview of Challenges in Modeling Heat and Mass Transfer for Living on Mars. Annals of the New York Academy of Sciences, 1077(1), 232–243. doi:10.1196/annals.1362.012
[er] Plaut, J. J., Safaeinili, A., Holt, J. W., Phillips, R. J., Head, J. W., Seu, R., ... & Frigeri, A. (2009). Radar evidence for ice in lobate debris aprons in the mid-northern latitudes of Mars. Geophysical research letters, 36(2).
[es] Wilson, P. W., & Leader, J. P. (1995). Stabilization of supercooled fluids by thermal hysteresis proteins. Biophysical journal, 68(5), 2098-2107.
[et] Banerjee, T., & Nayak, A. (2017). Competition Strategy and Geographical Proximity of Contract Research Organizations. Theoretical Economics Letters, 7(05), 1413.
[eu] Head, J., Dickson, J., Mustard, J., Milliken, R., Scott, D., Johnson, B., ... & Forget, F. (2015). Mars human science exploration and resource utilization: the dichotomy boundary Deuteronilus Mensae exploration zone. Geology, 36, 411-414.
[ev] Bussey, B., & Davis, R. R. (2015). First Landing Site/Exploration Zone Workshop for Human Mission to the Surface of Mars: October 27–30, 2015, Houston, Texas.
[ew] Ahmed, A., Fried, A., Roberts, J. J., & Limbachoya, M. (2004). Advantages and implications of high performance low density aircrete products for the UK Construction Industry. In 13th international brick and block masonry conference (IBBMaC), Amsterdam, Netherlands
[ex] British Board of Agrément, A. DUROX AIRCRETE BLOCKS SUPABLOC AND FOUNDATION. Agrément Certificate 00/3776Product Sheet 1
[ey] Ahmed, A., & Kamau, J. (2016). Properties of conventional cement and thin layer mortars. International Journal of Science, Environment and Technology, 5(5), 2837-2844.
[ez] Limbachiya, M. C., & Kew, H. Y. (2011). Physical properties of low density aircrete products. Cement Wapno Beton, (spec.), 66-69.
[fa] Fudge, C. A., & Limbachiya, M. C. (2006). Briefing: Aircrete.
[fb] Shahzad, A., & Pawar, S. S. (2017). Mortality (LC50) by standardization artificial soil test mathod for estimation of toxic effect of dieldrine and cythion on earthworm, Eisenia fetida. Int J Innov Res Adv Stud, 4(3), 148-152.
[fc] Gutteridge, S., & Pierce, J. (2006). A unified theory for the basis of the limitations of the primary reaction of photosynthetic CO2 fixation: Was Dr. Pangloss right?. Proceedings of the National Academy of Sciences, 103(19), 7203-7204.
[fd] Ndegwa, P. M., & Thompson, S. A. (2001). Integrating composting and vermicomposting in the treatment and bioconversion of biosolids. Bioresource technology, 76(2), 107-112.
[fe] Adhikary, S. (2012). Vermicompost, the story of organic gold: A review. Agricultural Sciences, 3(7), 905.
[ff] Hindman, J., Stehouwer, R., & MacNeal, K. (2008). Spent foundry sand and compost in blended topsoil: Availability of nutrients and trace elements. Journal of Residuals Science and technology, 5(2), 77-86.
[fg] Check, J., Sereda-Meichel, W., Man, L. S., & Macadam, E. Design Review Package for Vermicomposting Machine. WO2004039510
[fh] Grinias, J. P., Whitfield, J. T., Guetschow, E. D., & Kennedy, R. T. (2016). An Inexpensive, Open-Source USB Arduino Data Acquisition Device for Chemical Instrumentation. Journal of chemical education, 93(7), 1316-1319.
[fi] A Cost Estimation for CO2 Reduction and Reuse by Methanation from Cement Industry Sources in Switzerland
[fj] Aragona, I., Paudel, Y., Karanevich, P. (2012) MARS Return Fuel Production Using Table Top Electrolysis & Sabatier Reaction. COSGC Space Research Symposium 2
[fk] Puri, V., Nayyar, A., & Raja, L. (2017). Agriculture drones: A modern breakthrough in precision agriculture. Journal of Statistics and Management Systems, 20(4), 507-518.
[fl] Goetz, A. F., Vane, G., Solomon, J. E., & Rock, B. N. (1985). Imaging spectrometry for earth remote sensing. science, 228(4704), 1147-1153.
[fm] Wulder, M. A., Dymond, C. C., White, J. C., Leckie, D. G., & Carroll, A. L. (2006). Surveying mountain pine beetle damage of forests: A review of remote sensing opportunities. Forest Ecology and Management, 221(1-3), 27–41. doi:10.1016/j.foreco.2005.09.021
[fn] Fuller, R. M., Groom, G. B., & Jones, A. R. (1994). The land-cover map of great Britain: an automated classification of landsat thematic mapper data. Photogrammetric Engineering and Remote Sensing, 60(5), 553-562.

[fo] Croes, R. R. (2006). A paradigm shift to a new strategy for small island economies: Embracing demand side economics for value enhancement and long term economic stability. Tourism Management, 27(3), 453–465. doi:10.1016/j.tourman.2004.12.003
[fp] Fajardo, G. E. (2002). A struggle for identity.Caribbean Business,13,24–28.
[fq] Reddy, M. (1998). Energy consumption and economic activity in Fiji. The Journal of Pacific Studies, 22(s 81), 96.
[fr] Latimer, H. (1985). Developing-island economies — tourism v agriculture. Tourism Management, 6(1), 32–42. doi:10.1016/0261-5177(85)90053-6
[fs] Biopharmaceutical industry-sponsored clinical trials: impact on state economies. Pharmaceutical Research and Manufacturers of America (PhRMA); 2015. http://phrma-docs.phrma.org/sites/default/files/pdf/biopharmaceutical-industry-sponsored-clinical-trials-impact-on-state-economies.pdf.
[ft] Webster, C. J., & Woollett, G. R. (2018). Comment on "The End of Phase 3 Clinical Trials in Biosimilars Development?". BioDrugs : clinical immunotherapeutics, biopharmaceuticals and gene therapy, 32(5), 519–521. doi:10.1007/s40259-018-0297-y
[fu] Warren, S. R., Raisch, D. W., Campbell, H. M., Guarino, P. D., Kaufman, J. S., Petrokaitis, E., ... & Veterans Affairs Site Investigators. (2013). Medication adherence assessment in a clinical trial with centralized follow-up and direct-to-patient drug shipments. Clinical trials, 10(3), 441-448.
[fv] Ebert, M. A., Haworth, A., Kearvell, R., Hooton, B., Coleman, R., Spry, N., ... & Joseph, D. (2008). Detailed review and analysis of complex radiotherapy clinical trial planning data: evaluation and initial experience with the SWAN software system. Radiotherapy and Oncology, 86(2), 200-210.
[fw] Sahoo, U., & Bhatt, A. (2004). Electronic data capture (EDC)–a new mantra for clinical trials. Quality Assurance, 10(3-4), 117-121.
[fx] Henderson, L. (2013). Clinical Technologies. Applied Clinical Trials, 22(12), 18.
[fy] Lamberti, M. J., Kush, R., Kubick, W., Henderson, C., Hinkson, B., Kamenju, P., & Getz, K. A. (2015). An examination of eClinical technology usage and CDISC standards adoption. Therapeutic innovation & regulatory science, 49(6), 869-876.
[fz] Çalamak, S., Erdoğdu, C., Özalp, M., & Ulubayram, K. (2014). Silk fibroin based antibacterial bionanotextiles as wound dressing materials. Materials Science and Engineering: C, 43, 11-20.
[ga] Asakura, T. (2014). Biotechnology of silk (Vol. 5). T. Miller (Ed.). Dordrecht: Springer.
[gb] Frankopan, P. (2015). The silk roads: A new history of the world. Bloomsbury Publishing.
[gc] Sazzini, M., Garagnani, P., Sarno, S., De, S. F., Lazzano, T., Yang, D. Y., ... & Franceschi, C. (2015). Tracing Behçet's disease origins along the Silk Road: an anthropological evolutionary genetics perspective. Clinical and experimental rheumatology, 33(6 Suppl 94), S60-6.
[gd] Franceschi, F. (1995). Florence and Silk in the Fifteenth Century: the Origins of a Long and Felicitous Union. Italian History and Culture, 1, 3-22.
[ge] Cizakca, M. (1980). Price history and the Bursa silk industry: A study in Ottoman industrial decline, 1550–1650. The Journal of Economic History, 40(3), 533-550.
[gf] King, B. M. (2017). Silk and empire.
[gg] Altman, G. H., Diaz, F., Jakuba, C., Calabro, T., Horan, R. L., Chen, J., … Kaplan, D. L. (2003). Silk-based biomaterials. Biomaterials, 24(3), 401–416. doi:10.1016/s0142-9612(02)00353-8
[gh] Sakthivel, Nalliappan. (2014). Organic Farming in Mulberry: Recent Breakthrough.
[gi] Ellsworth, E. D. (2002). U.S. Patent No. 6,360,729. Washington, DC: U.S. Patent and Trademark Office.
[gj] LeCuyer, A. (2008). ETFE: technology and design. Walter de Gruyter.
[gk] Jones, M. R., & Giannakou, A. (2004). Thermally insulating foundations and ground slabs using highly-foamed concrete. Journal of ASTM International, 1(6), 1-13.
[gl] Bucklin, R. A., Leary, J. D., Rygalov, V., Mu, Y., & Fowler, P. A. (2001). Design parameters for Mars deployable greenhouses (No. 2001-01-2428). SAE Technical Paper.
[gm] Tinker, M., Hull, P. V., SanSoucie, M. P., & Roldan, A. (2006). Inflatable and Deployable Structures for Surface Habitat Concepts Utilizing In-Situ Resources. Earth & Space 2006. doi:10.1061/40830(188)72
[gn] Scherer, S., Rehder, J., Achar, S., Cover, H., Chambers, A., Nuske, S., & Singh, S. (2012). River mapping from a flying robot: state estimation, river detection, and obstacle mapping. Autonomous Robots, 33(1-2), 189-214.
[go] Cohen, M. M. (1998). Space habitat design integration issues (No. 981800). SAE Technical Paper.
[gp] Pedley, M. D., & Mayeaux, B. (2001). TransHab Materials Selection.

[gq] Pellenbarg, R. E., Max, M. D., & Clifford, S. M. (2003). Methane and carbon dioxide hydrates on Mars: Potential origins, distribution, detection, and implications for future in situ resource utilization. Journal of Geophysical Research: Planets, 108(E4).
[gr] Bolonkin, A., & Cathcart, R. (2007). Inflatable Evergreen Polar Zone Dome (EPZD) Settlements. arXiv preprint physics/0701098.
[gs] Suslov, S. (2011). Calculating of energy consumption of the sports hall. *****Lots of thermal math********
[gt] Duol Air Domes (2008) Why to Chose Duol Domes. Brezovica, Slovenija
[gu] Tsujino, J., Hongoh, M., Tanaka, R., Onoguchi, R., & Ueoka, T. (2002). Ultrasonic plastic welding using fundamental and higher resonance frequencies. Ultrasonics, 40(1-8), 375-378.
[gv] Grewell, D., & Benatar, A. (2007). Welding of plastics: fundamentals and new developments. International Polymer Processing, 22(1), 43-60.
[gw] Yano, K., Usuki, A., Okada, A., Kurauchi, T., & Kamigaito, O. (1993). Synthesis and properties of polyimide–clay hybrid. Journal of Polymer Science Part A: Polymer Chemistry, 31(10), 2493–2498. doi:10.1002/pola.1993.080311009
[gx] Hill, R. P. (1987). U.S. Patent No. 4,704,876. Washington, DC: U.S. Patent and Trademark Office.
[gy] Acharya, A., Arman, B., Olszewski, W. J., Bonaquist, D. P., & Weber, J. A. (2002). U.S. Patent No. 6,426,019. Washington, DC: U.S. Patent and Trademark Office
[gz] Little, W. A., & Sapozhnikov, I. (1998). U.S. Patent No. 5,724,832. Washington, DC: U.S. Patent and Trademark Office.
[ha] Filimonov, M. Y., & Vaganova, N. A. (2013). Simulation of thermal stabilization of soil around various technical systems operating in permafrost. Appl. Math. Sci, 7(144), 7151-7160.
[hb] Long, E. L. (1965). U.S. Patent No. 3,217,791. Washington, DC: U.S. Patent and Trademark Office.
[hc] Halfinger, J. A., & Haggerty, M. D. (2012). The B&W mPower™ scalable, practical nuclear reactor design. Nuclear technology, 178(2), 164-169.
[hd] Holtec International (2015) Essentials of SMR-160 Small Modular Reactor accessed: https://smrllc.files.wordpress.com/2015/06/htb-015-hi-smur-rev3.pdf
[he] Vujić, J., Bergmann, R. M., Škoda, R., & Miletić, M. (2012). Small modular reactors: Simpler, safer, cheaper?. Energy, 45(1), 288-295.
[hf] (2018) NuScale targets SMR staff costs below nuclear industry average. Nuclear Energy Insider
[hg] Dahlgren, E., Göçmen, C., Lackner, K., & van Ryzin, G. (2013). Small Modular Infrastructure. The Engineering Economist, 58(4), 231–264. doi:10.1080/0013791x.2013.825038
[hh] Clayton, D., & Wood, R. (2010). The role of instrumentation and control technology in enabling deployment of small modular reactors. In Seventh American Nuclear Society International Topical Meeting on Nuclear Plant Instrumentation, Control and Human-Machine Interface Technologies, NPIC &HMIT2010, Las Vegas, Nevada, November
[hi] Gartner, E. (2004). Industrially interesting approaches to "low-CO2" cements. Cement and Concrete research, 34(9), 1489-1498.
[hj] Kalogirou, S. A. (2002). Parabolic trough collectors for industrial process heat in Cyprus. Energy, 27(9), 813-830.
[hk] Zhang, Y., & Dao, S. (1996, November). Integrating direct broadcast satellite with wireless local access. In Proceedings of the First International Workshop on Satellite-based Information Services, New York (pp. 24-29).
[hl] Murray, J. P. (1999). Aluminum production using high-temperature solar process heat. Solar Energy, 66(2), 133-142.
[hm] Vasiliev, L. L. (2005). Heat pipes in modern heat exchangers. Applied thermal engineering, 25(1), 1-19.
[hn] Kemp, I. C. (2011). Pinch analysis and process integration: a user guide on process integration for the efficient use of energy. Elsevier.
[ho] Southworth, F. H., & Macdonald, P. E. (2003). The next generation nuclear plant (NGNP) project (No. INEEL/CON-03-01150). Idaho National Laboratory (INL).
[hp] Lee, W. J., Kim, Y. W., & Chang, J. H. (2009). Perspectives of nuclear heat and hydrogen. Nuclear Engineering and Technology, 41(4), 413-426.
[hq] Angulo, C., Bogusch, E., Bredimas, A., Delannay, N., Viala, C., Ruer, J., ... & Fütterer, M. A. (2012). EUROPAIRS: The European project on coupling of High Temperature Reactors with industrial processes. Nuclear Engineering and Design, 251, 30-37.

[hr] Megahed, M. M. (2001). Nuclear desalination: history and prospects. Desalination, 135(1-3), 169-185.
[hs] Edwards, J. H., Do, K. T., Maitra, A. M., Schuck, S., Fok, W., & Stein, W. (1996). The use of solar-based CO2/CH4 reforming for reducing greenhouse gas emissions during the generation of electricity and process heat. Energy conversion and management, 37(6-8), 1339-1344.
[ht] Canadian Nuclear Safety Commission (2019). Phase 1 Pre-Licensing Vendor Design Review Executive Summary: Ultra Safe Nuclear Corporation. Accessed: http://nuclearsafety.gc.ca/eng/reactors/power-plants/pre-licensing-vendor-design-review/executive-summary-ultra-safe-nuclear-corporation.cfm
[hu] Edwards, J. H., & Maitra, A. M. (1995). The chemistry of methane reforming with carbon dioxide and its current and potential applications. Fuel Processing Technology, 42(2-3), 269–289. doi:10.1016/0378-3820(94)00105-3
[hv] Nelson, M., & Dempster, W. F. (1996). Living in space- Results from Biosphere 2's initial closure, an early testbed for closed ecological systems on Mars. Strategies for Mars: A guide to human exploration(A 96-27659 06-12), San Diego, CA, Univelt, Inc.(Science and Technology Series., 86, 363-390.
[hw] Beloian, A. (1982). Use of a food consumption model to estimate human contaminant intake. Environmental monitoring and assessment, 2(1-2), 115-127.
[hx] Hall, K. D., Guo, J., Dore, M., & Chow, C. C. (2009). The progressive increase of food waste in America and its environmental impact. PloS one, 4(11), e7940.
[hy] Buzby, J. C., & Hyman, J. (2012). Total and per capita value of food loss in the United States. Food Policy, 37(5), 561-570.
[hz] Kant AK, Graubard BI (2005) Energy density of diets reported by American adults: association with food group intake, nutrient intake, and body weight. Int J Obes (Lond) 29: 950–956.
[ia] Javan, A., Grieco, W., Cumba, H., Weaver, H., & Lovas, B. (2012). U.S. Patent Application No. 13/265,525.
[ib] Putnam, J., Allshouse, J., & Kantor, L. S. (2002). US per capita food supply trends: more calories, refined carbohydrates, and fats. Food Review, 25(3), 2-15.
[ic] USDA (2014) Shelf-Stable Food Safety. Food Safety Information Sheet
[id] Pugliese, A., Cabassi, G., Chiavaro, E., Paciulli, M., Carini, E., & Mucchetti, G. (2017). Physical characterization of whole and skim dried milk powders. Journal of food science and technology, 54(11), 3433–3442. doi:10.1007/s13197-017-2795-1
[ie] Chávez-Servín, J. L., Castellote, A. I., Rivero, M., & López-Sabater, M. C. (2008). Analysis of vitamins A, E and C, iron and selenium contents in infant milk-based powdered formula during full shelf-life. Food chemistry, 107(3), 1187-1197.
[if] Itan, Y., Jones, B. L., Ingram, C. J., Swallow, D. M., & Thomas, M. G. (2010). A worldwide correlation of lactase persistence phenotype and genotypes. BMC evolutionary biology, 10(1), 36.
[ig] American Egg Board, Liquid Eggs to Dry: Conversion Worksheet. AEB.org/conversion
[ih] USDA. (2000). United States standards, grades, and weight classes for shell eggs. In AMS (Vol. 56, p. 210).
[ii] Koç, M., Koç, B., Susyal, G., Sakin Yilmazer, M., Kaymak Ertekin, F., & Bağdatlıoğlu, N. (2010). Functional and physicochemical properties of whole egg powder: effect of spray drying conditions. Journal of food science and technology, 48(2), 141–149. doi:10.1007/s13197-010-0159-1
[ij] Vignolles, M. L., Lopez, C., Madec, M. N., Ehrhardt, J. J., Méjean, S., Schuck, P., & Jeantet, R. (2009). Fat properties during homogenization, spray-drying, and storage affect the physical properties of dairy powders. Journal of Dairy Science, 92(1), 58-70.
[ik] Veglio, F., Passariello, B., & Abbruzzese, C. (1999). Iron removal process for high-purity silica sands production by oxalic acid leaching. Industrial & engineering chemistry research, 38(11), 4443-4448.
[il] Appelbaum, J., & Flood, D. J. (1990). Solar radiation on Mars. Solar Energy, 45(6), 353-363.
[im] Strategic Missions and Advanced Concepts Office, JPL. (2017) Solar Power Technologies for Future Planetary Science Missions. JPL D-101316
[in] Li, L., Kim, S., Wang, W., Vijayakumar, M., Nie, Z., Chen, B., ... & Liu, J. (2011). A stable vanadium redox-flow battery with high energy density for large-scale energy storage. Advanced Energy Materials, 1(3), 394-400.
[io] Hu, X., Zou, C., Zhang, C., & Li, Y. (2017). Technological developments in batteries: a survey of principal roles, types, and management needs. IEEE Power and Energy Magazine, 15(5), 20-31.
[ip] Davis, D. M., Nielsen, S. G., Magna, T., & Mezger, K. (2018, March). Constraints on the vanadium isotope composition of Mars. In Lunar and Planetary Science Conference (Vol. 49).

[iq] Sommer, W. C. (1970). U.S. Patent No. 3,488,067. Washington, DC: U.S. Patent and Trademark Office.
[ir] Brink, K., Olsson, M., & Bolmsjö, G. (1997). Increased autonomy in industrial robotic systems: a framework. Journal of Intelligent and Robotic Systems, 19(4), 357-373.
[is] Sun, Y., Coltin, B., & Veloso, M. (2013, June). Interruptable autonomy: Towards dialog-based robot task management. In Workshops at the Twenty-Seventh AAAI Conference on Artificial Intelligence.
[it] Gancet, J., & Lacroix, S. (2007). Embedding heterogeneous levels of decisional autonomy in multi-robot systems. In Distributed Autonomous Robotic Systems 6 (pp. 263-272). Springer, Tokyo.
[iu] Roberge, D. M., Ducry, L., Bieler, N., Cretton, P., & Zimmermann, B. (2005). Microreactor Technology: A Revolution for the Fine Chemical and Pharmaceutical Industries? Chemical Engineering & Technology, 28(3), 318–323. doi:10.1002/ceat.200407128
[iv] Brauch, S., van Berkel, S. S., & Westermann, B. (2013). Higher-order multicomponent reactions: beyond four reactants. Chemical Society Reviews, 42(12), 4948. doi:10.1039/c3cs35505e
[iw] Devlin, T., Kerr, R., Lyons, D., & Byrne, R. (2002). U.S. Patent No. 6,360,894. Washington, DC: U.S. Patent and Trademark Office.
[ix] Salter, J. D., & Wyatt, N. P. G. (1991). Sorting in the minerals industry: past, present and future. Minerals Engineering, 4(7-11), 779-796.
[iy] Lessard, J., de Bakker, J., & McHugh, L. (2014). Development of ore sorting and its impact on mineral processing economics. Minerals Engineering, 65, 88-97.
[iz] Santini, T. C., & Banning, N. C. (2016). Alkaline tailings as novel soil forming substrates: reframing perspectives on mining and refining wastes. Hydrometallurgy, 164, 38-47.
[ja] Dermont, G., Bergeron, M., Mercier, G., & Richer-Laflèche, M. (2008). Soil washing for metal removal: a review of physical/chemical technologies and field applications. Journal of hazardous materials, 152(1), 1-31.
[jb] Brantley, S. L. (2010). Weathering: Rock to regolith. Nature Geoscience, 3(5), 305.
[jc] McQueen, K. G. (2009). Regolith geochemistry. Regolith Science. CSIRO Publishing, Melbourne, 73-104.
[jd] Clark, B. C. (1993). Geochemical components in Martian soil. Geochimica et Cosmochimica Acta, 57(19), 4575–4581. doi:10.1016/0016-7037(93)90183-w
[je] Yong, R. N., Galvez-Cloutier, R., & Phadungchewit, Y. (1993). Selective sequential extraction analysis of heavy-metal retention in soil. Canadian Geotechnical Journal, 30(5), 834-847.
[jf] Banza, A. ., Gock, E., & Kongolo, K. (2002). Base metals recovery from copper smelter slag by oxidising leaching and solvent extraction. Hydrometallurgy, 67(1-3), 63–69. doi:10.1016/s0304-386x(02)00138-x
[jg] Goodfellow, R., & Dimitrakopoulos, R. (2017). Simultaneous stochastic optimization of mining complexes and mineral value chains. Mathematical Geosciences, 49(3), 341-360.
[jh] Quinkertz, R., Rombach, G., & Liebig, D. (2001). A scenario to optimise the energy demand of aluminium production depending on the recycling quota. Resources, Conservation and Recycling, 33(3), 217–234. doi:10.1016/s0921-3449(01)00086-6
[ji] McAuley, K. B. (1991). Modelling, estimation and control of product properties in a gas phase polyethylene reactor (Doctoral dissertation).
[jk] McAuley, K. B and Macgregor, J.F., (1991) https://doi.org/10.1002/aic.690370605
[jl] Boone, P. J., Canton, M. H., Niziol, S. F., Mapson, T. D., Cox, B. R., Kelly Jr, R. G., ... & Lichtenfiels, K. K. (1994). U.S. Patent No. 5,355,063. Washington, DC: U.S. Patent and Trademark Office.
[jm] Mason, L. S., Gibson, M. A., & Poston, D. (2013). Kilowatt-class fission power systems for science and human precursor missions.
[jn] Zhang, H. L., Baeyens, J., Degrève, J., & Cacères, G. (2013). Concentrated solar power plants: Review and design methodology. Renewable and sustainable energy reviews, 22, 466-481.
[jo] Freundlich, A., Ignatiev, A., Horton, C., Duke, M., Curreri, P., & Sibille, L. (2005, January). Manufacture of solar cells on the moon. In Conference Record of the Thirty-first IEEE Photovoltaic Specialists Conference, 2005. (pp. 794-797). IEEE.
[jp] Krebs, F. C., Gevorgyan, S. A., & Alstrup, J. (2009). A roll-to-roll process to flexible polymer solar cells: model studies, manufacture and operational stability studies. Journal of Materials Chemistry, 19(30), 5442-5451.
[jq] Valdivia, W. D. (2013). University start-ups: Critical for improving technology transfer. Center for Technology Innovation at Brookings. Washington, DC: Brookings Institution.

[jr] Xu, G. G. (2004). Information for corporate IP management. World Patent Information, 26(2), 149–156. doi:10.1016/j.wpi.2003.12.002
[js] Yang, Z., & Garimella, S. V. (2010). Thermal analysis of solar thermal energy storage in a molten-salt thermocline. Solar energy, 84(6), 974-985.
[jt] Leten, B., Vanhaverbeke, W., Roijakkers, N., Clerix, A., & Van Helleputte, J. (2013). IP models to orchestrate innovation ecosystems: IMEC, a public research institute in nano-electronics. California management review, 55(4), 51-64.
[ju] Eden Project (1992) Crop Yield Verification. Accessed: http://www.gardensofeden.org/04%20Crop%20Yield%20Verification.htm
[jv] Hochleitner, B., Desnica, V., Mantler, M., & Schreiner, M. (2003). Historical pigments: a collection analyzed with X-ray diffraction analysis and X-ray fluorescence analysis in order to create a database. Spectrochimica Acta Part B: Atomic Spectroscopy, 58(4), 641-649.
[jw] Cosentino, A. (2014). FORS spectral database of historical pigments in different binders. 53 TECHNICAL, 54.
[jx] Arruda, E. M., Cao, K., Siepermann, C. A. P., Thouless, M. D., Anderson, R. M., Kotov, N. A., ... & Waas, A. M. (2013). U.S. Patent Application No. 13/871,106.
[jy] Kitson, P. J., Symes, M. D., Dragone, V., & Cronin, L. (2013). Combining 3D printing and liquid handling to produce user-friendly reactionware for chemical synthesis and purification. Chemical Science, 4(8), 3099-3103.
[jz] Kitson, P. J., Glatzel, S., & Cronin, L. (2016). The digital code driven autonomous synthesis of ibuprofen automated in a 3D-printer-based robot. Beilstein journal of organic chemistry, 12(1), 2776-2783.
[ka] Gordeev, E. G., Galushko, A. S., & Ananikov, V. P. (2018). Improvement of quality of 3D printed objects by elimination of microscopic structural defects in fused deposition modeling. PloS one, 13(6), e0198370.
[kb] Kitson, P. J., Glatzel, S., & Cronin, L. (2016). The digital code driven autonomous synthesis of ibuprofen automated in a 3D-printer-based robot. Beilstein journal of organic chemistry, 12(1), 2776-2783.
[kc] Brychikhin, M. N., Chkhalo, N. I., Eikhorn, Y. O., Malyshev, I. V., Pestov, A. E., Plastinin, Y. A., ... & Toropov, M. N. (2016). Reflective Schmidt–Cassegrain system for large-aperture telescopes. Applied optics, 55(16), 4430-4435.
[kd] Graves, J. E., Northcott, M. J., Fleischer, S., Chang, R., & Zambon, P. (2015). U.S. Patent Application No. 13/799,923.
[ke] Thompson, K. L., Kuo, C. L., Yoon, K. W., & Ahmed, Z. (2018, July). Next-generation small CMB telescopes. In Ground-based and Airborne Telescopes VII (Vol. 10700, p. 107004K). International Society for Optics and Photonics.
[kf] Chkhalo, N. I., Malyshev, I. V., Pestov, A. E., Polkovnikov, V. N., Salashchenko, N. N., Toropov, M. N., ... & Rizvanov, A. A. (2018). Collimator based on a Schmidt camera mirror design and its application to the study of the wide-angle UV and VUV telescope. Journal of Astronomical Telescopes, Instruments, and Systems, 4(1), 014003.
[kg] Brychikhin, M. N., Chkhalo, N. I., Eikhorn, Y. O., Malyshev, I. V., Pestov, A. E., Plastinin, Y. A., ... & Toropov, M. N. (2016). Reflective Schmidt–Cassegrain system for large-aperture telescopes. Applied optics, 55(16), 4430-4435.
[kh] Mateja, J. (2016). Connecting a Spectrometer to a High Power Telescope in Order to Gather Light From Stars.
[ki] Naragani, D., Sangid, M. D., Shade, P. A., Schuren, J. C., Sharma, H., Park, J. S., ... & Parr, I. (2017). Investigation of fatigue crack initiation from a non-metallic inclusion via high energy x-ray diffraction microscopy. Acta Materialia, 137, 71-84.
[kj] Spencer, J. (2018). Harvesting the 'Hands-free Hectare'. Farmer's Weekly, 2018(18004), 52-53.
[kk] Khandait, P., Sariyam, S., Agarkar, D., Dhongde, V., Doble, S., Makhe, R., & Dhakate, R. (2018). Seed Sowing Robot. International Journal of Engineering Science, 16564.
[kl] Shinde, V. N., Sharma, N. S., & Kasar, M. S. (2018). Solar Based Multi-Tasking Agriculture Robot.
[km] Soundararajan, V., & Agrawal, A. (2016). U.S. Patent No. 9,244,147. Washington, DC: U.S. Patent and Trademark Office.

# 21: THE "TEAM BOLD" MARS COLONY

**Matt Wise**
matt@boldenergy.net
**Kyle Saffel**
**Patrick Fagin**
**Douglas Livermore**
**Kaitlin Davis**

## INTRODUCTION

This proposal will discuss a self-sustaining Mars colony capable of supporting 1,000 human colonists. This is more than a habitat for study and survival, this is a new life. This new home will allow humans to thrive and realize still greater dreams of progress, innovation, and exploration. The Mars colony will be exciting, enjoyable, and safe. Naturally the colony will not start out with 1,000 colonists from the beginning but will grow as more and more missions bring immigrants to this new home. This paper will describe the systems and lifestyle of a base that has reached 1,000 colonists, how it will function, and how it will be significant.

We must set our sights beyond mere survival to building and expansion. We have the capabilities and technology to not only land on Mars, but the live there and live well. While science is, unquestionably, a main purpose for humans on Mars, those efforts are best served by a colony base which is thriving. The first colony will be focused solely on optimal human survival. Once a colony of 1,000 people is firmly established on Mars the capabilities of strategically located research stations, laboratories, and expeditions to the farthest reaches of the planet will be much more robust. If we attempt to build the colony in a location which is ideal for scientific discovery, we risk shifting our odds of survival and generally making life harder than necessary. A thriving colony could produce materials, equipment, ferry food and supplies, and provide backside support for a multitude of outstations.

Production of spacecraft, fuels, food, and materials should be a key component in the colony's long-term plan. This colony will very likely service the nest steps in human exploration as we push out to the asteroid belt and other planets.

Ultimately, the colony must be sustainable requiring the minimal possible support from Earth. Creativity and broad skillsets of the initial colonists are key to the colony thriving. While specialized experts have a grasp on existing methods, creative genius and unconventional solutions will be required in a new world where we face new and unfamiliar challenges.

Positive social bonds and strong sense of purpose will propel this colony forward. At a population of 1,000 colonists this station is likely to be extremely diverse. I believe this step out into the universe will create the necessary momentum to further dissolve borders and unite the humans on earth as a planetary group rather than individual countries. Just as travelling abroad on earth tends to dissolve small-town rivalries and prejudices, becoming a multi-planetary species will elevate our perspective and make our petty differences irrelevant. People will unite behind a common cause and our common identity as humans.

## INITIAL ESTABLISHMENT

### Location

The right location is a fundamental key to the success of any enterprise. Great care will be taken to choose the optimal location for the first Mars Colony. The colony must be located near as many concentrated resources as possible to include water ice, chemical components, useable aggregate, and must be in a location which presents the best conditions for building colonial structures (a giant boulder field might be a bad idea).

We will build this first colony in either Arcadia or Tharsis. These regions are near the equator while still close to large water-ice deposits. The geography allows for suitable landing sites as well as good building sites for a permanent colony. Proximity to the equator provides ideal launch trajectory and maximum sunlight exposure for solar power and surface temperature ranges. The habitat site will be geologically protected to reduce the impact of dust storms, meteor strikes, and other natural phenomena. (Use of 'lava tubes' and natural geological formations should be considered when advantageous, but not relied on.) Building into slopes or hillsides offers the ability to use the ground itself to maintain pressurized habitats and mitigate temperature fluctuations.

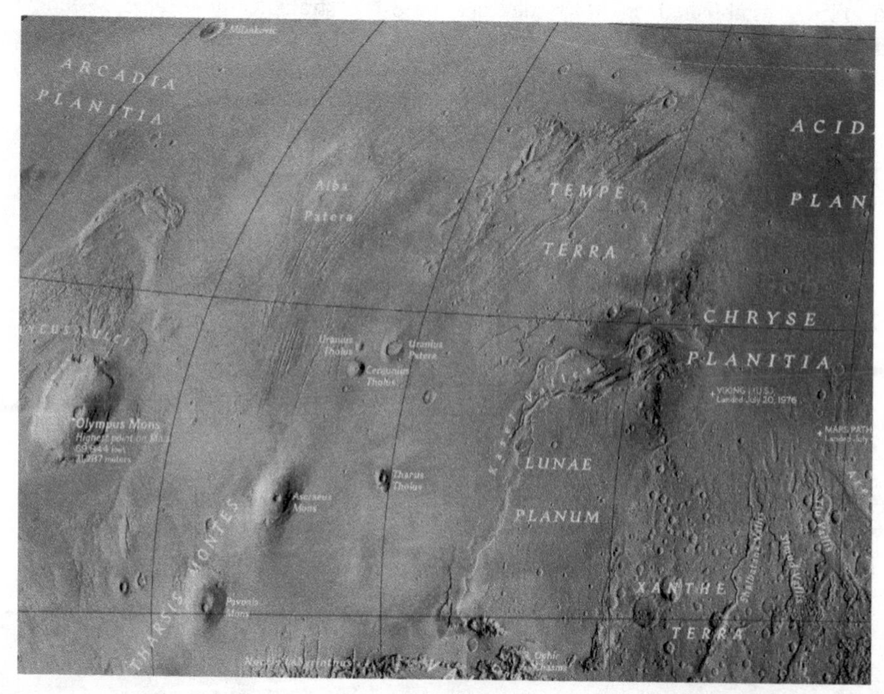

## TRANSPORT

Payload capacity per flight based on SpaceX Starship Super Heavy is 100,000+kg / 220,000+lb, at an average transport cost of $500/kg

| Cargo | kg | Transport Cost | Colonists |
|---|---|---|---|
| Power | 150,000 | $ 75,000,000 | 1,000 |
| Food Stores | 250,000 | $125,000,000 | |
| Medical | 35,000 | $ 17,500,000 | **Total Cargo** |
| Communications | 20,000 | $ 10,000,000 | 610,000kg |
| Fuel Depot | 10,000 | $ 5,000,000 | |
| Production Parts | 25,000 | $ 12,500,000 | **Flights** |
| Electronics | 10,000 | $ 5,000,000 | 16.1 |
| Data & Networking | | | |
| 3D Printer Components | | | **Cost (USD)** |
| Sensors | | | $305,000,000 |
| Circuitry | | | |
| Materials | 100,000 | $ 50,000,000 | |
| Metals, Plastics, etc | | | |
| Seed Stock | 5,000 | $ 2,500,000 | |
| Farm Materials/Supplies | 5,000 | $ 2,500,000 | |

*This list is by no means exhaustive. It's meant to show the transport cost only for the bare minimum requisite supplies for a complete colony and 1,000 colonists. Price to acquire supplies is additional.

**Landing/Launchpads**

These may be constructed using Martian aggregate to create a concrete or ceramic pad. Sulfur concrete will not likely withstand the heat, however a Portland cement equivalent can be produced. Any necessary lifts, launch towers, etc. will be constructed out of imported metal products until viable metals production has been established. Lower Mars gravity will allow for lower materials requirement.

**Transport tunnels to habitat**

A subterranean transport tunnel for a light rail system with pressurized passenger cars will run from a main port of entry to the launch terminal (approx 1km). Autonomous excavators will dig a trench for the light rail, 3D printers will create a support structure overhead, and then the excavators will back fill Martian regolith over the top for further protection. The route will initially be driven by self-driving transports until metals processing can create proper rails for a light rail train.

**The Terminal**
The upper level of the launch complex terminal will house launch operations including personnel and cargo logistics, refueling, flight control, and spacecraft maintenance. The lower level will contain passenger debarkation and inprocessing.

**POWER**

**Requirements**
Using the International Space Station (ISS) 75-90kw per day as a template for sustaining human life off-world, a basic habitat for 1,000 colonists will require a minimum 15,000kw. This would cover life support and general systems, but would be insufficient for larger-scale food and materials production.

Based on the following comparisons the optimal combination will be 8 nuclear fission generators and 3,000 solar panels. The nuclear reactors should provide adequate power while to solar panels will augment battery charging and expanded power requirements for industrial and construction applications. Additional wind power generators will be transported for diversification of power systems.

**a. Solar**
If solar were used exclusively, 15,000kw will require 6,000 solar panels at 250w per panel. The current, standard dimensions of a 250w solar panel are 64.5in x 39in. That's approximately 29 acres of panels just to cover the basics. Weight considerations at 18kg average per panel will net 108,000kg. That's just a bit more than a fully loaded Starship. This will be offset by the buildup of the colony over time, but it will still require a significant commitment of cargo space on each trip just for solar panels. Additionally, we'll require components, wire, and batteries. Conversely, diurnal cycle and atmospheric conditions could render a strictly solar power system unreliable. Solar should still be considered a primary element of the colony power grid but will have to be augmented by more reliable sources.

**b. Nuclear**
The Kilopower Project has been designing small 1kw-10kw nuclear power generators just for this purpose. While many may prefer to avoid any non-renewable, it's the best option for reliable energy. Annual dust storms can range planet-wide and significantly degrade energy from solar for as long as 2months in a worst-case scenario (1-3 days average).

Taylor Wilson at University of Nevada has also been developing safe, contained nuclear fission reactors which require minimal maintenance and refueling and capable of producing 5Mw of energy. These reactors are scalable and based on a heat exchange system rather than a steam turbine.

**c. Wind**

The 'Archimedes' wind power generators developed by Dutch MSc Marinus Mieremet require less wind and take advantage of a wider directional affect. Despite Mars' comparatively thin atmosphere these wind generators could still produce valuable power. Once materials are established it will also be possible to develop a production facility for these and other generators on Mars.

**d. Geothermal**

While there are no specifically known geothermal vents on Mars (yet), the temperature contrast from day to night (+70F to -100F) allows us to take advantage of thermodynamics in expansion and contraction to create energy.

**e. Power Storage**

Large battery banks strategically distributed in the colony will be a critical element. Distribution reduces the chance for critical failures and reduces impedance by shortening the distance energy travels over power lines. At minimum, the Mars colony will require a battery system capable of sustaining critical systems for up to 30 days. Disaster protocols will be in place in the case of extended power loss or catastrophic damage. These protocols will eliminate unnecessary sections and systems in order to conserve power.

We will develop efficient ways to manufacture batteries from locally sourced materials. Following traditional battery construction methods, anodes can be made of zinc, aluminum, lithium, cadmium, iron, and other materials known to exist on Mars. Cathodes could be produced from manganese, nickel, and other compounds which have been identified on the Martian surface.

**LIFE SUPPORT**

**Oxygen Requirements**

Each human requires an average of 1,230 Liters of oxygen per day on Mars (24hrs 37mins). Oxygen accounts for only about 23% of breathable air, and human lungs extract roughly ⅓ of the oxygen in each breath. A colony of 1,000 will require approximately 1,230,000 liters of oxygen every day or 2,000kg. Likewise, removal of $CO_2$ from the environment will be necessary as the average human produces approximately 500 liters of $CO_2$ per day.

**a. Recycling**

The air conditioning systems will include sabatier reactors and oxygen recyclers similar to that on the ISS. Breathable air is typically comprised of 78.09% nitrogen, 20.95% oxygen, 0.93% argon, 0.04% carbon dioxide, and trace amounts of other gases. The Air conditioning plants will also introduce the other necessary gasses into the environment.

**b. Water Ice Processing.**

At least two processing plants will be constructed for converting water ice for life support. Sabatier Reactors will be strategically located throughout the habitat to allow for redundancy and distribution. Water ice will be mined, transferred to the oxygen plant, a simple electrolyte and electric current are introduced which produces oxygen and hydrogen. The oxygen will be transferred to storage for life support and the hydrogen will be used to make fuel or recombined in the propellant depot. Multiple processing plants allow for greater efficiency cutting distance and energy cost and provide redundancy. Additionally, this same process will be employed for the fuel depot. Excess production may be shared between these systems while the fuel depot could be considered part of the contingency plan for emergency oxygen production.

**c. Greenhouse.**

The vertical farm facilities will likewise produce large amounts of oxygen and remove CO2 from the living environment. Each plant (average total leaf surface area of .5m$^2$) will produce an average of 250ml of oxygen per day under nominal conditions. Each plant removes an average of .52ml of CO2 per day. To satisfy total colony oxygen needs with plants alone this will require approximately 5,000 leafy green plants to be growing at any given time. If the average plant <1m tall requires .045m$^2$ the total dedicated growing space will be 227m$^2$ which will not include work spaces or utility areas. The air will be continuously cycled through all parts of the colonial air system to provide fresh oxygen. Facilities should be a minimum of 2 for redundancy and to minimize impact of catastrophic loss. Each operational greenhouse must be a minimum 150m$^2$. These figures assume all oxygen/CO2 management from these facilities without air recyclers or water-ice processing. Farming for food production will require much more growing space than what is necessary for oxygen production. This is very positive, given the capability to also recycle breathable air, minimal supplementation will be required Operation of the farming facilities is covered in more detail below.

**d. Cyanobacteria**

These oxygen-producing organisms may be introduced into the environmental processes to augment oxygen production. Further research will be required to understand what level of impact these organisms will have on the colony's oxygen requirements.

**Water Requirements**

The average human needs 2 liters of water per day. Our systems will need to provide a minimum of 2,000 liters of water every day to sustain basic needs. This doesn't include bathing, sanitation, food prep, farming, etc. It should be noted that in similar environments such as the ISS 70% - 90% of all water is recycled. That would lower the requirement of newly processed drinking water to around 600L/day. Properly maintained, the water systems in the colony will sustain quite well requiring only minimal periodic supplementation from mined water ice. Contingency storage should be a priority once colony needs are exceeded in case of shortage or emergency.

**a. Treatment recycle**

Water treatment facilities will be established for liquid waste. All water will be captured and recycled; waste water, humidity, and any other moisture is filtered, exposed to UV radiation, adjusted for mineral and PH, and then transferred to the holding tanks for use. 70% - 90% of all water needs can be fulfilled via efficient water recycling. A large 2-layer dome above the colony central square will be both functional and beautiful. The empty space between the two dome layers will be continually flooded with water. This will serve a dual purpose. Functional; The water layer blocks much of the harmful radiation, UV light kills unwanted microbes and bacteria in the water, and the constant flow keeps the water from becoming stale. Asthetic; The flowing water will create an optically pleasing display while creating ever changing patterns as the sunlight shines through the streams and plays on the floor.

**b. Rodriguez Well**

Once we identify suitable ice sheets for mining, we can employ a Rodriguez Well developed by the US Army and Cold Regions Research and Engineering Laboratory. This involves drilling through the sediment layer to the pure ice sheet below and creating a melting reservoir under the surface. One of the greatest challenges to this method is transporting the melted water back to the colony or propellant depot in sub-freezing temperatures. Heated lines or tanks require energy, however buried pipe may encounter a similar issue. This is also a task which can be accomplished by robots prior to human arrival. The Rodriguez Well typically takes an extended amount of time to develop a sustained reservoir so early site preparation will be advantageous. Heat from nuclear fission generators may also be useful for this application.

**c. Water Ice Processing**

Water-ice processing will be critical for drinking water. Allowing for sediment and other contaminating factors, it will take approximately 1.5 m3 of water-ice for every 1,000 liters of clean drinking water. If we’re able to consistently excavate below the sediment layer we may have access to a purer form of water-ice requiring much less processing. It’s possible to let the Martian atmosphere do some of the work for us by sublimation. The ice will turn directly to vapor which can then be captured and liquified. Ice will be transported to the water processing facility where it will be filtered and combined with recycled water before transferring into storage tanks for use in the colony.

**d. Internal Condensers.**

36% (ave.) of water consumed by humans is lost through breathing and perspiration. Condensers in the air conditioning system can add or remove necessary moisture and recycle the excess back in the water filtration systems and into reservoirs.

**Food Requirements**

An average, active human requires an average of 2kg of food per day which is approximately 745kg per earth year.

**a. Import**

The colony will import and maintain initial stock of rations and supplements while working to augment food stores through vertical farming techniques. Freeze-dried foods are the best option for preserved nutritional value, storage, and transport weight. (Example here: my personal favorite Mountain House Meals from Oregon Freeze Dry Inc). This gives us a wider menu of foods with greater long-term storability Food stores should be packaged in radiation shielded containers and cold stored to avoid nutrient degradation. Freeze dried foods also weigh 83% less than their fully constituted version. An average, active person requiring 2kg of food per day would only require 124kg per year in freeze dried meals.

| | **Needs (kg)** | **(lbs)** | **Needs (cals)** | **Freeze Dry wt (kg)** | **Freeze Dry wt (lbs)** |
|---|---|---|---|---|---|
| **Meal** | 0.68 | 1.50 | 833 | 0.11 | 0.25 |
| **Daily** | 2.04 | 4.50 | 2,500 | 0.34 | 0.75 |
| **Annual** | 745.02 | 1,642.50 | 912,500 | 124.17 | 273.75 |
| | | | | | |
| **1000 pax** | 745,025 | 1,642,500 | 912,500,000 | 124,170 | 273,750 |
| | | | | | |
| **Contingency*** | 750,000 | | 912,500,000 | 125,000 | |

*It's critical to build up a contingency stockpile of 1yr minimum in addition to principle 1yr supply. Food may be rationed to last longer in extreme circumstances, but the colony should be able to sustain in case of any interruption in resupply flights.

**b. Farming**

Requirements. To fully sustain a colony of 1,000 people, farming facilities will need to produce an average of 2,000kg of food per day. If 1 acre of farming space (4,047m$^2$) produces an average of 13.25kg of food daily it will require 75.5acres of growing area to produce the minimum about of food.

Greenhouses. If we consider vertical farming with each growing shelf comprised of 10 levels, that will still require farming space of 7.5 acres or 30,351m$^2$. For diversification and redundancy, the colony should create 10 greenhouses at approximately 3,000 m$^2$ each.

Inflatable spheres on the surface would be ideal for farming, especially if they can be built along a steep cliff or embankment. Surface domes will be a difficult due to pressurization and structural factors, but a spherical shape is ideal. A central column with a lift and stairs can provide access and successive decks along the inside of the sphere will provide growing space and exposure to natural sunlight. Lenses could be engineered in outer surface of the These would be easy to install, and multiple farming spheres could be operating in a very short time. In our design the service column in the center would continue underground where waste from the plants is used in mushroom farms which require no sunlight and produce a protein and nutrient rich food source.

Introduction of genetically optimized, high-yield plants, fish, and other food sources will dramatically reduce the burden on farming.

Hydroponic and enriched root spray techniques already perfected on Earth with be used in mass-production of fast growing, nutrient-rich plants. Leafy greens with fast growing cycles such as kale and spinach along with potatoes and other hearty, nutrient-dense foods.

Plants requiring soil will be grown in bins while leafy greens and others will be grown in trays and stacked in vertical shelves. Enriched root spray will likely be the most water-efficient process and allow the greatest control over plant growth factors.

In subterranean and low light areas LEDs will provide simulated sunlight. All the necessary parts may be created by 3D printer or extruder.

Individuals will also maintain a modest number of plants in their living spaces primarily for oxygen production, air cleaning, and psychological benefits. Plants shouldn't be confined to the farming spaces but should be intentionally distributed around the public and private areas of the colony.

**c. Animal Protein**

Eventually moving to a more robust hydroponic systems would allow for the introduction of fish and crustaceans. Large quantities of eggs could be packaged and transported from earth to be grown in the water troughs of a hydroponic system. Waste from the plants will help to feed the fish and nitrates from the fish will feed the plants.

Chickens and rabbits should also be considered as we expand food sources. Eggs are nutrient dense, chickens are relatively low-maintenance, and their droppings are excellent fertilizer. Likewise, both are hardy and have a relatively quick reproductive cycle.

**d. Soil Processing**

Facilities will process Martian soil with solid waste and other necessary nitrogen sources to create usable compost for horticulture applications. Soil processing facilities will process Martian soil with solid waste and other necessary nitrogen sources to create usable compost for horticulture applications Martian soil will be screened for usability and any toxic materials will be removed. Evidence suggests perchlorates are common in the Martian soil. Processing will seek to separate chlorine and oxygen molecules and remove toxins from the soil. Meanwhile waste from the colony will be processed to remove any harmful pathogens via enzyme processing or exposure to UV radiation. The two will then be blended together and transferred to containers for transport. Once processed, the soil will be stored underground until needed.

**e. Processing**

Food processing facilities will be adjacent to the farming facilities for ease of packaging, transport, and distribution.

**f. Shops**

As survival requirements are exceeded, fresh produce should be sold in convenience stores in the colony. This will provide Martian citizens a feeling of choice and the appearance of abundance within the colony.

**g. Synthetic**

A number of manufactured and 3D printable foods have been developed which use plant proteins to simulate meat dishes. These will be great options to expand menu options and provide greater variety to colonists.

**h. Supplements**

Dietary supplements will be necessary for complete nutritional health. Proteins, aminos, and Mars-specific nutritional requirements will be imported until more robust food production systems can sustain.

**i. Recreational Foods**

It's worth considering beyond survival requirements that the colony should produce foods and beverages which are enjoyable. Culinary experts will find ways to repurpose or recombine existing resources to create interesting texture and flavor pallets. Entrees, desserts, and tasty beverages are well within the realm of possibility. It might be possible to brew beer, ferment kombucha, or distill gin.

**HEAT**

**a. Direct**

Heat from electric-powered ceramic heaters throughout the colony air ducts. Ceramics may be produced on-planet if high efficiency methods can be implemented (fuels created on site, microwave fusion, etc)

**b. Recycle**

Launchpad may be fitted with a heat exchange system to pump super-heated liquids back to habitat. Radiant heat captured from electronics, batteries, and solar pumped through air system to maintain temperature.

**c. Insulation**

Protection against heat loss is key. Traditional insulation may not be feasible so using air-filled plastic film, and hexagonal wall structures should create empty spaces to buffer and maintain temperature.

## WASTE MANAGEMENT

**a. Human**
All biological waste will be recycled. Liquid waste will be treated and recycled while solid waste will be diverted to soil processing plant.

**b. Other**
All containers, products, etc will adhere to strict materials standards so as to be easily recycled. Plastics should be primary material.

## MEDICAL

All colonists will attend a basic medical course prior to arriving on planet. High preference may be given to medical professionals and specialists who also have advanced medical training (EMT-B and above).

One primary hospital and one clinic will be located on opposite sides of the colony for safety and to better care for colonists. All fitness and medical facilities will be co-located to provide the best overall care and monitoring of colonist health.

Nutritionists are focused on maintaining optimal nutrition to include regular testing, monitoring, and reports to Earth.

Fitness will be critical in the low gravity environment of Mars. Fitness facilities, trainers, regular exercise classes, and activities will be a regular part of the Martian lifestyle.

## HABITAT

Symmetry and strategic distribution of critical infrastructure within the colony will mitigate the chances of a single point of failure. Redundant systems and contingency stores will play a critical role in survivability.

**a. Subterranean Domes** (central communal)
In the initial stages of colonization, it will be ideal to build underground. Surface structures require more materials due to pressurization and are vulnerable to the effects of environmental hazards. However, it's not reasonable to go to Mars just to spend our entire time underground. If we're able to build into the side of a cliff or steep embankment we may have greater ability to see out while allowing the dirt and rock above to help contain our pressurized living space.

Living quarters and critical infrastructure should be built below ground. This allows for much larger spaces and protection. Greenhouses and observation decks may be built on the surface providing access to all colonists.

Structures will be built near the surface using telerobotic diggers to excavate. Aggregate will be transferred to a processing unit which is integrated with a 3d printer. Then a cylindrical dome habitat is inflated, supported by a tube frame structure. Mars aggregate then backfilled over dome as sulfur concrete. This provides insulation, protection from radiation and debris, and guards against rapid depressurization. Once initial subterranean structures are established, we'll be able to continue building downward without significant concerns of exposure and depressurization. Larger structures may be built by laying successive, fused layers of sulfur concrete over shaped aggregate. Martian soil will be excavated and piled in the desired shape, the building shell will be printed over the top of the regolith, then once the concrete has hardened robotic excavators remove the aggregate from underneath and backfill over the top of the structure.

Living quarters may be multiple levels below the surface with a central staircase and regional lifts. Each level will see hallways extend in 4 directions with multiple housing units along each hallway

**b. Dust Containment**

Decontamination areas in Airlocks and entry rooms with dust filtration, vacuum, and collection to minimize dust in living environments

Plants placed throughout corridors and living spaces to actively control dust and produce oxygen as well as providing a positive psychological benefit.

**FUEL**

**a. Propellant Depot**

One of the most mission-critical components of the colony will be the propellant depot to reliably refuel transports. The initial depot may be emplaced remotely and begin producing stores of liquid methane and liquid oxygen before humans ever set foot on the red planet. This depot will consist of two 2,500kg Sabatier reactors and storage tanks

**b. Sabatier Process**

Utilizing mined water-ice we will split the hydrogen and oxygen molecules and recombine with carbon from atmospheric CO2 via Sabatier process. Hydrogen, Oxygen, and Methane will be collected and liquified, then transferred into storage tanks for propellant.

**c. Atmospheric Methane**

It should be noted that trace amounts of methane have been detected in the Martianatmosphere by multiple sources (*NASA)*.

**COMMUNICATIONS**

**a. Long Distance/Interplanetary**

Satellite. It will be necessary to launch a small constellation of geosynchronous satellites from Mars orbit. The SpaceX Starlink satellites may be uniquely suited for this purpose. Satellites should be primarily communication systems, however they may be fitted with additional sensors to monitor planetary conditions. This system will provide continuous communications with earth and spacecraft in orbit without requiring direct LOS. It will also extend communications for remote mining sites, research stations, and other expeditions beyond the immediate colony area.

Data Gateway. Additionally, consideration should be given to a data gateway; either an orbital satellite such as the Mars Telecommunications Orbiter (MTO) proposed by NASA in 2009 or on one of Mars' moons. Data would be transmitted to the gateway via satcom and then to earth via high-speed/high-bandwidth laser. This should allow for greater speeds and data throughput while mitigating signal loss and data corruption due to atmospheric conditions.

**b. Local**

Cell Network. The current cell phone technology on Earth has been extensively developed and refined. Rather than reinvent the wheel, the Mars colony will use localized cell networks for the majority of communications. Rovers and EVA suits can be fitted with Bluetooth or wired connectivity, app development is very robust, and power consumption is relatively low compared to other communication systems currently available.

Every colonist will carry a smart phone and wear a smart watch. This will massively reduce any training or development time, provide instant communication between colonists, tracking capabilities, health monitoring, and endless potential for apps and efficiency. Tinder on Mars? Swipe right.

Multiple surface towers provide redundancy and geolocation with their own independent power sources (solar fields/batteries), while habitat will contain dispersed 'hotspots' which reduce power usage and signal deadzones.

Additionally, self-tuning HF radios should be used for backup and emergency communications. Mars has sufficient ionosphere to bounce a signal beyond line of sight. Similar radios currently exist on earth, transmit up to 10,000 miles, and are highly energy efficient.

**c. Internet**

A data network or the "Mars Wide Web" (MWW.colony.mars/ms/mar). Due to limited throughput from earth the internet on Mars will initially consist of a massive database connected to the local cell and wifi network. While science and operational data will always have priority, text messaging, and media updates will be uploaded from earth to the MWW. This will significantly boost morale by allowing colonists to maintain connection to loved ones, events, and popular culture back on Earth. The common connectivity will allow scientists and engineers to connect and collaborate on a wider scale.

Connectivity to earth will also create greater interest in the Mars colony. Awareness gathers support, and a steady flow of decentralized information will keep the pioneers at the front of everyone's mind. No advertising campaign could generate so much interest.

MWW Large RAID data storage with massive data libraries (technical/social/entertainment) creates an internet on Mars, accessible by all colonists via smartphone and wifi. Phones optimized for apps (available on the MWW) and power usage. Localized storage of common data will also reduce the bandwidth load on the interplanetary connection.

Updates Regular mass data dumps sent on every transport flight from Earth continually update Mars internet with content from home

**MATERIALS/PRODUCTION**

**a. Recycle/Reuse**

All materials should adhere to a strict guideline for reuse and recycling. Special attention will be paid to components for production equipment on initial supply flights.

**b. Local Production**

Energy efficient versions of 3D printers and extruders should be sufficient to produce the majority of parts needed in the colony. Recycling systems will produce additional raw material for production reducing the need for import materials

**c. Fabrics/Textiles**

Until a way is found for plant-based fibers to be mass produced, synthetic materials will have to be produced from ethylene and other plastics-based materials. 3D printers can produce most of the necessary production equipment, however some of the raw materials may still need to be imported from Earth. As technology advances, we may see more carbon-based materials and fabrics. A company which could establish themselves at the Mars Colony for such research and production could be relatively lucrative.

**d. Transparent Aluminum**

The raw elements are available on Mars to produce aluminum oxynitride (transparent aluminum) with mars-sourced aluminum, nitrogen, and oxygen. Structurally more secure at 20% mass than glass or plexiglass and does not block certain light spectrum as with glass. This material would be extremely useful for building domes and other pressurized structures.

**e. Metals Processing**

Resources for producing metals such as aluminum, titanium, and steel alloys are readily available on the Martian surface. As energy efficient methods are employed, the production of structural materials will be a major advantage.

Methods for steel-making would produce high grade steel and also produce oxygen, but may require massive amounts of energy. Scientists at Cambridge University announced a method for producing pure titanium directly from titanium dioxide, and the elements for making other important, and possibly new, metals are already on the planet.

## THE COLONY

The Mars colony will be laid out in a quartered format. A main corridor will run the length of the colony from the Port to a spacious main square in the center and on. The majority of structures in the colony will be subterranean even if only a few feet below the surface. Working areas and those needing access to the surface will occupy upper level domes while living spaces and recreational facilities will be located levels below ground. All levels will tie into the central corridor with additional redundant passageways in case of emergency.

Image Credit: Mars City Design

Each section of passageway will be fitted with a sealable hatch allowing for isolation of any part of the habitat in the event of depressurization, fire, or structural failure. Emergency stations dispersed throughout the colony will be fitted with oxygen, first aid, and containment apparatus.

### a. Society

Heavy focus will be placed on work-life balance and the importance of innovation, of working smart, not hard. Colony goals will be ambitious but tempered by a passion for enjoying the journey on this new planet. New colonists will be paired with compatible, experienced residents providing a social link and source of reference. New colonist will also be paired with a professional mentor; the more experienced teach those with less experience. In this way the flow of knowledge is promoted, and all colonist feel a sense of connectedness.

**b. Child Care**

In a colony of 1,000 it's unlikely to have more than a few children ...initially. Any long-term settlement of humans will very quickly have pregnant humans and therefore babies. The colony will have a safe child care area during the day which will focus on early development of social skills and critical reasoning. This child care center will also employ at least one Pediatrician to maintain continual monitoring of child health and development.

**c. Schools**

The curriculum will be a combination of lessons of history, engineering & science, and entrepreneurial spirit. Students will have access to a veritable maker's space of design and engineering equipment where their learning and their imaginations can interact hands-on with real world results.

Additionally, students of all ages will take advantage of college and vocational courses supported by universities on Earth through the **University of Mars at Tharsis**. Online and group coursework and lectures will be accessible via the MWW. Citizens will have access to everything from engineering, computer science, and advanced medical training to horticulture, dance classes, cooking classes and more.

**d. Social/Psychological**

Central Fireplace. It's been found that remote expeditions and people stranded in survival situations fare better when there's a central focal point in the camp. This keep morale high and strengthens the sense of community.

This colony must have a primary central 'town square'. The space could be a multipurpose meeting space, activity center, farming, movie theater, restaurants, etc.

Facilities are divided and colony will be symmetrical. A central, main corridor will lead to a central square or central dome. the majority of the colony will be underground for protection and logistics. However, some surface domes will be necessary both functionally and psychologically. Specially constructed observation areas will afford the colonists a view of the outside world.

Dress-up social. While exploring remote areas such as those in Africa, the British Army would require their men put on their dress uniform and have dinner in the evenings. They found this kept the men more civilized and 'prevented them from going wild'. A weekly or monthly social event would significantly boost the morale and camaraderie of the colonists.

Lighting. UV lighting, IR saunas, and day/night cycles to prevent Seasonal Affectedness Disorder, provide Vitmin D, and keep everyone happy.

**e. Fitness**
Fitness will be a critical component of colony life. Maintaining muscle and cardiovascular health in Martian gravity will require daily effort. Resistance machines, cardio, and specialty exercise will be available as well as organized team sports, group classes, yoga, and more. A wide array of multipurpose exercises and engaging programs to keep fitness and motivation high. As mentioned earlier, these facilities will be located near the medical areas to facilitate health monitoring and a ‘preventative health-care’ approach. Fitness trainers, nutritionists, and physical therapists will also be on staff at the exercise facilities for a holistic approach to fitness and well-being.

Each colonist will have a personal health profile where they can monitor fitness data, chart their workouts, and set personal goals. This profile will also allow health care personnel to track health and fitness trends throughout the population. All exercise equipment will be fitted with power generation modules to make use of the kinetic energy expended. All moving exercise equipment will be fitted with devices to generate electricity. Colonist will have the option of charging personal electronics via exercise.

All moving exercise equipment will be fitted with devices to generate electricity. Colonist will have the option of charging personal electronics via exercise.

**f. Social**
A vibrant social life will be an important component of the Martian colony. Time to interact with fellow colonist, particularly those working in different sections, will foster human connection between residents and provide a psychological boost. Access to social media and ease of communication via smartphone and the MWW will help to maintain connectivity. Well-organized social events will relieve tensions of living in the Martian habitat so far from Earth and ease the stresses of the common work day. Group gatherings and interaction will remind us what it is to be human and why we came here in the first place. Work-life balance will be a high priority, and the goal will always be to build a strong and vibrant community.

**g. Currency & Banking**
As the colony grows it will become necessary to pay for goods and services particularly as we push toward economic viability. Transactions between Earth and Mars, and commerce will require a cohesive payment system. Especially as contract workers arrive, asteroid mining ramps up, real estate sales occur, and further expansion of the colony there must be a reliable financial system. The logical choice is to use a digital currency on Mars. Printing or minting physical currency is both archaic and consumes resources, but a blockchain crypto-currency such as Dash, Bitcoin, or Ethereum will likely be the best options for utility and for broader banking applications. Additionally, the use of a blockchain-

based digital currency allows for interplanetary transactions, portability, and secure transfers.

## POLITICAL/ORGANIZATIONAL

### a. Jurisdiction

It's possible that legal jurisdiction may fall under the laws of the country with the greatest interest in the colony or, possibly, presided by the UN. The legal agreements we use to govern the ISS won't work for a Mars colony however. Those agreements give each country jurisdiction over it's nationals and components of the station registered by that country. While Mars may have laboratories or other facilities dedicated to the work of one particular country back on Earth, the common spaces and overall interdependence of the colony will not hold up under a fractured legal framework. The end result will be political turmoil. With an eye on history, a move toward independence should be anticipated. Sooner or later, a self-sustaining colony will likely push toward self-governance.

### b. Parliament

One potential solution is representatives from each country with a vested interest in the colony form a council. Each country maintains an ambassador on the Mars Colony who, collectively, act as an independent parliament governing the Colony. Government should maintain broad ties to Earth as humans and avoid any sort of isolated, cellular structure which could lead to opposing sides. Council members will be able to represent their home government's interests while allowing the colony to maintain a relatively unfettered sort of independence. Politicians will inevitably attempt to hijack the bold and pioneering spirit for personal or political gain. This representative council should alleviate some of those issues.

### c. Community Structure

#### i. Leadership

Each fundamental section of the colony will have a primary representative followed by top-line supervisors who may or may not be chosen by that representative. Lead representatives answer to the Parliament as a whole. Supervisors will be independently responsible for cellular 'teams'. Teams should be creative and functional, able to act independently, requiring only minimal guidance. Ultimately there must be one coordinating director, but the majority of operational decisions can be made by section chiefs. A physician will head the medical, a botanist should head the farming section, a Green Beret should handle security, and engineers will head the propellant depot, waste processing, and construction teams, and so on. While each section will have a supervisor, their responsibility should be solely focused on strategic coordination. If there are 2 water-ice processing facilities, each team should be able to function independently with minimal guidance.

#### ii. Security

This is the new frontier. Every old west town has a sheriff if for no other reason than to maintain order and Mars will be no different. As long as there are humans there will be disagreements. Moreover, security personnel will be present to ensure the safety of colonists, perform rescue operations, and augment medical personnel when necessary. (They also fight off the Klingons should negotiations break down and things get tense.) US Army Green Berets are uniquely suited to this position. As warrior diplomats they tend to be exceptionally physically fit and possess above-average calm and intelligence. They are adaptable and are accustomed to working in the most hostile places Earth has to offer while keeping a great, albeit warped, sense of humor. Security personnel shall possess above average maturity and judgment. Mediating disagreements, and maintaining long-term peace will be utmost priority, superceded only by a commitment to the safety of all colonists.

**ECONOMIC**

Mars must develop value beyond science and exploration to be sustainable. The production of fuels and energy provide the muscle to make use of the vast mineral resources which are so accessible on Mars.

**a. Advertising**

If companies such as Google and Facebook have taught us anything, it's that massive revenues can be generated purely by drawing consumer attention. The advertising dollars these companies receive annually outpaces the GDP of many countries. We will set up a company devoted to everything Mars. 24hr streaming social, entertainment, economic, and scientific content from the red planet will inspire and captivated every mind on earth. Properly positioned and ethical marketing will take advantage of this massive opportunity for the financial benefit of the Martian colony. Consider how many people watched the first moon landing in 1969; licensing of content and ad placement will generate significant revenue and fan the flames of inspiration by allowing everyone on Earth to watch history unfold in real time.

**b. Data/Intellectual**

The potential for advancements in chemistry, engineering, and data systems development are extensive. Corporate research facilities on Mars can develop new methods and new products to be implemented in mass production on Earth.

**c. Ceramics**

Mars aggregate could be extremely useful in the production of advanced ceramics for use in space vehicles and right here at home, on Mars.

**d. Rocket production**

I propose rocket companies like SpaceX establish rocket factories on Mars. Raw materials are available and the reduced gravity aids in the production of large-scale space vehicles. The propellent depots provide fuel and launches from the Martian surface will be less expensive than those on earth. Engineering firms will locate offices here to develop new machinery and spacecraft. This colony could very well be the 'ship yards' of the new age.

**e. Mineral Exports**

Companies interested in asteroid mining should make significant investments into infrastructure and production plants. As Mars Transport flights unload colonists and supplies, it seems prudent not to send them home empty. Valuable metals and rare minerals can be transported back to Earth for approximately $200/kg. One cargo load of precious metals would be worth billions. This should be enticing enough for an earth company to invest in propellant plants, launch systems, telerobotic miners, and other infrastructure to take advantage of Mars-based exports.

**f. Real Estate**

The potential for the real estate market is huge. As viable transport, private contracting, and massive interest grow, there will be large numbers of humans waiting to get to Mars.

Register. First, a method of record-keeping is necessary and a blockchain property record would be ideal. Blockchain record systems for property and identity are already in full-scale use by various countries on Earth. The only question is, "How do I know what I'm registering?"

As we have more detailed imagery of the planet, better surveys, and established transport to Mars we will be able to register a land claim. The claim will be entered into the Mar Land Registry and you may develop that land. The highest value will likely go to land which is already developed; condominiums, farm spheres, and other habitable areas. An enormous market will emerge for land development, from telerobotic excavators, engineers, and construction, to agency, furnishings, and supplies.

## CONCLUSION

The Mars colony concept is not only viable, it is imperative for our growth as a species. The impact to human society on earth and our expanded understanding of the universe is immeasurable. We have the technology, ingenuity, and resolve to not only survive on the red planet, but to thrive and continue beyond. Not only is there endless opportunity for science, but massive opportunity for private enterprise. Mars is not just economically viable, but in the coming years will show herself to be a golden opportunity. Humanity is on the brink of its next great leap, the threshold of the next New World. What if Christopher Columbus had decided to stay home? Establishment of a thriving Colony on Mars will change the world on the order that colonizing the Americas changed the 'known' world to what it is now.

We can and we must be bold.

# 22: TWARDOWSKY COLONY

**Amanda Solaniuk, Anna Wojcik, Joanna Kuzma, Natalia Cwilichowska, Katarzyna Lis, Slawek Malkowski, Dariusz Szczotkowski, Szymon Loj, Orest Savystskyi, Dominik Liskiewicz, Wojciech Fikus, Jakub Nalewaj, Anna Jurga, Leszek Orzechowski, Bartosz Drozd, Pawel Gorniak, Krzysztof Ratajczak, Pawel Piszko, Maciej Piorun**
Wroclaw University of Science and Technology
Poland
orzechleszek@gmail.com

## 1. LOCATION

Primary restrictions for a Mars colony location, were latitude and elevation limits. Low elevation minimizes the challenge of landing large mass- es on Mars, and reduces the radiation exposure. Latitude is restricted due the launches and orbital mechanics. Furthermore the level of heavy metals in the soil and strong radiation make a location unsuitable for establishment. Numerous (>40) potential locations sites has been considered. Each one of them included a 100 km radius circle Exploration Zone (EZ), with centralized landing site, which defines its boundaries.

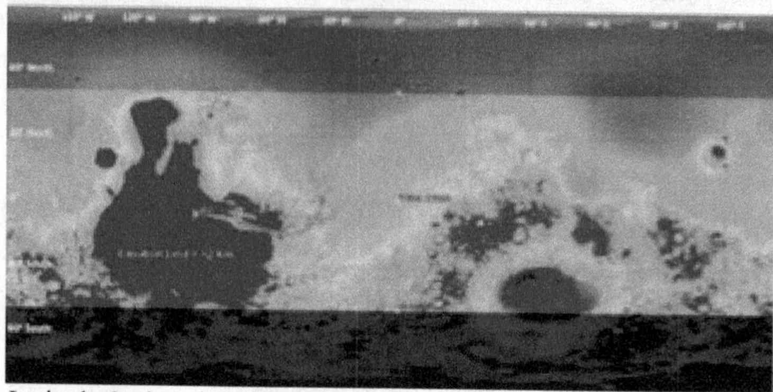

Fig.1. Latitude & elevation restriction for Mars settlement location

Figure 2 shows the Main Areas contained within Exploration Zone (black circle):

1. Region of scientific interest (red circles)
2. Habitation Area (yellow dot)
3. Landing Zone (blue dot)
   (1) Primary
   (2) Secondary
4. Resources regions (green circles)
   (1) In-Situ Resource Utilization Zone
   (2) Mining Zone

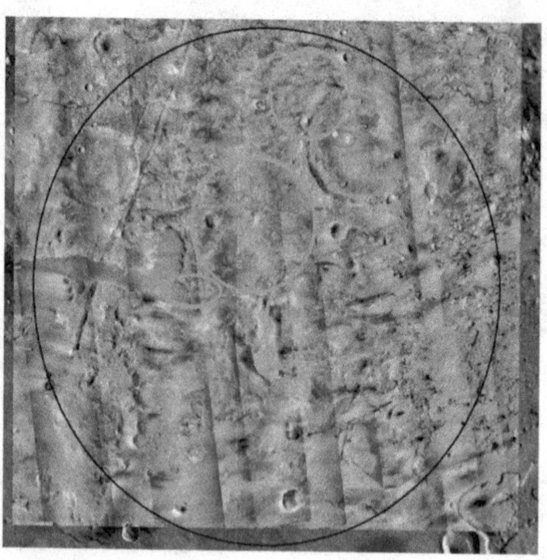

Fig.2. Detailed description of specific areas of interest in Jezero Creater and Syrtis

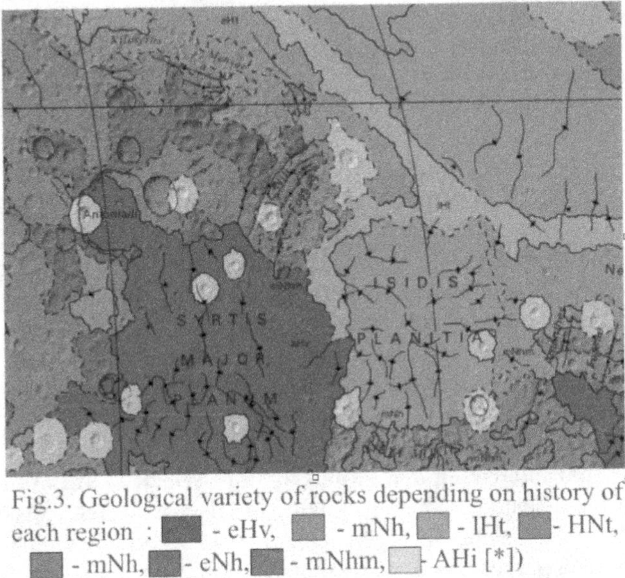


Fig.3. Geological variety of rocks depending on history of each region : - eHv, - mNh, - lHt, - HNt, - mNh, - eNh, - mNhm, - AHi [*])

After the in-depth analysis of all potential regions accessible data, Jezero Crater was chosen. It is a well-studied region with an extensive literature supporting an exceptional geological formations with astrobiological significance. It is located, between early Hesperian volcanic units (eHv) and late Hesperian lowlands (lHl) from the south, highland units (mNhm, mNh) from northwest and transition units (eHt, lHt) from northeast, on Hesperian and Noachian transition unit (HNt). Geological diversity increase the probability of finding minerals and metallic ores undetected yet.

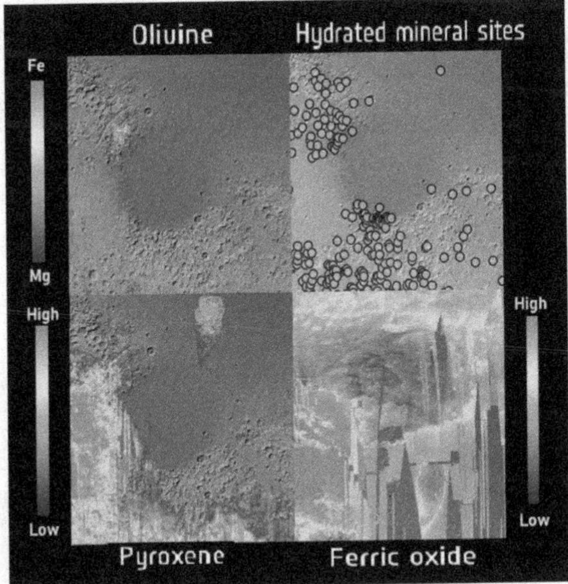


Fig.4. Location & quantity of specific material necessary for Mars settlement

The additional advantage is easily found there - natural resources. The most important minerals detected in surroundings are pyroxene, magnesium and iron rich olivine, amphiboles, feldspar, sulfates, carbonates and phosphate. Occurrences of hydrated minerals were observed north and south of selected location. In the mentioned there is a high probability of encountering ice; ferric oxide is a common compound across the area.

Additional reason for selecting this region is the possibility to acquire large amount of data from Mars 2020 mission, for which both Syrtis and Jezero regions-marked as excellent in terms of suitability - are supposed to be main areas of analysis. The rover will not only seek signs of potential habitability – and past microbial life – but will also collect rock and soil samples and store them in a cache on the planet's surface, which would become a huge burden to retrieve back to Earth, without a Mars colony settled nearby.

**1.1 - The parametric aided analysis of selected Habitation Area**
We utilise a multi-scale parametric design and analysis to aid decision-making process. This gives us the ability to create custom tools with high level of flexibility. For this purpose we use Grasshopper 3D software.

Grasshopper's Ladybug add-on component is used to calculate and visualize the number of hours of direct sunlight received by localization mesh during the Mars year. Sun angle data is extracted from Mars Trek and then converted from azimuth and elevation degrees to vectors. Data contains hourly sampling on particular Sol each month through the Martian year. The geometric center of each face of the terrain model mesh is used as a test point. The total quantity of connected vectors is 290.

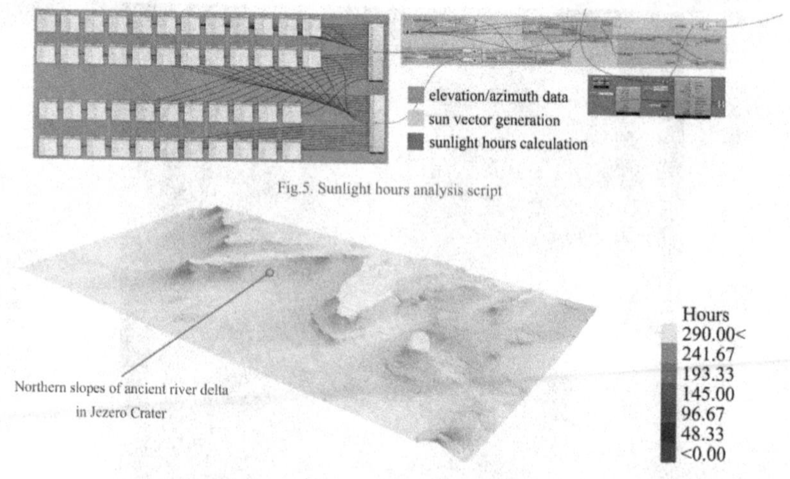


Fig.5. Sunlight hours analysis script

Fig.6. Sunlight hours analysis output. A colored mesh representing the number of hours of direct sunlight received by localization, out of the total number of connected sun vectors

Final localization conclusions are based on analysis of existing data and parametric calculations:

- Based on geological and geographical criteria Jezero crater was selected as Habitation Area,
- Jezero crater being the lowest point in the area is preferable localization of the Space Port
- Ancient river delta in Jezero Crater with its landscape allows for easiest access to nearby plateau rich in Resources Regions of Interest and Science Regions of Interests as well as access to the Space Port -
- Based on MarsTrek data analysis northern slopes of ancient river delta in Jezero crater receives least sunlight hours, therefore, less potential solar flares exposure. Usage the landform of the delta to our advantage is an example of landscape-based radiation protection for the colony.
- Locating residential area on northern slopes allows for verticality of architecture and panoramic views of the landscape without sacrifice in radiation protection.
- Verticality of architecture allows for reflected natural martian sunlight being used in a residential area.

**Verdict**: The first colony will be located northern slopes of ancient river delta and will be called Twardowsky, in the name of the sorcerer Jan Twardowski, a character derived from the old Polish folklore legend who was the first polish space traveler.

Fig.7. Twardowsky - Artist's vision

## 2. ARCHITECTURE

Following the decision about the colony's location, the position of Space Port is determined. To prevent the possibility of the harmful impact caused by rocket accident during the landing, the 3-5 km safe zone radius is set. A very flat area in the south-east direction from the colony within the allowed distance was considered best from multiple possible choices. Finally, it was decided that Space Port will be located 5km south east from the base for extra precaution. It will consist of 5 landing pads as well as an administrative building and a warehouse. The fuel will be produced and transported to the Space Port from the colony.

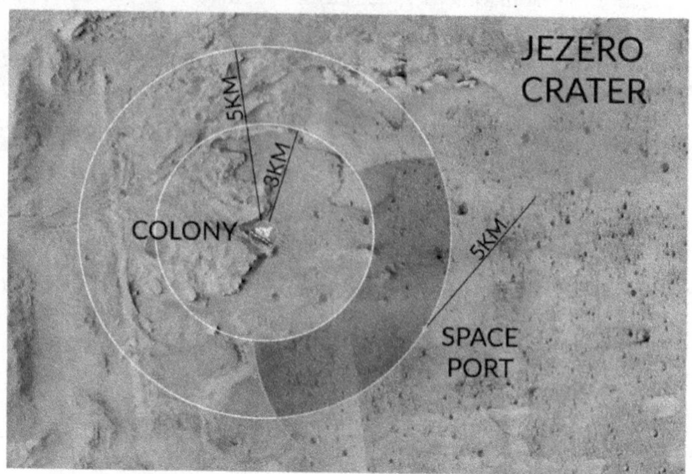

Fig.8. Twardowsky - chosen localisation

The next designing step was to determine how to provide the colony with as much natural light as possible by using a system of 6000 mirrors. The light reflected by mirrors will be a compensation to the artificial sources of light used in the base and will be essential to the growth of the plants.

Fig.9. Mirror array overlapping analysis

The number and geometry of mirrors arrays are derived from parametric design. Mirror geometry is generated on the selected area, concentrically around chosen BIOMe. Orientation of each mirror is generated by calculation of plane normal from sunlight and reflection vectors. Overlapping is analyzed by checking if rays, reflected from mirror vertices and center, aren't blocked by other mirrors. Then geometry is colored accordingly to the amount of collisions. The distance between rows and mirrors inside each row can be adjusted till the results of analysis is acceptable. The graph depicted on fig. 10 represents the number and type of scripts used.

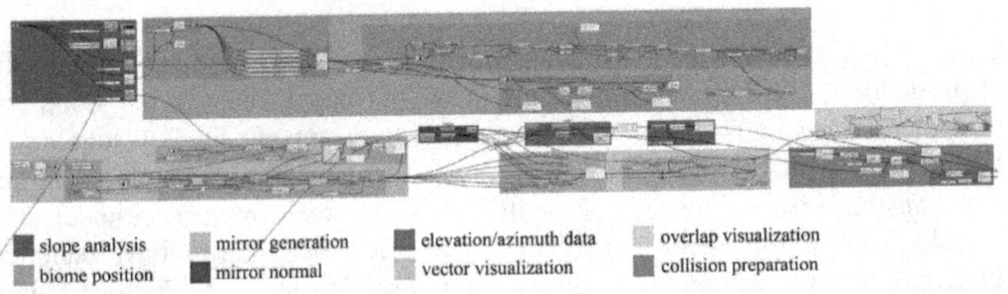


Fig.10. Mirror array overlapping analysis and BIOMs placement scripts

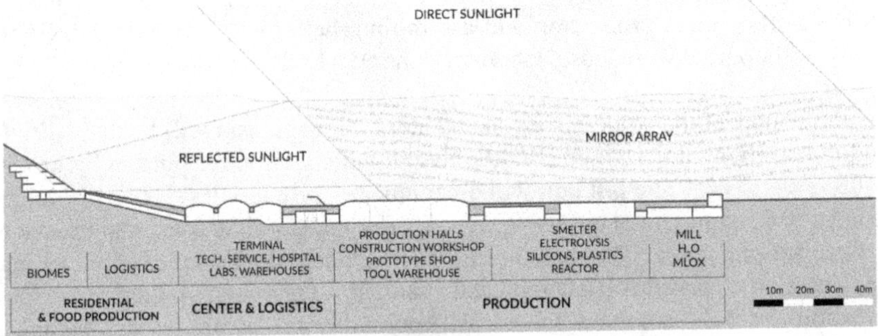


Fig..11 Twardowsky - section

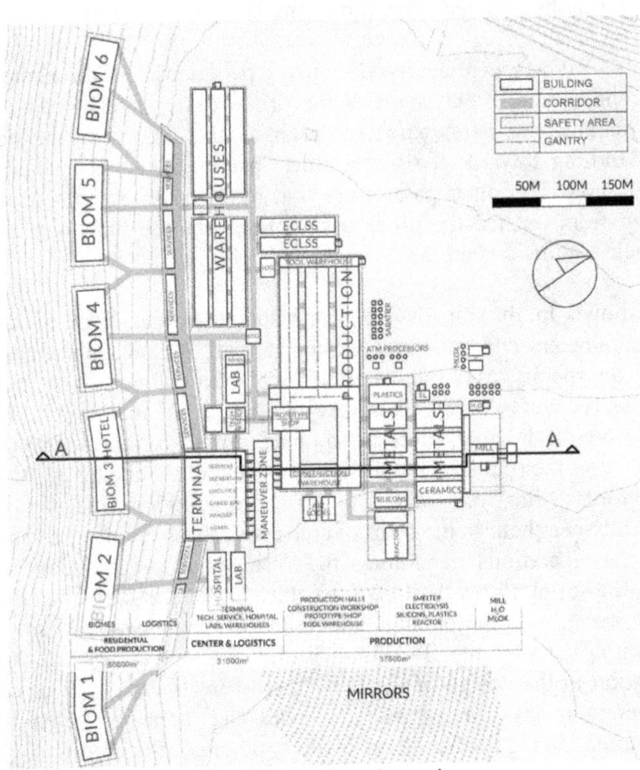


Fig.12 Twardowsky - plan

The colony's structure will be divided into three main sections: residential with food production, logistics and production. The first part located in the slope will consist of 6 living units – biomes – each providing housing for 200 inhabitants and integrated with various food production structures. Next two partitions are situated on flat terrain. The logistics will include the main hangar, connected by air- locks with a maneuver zone, passengers' terminal, cargo bay, technical service and multiple warehouses. The producing section will have its own hangar and will involve all the production processes. The most crucial parts being: a mill for the regolith pre-processing, a smelter connected with silicon and plastics acquiring, three main production and assembling halls, a construction workshop, prototype shop, tool warehouse, laboratory and two ECLSS dedicated halls.

The population of the colony is estimated at 1100 people and will be divided into 6 individual living units called BIOMs, five of which will be populated by groups of 200 settlers each. The last unit will be unique because it will serve as a hotel for around 100 potential hotel guests as well as offer many public spaces and services for their entertainment. The structure of each standard BIOMs were divided into three sections: private (housing and sanitary facilities), public (services, recreation, kitchen, common rooms, offices and sport facilities) and technical maintenance spaces (technical rooms, airlocks) including food production related areas (ECLSS warehouses, technical rooms, planting seedlings etc.).

Housing units - BIOMs were designed by using parametric design methods, based on analysis and optimizations made at the urban design stage, from which the appropriate parameters were extracted, such as the length of the object, the direction of turning towards the world sides, positioning in the slope space of individual BIOMs. The other parameters that make up the whole structure, but those coming from outside the urban assumptions are the functions included in each unit, their number and areas.

The script shown in the Fig.13. takes the inputs in the form of the above-mentioned parameters and generates the appropriate function distribution in the object. We can specify any number of different functions, assign them to the appropriate story, and we can also specify their area and height. The script automatically breaks down all functions by placing them on the appropriate story, and also calculates their areas and dimensions in relation to the length of the entire BIOM structure. After the function has been divided, the script creates connections between them in the form of communication and shared spaces. After performing the operations generating the distribution of functions and their connections, the script places the entire structure in the urban space defined in the urban design stage, with the optimal orientation towards the world sides and the optimal opening angle of BIOMs wings. The advantage of using the parameterization in the design of the BIOM structure is that you can change the input parameters at any time or add new ones and, as a result, get the output geometry without having to change the entire structure of the object.

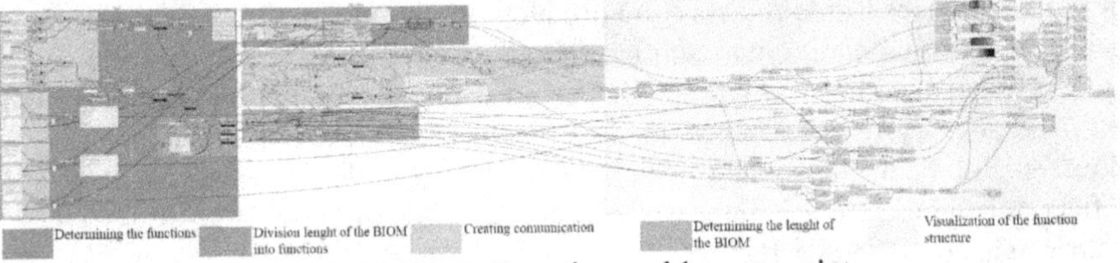


Fig.13. BIOMe's Function and layout scripts

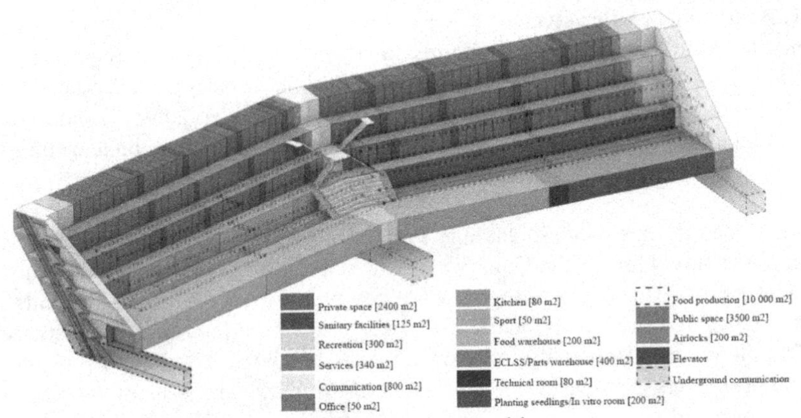


Fig.14. BIOMe: functions and layout

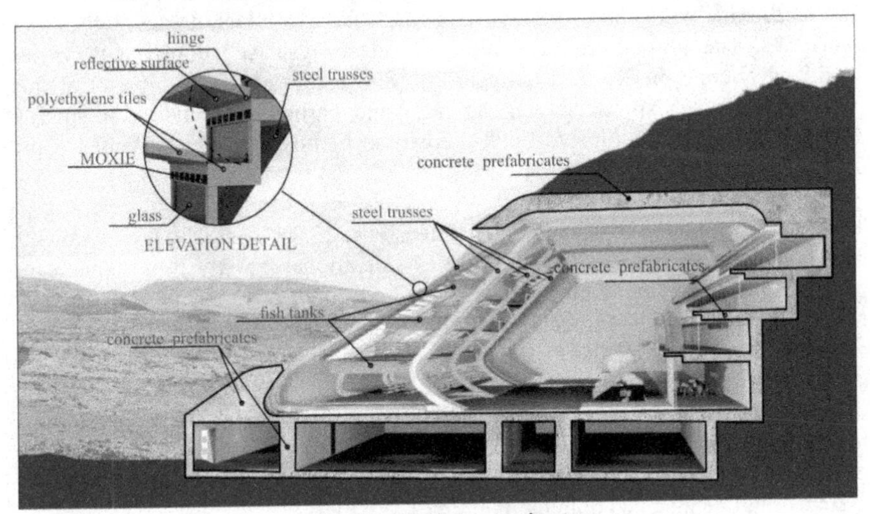


Fig.15. BIOMe: section

## 2.1 Structure

BIOMe main structure will be made of concrete prefabricates and steel trusses created in the construction workshop and assembled on site. Landmass excavated for BIOMe would serve as a raw material deposit for constructing its components.

Inside free spaces inside trusses food production with aquaponics water tanks will be placed creating additional radiation shielding. On the trusses external elevation will be placed with 25 cm polyethylene tiles radiation shielding together with MOXIE sub- system described in detail in the next section. Above windows, shutters will be installed that will both additionally shield from radiation and reflect mirrored sunlight into the colony. In case of danger i.e. dust storm, dethermalization or solar flare shutters could be closed.

## 3. RECLSS

### 3.1. Atmosphere subsystem

One of the most crucial life-sustaining challenges in the Mars Colony project is to mimic an Earth's atmosphere. To achieve this ultimate goal, one has to generate and maintain sufficient levels of nitrogen (78.08%), oxygen (20.95%) and other essential components of Earth's atmosphere in the same ratio and balance the pressure and temperature of the colony.

The issue is not trivial, taking into consideration the Mars atmosphere composition, volume of the Colony's BIOMes and the requirement of the constant generation of oxygen and nitrogen while getting rid of excess carbon dioxide and other unwanted gas components. To tackle issue the Massachusetts Institute of Technology and University of Copenhagen collaborated with NASA on the concept of Mars Oxygen ISRU Experiment or MOXIE. The system is aimed to be tested onboard the rover during Mars 2020 mission. This is a device capable of generating oxygen from $CO_2$, in the electrochemical-based catalytic process. Carbon dioxide will come from the Mars atmosphere of which it constitutes more than 95%. The process utilise solid oxide electrolysis cell doped with yttria-stabilized zirconia in elevated temperature (800 degrees C). The product of this process is oxygen (produced on cathode) and carbon monoxide (on anode). Afterwards one has to filter through the carbon monoxide using the outtake and pump oxygen into the colony's BIOMes.

The current prototype of MOXIE has dimensions: 23.9 x 23.9 x 30.9 cm, weights approximately 15 kg and has 300 W of operational power. It is capable of generating about 10 g of pure oxygen which is equivalent of 7.52 $dm^3$ in NTP (293.15 K, 1013 hPa). According to Earth's atmosphere, the $O_2$ level make 21% of volume so i.e. to fill the necessary amount with oxygen for 40 x 40 x 10 m (living quarter) MOXIE would take couple of months. However, the scientists are currently trying to drastically elevate the efficiency of the process and the next iterations of MOXIE could be incorporated on the dome of the colony in racks consisting of multiple units to maintain the sufficient level of oxygen. Multiple systems shall be installed onto the building's scaffolds.

The nitrogen will be generated in larger quantities while treating the biological waste in the Colony as described in the ECLSS chapters of this report. The pressure and temperature will be constantly con- trolled with a system utilising $O_2$, $CO_2$, CO and $N_2$ sensors and a set of pumps, gas-specific filters and heat exchangers for cleansing and equalizing the atmospheric conditions and to detect any anomalies.

**3.2. Water subsystem**

Water subsystem will distribute water in the water network developed in the Base, for crew purposes such as potable water and hygiene water. It is estimated that over 24 $kg \cdot crewmember^{-1} \cdot day^{-1}$ water will be used for human needs. As system will be working in closed loop its aim will be to recover water and nutrients from wastewater streams created in the Base. There will be three main wastewater streams: urine, condensate (as the crew condensate and transpiration water from biomass cultivation) and grey water. Hybrid wastewater processing is chosen for wastewater treatment. It includes application of both, physicochemical and biological treatment. Urine processing will start in residential units. There, the streams of urine and feces will be separated in the diversion toilet. Feces will be distributed to the solid waste subsystem and subjected to further processing. Urine pre-treatment is provided by monopersulfate oxidation and solid sulfuric acid. Then, urine will be combined with other two wastewater streams: grey water and condensate from crew. Combined wastewater streams will go to the biological treatment unit. Dilution of urine, which is nitrogen rich, prevents the process from being inhibited due to the high concentration of $NH_3$. In this stage nitrification and/or denitrification is performed and recovery of nutrients occurs. This stream could partially feed aquaponic cultivation of plants (as nitrogen in form of $NO_3$ is the most efficient to absorb by plants) . After biological treatment, wastewater will be sent to the unit of reverse osmosis. In this stage two streams will be created: permeate (treated wastewater) and brine (highly concentrated mixture). The final step will be catalytic oxidation and water supplementation with minerals. Water storage and suitable mineral components supplementation will be realized in pressure tank. Supplementation in water tank will be controlled automatically and supported by mechanical stirrer.

Water transpired from plants and cultivation bed in aquaponic system will be captured by rotor ex- changer. Water is condensed and returned back to the system. Whole water subsystem will be equipped with water quality and quantity monitoring (parameters such as: conductivity, temperature, pH, iodine level, TOC, odor control). System will provide microbiological control and mineral components analysis derived by conductivity measurements.

**3.3. Biomass production subsystem**

According to accepted standards amount of calories, which must be delivered per person is 3072 kcal per day. It means that the total amount of produced calories should be approximately 3,800,000 kcal a day. This amount can cover the daily demand and create a supply, that will allow the colony to grow, feed the tourists and help survive in the case of failure of any of the nutritional modules. About 40% of colonists' calorific demand will be provided by plants. In each residential module there will be 4 greenhouses each with a volume of 2,500 $m^2$. In order to provide a varied diet in greenhouses, 18 species of plants will be grown (Cabbage, Carrot, Chard, Celery, Green Onion, Lettuce, Onion, Pea, Radish, Red Beet, Rice, Soybean, Spinach, Strawberry, Sweet Potato, Tomato, Wheat, White Potato).

Plants will be divided into 4 greenhouses per BIOMe according to the length of the growth cycle and the possibility of feeding with salt water as a nutrient solution. The use of 4 greenhouses will be aimed at limiting damage in the case of a system failure or the appearance of plants disease which might damage crops. Plants will be grown in the aquaponic system. This system was chosen because of the smallest demand for nutrients supplied outside the system. Fish will provide another source of calories. 3 greenhouses in each living module will contain freshwater aquaponics, and one saltwater aquaponics. Beehives will be placed in the area of cultivation to provide honey and to pollinate plants. Fish farmed in freshwater systems will be: tilapia, carp, blue gill and catfish. Organisms grown in saltwater systems will be: prawns, saltwater bivalves, urchins and algae. Algae, due to the high content of micronutrients and vitamins, can complement the diet of colonists.

In addition, they can be a source of food for fish at the early stages of plant growth. Another source of food is meat grown in vitro, in laboratories. In order to ensure the most varied diet pork, poultry, beef and lamb will be grown. For a security reasons, greenhouses will be closed to residents. However, small ornamental hydroponic installations will be located in the colony and in apartments on colonist's demand. In the public area will be grown bamboos as a raw material, dandelion, ficus elastica, and rubber tree in order to obtain a natural rubber. Dandelions are not very efficient, but will provide access to the rubber at an early stage of development, a ficus that will take over the role of dandelions and rubber trees, which will have been the main source of rubber after about 10 years and will have replaced the dandelions and ficus. Ficus will remain as a decorative plant.

**3.4. Waste management subsystem**

Due to the daily production of organic waste, at the level of more than 5 tons per BIOMe, designing a bio-refinery that would be able to convert biomass into valuable products was very crucial. All technological processes, used in this concept, are selected in order to availability substrates or catalysts on Mars. What is more rare chemical elements, used as catalysts, can be gathered from the main LFTR. These elements are waste products of conducted reaction. Non edible part of plants, which will be main part of all organic wastes are very rich in valuable substances, which can be extracted and processed to obtained high valuable products. These products can be use in all aspects of colony's operations. What is more obtained products are characterized by biodegradability, thanks to which we eliminate the problem of waste storage. In the proposed colony model, 3 refineries would be created. Each of them will have a universal apparatus, easily adjustable for current needs of products. The first stages of processing, for each installations would be identical: biomass grinding, and subsequent placing the material on the filter press and separation into 2 products: green juice-rich in nutrients, such as sugars, proteins, minerals and press cake, mainly containing lignin, cellulose and hemicellulose. Due to its composition, juice would have to be purified on chromatographic columns or filtration technique to remove undesirable compounds, and can be used as a dietary supplement for residents. The filter cake creates many more possibilities of development, hence the need to build 3 installations.

Celluloses and hemicelluloses are polymers composed of glucose monomers. Enzymatic hydrolysis allows us to achieve sugar platform C6, C5/C6 for further synthesis. For instance it is possible to obtain ethyl alcohol using microorganisms and sugar - glucose (under anaerobic conditions). Due to the high water content in product of fermentation, distillation is a necessary step, and recycled water can be used for the next technological cycle. Obtained ethanol can be used as a disinfectant, fuel, and also for commercial purposes - for the production of Martian vodka. Ethanol can be used for polyethylene production. However it is possible to achieve methane from another type of fermentation and use it as substrate along with chlorine to obtain polyvinyl chlorine. Due to the presence of significant amounts of cellulose, it is possible to produce paper, using small scale installation, due to low demand. Cellulose is also a raw material for the production of biodegradable textiles - modal, lyocell, viscose, which will be used in the production of everyday-use clothing. Textile production together with the sewing room represent second installation. The third installation makes possible to obtain an additional source of energy from biomass and semi-finished products for further syntheses. Pyrolysis of biomass, in which synthesis gas is produced, and the Fisher-Tropsch reaction, allows to obtain liquid fuels and water as a by-product.

## 4. HUMAN HEALTHCARE SYSTEM

### 4.1. Medical Care

Every habitat will have their own medical point which will provide fundamental aid. Telemedicine will displace standard medical visits; all prescribed medicaments and supplements will be transported to individual patients by courier. It is crucial to monitor health of habitants to predict abnormalities or future diseases. Measurements of human physiology can be done by using wearable sensors, which results will be stored in medical database for further analysis. Complete database of medical examination will be fundamental element in earlier detection of not significant pathological changes. That should be done once a week, to find the balance between the fastest possible reaction to negative physiological changes and decreasing a sensation of no privacy.

Once per 3-6 months should be made, listed below, medical imaging which will give us valuable in- formation for research purposes.

- Metabolomic NMR - obtaining information about the metabolic profile of body fluids, tissue extracts, tissues.
- Hybrid SPECT/MRI NP – will allow imaging of structure and function organs at subcellular levels and for targeted imaging of the disease.
- Brain Diffusion Tensor Imaging, Diffusion Weighted Imaging & Tractography (Fiber Tracking) – will allow to tract-specific localization of white matter (WM) lesions, localization of tumours in relation to the WM tracts and localization of the main WM tracts for neurosurgical planning.

In case of surgical intervention necessity in hospital operating theatre will be two surgical robots, for abdominal surgeries and orthopaedical one. Surgical robots have greater range of motion than human hands, magnifies view of the procedure and translates surgeon's movements into smaller, more precise robotic movements. Besides that, rehabilitation time is shorter due to reduced invasiveness of the surgery in comparison to traditional methods.

### 4.2. Decompression sickness

Providing adequate alveolar $O_2$ pressure in the habitat and suit, to reduce the risk of fire and to prevent decompression sickness from, is one of the most significant factors. The presence of Ar significantly increased a risk of DCS; it has comparable solubility to $O_2$. The proper $N_2$-Ar concentration ratio in habitat is 1.68 $N_2$/1.0 Ar.

Before wearing low-pressure (3.75 psi; ~259 hPa) suit it is necessary to use pre-breathe cabin (100% $O_2$) for several hours. Nonetheless those conditions do not reduce the risk of DCS completely, but some actions would be necessary to reduce the risk of DCS to an acceptable level like several hours of pre-breathing 100% $O_2$, a higher suit pressure, or exercise during pre-breathing.

Tab.1 The best option evaluated for a martian habitat atmosphere

| pressure | Partial pressure as % of total pressure | | |
|---|---|---|---|
| [psia] | $O_2$ | $N_2$ | Ar |
| | [%] | [%] | [%] |
| 8.0 | 32 | 42.6 | 25.2 |

**4.3. Medicines**

Experiments on the ISS showed that the pharmacokinetics of drugs are changed in microgravity. Due to the NASA research, it was found out that some pharmaceuticals have a shorter shelf-life and de- graded potency in space. It is required to achieve higher stability of pharmacotherapeutics with longer shelf-life or with higher resistance to unique ambient conditions. It can be provided inter alia by pharmacokinetics and pharmacodynamics research, specialized therapeutic monitoring for spaceflight-related pathophysiology. New alternative formulations and innovative packaging will provide slower degradation as well. However, if it is not in our reach to prolong shelf-time of some medicaments it can produce them on Mars.

4.4. **Gravitational Issues**

The most efficient way, to prevent muscle atrophy and bone loss, is everyday exercising. What is more, lower gravitational forces cause that fluids are shifting to the upper part of the body, so it is highly recommended to wear soft mechanical counter pressure suit not only during exercising but for a few hours during normal day activities as well. Such repeatable activity can be diversified by VR actions which will help for boredom and entertainment. Some of inhabitants will have aggravating symptoms of space sickness and the only working treatment are pharmaceuticals.

**5. PRODUCTION**

**5.1. Production and distribution of energy**

Energy system for colony will consist of a several technologies combined to achieve high redundancy. Liquid Fluoride Thorium Reactor (LFTR) will be used as a main energy source and it will be capable of producing output power up to 100 MW of electric energy. Advantages of using LFTR for Mars colony:

- Reduced dimensions and weight of the reactor system,
- Reactor vessel operates on low pressure which minimize the size of the safety installation,
- Thorium (Th) fuel cycle - Th can be obtained from the martian soil,

- Fertile material (Th) can be added to the reactor liquid salts solution online. Only starting fuel batch needed, estimated 100 kg of thorium for one year of operation,
- Reactor salts installation has long lifetime without significant renovation, dependent of energy output density,
- High operational temperature enables to use supercritical CO2 turbomachinery in closed loop Brayton cycle - reduced weight and dimensions of the power conversion system, improved power conversion efficiency.

Waste heat generated by reactor will be recycled. The 100 MW electric power reactor will be capable to generate up to 250 MW of heat which will be used in industry and agriculture. In normal situation, most of the energy consumption will be appeased by the LFTR but the critical subsystems like life support for habitats, communication and control will be supplied by Scattered Kilopower Reactor power grid, where single module will be able to generate 10 kW of electrical power. In an emergency where LFTR is not able to produce output, Kilopower grid will maintain the colony in emergency mode until the LFTR malfunction is fixed. The electrical and thermal energy production will be driven by energy control system where optimization algorithm will be able to adapt electrical and thermal energy production for each module actual request, where distribution of Kilopower modules will depend on each sector's energy consumption potential. In case of sector breakdown, main control system is able to redirect power stream to a specific location using power lines, which will be placed in pipes parallel to the base connection tunnels.

Based on research of current designs of molten salt type reactors and supercritical Brayton cycle power conversion systems, 3D model of Twardowsky base LFTR power plant has been created. It has been divided into five modules: reactor system, power conversion, heat exchanger, electric converters and nuclear laboratory connected with reactor's salt processing unit.

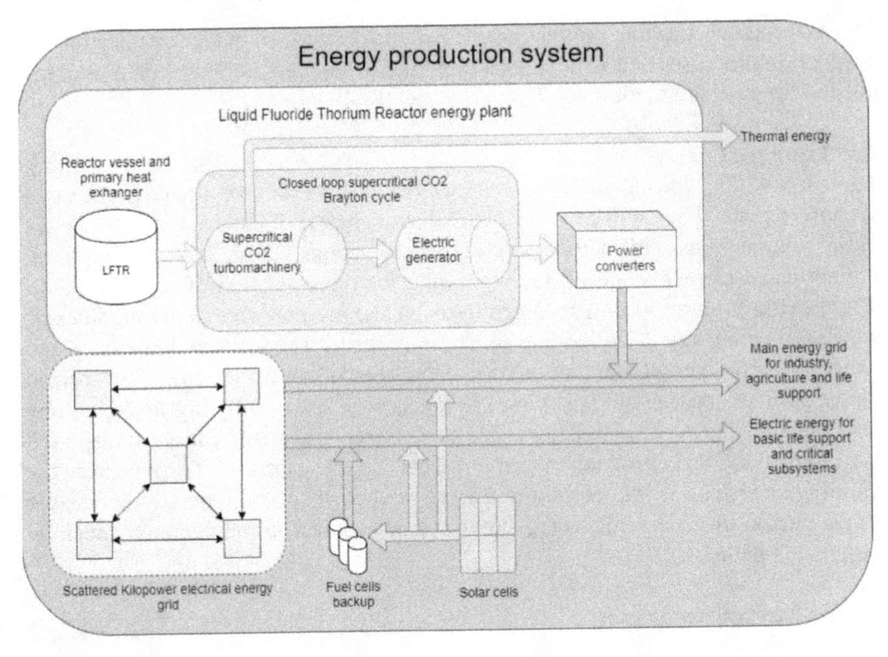


To achieve a higher redundancy, power system will have backup energy in separate hydrogen and oxygen stored in regenerative fuel cells (RFCS) by process of electrolysis. In case of an emergency it could be combined into water. Energy up to few kilowatts per unit will be extracted and transferred into the grid. Solar panels will be used to power up process of electrolysis and as a support energy source for base to reduce Kilopower necessary constant output to extend modules lifetime. It will be placed flat on the top of plateau in form of long, modular stripes that are able to roll on reel in case of sandstorms. Solar panels will be constantly produced in base's factory.

Synergy of LFTR, Scattered Kilopower Reactor and backup fuel and solar cells will provide not only significant power output for the colony fast growth, and also provide a high safety and serve as a backup in every scenario.

**5.2. Open pit mine and atmosphere processor**

The Raw materials for production in the colony will be taken from two sources - the ground and the atmosphere. Acquisition of compounds for further processing into metal, ceramic and polymer semi-products will be carried out by means of open pit mines from the colonies, from which spoils will be transported using an autonomous transport system. After that, the spoil minerals will be thrown into the container and after passing through the airlock unloaded onto the pre-processing line. In the industrial part of the colony, water will be extracted from pre-prepared (i.e. ground / sifted) spoil (in the first place), and then the spoils will be transported to obtain the desired compounds from it. The second source of resources is Martian atmosphere, from which carbon dioxide will be taken, as well as nitrogen and argon. Those will be mixed with oxygen and pumped into the life support system. Carbon dioxide itself will have a much wider application in production processes and will be used to produce, among others, carbon monoxide needed for steel production.

### 5.3. Transport

Transport will be carried out using vehicles that will be fully adaptable to the tasks set before them. They will consist of a universal chassis module and one of the six basic functional modules: transport of people, goods, spoil, liquid substances, prefabricated elements and a special module for the collection of regolith water. There exists a possibility of redesigning or designing new functional modules as well as adapting the drive module to existing needs. The airlocks have the same dimensions in the whole transport system, thanks to which the standardized dimensions of the vehicle will allow free access to every building in the colony regardless of the type of functional module. Large-size items can be transported between individual production departments using gantries. To improve the mobility, residents of the colony will have at their disposal generally accessible single-person electric vehicles (eg. Segways or scooters) to navigate on specially designated paths.

### 5.4. $H_2O$ production

One of the most important raw materials necessary to create any human colony is water. On Mars, there exists a possibility of acquiring water in three ways depending on the degree of its occurrence on and under the surface.

- The first method consists in separating frozen water by evaporating it from regolith during pre-processing of regolith, which is extracted in open-cast mines. This is the most reliable and promising source of water that can be exploited while extracting other needed resources. It is a method that due to the low water content (or unknown - presence of water is certain, but exact percentage depends on the place) becomes efficient only when processing a large amount of spoil.
- The second method is intended for use in areas with a slightly greater share of frozen water in the ground. It involves drilling a hole and introducing a source of heat, evaporated water is collected in tanks of autonomous vehicles. Boreholes would be made near the colonies in specially designated zones with higher frozen water share.
- The third method is Rodwell's method, which consists in drilling a well where a device is then introduced to melt the water and create a so-called. "melt pool", this water is then pumped to the tank on the surface and transported to the colony. This method only makes sense when large clusters of water are found in the form of relatively pure ice.

All these methods can complement each other, it all depends on how much water will close the soil near the colony and how much it will be concentrated. In the project, we assume as the main source of water extracting it directly from regolith as one of the first stages of the regolith processing.

**5.5. Sabatier reaction and Electrolysis**

In order to obtain oxygen and hydrogen, it will carry out electrolysis of water. The oxygen obtained will serve, among others for the oxidizer in the arc furnace in the production of metals and will be used to produce air for the life support system. Hydrogen will also be used in the RWGS process and will be a component of the gas mixture allowing the deposition of iron on the chamber walls in the DRI process. Thanks to the use of Sabatier's reaction, it will also be used for the production of methane, which in turn can be subjected to a pyrolysis process in order to obtain coal, eg. for the production of electrodes.

**5.6. Rocket propellant**

The most common rocket propellant solutions are mixtures of methane and oxygen as well as hydrogen and oxygen. Storage of hydrogen in liquid form is problematic due to the very low temperature of 20 K (-253 °C). There was made a decision that production of fuel will be based on methane and oxygen. Methane is easier to store than hydrogen, it only requires lowering its temperature to 111 K (-162 °C). Methane will be obtained thanks to the methanation reaction ($CO_2 + 4H_2 = CH_2 + 2H_2O$) taking place in the Sabatier Reactors. Ps. SpaceX engines are driven by this mix.

**5.7.Plastics**

Plastics are a group of materials necessary for the production of: housings, casings, packaging, furniture, insulation of electrical wires, clothing, thermal insulation, seals, and partly for anti-radiation protection (PE). The production of plastics will be based mainly on raw materials taken from the atmosphere and organic waste from colonies in the form of biomass (ECLSS). At the initial stage of colony development, production will be limited to only a few types of materials, which will include : PE, PA, PvC, PHB as well as resins and silicones. Additives for polymers that will be used in quantities below 10 kg per year will be imported from Earth, but in the future we assume the total self-sufficiency of the colony.

**5.8. Metals**

The production of steel in the colony can be carried out by the mean of a direct iron reduction process. In the furnace for direct reduction, iron oxides are reduced at a temperature below the melting point of iron, which results in the lack of formation of slag. This process does not require the addition of fluxes. The free iron atoms then combine with the carbon monoxide co-pressed with hydrogen as a mixture known as syngas and formed iron pentacarbonyl is forced into a chamber in which it dissociates on walls heated to about 120°C leaving a layer of almost pure iron (> 97.5%). The released carbon monoxide is forced back into the first chamber and the cycle is repeated. The process works best for ore with high iron content (> 60%).

The content of aluminum oxide in the regolith is around 8%. It occurs in the form of corundum which can be dissolved in a cryolite and then electrolyzed with consumable carbon electrodes. As a result, we get the reaction $Al_2O_3 + 3C \rightarrow 2Al$

+ 3CO. This process requires large energy inputs, as much as 70 MJ of energy to produce 1 kg of aluminum. It will not be used in the construction, but rather in elements in which a small mass will be crucial, and in the case of problems with obtaining copper in electrical wires and other parts of the energy consumption and transmission system.

## 5.9. Manufacturing

The independence of the Colony from deliveries from the Earth is a key issue. However, at such an early stage of development as a colony with only 1,000 people, it is not possible to produce all the things necessary for the functioning of the colony and to ensure development opportunities. Materials such as electronics, some medications and substances for which it is not profitable to build a special infrastructure due to small quantities of demand will be transported from Earth. The production part will therefore be adapted to facilitate the easy retooling of workstations in order to produce as much detail as possible. The use of state-of-the-art manufacturing techniques will be necessary for production, this is to reduce the number of necessary technologies needed for production. The most modern production techniques include technologies such as incremental technologies (3D printing) and laser technologies (cutting, finishing, layering, joining materials). However, it is necessary to use conventional machines for processing plastics and metals, such technologies as thermoforming, extrusion or turning or milling. We do not assume full automation of production because it is an expensive, complicated and insufficiently flexible solution. It is not possible to create so many different detailed parts without using dedicated lines and devices.

Production begins with the initial processing of input raw materials in the form of excavated coal from the opencast mine, Martian atmosphere and waste from colonies. In the case of spoil from the mine, it is poured into a container, which then passes through the lock. The next step is a milling / grinding ore and rinsing it to remove the perchlorate, the next step is to transport the feed through conveyor system (belt) to the selected four hangars, including steel mills (each shed has an instrumentation dedicated to recovering / extracting the substance or group of substances such as steel, aluminum, ceramics and others). In the case of metals, it is necessary to transport them to further halls in order to re-melt them and cast and / or plastic work, the result of these processes are semi-finished products, among others in the form of sheets, profiles or powders needed for 3D printers.)
The materials will be transported on production in the form of eg granules or "dough" (silicone). The infrastructure allows for the production in significant quantities of glass and materials intended for the construction of prefabricated buildings. For the production of photovoltaic panels and lighting, the colony will produce and process semiconductors. Unfortunately, at this stage of colonization, we have a lot of restrictions related to the availability of selected technologies. There will be no large variety of products and alloy additions will have to be imported from Earth.

The largest parts of the entire production system are three large production halls, in which production machines are located from selected manufacturing technologies. These halls are connected with each other by a transport system, which consists of transport trolleys and overhead traveling gantries. The gantries connect the production halls and the logistics part, which in turn connects with the construction workshop, prototype and hangars.

Materials in the form of semi-finished products from metals, ceramics and plastics go to the afore- mentioned logistic part, from which they are then transferred by cranes to the indicated work stations.

**5.10. Science and Development**

The workshop is one of the most important and one of the first objects created in the colony, its main task is to provide the colony with the possibility of continuous development and adaptation to the existing needs by introducing necessary modifications to existing systems. It is divided into such departments as prototyping, construction workshop, hangar, and laboratory & administration. Solutions developed on Earth may become insufficient in the Martian reality, the solution to this problem is to develop new solutions and improve existing ones. All repairs, improvements and production of vehicles and research equipment takes place in a prototype shop, which has dedicated workstations where work machines are placed. You can distinguish the position of 3D printers (in various technologies and with a different working field, which, for example, allows you to print entire vehicle frames), you can also extract positions with laser technologies, welding workshop and assembly part. In order to facilitate the handling of large components, overhead cranes, hoists and platforms / trolleys will be used. In addition to the prototype lab, a laboratory is located where research is conducted on the processes used in production, the design of new construction solutions and the development of methods for their production (laboratories for constructors and technologists). In addition, the laboratory is equipped with chemical laboratories, laboratories for materials research and computer laboratories. During the de- sign, technologies such as vR and AR are widely used to visualize and to better look at the problems. Hangar is a common part of the prototype and construction workshop, it performs mainly logistics and service functions, it is also a complement to the transport system. The last object is a construction workshop whose main goal is the production of precast construction elements.

**5.11. Prefabrication**

Precast concrete and steel for the construction of new structures will be manufactured in a specially designed construction workshop with hangars, which at the same time serve as air locks on the surface of Mars. This allows the rapid transport of these elements to the construction site without interfering with the basic transport system, which are part of the hangars used by the main terminal. The transport will take place using vehicles with a functional module for transporting prefabricated elements. In the case of relatively heavy components, running modules with doubled wheels will be used.

**5.12. Warehouse**

All colony warehouses are part of a large smart storage system that remotely controls the flow of materials to and from warehouses and within them. Goods are coded and entered into the system by the operator, which are then transported by special carts to their destination. The transport is fast and collision free because the individual carts are contacting with each other. They operate within the whole colony, where only transport routes are designated for them.

## 6. MARTIAN SOCIETY OVERVIEW

The fundamental problem will be to avoid the social anomie. The anomie leads to the break-ups of societies, enforces undesirable behaviors like suicides or rebellion. Because of that, it is necessary to create a society that reflects the earth society as closely as possible however, taking into account the specificity of the colony. The Martian society will live in a political system which will have its democratic representation in decision-making institutions and organizations. We call these castes guilds, because their members will be selected based on their professional qualifications and competences. On the other hand the economic system of the Martian colony will be a free market economy excluding some rare goods ex real estate, which turnover will be regulated by the administration of the colony. Martian society will be open, however, first colonizers, should show tolerance for other cultures, who live and accept the lifestyle of a globalized and unified society as much as possible. The aim of such a model is the possible minimization of social conflicts at least at the initial stages of colonization. It reduces the risk of conflicts: at the political, economical and social level. The overall goal of the colony is to expand, and basic is to survive.

## 7. THE SYSTEMIC CONCEPT OF THE MARTIAN COLONY

### 7.1. Colony status relative to the Earth.

To determine to determine the status of the colony, three establishment was adopted :

- Together with the rapid development of space technology, nowadays cosmic law is outdated and therefore the colony status cannot be determined on its basis
- On Earth, there were no changes in geopolitical forces in military and economic terms
- The richest countries (G8 group + representation of the European Union but also corporations) will finance the colony at the early stages.

Bearing in mind the above assumptions, the optimal solution is to provide the Martian colony with an autonomous status, while guaranteeing its status and maintaining basic economic functions. Given the common position of G8 and their economical and political position and military strength , they are able to bear the necessary costs of functioning of the colony on early stages, and above all, ensure its security against possible aggression. At the same time, the Martian colony will be obliged to pay a fixed interest tax to the financing countries from the moment the net profit is obtained, until the investors guarantee a rate of return established at the time of signing the contract adjusted for deflation or inflation rate.

### 7.2. The systemic concept of the Martian colony

In the colony, civil liberties and democratic decision-making mechanisms should be provided. However the knowledge of experts from various fields will have an important role for the survival of settlers. In order for the position of these experts to be ensured, and the area of their knowledge will not be the subject of voting, a guild system will be established. The guilds will delegate their representatives to the Martian Council of colony - which is a collegial law body. At the same time, narrow ranges of specialization will be resolved only by guilds. In the event of a crisis caused by a threat, a specific guild will be able to take over most of the power during this crisis - but only with the consent of the Council. In addition, guilds will be responsible for organizing work, generating profit, and also education (secondary school and university level).

01 Supporting Life and Healthcare Guild — Responsible for providing basic living needs: water, food, air or medical care.

02 Engineers' Guild — Responsible for construction and architecture, energy, creating transmission installations etc.

03 Logistic Guild — Responsible for managing the production of goods intended for the colony to trade with the Earth, as well as import & export of goods between Earth and Mars.

04 Scientists Guild — Responsible for exploring and exploring Mars, inventions, which can also be export products of Cologne, and improving the quality of life in the colony.

05 Administrative and Social Guild — Responsible for associating specialists in the humanities and social fields as well as responsible for administering the colony, guarding the law, organizing sociological and political life and distribution of culture.

Guilds will exercise executive power in colony. The Council, when adopting resolutions, will have to indicate in them the guild responsible for implementing the law in force. In addition, the colony will be headed by a Presidium composed of the President of each Guild. The Presidium will represent the colony in dealing with the Earth or be, for example, an instance of appeal against each of the colony's decisions, manage and carry out elections to the Council, etc. In the first phase of colony existence, 5 guilds will be created for the survival, development and functioning:

### 7.3. Age structure and sex of colonizers

The unstable social structure was chosen, but in a way that it would naturally become a stationary structure after certain amount of time.

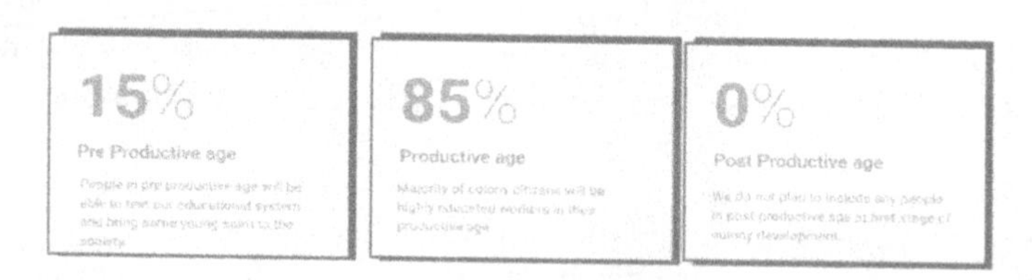


The same number of women and men is assumed. This will ensure the possibility of pairing between colonizers, which in turn will provide sexual needs, self-fulfillment, etc. There are no sociological premises for the sex life of the Martian colony to look differently than on Earth. Taking into account the fact that sexual needs are one of the basic human needs, and to reduce the stress level and possibility of conflict occurring in the sexual abstinent societies. As for the fulfilling of the basic sexual needs the dating application ErOS (Erotic Operating System) will be created. This kind of idea is way easier and more efficient than classical prostitution because it doesn't require special space and staff. Algorithms used in the application will help to find the best available partner and help in arranging the meet- ing. The application can be also extended into other modules that will not consider sexual activities but will be plainly romantic or platonic just as the application used right now on Earth. These will help establishing new relationships inside the colony or with the people arriving to the colony as tourists.

**7.4. Education and competences**

Education and skills factor is much more important than age and nationality or sexual orientation of colonizers. Harsh conditions of extraterrestrial colony demand specific, usually specialist knowledge and set of skills. Colonizers experience and education character will directly affect on described previously guild system, because it will be census based on which they will be assigned to different guilds.

It is important to take into consideration, that 15% of population will be children and teenagers. The most numerous Guild is the "Life sustaining and health care" guild. These are specialists who pro- vide the basic needs of the colony - without their satisfaction, development and proper functioning are impossible. It is assumed that specialists will constitute 30% of the whole society. Around the same population percentage will be occupied by the Engineers' Guild. Right after that we have Logistics and Scientific Guilds both with 15% of population and lastly the Administrative Guild.

## 8. ECONOMY AND MARKET

The main goal is to create a colony economically independent of Earth investors, which will develop gradually through the production and development. To this purpose, a business model based on the free market was set up. Guilds can both trade with each other and sell their products to other colonists, but mainly, the customer of the colony will be buyers from Earth.

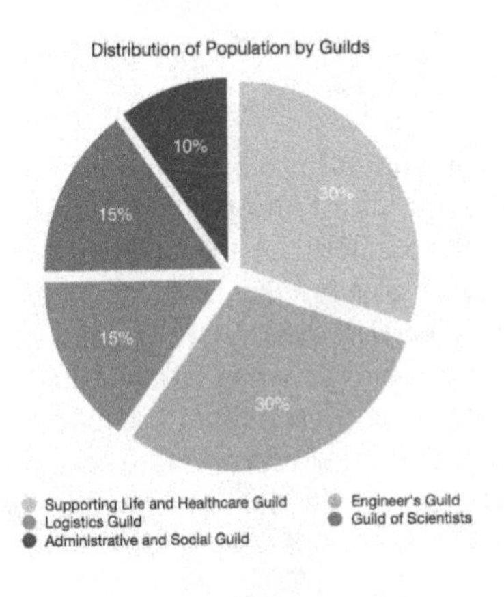

**8.1. Earth companies as the investors and service operators.**
On Mars, it will be possible to conduct world corporate investments. Apart from income, it will provide two benefits: the investor will be responsible for his investment, and the presence of corporations and their employees will contribute to the development of the colony. Existence of Earth companies will positively influence the civilians of a colony on Mars because they will know them - it is connecting them with their life on Earth. Providing service space to terrestrial companies and conducting tenders for the provision of services will ensure the participation in the profits of operators and reduce the costs of running them. Thanks to this, the company that wins the contracts will bear the costs, not the colonists.

At the level of the report, it is difficult to determine what profit we are talking about, however, with the estimated assumptions of:

| Number of spots | Turnus length | Cost of spot per turnus | Turnus revenue |
|---|---|---|---|
| 100 | 14 sols | $ 6,000 | $ 620,000 |

We can predict an full annual income of $ 16.1 million which means that the colony that provided the operator with access to the investment on the 10% revenue rate , obtains a total profit of $ 1,612,000 a year.

However, the cost depend on the standard of apartment and the hotel is only one tourist object and it is an example of profit from the service license itself. In addition, the hotel must pay - as an external entity, for example, purchase of energy from the colony and purchase of other goods. The hotel will be equipped with a range of world-wide entertainment such as spa, wellness, gyms, bowling alleys, and massage rooms.

**8.2. Export of goods to earth and its production on Mars.**
Because of uncompetitive costs of good's transport to and from Earth, a strategy should be applied that will minimize the consequences of the law of supply and demand. In this case one of the sectors of activity will be the production of very limited products, eg Martian vodka with filings of cosmic rocks. For this kind of goods important value resulting from their acquisition is not the material value of the object and the resulting prestige.

However, the production of limited products may not provide sufficient income for colonies. In the initial phase of the colony's existence, a thorough market analysis will be carried out to determine the niche of Martian products and their export to the ground. We can assume that products, created by both the logistics guild and minor activities carried out by the residents, will meet the optimal conditions for export profitability when they meet several establishments. Standard products exported from Mars should be, relatively light, small and their production should be relatively inexpensive and, if possible, take place using the maximum number of resources obtained on Mars and in the colony.

Based on the chart above, we can try to pre-estimate and generally estimate the possible income from the above products annually in the first phase of colony development. Among the products that meet these expectations there are:

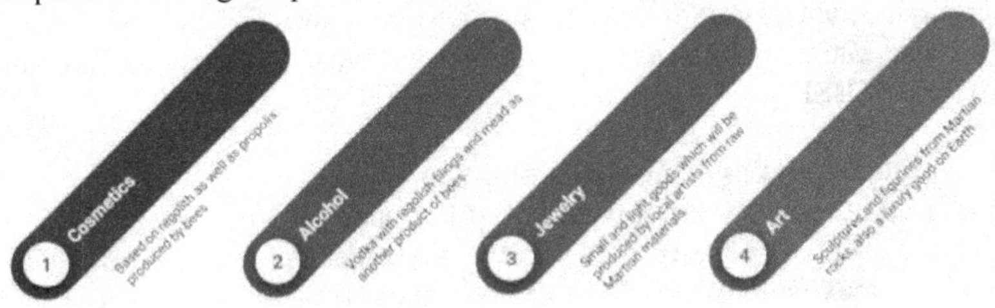


**8.2.1. Cosmetics:**

It is difficult to estimate the level of production and export of cosmetics at the level of the report. We are planning to prepare a production line ready for the production of approx. 100,000 tons of cosmetics. Each ton should bring about $ 10,000 in profit. This means that a profit of $ 1 billion is planned in the first year of production. These data are not overestimated at all, currently the cosmetics market is very large. The increase in the value of the market of cosmetics for skin care in the years 2014 - 2019 amounted to over $ 20 billion, at the same time to hair care another 7 billion, mouth 7 next. Thus, Martian production despite high profitability will initially constitute a small share of global cosmetics production

**8.2.2. Alcohol**

We will export two types of alcohol: vodka, which will be an expensive, luxurious product and mead, which will be a cheaper mass product. We estimate costs of production, transport, packaging and advertising to calculate cost of production of one bottle of alcohol (vodka or mead).

| Production costs | Transport costs | Packaging costs | Advertising costs | Sum |
|---|---|---|---|---|
| $ 2 | $ 100 | $ 2 | $ 5 | $109 |

With the estimated production scale of around 10,000 liters (20,000 bottles ) we can calculate:

| Alcohol | Price per bottle | Income | Costs | Profit |
|---|---|---|---|---|
| Vodka | $ 1,500 | $ 1,609,000 | $ 59,000 | $ 1,540,000 |
| Mead | $ 10 | $ 2,380,000 | $ 2,180,000 | $ 200,000 |

The scale of mead production should increase by ca. 10% each year, therefore the profit in the following years from the year of commencement ( the n-th year ) of production

| Year | n | n+1 | n+2 | n+3 | After first four years |
|---|---|---|---|---|---|
| Profit | $ 200,000 | $ 220,000 | $ 242,000 | $ 266,200 | $ 928,200 |

The scale of mead production should increase by ca. 10% each year, therefore the profit in the following years from the year of commencement ( the n-th year ) of production To sum up - from the above estimates - the profit from alcohol over first four years of production will be around $ 7,088,200.

**8.2.3. Jewelry**

Jewelry is a good whose value is estimated primarily on the basis of the value of the metal from which it is made, in this case, material coming from another planet. That's why we can bet a big price for this jewelry. It is estimated that the colony should produce about 10,000 kilograms of jewelry a year, with a profit of around 5,000 $ per kilogram of consumed minerals. Annual profit should amount to approximately USD 50,000,000 million.

### 8.2.4. Sculptures and figurines from Martian rocks

Regolith sculptures are a difficult thing to price, which is why we accept the price for 1 kg the same as in the case of jewelry. A standard sculpture of 20/20/20 cm will weigh a maximum of 12 kg - it means a transport cost of 2,400 $, including the value of creative work should be priced at about $ 1000. The assumed market value of the regolith will be around $ 60,000. So the minimal cost of the figure is: 63 400 $. It is estimated that the sculptures will be about 500 pieces a year, which gives us income of $ 31.7 million and a profit of 30,000,000.

## 8.3. Small and medium companies

Residents of the colony will also be able to conduct their own activities: production, service or artistic. Such action will allow not only additional income, but also colony development. Own activities will be financed only from the resources of the person running it and will be taxed.

## 8.4. Regulated goods

The free market on Mars will not be like on Earth, because as a result of capital accumulation of a given good there will always be more space for development, for example. The acquisition of real estate by individuals will be regulated by the administration. Space and energy are the most expensive goods on Mars. Among the rationed goods will also be alcohol In addition, only the logistics guild will be entitled to import goods from land to the colony.

## 8.5. Currency

We assume that the Martian society will create a system of exchange of goods itself, because it is a natural process of community development. Nevertheless, the proposed currency in the first stage will be the exchange of goods for electricity units that one can use, for example for entertainment devices (due to lack of bullion and unprofitable gold import we want avoid fiat money).

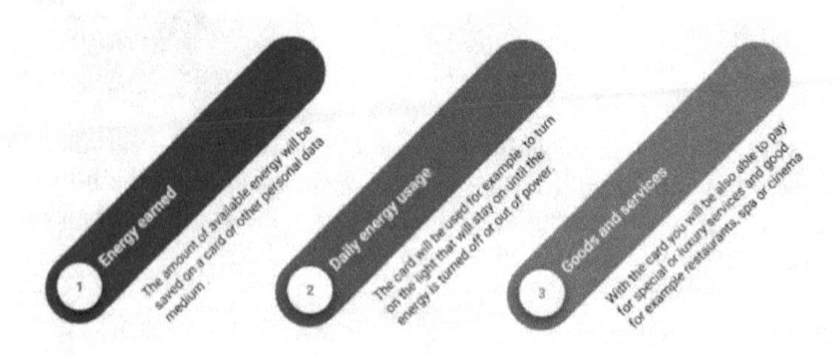

The payment will take place by means of downloading the energy limit to be used
The basic benefits of energy-based currency:

- basing the currency on a physical economic good will allow real valuation of work and avail- able products
- based on the used energy, the colony authorities will know the demand for it and manage its quantity at the same time ensuring the minimum necessary for the functioning of the colony with the necessary reserves
- basing the currency on energy allows easy exchange of different Earth currencies by tourists - they simply buy and top-up a payment card
- colony development leads to an increase in energy resources - therefore the richer the colony the richer its inhabitants will be.

However, the creation of the currency will be necessary to preserve commercial relations with the Earth, and first of all to be able to pay off the investment costs of building a colony. Another important economic element is the presence of a colony on the Earth's stock exchange, at least at the initial stage of development, until the investment costs are repaid. Such stock exchange shares would be very at- tractive for investors from Earth.

**8.6. Economic independence**

To sum up, the colony can get financial independence because it will continue to be from the sources of profit from production and export, trade in the internal market, licensing and taxing operators and taxes on small and medium enterprises. At the report level, it is only possible to estimate production and export profits.

**9. SOCIAL LIFE**

**9.1. Education**

Martian education will be divided into two stages: basic and guild. At the first stage, children and adolescents learn principles of social life, basic science and technology and take part in practicing in each guild (ex semi-annual internships). Then, after completing this stage, the youth will choose a guild in which one wants to learn the profession. It is worth noting that specialist knowledge will be taught in the last year of education which will end with a special exam checking the candidate adequacy to perform in specified profession. Thanks to this fact we ensures continuity of the guild system and educating young people in occupations that are necessary for the survival of the Martian colony.

**9.2. Culture & art**

The issue of the distribution of culture and art will be left to the market - it will be based on the law of supply and demand regulating what culture and at what price will be purchased by the inhabitants of Mars. However, in order to ensure the contact with the Earth's culture, free digital collections will be made available: with films, recordings of performances, music and literature. The formation of a separate Martian culture will naturally come after with new generations, however it will be a long and unregulated process.

### 9.3. The Penal Code and the law enforcement authorities

The social and administrative guild will be responsible for the social order in the colony. However, punishing dangerous criminals can be a problem in the form of a waste of human resources. Therefore, for offenses of a lower order their perpetrators will be punished by additional work or material fines. In turn, serious crimes such as rape or murder means confiscation of the colonial property of the perpetrator and deportation to the ground at his expense, where will be judged according to the law of the country of origin. In this case, it should be taken into consideration that someone does not have enough money to allow such transport. The solution to this problem is "crime insurance" operating as an ordinary Civil Liability insurance. Having such insurance will be a condition of consent to the colony. In addition, this insurance will allow to return to Earth people who will stop wanting to live on Mars.

## 10. QR CODES

Hi-res cover art

Literature link

Hi-res panorama

## 11. ACKNOWLEDGEMENTS

Special thanks: to Gordon Wasilewski and Pawel Rudzki

CPSIA information can be obtained
at www.ICGtesting.com
Printed in the USA
BVHW072009230920
589459BV00001B/12